Recent Achievements and Perspectives in

Nuclear Physics

Proceedings of the

5th Italy-Japan Symposium

Recent Achievements and Perspectives in **Nuclear Physics**

Proceedings of the
5th Italy-Japan Symposium

Naples, Italy
3 – 7 November 2004

editors

G. La Rana
University of Naples & INFN
Italy

C. Signorini
University of Padova & INFN
Italy

S. Shimoura
CNS University of Tokyo, Japan

World Scientific

NEW JERSEY • LONDON • SINGAPORE • BEIJING • SHANGHAI • HONG KONG • TAIPEI • CHENNAI

Published by

World Scientific Publishing Co. Pte. Ltd.
5 Toh Tuck Link, Singapore 596224
USA office: 27 Warren Street, Suite 401-402, Hackensack, NJ 07601
UK office: 57 Shelton Street, Covent Garden, London WC2H 9HE

British Library Cataloguing-in-Publication Data
A catalogue record for this book is available from the British Library.

RECENT ACHIEVEMENTS AND PERSPECTIVES IN NUCLEAR PHYSICS
Proceedings of the 5th Italy–Japan Symposium

ISBN 981-256-523-X

Printed in Singapore by Mainland Press

FOREWORD

The 5th Italy-Japan Symposium 'Recent Achievements and Perspectives in Nuclear Physics', was held in Naples from November 3 to November 7, 2004, within the framework of the RIKEN-INFN Research Collaboration Agreement. The aim of this Agreement is to promote cooperation between the two communities, and thereby to enhance the development of nuclear physics in the two countries. The first meeting of this series was held in Catania in 1992, the second at Saitama in 1995, followed by the third in Padova in 1997 and by the fourth in RIKEN and University of Tokyo in 2001. This Symposium was intended to stimulate discussions and exchange of ideas between the two communities on recent achievements in theoretical and experimental nuclear physics and future projects.

Cooperation between the two communities has grown up in the last ten-fifteen years; collaborations on different subjects of theoretical and experimental heavy-ion nuclear physics, nuclear astrophysics, hypernuclei and exotic atoms, are currently well established.

Considering the strong engagement of the two communities in many domains of nuclear physics, these collaborations appear to be still limited to a relatively small area leaving a wide room for future collaborations. Concerning the heavy-ion physics, which is one of the main subjects of this Symposium, both communities are very active in the domain of gamma-ray spectroscopy and reaction dynamics. The considerable development of radioactive ion (RI) beam facilities makes the research in Japan mainly focused at exploring exotic nuclei. In contrast, the study of heavy-ion physics with stable beams is very active in Italy. These studies, carried out by the two communities with emphasis on different aspects of nuclear phenomena, are often complementary for the understanding of the behaviour of nuclei going from ordinary to extreme physical conditions. In this respect, mutual participation to these projects and blending of the different methods and approaches would be very important for the development of heavy-ion nuclear physics. Similar considerations could be made for other areas of common interest like the Quark-Gluon-Plasma physics as well as applications of nuclear physics.

Nuclear studies of exotic nuclei represent one of the main perspectives of nuclear physics in the world. The development of new facilities, which will provide high-intensity beams, is in progress or under discussion in Europe, Asia and North America. At this Symposium we had a comprehensive overview of the projects for future accelerators of RI beams in Japan and in Italy: RIBF and SPES, respectively, as well as of the innovative instrumentation which will enhance substantially our detection capability. Cooperation between the two communities on these subjects can play an important role in the future of nuclear physics.

The current Italy-Japan collaborations have been the starting point for the program of the Symposium, which has been then extended to other topics of common interest. As usual, the program consisted of invited talks. The speakers covered eight main topics: i) Nuclear Structure; ii) Nuclear Dynamics and Hot Nuclei; iii) Ultrarelativistic Heavy-Ion Collisions; iv) Hypernuclei and Exotic Atoms; v) Nuclear Astrophysics; vi) Applications of Nuclear Physics; vii) Innovative Instrumentation: Detectors and Electronics; viii) RI Beams. The Symposium had about 100 participants including as many as 23 physicists from Japan.

I would like to thank all the participants for the active contribution they had given to the meeting, especially those from Japan. We are confident that this Symposium has stimulated friendly collaborations between the participants. I would also like to express my gratitude to the International Advisory Committee and the Local Organizing Committee as well as to the staff of the INFN and of the Dipartimento di Scienze Fisiche who worked on this Symposium

I gratefully acknowledge the financial support of the Istituto Nazionale di Fisica Nucleare, Dipartimento di Scienze Fisiche, Università di Napoli 'Federico II' and CAEN-S.p.A.

Giovanni La Rana
The Chairman of the Symposium

LOCAL ORGANIZING COMMITTEE

L. Gialanella
M. La Commara
G. La Rana, *Chair*
M. Romoli
E. Rosato
M. Trotta
E. Vardaci

INTERNATIONAL ADVISORY COMMITTEE

E. Chiavassa (*Torino*)
T. Motobayashi (*RIKEN*)
T. Nomura (*KEK*)
G. Ricco (*Genova*)
M. Sandoli (*Napoli*)
S. Shimoura (*CNS*)
C. Signorini (*Padova*)

5th Italy-Japan Symposium

Recent Achievements and Perspectives in Nuclear Physics

Naples, November 3-7, 2004

PROGRAMME

Wednesday, November 3

9:30 Opening address

Chairperson: G. La Rana (Napoli)

L. Merola (Napoli) Director of the Sezione INFN of Napoli

C. Signorini (Padova)

Nuclear Structure

Chairperson: H. Ikezoe (JAERI)

10:00 **S. Shimoura** (CNS): In-beam spectroscopy of exotic nuclei using direct reactions

10:30 **A. Covello** (Napoli): Structure of exotic nuclei around closed shells

11:00 **Y. Utsuno** (JAERI): Shell structure and correlation studied by large-scale shell-model calculations

11:30 *Coffee break*

Chairperson: G. De Angelis (LNL)

12.00 **K. Morita** (RIKEN): Superheavy elements

12:30 **P.G. Bizzeti** (Firenze): Past and future researches at the GASP array

13:00 Session close

Chairperson: S. Kubono (CNS)

14:30 **S. Lunardi** (Padova): Euroball results on nuclei at high spins and far from stability

15:00 **A. Bonaccorso** (Pisa): Reaction mechanisms for structure study

15:30 **H. Ikezoe** (JAERI): Fusion reaction of deformed nuclei and effect of closed shell structure

16:00 **S. Kimura** (LNS): Chaos driven fusion enhancement factor at astrophysical energies

16:20 **E. Fioretto** (LNL): The heavy-ion magnetic spectrometer PRISMA at LNL

16:40 **A. Gadea** (LNL): First results of the CLARA-PRISMA setup

17:00 *Coffee break*

Chairperson: M. Sandoli (Napoli)

17:30 **H. Sakai** (Tokyo): Construction of polarized proton target and first experiment with ^{6}He

17:50 **A. Bracco** (Milano): The gamma decay of the giant dipole resonance: New measurements and perspectives

18:20 **B. Martin** (Napoli): Evolution of the prompt dipole γ-ray emission with incident energy in fusion reactions

18:40 **G. Benzoni** (Milano): Selection rules on K-quantum number in warm rotating nuclei

19:00 Session close

Thursday, November 4

Nuclear dynamics and hot nuclei

Chairperson: S. Shimoura (CNS)

8:50 **A. Vitturi** (Padova): Low-energy nuclear reactions with unstable beams

9:20 **D. Pierroutsakou** (Napoli): Elastic scattering and reaction mechanisms of weakly bound and exotic nuclei at near-barrier energies

9:50 **M. Mazzocco** (Padova): Elastic scattering for the system ^{11}Be + ^{209}Bi at coulomb barrier energies with the EXODET apparatus

10:10 **E. Vardaci** (Napoli): Fission dynamics in systems of intermediate fissility

10:40 **N. Gelli** (Firenze): Search for isospin effects on the level density of nuclei involved in the decay of ^{139}Eu

11:00 *Coffee break*

Chairperson: G. Poggi (Firenze)

11:20 **M. Di Toro** (LNS): Isospin dynamics in heavy ion collisions: EOS-sensitive observables

11:50 **M. Bruno** (Bologna): Thermodynamical aspects in heavy ion reactions

12:20 **A. Pagano** (LNS): Heavy ions studies with the detector CHIMERA at the LNS in Catania

12:50 **D. Santonocito** (LNS): Disappearance of collective motion in hot nuclei

13:10 Session close

Ultrarelativistic heavy-ion collisions

Chairperson: T. Ishikawa (KEK)

14:40 **E. Scomparin** (Torino): Dimuon and charmonium production in heavy-ion collisions at the CERN SPS

15:10 **M. Asakawa** (Osaka): What is Quark-Gluon Plasma? – Recent progress

15:40 **E. Nappi** (Bari): Design and performance of the ALICE experiment at LHC

16:10 **H. Hamagaki** (CNS): Is QGP found at RHIC?

16:40 *Coffee break*

Chairperson: H. Tamura (Tohoku)

17:10 **C. Guaraldo** (LNF): Measurement of kaonic hydrogen with DEAR at DAFNE and future perspectives

17:40 **Y. Nagame** (JAERI): Atom-at-a-time chemistry of the transactinide element, Rutherfordium (element 104) – towards experimental verification of relativistic effects in chemical properties

18:10 **C. Petrascu** (LNF): The kaonic helium case

18:30 Session close

Friday, November 5

Hypernuclei and exotic atoms

Chairperson: A. Feliciello (Torino)

8:50 **H. Tamura** (Tohoku): Gamma spectroscopy of Hypernuclei

9:20 **T. Bressani** (Torino): Hypernuclear physics with FINUDA now and in the future

9:50 **M. Iwasaki** (RIKEN): Strange multi-baryon and kaonic atoms

10:10 **F. Garibaldi** (ISS): Recent results on High Resolution Hypernuclear Spectroscopy by Electroproduction at Jefferson lab, Hall A

10:30 *Coffee break*

Nuclear astrophysics

Chairperson: T. Motobayashi (RIKEN)

10:50 **F. Terrasi** (Napoli-II Univ.): Recent developments in Nuclear Astrophysics in Italy

11:20 **S. Kubono** (CNS) Study of stellar reactions relevant to explosive hydrogen burning with CRIB

11:50 **S. Cherubini** (LNS): The Trojan Horse Method in Nuclear Astrophysics

12:20 **H. Ishiyama** (KEK): Direct measurement of the astrophysical $^{8}Li(\alpha,n)^{11}B$ reaction

12:40 **P. Prati** (Genova): Perspectives in Nuclear Astrophysics

13:10 Session close

Applications of nuclear physics

Chairperson: A. Foti (LNS)

14:40 **M. Torikoshi** (NIRS): Medical applications of Heavy Ion Accelerators

15:10 **A. Andrighetto** (LNL): Thin UC_2 target for SPES project

15:30 **F. Lucarelli** (Firenze): Applied nuclear physics for cultural heritage and environmental studies: the new LABEC laboratory in Florence

15:50 **K. Satou** (Tsukuba): Hydrogen analysis for geo-science by using nuclear micro-beam

16:10 *Coffee break*

Innovative instrumentation: detectors and electronics

Chairperson: C. Signorini (Padova)

16:40 **A. Cunsolo** (LNS): New lines of research with the MAGNEX large-acceptance spectrometer

17:10 **T. Fukuchi** (CNS): Development of Position Sensitive Ge Detector

17:40 **D. Bazzacco** (Padova): Status of the AGATA project

18:10 **G. Prete** (LNL): Development of charged particle detectors and ancillary for gamma arrays

18:30 **E. Vardaci** (Napoli): EXODET: a new detector array for charged particles with integrated electronics for the position read-out

18:50 Session close

Saturday, November 6

Chairperson: G. Prete (LNL)

9:00 **L. Bardelli** (Firenze): Hardware and software development for fast digital processing of signals from nuclear physics detectors

9:20 **Y. Watanabe** (RIKEN): DAQ system for RI beam facilities – present and future

9:50 **E. Rosato** (Napoli): 3-D tomography of nTD silicon detectors.
The 3 MV tandem accelerator in Naples

10:20 **G. Grieco** (CAEN Spa): CAEN Electronics for Nuclear Physics

10:40 *Coffee break*

RI beams

Chairperson: T. Nomura (KEK)

11:10 **E. Chiavassa** (Torino): Experimental Nuclear Physics in Italy

11:40 **M. Menna** (LNS): The EXCYT project

12:10 **H. Miyatake** (KEK): Present status of the KEK-JAERI joint RNB project

12:40 Session close

Chairperson: T. Nomura (RIKEN)

14:20 **T. Kubo** (RIKEN): Present status of BIGRIPS separator project at RIKEN RI- Beam Factory

14:50 **A. Pisent** (LNL): The SPES and EURISOL projects

15:20 **T. Motobayashi** (RIKEN): Research program in RIKEN RIBF project and orientation for future collaborations

16:00 Conclusive remarks

16:10 Session close

CONTENTS

Section III
ULTRARELATIVISTIC HEAVY-ION COLLISIONS

Section IV
HYPERNUCLEI AND EXOTIC ATOMS

Section VII
INNOVATIVE INSTRUMENTATION: DETECTORS AND ELECTRONICS

SECTION I

NUCLEAR STRUCTURE

IN-BEAM SPECTROSCOPY OF EXOTIC NUCLEI USING DIRECT REACTIONS*

S. SHIMOURA

Center for Nuclear Study (CNS), University of Tokyo,
Wako branch at RIKEN, 2-1 Hirosawa, Saitama 351-0198, Japan
E-mail: shimoura@cns.s.u-tokyo.ac.jp

Direct reactions of exotic nuclear beams populating their excited states are discussed from the aspects of the inverse kinematics by measuring in-flight decay of the beam-like nucleus and the selectivity of the reactions. Emphasis is made for inelastic scatterings and nucleon transfer reactions of α using a liquid helium target. As an example of the α inelastic scattering, recent experimental results on the isoscalar electric responses of the ^{14}O are presented and discussed on the continuum effects on the isoscalar E0 and E1 responses. Recent experiments on γ-ray spectroscopy of ^{13}B* and ^{23}F* populated by the proton-transfer reaction (α,t) at 30–55 A MeV are also demonstrated for studies of single particle states in exotic nuclei.

1. Introduction

Direct reactions have been commonly used to populate excited states having characteristic quantum numbers depending on the probes such as proton, deuteron, alpha, Coulomb field, and so on. In RI-beam experiments, these probes are prepared as experimental target and excited states of beam or beam-like nucleus are identified by measuring all the decay products including γ-rays where the invariant masses are reconstructed to determine the excitation energies. It is noted that the Doppler-shift correction of the γ-ray from the moving nucleus is equivalent to reconstruct the invariant mass of the γ and the residual nucleus. The combination of the inverse kinematics and the invariant mass spectroscopy has the following advantages to overcome poor energy resolutions and weak intensities of RI beams:

*This work is partially supported by the Grant-in-Aid for Scientific Research (Nos. 11440081 and 15204018) by the Ministry of Education, Culture, Sports, Science and Technology.

- Because of the kinematical focus of the decaying particles, a large acceptance can be performed by using particle detectors at forward angles.
- Physical probe can be changed by replacing the experimental target without changing the configuration of the detector system. Therefore, the excitation spectra with different probes can be directly compared with the same acceptance.
- Since an RI beam contains some nuclides, which is usually considered to be contaminants, different reaction processes from different nuclei produce the same nucleus with a target. This helps us to characterise excited states from differences in population probabilities via different reaction processes with the same detector system.
- Resolution of the excited energy is determined by the resolution of detectors independent of the resolution of the beam. For the case of particle decay, a few hundred keV resolution for a decay energy of a few MeV is typically achieved whereas the resolution of the RI beam is more than several MeV. For the case of γ-decay, 100–200 keV resolution can be achieved by using an array of NaI(Tl) scintillators. This magnitude becomes more than one order of magnitude better by using an array of position sensitive Ge detectors.
- Information on the spin and the alignments of the excited states can be obtained by analysing the angular correlation in the decay process as well as the angular distribution of the reaction characterising the transferred angular momentum.

Based on these considerations we have been performing in-beam spectroscopy of unstable nuclei via several direct reactions of α particles by using a liquid helium target. Among them, two kinds of our recent experimental results on the inelastic scattering and the proton transfer reaction of α taken at RIPS beamline in RIKEN accelerator research facility are presented in this report.

2. Nuclear Responses in Unstable Nuclei

For the studies of asymmetric nuclear matter, responses of compressional modes in unstable nuclei are of interests. The α inelastic scatterings above 50 A MeV have been used for the isoscaler monopole (ISE0) and isoscalar dipole (ISE1) responses in stable nuclei, since the lowest orders of these responses are compressional modes[1]. In order to extend this idea to unstable nuclei, we have measured the α inelastic scattering on the ^{14}O nucleus at

60 MeV per nucleon. As described before, the inverse kinematics and the invariant mass method are fully used since all the excited states in the ^{14}O are unbound for particle emission. Although the ^{14}O is considered to be not so exotic, where first excited state is 1^- at 5.2 MeV indicating a semi magic nucleus, the lower particle (proton) threshold may affect the response function in the continuum. Experimentally, in order to determine the excitation energies, invariant masses of ^{13}N+p, ^{12}C+2p, ^{12}C+γ+2p, ^{10}C+α, and ^{10}C+γ+α channels were reconstructed. The transition strength of a certain multipole at a given excitation energy was deduced by a multipole decomposition analysis (MDA) using DWBA predictions of the angular distributions. Optical and transition potentials were obtained by folding models using the density distribution and the collective transition densities[2,3,4].

Figure 1 shows the obtained E0 and E1 transitions as a function of the excitation energy. Scattered strength distributions are seen as well as transition to the known resonant states at 5.9 MeV (0_2^+) and 5.2 MeV (1_1^-). The broad bumps at 6 – 8 MeV may be not a resonance but a continuum effect, since this region corresponds to the bound states in the mirror nucleus ^{14}C and neither 0^+ nor 1^- states exists in the ^{14}C. Qualitative tendencies are recently predicted by an RPA calculation taking into account the particle thresholds.

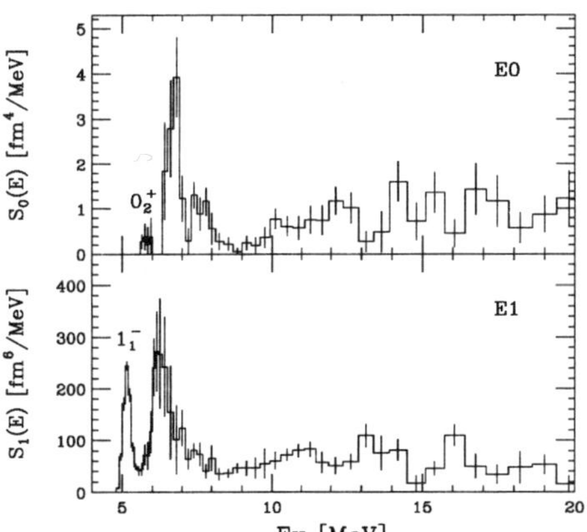

Figure 1. Strength distributions of the isoscaler monopole and isoscaler dipole response of the ^{14}O nucleus

3. Proton Single Particle States in Neutron-Rich Nuclei

The single particle states in a nuclear mean-field have been studied through nucleon transfer reactions such as (d,p) and (d,n) around 10 MeV per nucleon because of the kinematical matching conditions. We found that for the higher incident energies, the cross sections of the (α,t) and (α,^{3}He) reactions are rather large, because of a large negative Q-value and a wide momentum distribution of a nucleon in the α particle.

In order to study the proton single particle states in neutron-rich nuclei, which may affected by adding neutrons to the stable nuclei, we have performed spectroscopy of the ^{13}B* and the ^{23}F* nuclei populated by the (α,t) reactions on the ^{12}Be and the ^{22}O nuclei, respectively. For the measurement of the ^{13}B*, we introduced an array of position-sensitive Ge detectors (GRAPE[5]) for Doppler-shifted γ rays to distinguish decays from nearly degenerate excited states. We have found that the strongly populated 4.8-MeV state (see Fig. 2) has spin and parity of $1/2^+$ by a DWBA analysis of the angular distribution (Fig. 3). This state may be a candidate of of the proton single particle state in ($2s_{1/2}$) (spherical) or $\Omega^\pi = 1/2^+$ (deformed).

For spectroscopy of ^{23}F*, we measured the ^{4}He(^{22}O,^{23}Fγ) (proton transfer), the ^{4}He(^{23}F,^{23}Fγ) (inelastic excitation), and the ^{4}He(^{24}F,^{23}Fγ) (neutron knockout), simultaneously, because of the cocktail RI beam. We have found several new excited states by analysing the γ and γ-γ spectra. Comparison among the population probabilities for three kinds of reactions makes it possible to identify proton single-particle states, although the analysis is still going on.

4. Summary

Direct reactions using various probes combined with invariant-mass/γ spectroscopy are powerful tools to investigate excited states in exotic nuclei. Among them, alpha inelastic scatterings in inverse kinematics can provide information on isoscalar modes, especially for isoscalar monopole (ISE0) and dipole (ISE1). We have deduced ISE0 and ISE1 responses in the ^{14}O nucleus and found substantial strengths just above the proton threshold which may correspond to threshold effects. Alpha induced nucleon transfer reactions at 30-50 A MeV are useful for studying single-particle states. We have applied the (α,t) reactions on ^{12}Be and ^{22}O in the inverse kinematics and deduced the transferred angular momentum by means of a DWBA analysis. We plan to study other exotic nuclei such as nuclei around ^{32}Mg by using α induced reactions, which will give information on exotic structures in neutron-rich nuclei.

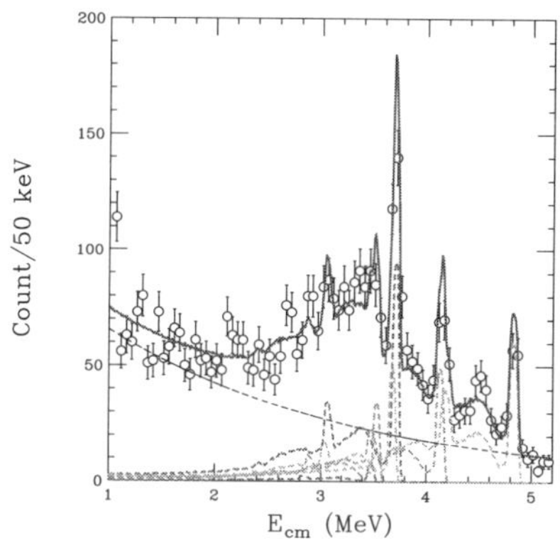

Figure 2. Doppler-shift corrected γ-ray spectra in ^{4}He(^{12}Be,^{13}Bγ) reaction taken by a Ge array GRAPE. Lines denotes simulated response functions of the detector system assuming emission of the known γ-rays.

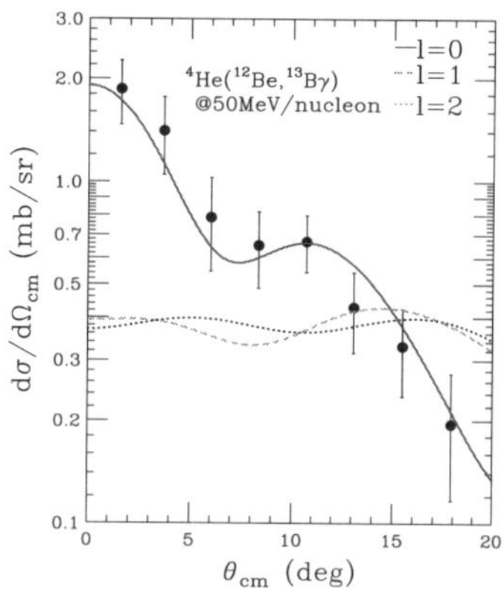

Figure 3. Angular distribution of the ^{12}Be(α,t)^{13}B*(4.8 MeV) reaction. Solid, dashed, and dotted lines denote the predictions of DWBA calculation assuming transferred angular momenta of 0, 1, and 2, respectively.

Acknowledgments

The author would like to thank the members of the CNS-Kyoto-RIKEN-Tokyo-Rikkyo collaboration for the present works and projects.

References

1. B. John et al., *Phys. Rev.* **C68**, 014305 (2003); D. H. Youngblood, Y.-W. Lui, and H.L. Clark, *Phys. Rev.* **C65**, 034302 (2002); D. H. Youngblood, Y.-W. Lui, and H.L. Clark, *Phys. Rev.* **C63**, 067301 (2001); D. H. Youngblood, Y.-W. Lui, and H.L. Clark, *Phys. Rev.* **C60**, 014304 (1999).
2. G.R. Satchler, *Nucl. Phys.* **A472**, 215 (1987).
3. A. Kolomiets, O. Pochivalov and S. Shlomo, *Phys. Rev.* **C61**, 034312 (2000).
4. M.N. Harakeh and A.E.L. Dieperink, *Phys. Rev.* **C23**, 2329 (1981).
5. S. Shimoura, *Nucl. Instr. Meth.* **A525**, 188 (2004).

STRUCTURE OF EXOTIC NUCLEI AROUND CLOSED SHELLS

A. COVELLO, L. CORAGGIO, A. GARGANO AND N. ITACO
Dipartimento di Scienze Fisiche, Università di Napoli Federico II,
and Istituto Nazionale di Fisica Nucleare,
Complesso Universitario di Monte S. Angelo, Via Cintia, 80126 Napoli, Italy
E-mail: covello@na.infn.it

We have studied several exotic nuclei around doubly magic ^{100}Sn and ^{132}Sn in terms of the shell model employing realistic effective interactions derived from the CD-Bonn nucleon-nucleon potential. The short-range repulsion of the bare potential has been renormalized by constructing a smooth low-momentum potential, $V_{\rm low-k}$, that is used directly as input for the calculation of the effective interaction. In this paper, after a brief survey of the theoretical framework, we present some selected results of our studies. Comparison shows that our results are in very good agreement with the available experimental data. This may stimulate experimental efforts to gain more information on nuclei in these regions of shell closures off stability.

1. Introduction

The study of exotic nuclei in the regions of shell closures is currently drawing much attention. In this context, of special interest are nuclei in the close vicinity to doubly magic ^{100}Sn and ^{132}Sn. The experimental study of these nuclei is very difficult, but in recent years new facilities and techniques have made some of them accessible to spectroscopic studies. Today, the development of radioactive ion beams opens up the prospect of gaining substantial information on ^{100}Sn and ^{132}Sn neighbors, which should allow to explore the evolution of the shell structure when approaching the proton and neutron drip lines. This makes it very interesting and timely to perform shell-model calculations not only to try to explain the available data, but also to make predictions which may stimulate, and be helpful to, future experiments.

On these grounds, we have recently started[1,2] a comprehensive study of nuclei around ^{100}Sn and ^{132}Sn within the framework of the shell model employing realistic effective interactions derived from the CD-Bonn nucleon-

nucleon (NN) potential.[3] The main aim of this paper is to report on some selected results of our calculations and show that they explain very well the existing data, which indicates that there is no evidence so far of shell structure modifications in these regions.

In Sec. 2 we give an outline of the theoretical framework in which our realistic shell-model calculations have been performed. In Sec. 3 we present our results for some nuclei around ^{100}Sn and beyond ^{132}Sn, and compare them with experiment whenever possible. Sec. 4 presents some concluding remarks.

2. Outline of theoretical framework

As already mentioned in the Introduction, in our shell-model calculations we have made use of a realistic effective interaction derived from the CD-Bonn free NN potential.[3] This high-quality potential, which is based upon meson exchange, fits very accurately (χ^2/datum ≈ 1) the world NN data below 350 MeV available in the year 2000.

The shell-model effective interaction $V_{\rm eff}$ is defined, as usual, in the following way. In principle, one should solve a nuclear many-body Schrödinger equation of the form

$$H\Psi_i = E_i\Psi_i, \tag{1}$$

with $H = T + V_{NN}$, where T denotes the kinetic energy. This full-space many-body problem is reduced to a smaller model-space problem of the form

$$PH_{\rm eff}P\Psi_i = P(H_0 + V_{\rm eff})P\Psi_i = E_iP\Psi_i. \tag{2}$$

Here $H_0 = T + U$ is the unperturbed Hamiltonian, U being an auxiliary potential introduced to define a convenient single-particle basis, and P denotes the projection operator onto the chosen model space.

A main difficulty one is confronted with in the derivation of $V_{\rm eff}$ from a modern NN potential, such as CD-Bonn, is the existence of a strong repulsive core which prevents its direct use in nuclear structure calculations. This difficulty is usually overcome by resorting to the traditional Brueckner G-matrix method. Here, we have made use of a new approach[4] which provides an advantageous alternative to the use of the above method. It consists in constructing a low-momentum NN potential, $V_{\rm low-k}$, that preserves the physics of the original potential V_{NN} up to a certain cut-off momentum Λ. In particular, the scattering phase shifts and deuteron binding energy calculated by V_{NN} are reproduced by $V_{\rm low-k}$. The latter is a smooth potential that can be used directly as input for the calculation of shell-model effective interactions. A detailed description of our derivation of $V_{\rm low-k}$ can be found in Ref. 4, where a criterion for the choice of the cut-off parameter Λ is also given. We have used here the value $\Lambda = 2.1$ $\rm fm^{-1}$.

Once the $V_{\rm low-k}$ is obtained, the calculation of the matrix elements of the effective interaction is carried out within the framework of a folded-diagram method,[5] including $V_{\rm low-k}$ diagrams through second order.[6]

3. Results

In this section, we report some results of our study of exotic nuclei in the close vicinity to ^{100}Sn and ^{132}Sn. As regards the former, we consider the three odd-odd nuclei ^{98}Ag, ^{102}In, and ^{100}In, focusing attention on the particle-hole (ph) multiplets, which are a direct source of information on the neutron-proton effective interaction. Clearly, the most suitable system for this study is ^{100}In with only one neutron valence particle and one proton valence hole. Although, no experimental information is yet available for this highly important nucleus, we have found it interesting to predict some of its spectroscopic properties in the hope that they will be verified in a not too distant future. Some experimental information is instead available for ^{98}Ag and ^{102}In, which have two additional proton holes and neutron particles with respect to ^{100}In, respectively. The study of the ph multiplets in these nuclei not only is interesting in its own right, but also provides a rather stringent test of the reliability of our predictions for ^{100}In.

In the ^{132}Sn mass region, we consider the two nuclei ^{134}Sn and ^{135}Sn, which are of great interest with regard to a possible appearance of shell quenching in neutron-rich nuclei. While for ^{134}Sn some experimental information has recently become available, the nucleus ^{135}Sn has not been accessible to spectroscopic studies so far. We report here our predictions for its low-energy spectrum.

The results of our calculations are presented separately for each mass region in the two following subsections.

3.1. ^{100}*Sn region*

We assume that ^{100}Sn is a closed core and let the valence neutrons occupy the five levels $0g_{7/2}$, $1d_{5/2}$, $1d_{3/2}$, $2s_{1/2}$, and $0h_{11/2}$ of the 50-82 shell, while for the proton holes the model space includes the four levels $0g_{9/2}$, $1p_{1/2}$, $1p_{3/2}$, and $0f_{5/2}$ of the 28-50 shell. The neutron single-particle and the proton single-hole energies cannot be taken directly from experiment, since no spectroscopic data are yet available for ^{101}Sn and ^{99}In. We have determined them by an analysis of the low-energy spectra of the odd Sn isotopes for the former and of the $N = 50$ isotones for the latter. Our procedure is discussed in detail in Ref. 2, where the adopted values are also reported. As regards the two-body matrix elements of the effective interaction, ^{100}In requires only the neutron-proton matrix elements in the *ph* formalism, while for ^{98}Ag and ^{102}In we also need proton-proton and neutron-neutron matrix elements in the hole-hole (*hh*) and particle-particle (*pp*) formalism, respectively. A description of our derivation of the effective interaction in the *pp* and *hh* formalism can be found in Refs. 7 and 8, respectively, while the matrix elements in the *ph* formalism have been explicitly derived in Ref. 1.

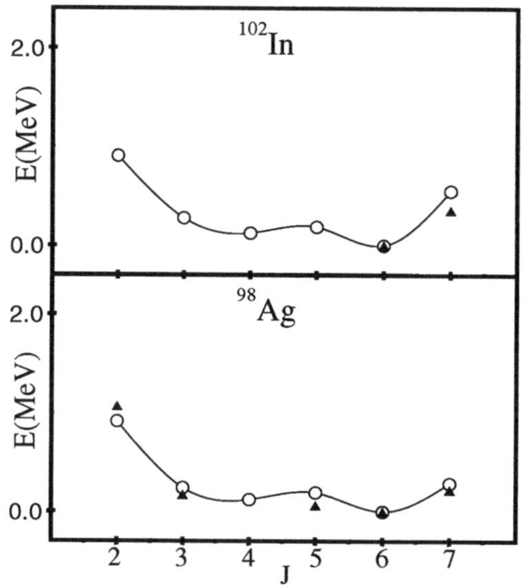

Figure 1. Proton hole-neutron particle $\pi g_{9/2}^{-1} \nu d_{5/2}$ multiplet in ^{98}Ag and ^{102}Ag. The calculated results are represented by open circles while the experimental data by solid triangles. The lines are drawn to connect the points.

In Fig. 1 we report the results of our calculations for the $\pi g_{9/2}^{-1} \nu d_{5/2}$ multiplet in ^{98}Ag and ^{102}In and compare them with the experimental data.[9] The members of the calculated multiplet in both nuclei have been identified as those dominated by this configuration with the two remaining valence nucleons forming a zero-coupled pair. We see that the calculated energies are in good agreement with the experimental ones, the largest discrepancy being about 200 keV for the 7^+ state in ^{102}In. A detailed discussion of the structure of the calculated states can be found in Ref. 2.

Let us now come to ^{100}In. In Fig. 2 we show our predictions for the $\pi g_{9/2}^{-1} \nu d_{5/2}$ multiplet. In this case there is little configuration mixing, the weight of the leading component in each member being at least 77%.

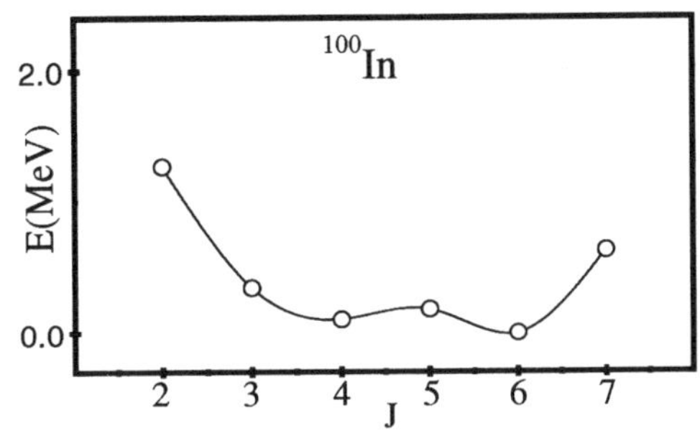

Figure 2. Same as Fig. 1, but for ^{100}In.

From Figs. 1 and 2 we see that a main feature of the calculated multiplets is that the states with the minimum and maximum J have the highest excitation energy, while the state with next to the highest J is the lowest one, the latter feature being in agreement with the predictions of the Brennan-Bernstein empirical coupling rule.[10] This pattern reproduces very well the experimental data in ^{98}Ag and is quite similar to that of the multiplets in the proton particle-neutron hole nuclei around doubly magic ^{132}Sn and ^{208}Pb.[1,11]

3.2. ^{132}*Sn region*

We consider ^{132}Sn as a closed core and let the valence neutrons occupy the six single-particle levels $0h_{9/2}$, $1f_{7/2}$, $1f_{5/2}$, $2p_{3/2}$, $2p_{1/2}$, and $0i_{13/2}$ of the 82-126 shell. As regards the single-neutron energies, they have been taken from the experimental spectrum[12] of ^{133}Sn , except that relative to the $i_{13/2}$ level which has not been observed. The latter has been taken from Ref. 13, where all the adopted values are also reported. In this connection, it should be mentioned that the calculations performed in this previous work differ from the present ones in that there we used a G-matrix folded-diagram formalism and included diagrams up to third order in G. From Fig. 3 we see that the calculated spectrum of ^{134}Sn is in good agreement with that proposed in the experimental studies,[14,15] confirming the interpretation given in these studies.

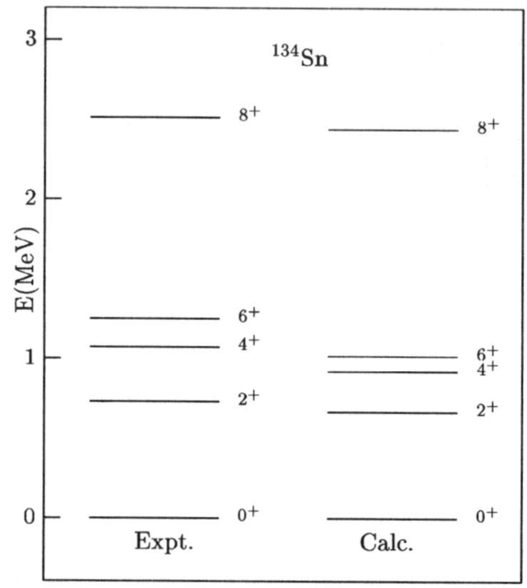

Figure 3. Experimental and calculated spectrum of ^{134}Sn.

Very recently,[16] the $B(E2; 0^+ \rightarrow 2_1^+)$ values in ^{132}Sn and ^{134}Sn have been measured using Coulomb excitation of neutron-rich radioactive ion beams. We have calculated this $B(E2)$ for ^{134}Sn with an effective neutron charge of $0.70e$, according to our early study.[13] The experimental and theoretical values are shown in Fig. 4, where we also compare with experiment[17] our $B(E2; 0^+ \rightarrow 2_1^+)$ values for ^{128}Sn and ^{130}Sn. The experimental $B(E2)$ value for ^{132}Sn is also reported to show the trend across the $N = 82$ magic number. The neutron single-hole energies and the neutron-hole effective charge used in our calculations for ^{128}Sn and ^{130}Sn are the same as those adopted in our recent study[18] of ^{129}In and ^{129}Sb. From Fig. 4 we see that our values reproduce remarkably well the experimental ones.

Let us now come to ^{135}Sn. This nucleus is expected to have a $\frac{7}{2}^-$ ground state[19,20] with a binding energy of 8.393 ± 0.401 MeV, as derived from the systematic trend.[21] No other experimental information on its spectrum is presently available. In Table 1 we report the calculated excitation energies up to about 1.5 MeV.

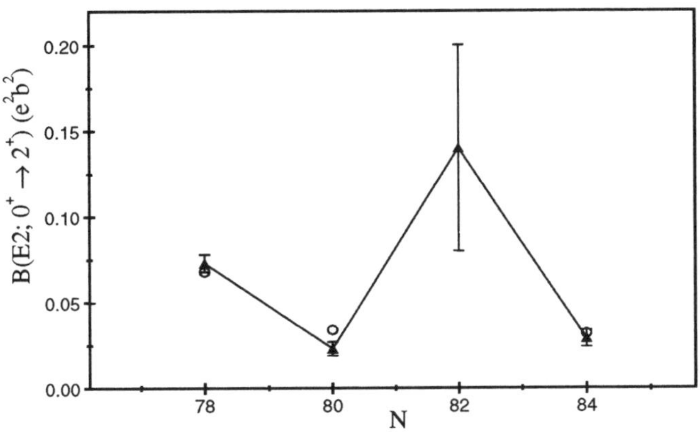

Figure 4. Experimental (solid triangles) and calculated (open circles) values of $B(E2;0^+ \to 2_1^+)$ for even-even Sn isotopes around neutron number $N = 82$.

Table 1. Calculated excitation energies for states of ^{135}Sn up to about 1.5 MeV.

J^π	E(MeV)
$\frac{7}{2}^-$	0.0
$\frac{5}{2}^-$	0.285
$\frac{3}{2}^-$	0.429
$\frac{11}{2}^-$	0.620
$\frac{3}{2}^-$	0.656
$\frac{9}{2}^-$	0.690
$\frac{15}{2}^-$	0.877
$\frac{9}{2}^-$	1.159
$\frac{7}{2}^-$	1.229
$\frac{3}{2}^-$	1.250
$\frac{1}{2}^-$	1.263
$\frac{5}{2}^-$	1.296
$\frac{7}{2}^-$	1.339
$\frac{5}{2}^-$	1.455
$\frac{9}{2}^-$	1.493

Our calculations confirm the $\frac{7}{2}^-$ nature of the ground state and predict for the binding energy the value 8.481 ± 0.078 MeV. A detailed discussion of the structure of the calculated states can be found in Ref. 13.

4. Concluding remarks

We have presented here some results of a shell-model study of exotic nuclei in the close vicinity to doubly magic ^{100}Sn and ^{132}Sn. The effective interaction has been derived from the CD-Bonn nucleon-nucleon potential, making use of a new method to renormalize the short-range repulsion of the bare potential without first calculating the Brueckner G matrix.

In the ^{100}Sn region we have focused attention on the three odd-odd nuclei ^{98}Ag, ^{102}In, and ^{100}In. We have found that the proton hole-neutron particle multiplets in these nuclei exhibit the same behavior as that of the multiplets in the proton particle-neutron hole nuclei around ^{132}Sn and ^{208}Pb. As regards the comparison with experiment, the data available for ^{98}Ag and ^{102}In are well reproduced by our calculations, providing confidence in our predictions for ^{100}In.

In the ^{132}Sn region we have considered the two very neutron-rich nuclei ^{134}Sn and ^{135}Sn. We have shown that our results for ^{134}Sn are in very good agreement with the available experimental data. In particular, our calculations reproduce quite accurately the $B(E2; 0^+ \rightarrow 2_1^+)$ value that has been very recently measured through Coulomb excitation of radioactive ion beams.[16] As is the case for ^{100}In, this supports confidence in our predictions for ^{135}Sn.

In summary, we may conclude with the following remarks.

(i) At present there is no real evidence of shell modifications in the immediate neighbors of ^{100}Sn and ^{132}Sn.

(ii) It is of key importance to gain more information on nuclei in the regions of shell closures off stability. This is certainly a main challenge of nuclear structure experiments with radioactive ion beams.

Acknowledgments

This work was supported in part by the Italian Ministero dell'Istruzione, dell'Università e della Ricerca (MIUR).

References

1. L. Coraggio, A. Covello, A. Gargano, N. Itaco and T. T. S. Kuo, *Phys. Rev.* **C66**, 064311 (2002).
2. L. Coraggio, A. Covello, A. Gargano and N. Itaco *Phys. Rev.* **C70**, 034310 (2004).
3. R. Machleidt, *Phys. Rev.* **C63**, 024001 (2001).

4. S. Bogner, T. T. S. Kuo, L. Coraggio, A. Covello and N. Itaco, *Phys. Rev.* **C65**, 051301 (2002).
5. T. T. S. Kuo and E. M. Krenciglowa, *Nucl. Phys.* **A342**, 454 (1980).
6. J. Shurpin, T. T. S. Kuo and D. Strottman, *Nucl. Phys.* **A408**, 310 (1983).
7. M. F. Jiang, R. Machleidt, D. B, Stout and T. T. S. Kuo, *Phys. Rev.* **C46**, 910 (1992).
8. L. Coraggio, A. Covello, A. Gargano, N. Itaco and T. T. S. Kuo, *J. Phys.* **G26**, 1697 (2000).
9. Data extracted using the NNDC On-line Data Service from the ENSDF database, file revised as of December 28, 2004.
10. M. H. Brennan and A. M. Bernstein, *Phys Rev.* **120**, 927 (1960).
11. A. Covello, L. Coraggio, A. Gargano and N. Itaco, *Yad. Fiz.* **67**, 1637 (2004).
12. P. Hoff *et al.*, *Phys. Rev. Lett.* **77**, 1020 (1996).
13. L. Coraggio, A. Covello, A. Gargano and N. Itaco, *Phys. Rev.* **C65**, 051306(R) (2002).
14. C. T. Zhang *et al.*, *Z. Phys.* **A358**, 9 (1997).
15. A. Korgul *et al.*, *Eur. Phys J.* **A7**, 167 (2000).
16. J. R. Beene *et al. Nucl. Phys.* **A746**, 471c (2004).
17. D. C. Radford *et al.*, in *Frontiers of Collective Motion* , edited by H. Sagawa and H. Iwasaki (World Scientific, Singapore, 2003), p. 319.
18. J. Genevey, J. A. Pinston, H. R. Faust, R. Orlandi, G. S. Simpson, I. Tsekhanovic, A. Covello, A. Gargano and W. Urban, *Phys. Rev.* **C67**, 054312 (2003).
19. A. Korgul, H. Mach, B. Fogelberg, W. Urban, W. Kurcewicz and V. I. Isakov, *Phys Rev.* **C64**, 021302(R) (2001).
20. J. Shergur *et al.*, *Nucl. Phys.* **A682**, 493c (2001).
21. G. Audi and A H. Wapstra, *Nucl. Phys.* **A595**, 409 (1995).

SHELL STRUCTURE AND CORRELATION STUDIED BY LARGE-SCALE SHELL-MODEL CALCULATIONS

YUTAKA UTSUNO

Department of Materials Science, Japan Atomic Energy Research Institute, Tokai, Ibaraki 319-1195, Japan
E-mail: utsuno@nsbox.tokai.jaeri.go.jp

TAKAHARU OTSUKA

Department of Physics and Center for Nuclear Study, University of Tokyo, Hongo, Tokyo 113-0033, Japan
RIKEN,
Hirosawa, Wako-shi, Saitama 351-0198, Japan

MICHIO HONMA

Center for Mathematical Sciences, University of Aizu, Tsuruga, Ikki-machi Aizu-Wakamatsu, Fukushima 965-8580, Japan

TAKAHIRO MIZUSAKI

Institute of Natural Sciences, Senshu University, Higashimita, Tama, Kawasaki, Kanagawa 214-8580, Japan

NORITAKA SHIMIZU

Department of Physics, University of Tokyo, Hongo, Tokyo 113-0033, Japan

Recent advances in the large-scale shell-model calculation performed by the authors are presented. Since the shell model, a powerful tool to study the nuclear structure, is often confronted with the computational limitation when applied to a heavier region, the Monte Carlo shell model is proposed to overcome the limitation. Owing to the development of the large-scale shell-model calculation, we can obtain a systematic description of some regions. From this the relation between the effective interaction and the nuclear structure is clarified, as exemplified in the $N \sim 20$ region, the pf-shell region, and the medium-heavy mass region.

1. Introduction

The goal of the nuclear-structure theory is to understand the nuclear many-body system whose character appears in the energy levels, electromagnetic transitions and moments, *etc.* In microscopic calculations, a nucleus is treated as a quantum many-body system consisting of protons and neutrons. The task of the microscopic calculation is then solving the Schrödinger equation, once the nucleon-nucleon interaction is given. However, it is still impossible to solve the Schrödinger equation exactly or rather precisely, except for very light nuclei with $A \lesssim 10$. Thus, most of nuclei must rely upon models in which some of essential facets in the structure are extracted.

Several microscopic models are often adopted such as the mean-field model, the shell model, the cluster model, *etc.* Each model has a different advantage over the others, i.e., the flexibility of the single-particle wave function in the mean-field model, for instance. The advantage of the shell model is the *correlation* or the configuration mixing, whose importance to the magnetic moment was for the first time pointed out by Arima and Horie[1]. While the target of the shell model was at its early stage restricted to nuclei having a few nucleons (or holes) on an inert core, it has been recently extended to deformed nuclei. As a result, a systematic shell-model description in several regions has been possible, and we can now trace the nuclear structure back to the effective interaction from the large-scale shell-model results.

In the present symposium, we review recent advances and applications of the large-scale shell-model calculation performed by the authors, and focus upon how the shell structure varies from stable to unstable nuclei influenced by the effective interaction. In Sec. 2, we first discuss the need for the large-scale calculation, and show some computational methods and their present status. Some results are presented in Sec. 3 with emphasis on the similarity of the evolving shell structure among different regions. In Sec. 4, the summary is given.

2. Present status of large-scale shell-model calculations

In the shell-model calculation, one first assumes an inert core, and then the Schrödinger equation is solved for the rest nucleons (or holes) within appropriate single-particle orbits (called valence orbits). To perform a shell-model calculation, we should diagonalize the Hamiltonian whose bases consist of all possible many-body states. The number of the bases is then

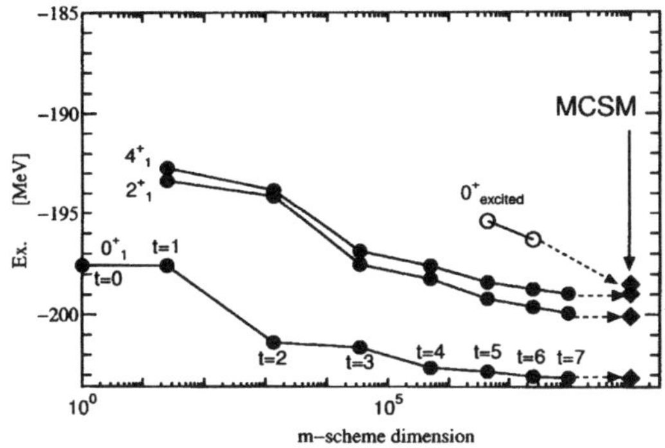

Figure 1. Convergence of the energies in ^{56}Ni as a function of the truncation level t in the conventional method in comparison with the MCSM energies. Taken from Ref. 7.

explosively increasing for systems with a number of the valence orbits and nucleons, which often prevents us from the actual computation. This is the main reason why the shell model cannot be presently applied to all nuclei in the nuclear chart.

Let us next consider how large scale calculation is required in the shell model. For this purpose, the shape coexistence of a doubly magic nucleus ^{56}Ni is taken as an example. Its ground state is mainly comprised of the fully occupied $(0f_{7/2})^{16}$ configuration, and not so deformed. On the other hand, a strongly deformed band has been recently observed[2] in this nucleus starting at ~5 MeV, regarded as 4-particle 4-hole ($4p4h$) excited states across the $N = Z = 28$ shell gap in terms of the Nilsson diagram[2].

From the shell-model viewpoint, it seems to be a good approximation to truncate the number of bases by restricting the number of nucleons excited across the $N = Z = 28$ shell gap defined as t. Figure 1 presents the convergence of the energy levels as t increases. The ground-state energy is almost convergent at $t \simeq 5$, whereas the energy of $0^+_{\rm excited}$, i.e., the band head of the deformed band, is not, although both states are equally comprised of the pf-shell configurations. This indicates that a very large scale calculation is often needed to describe low-lying states in a systematic way. Full $0\hbar\omega$ calculations are often performed for a unified description of a certain region. While the maximum $M = 0$ basis dimension[a] of the full sd

[a]Since the Hamiltonian conserves the angular momentum, only bases with a designated M (i.e., z component of the total angular momentum) should be taken into consideration, which is called the m-scheme calculation.

shell is about 10^5, that of the full pf shell exceeds 10^9. For heavier nuclei, the dimension is prohibitively large except for ones around the closed shells.

In a conventional way, the shell-model calculation is made by diagonalizing the Hamiltonian with the Lanczos algorithm. Up to about 10^9 m-scheme bases are manageable with some recent codes[3,4]. Beyond this dimension, the numerically exact calculation is impossible, and we must rely on alternative methods. Within the conventional technique, the model space is often truncated, for instance, by restricting the number of nucleons occupying a certain orbit. The shell-model Monte Carlo (SMMC) method[5] is an application of the path integral Monte Carlo technique to the shell model, evaluating the ground-state and thermal properties of a many-body system. Note that for systems with a realistic Hamiltonian, the SMMC suffers from the minus-sign problem. In the so-called VAMPIR method and its families[6], the many-body state is represented by a single or a few quasiparticle basis state(s), optimized by the variational principle, together with the symmetry restoration.

In the Monte Carlo shell model (MCSM)[7] calculation, adopted in the present study, the many-body state is described by a superposition of several tens of the deformed Slater determinants with the symmetry restoration based on the quantum Monte Carlo diagonalization (QMCD) method (See Ref. 8 for details). Candidates of the basis are stochastically generated by a quantum Monte Carlo technique similarly to the SMMC method, and then we select favorable ones among them so as to make the eigenvalue of the Hamiltonian as low as possible. As the selected bases are of high importance, the MCSM can be characterized as the *importance truncation*[7]. The feasibility of the MCSM is confirmed in the pf-shell region, even if the m-scheme dimension becomes huge (see Fig. 1 and Ref. 9 for the comparison with the truncation).

3. Applications

In this section, we discuss how these large-scale calculations, surveyed in the last section, clarify the nuclear structure and the physics behind it, showing some applications. We discuss in Sec. 3.1 the role of the shell structure varying from stable to unstable nuclei in the structure of $N \sim 20$ nuclei, and show in Sec. 3.2 the similar behavior in the pf-shell region. In Sec. 3.3, the valance of the p-p, n-n and p-n interactions is argued on the basis of a systematic description of the medium-heavy nuclei around ^{132}Sn. For detailed discussions of each subject, refer to recent publications

(Refs. 10, 11 and 12 for Secs. 3.1, 3.2 and 3.3, respectively).

3.1. *Exotic nuclei around* $N = 20$

It is known that the $N = 20$ magic structure disappears in neutron-rich nuclei around ^{31}Na, where the ground state is predominantly comprised of the $2p2h$ excitation across the $N = 20$ shell gap. As the mechanism of this disappearance, we have proposed that the gap becomes narrower towards smaller Z, affected by a strongly attractive $T = 0$ monopole interaction between the $j_>$ and $j_<$ orbits[13]. Here, the monopole interaction is a "mean" interaction between two orbits whose angular dependence is averaged out[14], and gives rise to the *effective single-particle energy*[15] depending on the proton and neutron number. In this section, we investigate the importance of the narrowing shell gap related to the nuclear structure calculated by the MCSM with the model space and the Hamiltonian described in Ref. 15.

3.1.1. *Onset of intruder ground state in exotic Na isotopes*

In order to establish the region where the ground state is dominated by the intruder (i.e., $2p2h$) state, we compare the quadrupole moments and the magnetic moments of Na isotopes between the shell-model calculations with recent experimental data by Keim *et al.*[16] in Fig. 2. At $N = 16$ and 17, the moments are well reproduced by the *sd*-shell model, indicating negligible *pf*-shell components in their ground states. The experimental quadrupole moment somewhat deviates from that of the *sd*-shell model at $N = 18$, and finally the experimental moments fully disagree with the *sd*-shell ones at $N = 19$ and 20. On the other hand, the overall agreement with the MCSM is seen. As presented in Fig. 2 (c), this result indicates that the transition to the intruder-dominant ground state occurs already at $N = 19$, contrary to the conventional picture of the so-called "island of inversion"[17].

The mechanism of the disappearance of the magic structure becomes accessible by the clarification of the onset of the intruder ground state at $N = 19$, where the competition between the normal and intruder states severely occurs. In the present Hamiltonian, the shell gap of the Na isotopes is around 3.5 MeV, depending weakly on the neutron number. Note that the shell gap of the doubly magic nucleus ^{40}Ca is as large as about 6 MeV. When we widen the shell gap by shifting the monopole interaction, the ground state of this nucleus moves drastically to the state dominated by the normal state. This transition occurs at the shell gap ~ 4 MeV, where the ground state of the $N = 20$ isotope ^{31}Na is still dominated by the intruder

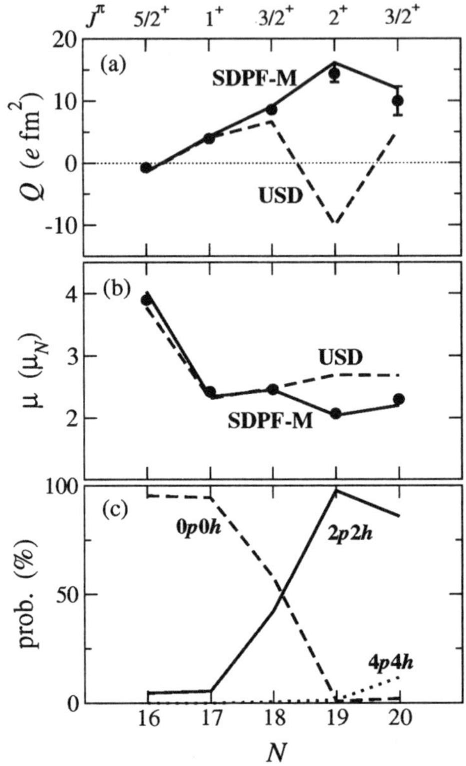

Figure 2. (a) Quadrupole moment, (b) magnetic moment, and (c) $npnh$ ($n = 0, 2, 4$) probabilities in the ground state of Na isotopes compared between the experiment, the MCSM calculation with the SDPF-M interaction, and the sd-shell calculation with the USD interaction. Taken from Ref. 10.

state. Namely, the present result clearly indicates that only the narrow shell gap is allowed to describe the ground state at $N = 19$, supporting the evolution of the shell structure proposed in Ref. 13.

3.1.2. *Stability of the deformation in exotic Mg isotopes*

The magic structure disappears also in some Mg isotopes including $N = 20$ and 22 isotopes ^{32,34}Mg, which is deduced from the 2_1^+ level and the $B(E2)$ value. Even though the large deformation is similarly seen in ^{32}Mg and ^{34}Mg, it has been recently clarified that the ratio of the $E_x(4_1^+)$ to the $E_x(2_1^+)$ is rather different between these two nuclei[18,19]: the ratios are 2.6 and 3.2 for the former and the latter, respectively. This difference attracts theoretical interest, and we have found[20] that this informs us much of the purity of the intruder state and the shell structure as briefly surveyed below.

First, we confirm that the energy levels and the $B(E2; 0_1^+ \to 2_1^+)$ values of ^{32,34}Mg are well reproduced by the MCSM. Both the ground states are dominated by the intruder state, having the large deformation. In order to understand the difference of the energy levels between them, we calculated the potential energy surfaces with the constrained Hartree-Fock calculation. As a result, the minimum of the energy in ^{32}Mg is rather unstable along the γ direction, whereas the prolate deformation in ^{34}Mg is very stable. Indeed, compared with the Davydov-Filippov model[21], the energy levels and the $B(E2)$ values of ^{32}Mg calculated by the MCSM are well described by an asymmetric rotor with $\gamma \sim 20°$, whereas those of ^{34}Mg show a typical rigid symmetric rotor.

Let us further investigate the implication of the stability of the deformation in ^{34}Mg. The $E_x(4_1^+)$ to $E_x(2_1^+)$ ratio of ^{34}Mg is an extraordinarily large value in light nuclei. This is achieved by the stability along not only the γ direction but also the β direction. In ^{34}Mg, another local minimum exists at a smaller β value, dominated by the normal state. Since the energy difference between the two local minima is as large as $\sim$ 2.5 MeV, the resulting ground state by the MCSM is almost a pure intruder state. As the shell gap is narrowed, those two minima compete with each other and the stability in the β direction is disturbed. We confirm that in this situation the $E_x(4_1^+)$ to $E_x(2_1^+)$ ratio deviates from the experimental value. Thus, the experimental result indicates not only that the ground state of ^{34}Mg is dominated by the strongly prolate deformed state but also that the minimum is very stable along the β direction caused by the narrow $N = 20$ shell gap for Mg isotopes.

3.2. *Systematic description of pf-shell nuclei*

It is one of the most important problems in the shell model to construct the *unified* effective interaction by which the nuclear structure in a certain region is systematically described. In the *p*- and *sd*-shell regions, the Cohen-Kurath[22] and the USD[23] interactions are successfully applied, respectively. On the other hand, the unified interaction in the *pf*-shell region was missing, partly because the full shell-model calculation was difficult in this region.

Recently, Honma *et al.* have constructed a new effective interaction in the *pf* shell named GXPF1[11,24]. The total 195 two-body matrix elements

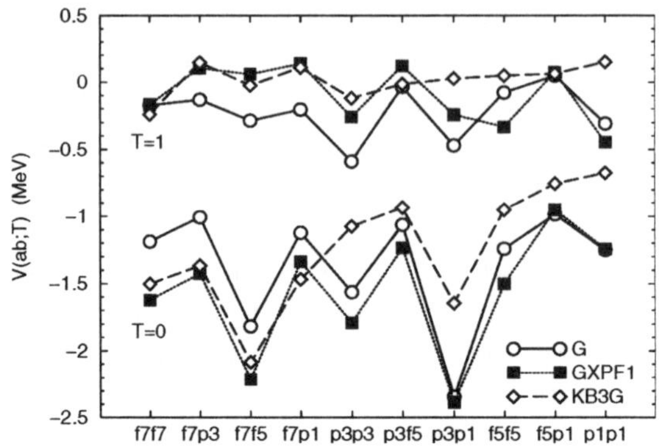

Figure 3. Strength of the monopole interaction of the pf shell compared among the G[25], GXPF1, and KB3G[26] interactions. The label "f5p3" in the horizontal axis denotes $a = f_{5/2}$ and $b = p_{3/2}$, for instance. Taken from Ref. 11.

of the interaction are determined to fit 699 experimental energies of the pf-shell nuclei in a similar way to the construction of the USD interaction[23] in the sd shell. The resulting GXPF1 interaction succeeds in describing not only the yrast levels and the masses but also many non-yrast levels and the electromagnetic properties (see Ref. 11 for details). A significant improvement from existing interactions is achieved in the description of the $N = Z = 28$ magic structure, and it is thus likely to have a high predictive power in the description of the shell structure over the region with appropriate strengths of the monopole interaction.

Figure 3 compares the strength of the monopole interaction between some interactions. The $T = 0$ monopole interactions are more strongly attractive than the $T = 1$ ones as a whole, and show strong orbital dependence: the ones of $f_{7/2}$-$f_{5/2}$ and $p_{3/2}$-$p_{1/2}$, i.e., $j_>$-$j_<$ in general, are the strongest. This situation is the same as the case of the sd-shell interaction, which constitutes one of the major sources of the appearance and disappearance of the $N = 16$ and 20 magic numbers discussed in the last section. In a similar way, the present feature of the monopole interaction predicts the appearance of a new $N = 34$ magic number around $Z = 20$.

3.3. *Medium-heavy nuclei around* ^{132}*Sn*

In heavier nuclei, the number of the valence orbits increases, and the pairing correlation thus becomes more important. It is then efficient to adopt a basis state taking the pairing correlation into account in the MCSM. A pair-condensed state has been taken as the basis state for medium-heavy nuclei on the top of the ^{132}Sn core[27]. Accordingly, it is possible to systematically describe this region with a precise shell-model calculation.

One of the most intriguing features in this region is the transition from the non-collective to collective structure. Using a schematic interaction consisting of the monopole- and quadrupole-pairing interaction and the quadrupole interaction, it is demonstrated that this transition is reproduced in the chain of Ba isotopes within a single framework[27]. With the same interaction, an anomalously small $B(E2; 0^+ \rightarrow 2_1^+)$ value[28] in ^{136}Te is reproduced also[12]. Analyzing its wave functions, it turns out that the 2_1^+ state is predominantly comprised of the neutron quadrupole pair, affected by the difference in the strength of the interactions between p-p and n-n. On the other hand, the Ba isotopes with several valence neutrons show the large quadrupole collectivity dominated by the proton-neutron quadrupole interaction. Thus, the present result not only shows the feasibility of the shell model in this region, but also gives useful information on the valance of the effective interaction between p-p and n-n and p-n.

4. Summary

We present that the shell model has been developing in terms of both the methodology and the application to realistic systems. While the limit of the basis dimension has been raised in the framework of the conventional diagonalization method, the Monte Carlo shell model calculation can be applied to more and more nuclei with a sufficient precision. As a consequence, a systematic description is obtained not only in light regions near stable nuclei but also heavier regions including exotic nuclei. We pay an attention to the varying shell structure from stable to exotic nuclei caused by the monopole interaction, and point out its importance to understand the nuclear structure from a more unified viewpoint. This picture can be drawn only on the basis of the calculation properly including the correlation, whose role is clearly seen in the medium-heavy nuclei, too.

Acknowledgments

This work was supported mainly by Grant-in-Aid for Specially Promoted Research (13002001) from the MEXT and by the RIKEN-CNS collaboration project on large-scale nuclear structure calculation. One of the authors (Y.U.) acknowledges the support in part by Grant-in-Aid for for Young Scientists (14740176) from the MEXT, and the Helios computer system in JAERI.

References

1. A. Arima and H. Horie, *Prog. Theor. Phys.* **11**, 509 (1954).
2. D. Rudolph *et al.*, *Phys. Rev. Lett.* **82**, 3763 (1999).
3. E. Caurier, code ANTOINE (Strasbourg, 1989).
4. T. Mizusaki, code MSHELL, *RIKEN Accel. Prog. Rep.* **33**, 14 (2000).
5. S. E. Koonin, D. J. Dean and K. Langanke, *Phys. Rep.* **278**, 1 (1997).
6. K. W. Schmid, *Prog. Part. Nucl. Phys.* **52**, 565 (2004).
7. T. Otsuka, M. Honma, T. Mizusaki, N. Shimizu and Y. Utsuno, *Prog. Part. Nucl. Phys.* **47**, 319 (2001).
8. T. Otsuka, M. Honma and T. Mizusaki, *Phys. Rev. Lett.* **81**, 1588 (1998), and references therein.
9. T. Mizusaki, T. Otsuka, Y. Utsuno, M. Honma and T. Sebe, *Phys. Rev. C* **59**, R1846 (1999).
10. Y. Utsuno, T. Otsuka, T. Glasmacher, T. Mizusaki and M. Honma, *Phys. Rev. C* **70**, 044307 (2004).
11. M. Honma, T. Otsuka, B. A. Brown and T. Mizusaki, *Phys. Rev. C* **69**, 034335 (2004).
12. N. Shimizu, T. Otsuka, T. Mizusaki and M. Honma, *Phys. Rev. C* **70**, 054313 (2004).
13. T. Otsuka, R. Fujimoto, Y. Utsuno, B.A. Brown, M. Honma and T. Mizusaki, *Phys. Rev. Lett.* **87**, 082502 (2001).
14. For instance, A. Poves and A. Zuker, *Phys. Rep.* **70**, 235 (1981).
15. Y. Utsuno, T. Otsuka, T. Mizusaki and M. Honma, *Phys. Rev. C* **60**, 054315 (1999).
16. M. Keim, *AIP Conf. Proc.* **455**, p. 50 (AIP, New York, 1998); M. Keim *et al.*, *Euro. Phys. J. A* **8**, 31 (2000).
17. E. K. Warburton, J. A. Becker and B. A. Brown, *Phys. Rev. C* **41**, 1147 (1990).
18. F. Azaiez *et al.*, *AIP Conf. Proc.*, 481, p. 243 (AIP, New York, 1998).
19. K. Yoneda *et al.*, *Phys. Lett. B* **499**, 233 (2001).
20. Y. Utsuno, T. Otsuka, T. Mizusaki and M. Honma, in preparation.
21. A. S. Davydov and G. F. Filippov, *Nucl. Phys.* **8**, 237 (1958).
22. S. Cohen and D. Kurath, *Nucl. Phys.* **73**, 1 (1965); *Nucl. Phys. A* **101**, 1 (1967).
23. B. A. Brown and B. H. Wildenthal, *Annu. Rev. Nucl. Part. Sci.* **38**, 29 (1988).
24. M. Honma, T. Otsuka, B.A. Brown and T. Mizusaki, *Phys. Rev. C* **65**, 061301(R) (2002).
25. M. Hjorth-Jensen, T. T. S. Kuo and E. Osnes, *Phys. Rep.* **261**, 125 (1995).
26. A. Poves, J. Sánchez-Solano, E. Caurier and F. Nowacki, *Nucl. Phys. A* **694**, 157 (2001).
27. N. Shimizu, T. Otsuka, T. Mizusaki and M. Honma, *Phys. Rev. Lett.* **86**, 1171 (2001).
28. D. C.. Radford *et al.*, *Phys. Rev. Lett.* **88**, 222501 (2002).

STUDY OF SUPERHEAVY ELEMENTS AT RIKEN: EXPERIMENTS ON THE SYNTHESIS OF ELEMENT 113 IN THE REACTION ^{209}BI(^{70}Zn, n)^{278}113 *

K. MORITA

RIKEN (The Institute of Physical and Chemical Research),
Wako-shi, Saitama 351-0198, Japan
E-mail:morita@rarfaxp.riken.jp

The event assigned to the convincing candidate of an isotope of the 113th element, ^{278}113, and its daughter nuclei, ^{274}Rg and ^{270}Mt, were observed in the reaction ^{209}Bi + ^{70}Zn at a beam energy of 349.0 MeV with a total dose of 1.7×10^{19}, for the first time. Alpha decay energies and decay times of the candidates, ^{278}113, ^{274}Rg, and ^{270}Mt were (11.68±0.04 MeV, 0.344 ms), (11.15±0.07 MeV, 9.26 ms), and (10.03±0.07 MeV, 7.16 ms), respectively. The production cross-section of the isotope were deduced to be $55\ ^{+150}_{-45}$ fb (10^{-39} cm^2).

1. Introduction

It is one of the exciting topics to find out the new element both for nuclear physics and nuclear chemistry. For the heaviest atomic number region (Z≥110), the production cross-sections became extremely small down to the level of sub pico-barn. This fact makes the production of isotopes of the heaviest elements very difficult. At the Institute of Physical and Chemical Research (RIKEN), Japan, we have developed a system to study the heaviest elements using a gas-filled recoil ion separator (GARIS) at the RIKEN Linear Accelerator (RILAC) Facility. Utilizing the high intensity heavy ion beams from the RILAC and the high efficiency, high separating power separator GARIS, we have performed a series of experiments to produce the heaviest elements Z = 108 (Hs), 110 (Ds)[1], 111 (Rg)[2], and 112[3], and measured the α-decay chains of those nucleus successfully. The reactions used

*A part of contents in this report is published in Journal of the Physical Society of Japan **73**, 2593-2596 (2004) by K. Morita, K. Morimoto, D. Kaji, T. Akiyama, S. Goto, H. Haba, E. Ideguchi, R. Kanungo, K. Katori, H. Koura, H. Kudo, T. Ohnishi, A. Ozawa, T. Suda, K. Sueki, H.-S. Xu, T. Yamaguchi, A. Yoneda, A. Yoshida, and Y.-L. Zhao were ^{208}Pb(^{58}Fe, n)^{265}Hs, ^{208}Pb(^{64}Ni, n)^{271}Ds, ^{209}Bi(^{64}Ni, n)^{272}Rg, and, ^{208}Pb(^{70}Zn, n)^{277}112, respectively. Those reactions were firstly studied by the group at the Gesellschaft für Schwerionenforschung (GSI), Germany, and those elements were discovered by their work[4–9]. The results of our experiments reproduce well the results obtained at GSI, providing confirmations for their discovery of those elements.

As an extension of our previous work, we performed experiments aimed at synthesizing an isotope of larger atomic number, $Z = 113$, using the ^{209}Bi + ^{70}Zn reaction. During irradiation, we observed one α-decay chain that can be assigned to subsequent decays from ^{278}113, using the genetic correlation of α-decays connected to the known nuclides, ^{266}Bh and ^{262}Db [10].

The production of element 113 was first reported by Oganessian *et al.* [11] in 2004 using the ^{243}Am(^{48}Ca, xn),(x=3, 4) reaction at the Flerov Laboratory of Nuclear Reaction (FLNR) of Joint Institute of Nuclear Research (JINR), Russia. They reported that the observed decay chains were consistent with the subsequent decays from ^{288}115 produced by the 3n evaporation channel and from ^{287}115 produced by the 4n evaporation channel. In these chains, three atoms of ^{284}113 and one atom of ^{283}113 were consequently assigned to be α-decay daughters of ^{288}115 and ^{287}115, respectively. The chains ended by spontaneous fission of previously unknown dubnium isotopes, ^{268}Db and ^{267}Db. To confirm the atomic number of these isotopes, experiments on chemical separation of long-lived dubnium isotopes[12] were performed at FLNR.

The results of the present work strongly indicate the synthesis of the 113th element, even though the number of chains observed was only one, because the chain ended with known nuclides. More chains are expected to be observed in further experiments.

2. Experimental procedure

A ^{70}Zn ion beam of 352.6 MeV was extracted from RILAC. The absolute accuracy was ± 0.6 MeV. The typical beam intensity on the target was 2.4×10^{12} s^{-1}.

Targets were prepared by vacuum evaporation of metallic bismuth onto carbon backing foils of 30 μg/cm^2 thickness. The thickness of the bismuth layer was about 450 μg/cm^2. The targets were covered by 10-μg/cm^2-thick carbon to protect them from sputtering. Beam energy at the half-depth of the targets was estimated to be 349.0 MeV. Sixteen targets were mounted on a rotating wheel of 30 cm diameter. The wheel was rotated during irradiation at 3000 rpm.

The reaction products were separated in-flight from the beam using a gas-filled recoil ion separator, GARIS [1], and were guided into a detector box placed at the focal plane of GARIS. The separator was filled with helium gas at a pressure of 86 Pa. The value of the magnetic rigidity (Bρ) of GARIS for evaporation residue measurement was set at 2.09 Tm.

Evaporation residues (ERs) are implanted in the position sensitive silicon semiconductor detector (PSD) with 60 mm$\times$60 mm in size after passing through the two timing counters. The timing detector is an assembly of microchannel plates (MCP) that detect secondary electrons emitted from a thin foil by impact of ions passing through the foil. The timing signal was used for two purposes. One is to measure the time of flight (TOF) of incoming particles for rough estimation of their mass number together with energy signals from the PSD. The second purpose is to identify decay events originating from implanted nuclei in the PSD. The events without signals from either timing detector are regarded as the decay events. There is a possibility that decaying particles (α or spontaneous fission decay products) emitted in backward direction are escaped from the PSD because the ranges of decaying particles are larger than those of implanted evaporation residues. In order to detect such decaying particles, we set four semiconductor detectors (SSDs). The size of each SSD was 60 mm$\times$60 mm, and has no-positional sensitivity.

3. Results and discussion

We observed one event of implantation of an evaporation residue (ER) in the PSD followed by four consecutive α-decays terminated by a spontaneous fission decay. The measured positions of all six sequential events were within the spatial resolution of the PSD [1]. The observed energies, time differences between events, and positions of each decay are summarized in Table I, together with those of the implantation event. The energies have been calibrated for α-particles. Therefore, the listed energies for the ER and spontaneous fission are not accurate because possible effects of pulse height defect are not taken into account. The singles counting rate of the PSD was 2 s^{-1} at the typical beam intensity.

The total dose of the ^{70}Zn projectile was 1.7$\times$10^{19}. The production cross section of this specific event was deduced to be 55^{+150}_{-45} fb (10^{-39}cm^2) using the transmission efficiency of GARIS of 0.8 [2]. Indicated uncertainty in the cross section is only statistical one with 1 σ (68% confidence level).

Table I. Observed events. ΔE_{sum}: Energy resolution in FWHM. See text. ΔT: Time differences between events. Position: measured from the bottom of the detector.

	E_{PSD} (MeV)	E_{SSD} (MeV)	E_{sum} (MeV)	ΔE_{sum} (MeV)	ΔT	TOF (ns)	Position (mm)
ER	36.75	-	36.75	-	-	44.6	30.33
α_1	11.68	-	11.68	0.04	0.344 ms	-	30.49
α_2	6.15	5.00	11.15	0.07	9.26 ms	-	30.40
α_3	1.14	8.89	10.03	0.07	7.16 ms	-	29.79
α_4	9.08	-	9.08	0.04	2.47 s	-	30.91
SF	204.1	-	204.1	-	40.9 s	-	30.25

The experiment was designed to produce the isotope, $^{278}113$, by the one-neutron evaporation channel in the ^{209}Bi + ^{70}Zn complete fusion reaction. On the basis of a systematic study of the most probable reaction energies for the one-neutron evaporation channel in the ^{208}Pb(^{64}Ni, n)^{271}Ds [1], ^{209}Bi(^{64}Ni, n)^{272}111 [2], and ^{208}Pb(^{70}Zn, n)^{277}112 [3] reactions, a reaction energy of 349.0 MeV at the half-depth of the targets was adopted to maximize the relevant cross section for producing $^{278}113$. The corresponding energy of the center of mass frame is 261.4±2.0 MeV, where ±2.0 MeV indicates the range of reaction energy due to the energy loss of the beam in the bismuth targets. The excitation energy (E^*_{CN}) of the compound nucleus, $^{279}113$, is calculated to be 14.1±2.0 MeV using the predicted mass for a compound nucleus [13] and the experimental masses of the beam and target.

If we assume that the implanted nucleus followed by the decay chain observed in the present study is the product of the ^{209}Bi(^{70}Zn, n)^{278}113 reaction, the members of the decay chain are assigned to be ^{278}113, ^{274}111, ^{270}Mt, ^{266}Bh, and ^{262}Db.

The isotope ^{262}Db is known to decay by α and spontaneous fission and branching ratios of ≈64% and ≈33%, respectively, with a half-life ($T_{1/2}$) with the 1σ error (68% confidence level) of 34±4 s [14]. The measured decay time of 40.9 s for SF, corresponding to $T_{1/2} = 28^{+140}_{-13}$ s, is well consistent with the reported value.

The α-decay of ^{266}Bh produced by the ^{249}Bk + ^{22}Ne reaction was reported by Wilk *et al* [15]. This decay was followed by the α-decays of ^{262}Db and ^{258}Lr. The decay energy (E_α) and decay time for ^{266}Bh were 9.29 ($\pm \sim$ 0.10) MeV and 0.87 s, respectively. The decay time, 0.87 s, corresponds to $T_{1/2} = 0.6^{+2.9}_{-0.3}$ s. The measured decay time of α_4 in the present work, 2.47 s, corresponding to $T_{1/2} = 1.7^{+8.2}_{-0.8}$ s, agrees within the 1σ error with the value reported by Wilk *et al.* Considering the rather wide distribution of α-decay energies from odd-odd nuclei, as shown in the case of ^{272}111[2], our present value for α_4 ($E_\alpha = 9.08 \pm 0.04$ MeV) dose not contradict the value reported in the reference [15]. It should be noted that only one α-decay event was reported in the reference [15].

The possibility of a radiative capture process (0-neutron evaporation channel) to form the observed decay chain could not be excluded purely on the basis of the property of the measured SF. However, this channel is excluded by considering the excitation energy of the compound nucleus. The calculated value, $E^*_{CN} = 14.1 \pm 2.0$ MeV, is too high to populate the ground state of the compound nucleus by only γ-ray emission, because the calculated one-neutron separation energy of the compound nucleus is only 7.5 MeV, using the same mass prediction [13].

The two-neutron evaporation channel is also excluded in this excitation energy because the phase volume for the channel is small compare to the one-neutron evaporation channel.

The possibility of a one-proton evaporation channel leading to the product of ^{278}112 is excluded by comparing the observed decay time of SF with that of ^{262}Rf which decays by 100% spontaneous fission with a 47±5 ms half-life [16], greatly shorter than the observed value, $T_{1/2} = 28^{+140}_{-13}$ s.

In conclusion, the reaction product, followed by the decay chain observed in our experiment, was considered to be most probably due to the ^{209}Bi(^{70}Zn, n)^{278}113 reaction. As a result, the members of the decay chain were consequently assigned as ^{278}113, ^{274}Rg, ^{270}Mt, ^{266}Bh, and ^{262}Db. The decay energies and times are listed in Table I.

Acknowledgments

The experiments were carried out in collaboration with the following coworkers. K. Morimoto, D. Kaji, H. Haba, R. Kanungo, K. Katori, T. Ohnishi, T. Suda, A. Yoneda, A. Yoshida (RIKEN), T. Akiyama, T. Yamaguchi (Saitama Univ.), S. Goto, H. Kudo (Niigata Univ.), E. Ideguchi (Center for Nuclear Study, Univ. of Tokyo), H. Koura (JAERI Tokai), A. Ozawa, K. Sueki (Univ. of Tsukuba), H.-S. Xu (IMP Lanzhou, Chinese Academy of Sci.), and Y.-L. Zhao (IHEP Beijing, Chinese Academy of Sci.).

The author thanks all accelerator staff members for excellent operation for very long period of beamtime.

References

1. K. Morita, K. Morimoto, D. Kaji, H. Haba, E. Ideguchi, R. Kanungo, K. Katori, H. Koura H. Kudo, T. Ohnishi, A. Ozawa, T. Suda, K. Sueki, I. Tanihata, H. Xu, A. V. Yeremin, A. Yoneda, A. Yoshida, Y.-L. Zhao, and T. Zheng, *Eur. Phys. J.* **A21**, 257(2004).
2. K. Morita, K. Morimoto, D. Kaji, H. Haba, E. Ideguchi, J. C. Peter, R. Kanungo, K. Katori, H. Koura H. Kudo, T. Ohnishi, A. Ozawa, T. Suda, K. Sueki, I. Tanihata, H. Xu, A. V. Yeremin, A. Yoneda, A. Yoshida, Y.-L. Zhao, T. Zheng, S. Goto, and F. Tokanai, *J. Phys. Soc. Jpn.* **73**, 1738 (2004).
3. K. Morita et al; to be submitted in *J. Phys. Soc. Jpn.*.
4. S. Hofmann, *Rep. Prog. Phys.* **61**, 639 (1998).
5. S. Hofmann, V. Ninov, F. P. Heßberger, P. Armbruster, H. Folger, G. Münzenberg, H. J. Schött, A. G. Popeko, A. V. Yeremin, A. N. Andreyev, S. Saro, R. Janik and M. Leino, *Z. Phys.* **A350**, 281 (1995).
6. S. Hofmann, V. Ninov, F. P. Heßberger, P. Armbruster, H. Folger, G. Münzenberg, H. J. Schött, A. G. Popeko, A. V. Yeremin, S. Saro, R. Janik and M. Leino, *Z. Phys.* **A354**, 229 (1996).
7. S. Hofmann and G. Münzenberg, *Rev. Mod. Phys.* **72**, 733 (2000).
8. S. Hofmann, F.P. Heßberger, P. Armbruster, G. Münzenberg, S. Antalic, P. Cagarda, B. Kindler, J. Kojouharova, M. Leino, B. Lommel, R. Mann, A. G. Popeko, S. Reshitko, S. Śaro, J. Uusitalo and A. V. Yeremin, *Eur. Phys. J.* **A14**, 147 (2002).
9. S. Hofmann, *J. Nucl. Radiochem. Sci.* **4**, R1 (2003).
10. K. Morita, K. Morimoto, D. Kaji, T. Akiyama, S. Goto, H. Haba, E. Ideguchi, R. Kanungo, K.. Katori, H. Koura, H. Kudo, T. Ohnishi, A. Ozawa, T. Suda, K. Sueki, H. Xu, T. Yamaguchi, A. Yoneda, A. Yoshida, and Y.-L. Zhao, *J. Phys. Soc. Jpn.* **73**, 2593 (2004).
11. Yu. Ts. Oganessian *et al.*, *Phys. Rev.* **C69**, 021601(R) (2004).
12. S. N. Dmitriev *et al.*, *Proceedings of International Symposium on Exotic Nuclei (EXON2004), Peterhof, Russia, July 2004, World Scientific Publishing.*
13. W. D. Myers and W. J. Swiatecki, *Nucl. Phys.* **A601**, 141 (1996).
14. R. B. Firestone and V. S. Shirley, *Table of Isotopes, 8th ed.* (John Wiley & Sons, New York, 1996).
15. P. A. Wilk, K. E. Gregorich, A. Türler, C. A. Laue, R. Eichler, V. Ninov, J. L. Adams, U. W. Kirbach, M. R. Lane, D. M. Lee, J. B. Patin, D. A. Shaughnessy, D. A. Strellis, H. Nitsche, and D. C. Hoffman, *Phys. Rev. Lett.* **85**, 2697 (2000).
16. L. P. Somerville, M. J. Nurmia, J. M. Nitschke, and A. Ghiorso, E. K. Hulet and R. W. Lougheed, *Phys. Rev.* **C31**, 1801 (1985).

DYNAMICAL SYMMETRIES FROM ISOSPIN TO X(5): PAST AND FUTURE RESEARCHES AT THE GASP ARRAY

P. G. BIZZETI

Dipartimento di Fisica, Universitá di Firenze
I.N.F.N., Sezione di Firenze
Via G. Sansone 1
I 50019 Sesto Fiorentino (Firenze), Italy
E-mail: bizzeti@fi.infn.it

A few examples of recent experimental and theoretical works performed in the frame of the GASP collaboration are reported. Selected examples concern recent results on dynamical symmetries and phase transitions in nuclear shape: in particular, the isospin symmetry in the $A = 35$ isobaric doublet, the X(5) symmetry at the critical point of phase transition in ^{178}Os and the new field of phase transitions in the octupole mode.

1. Introduction

After the last Italy–Japan Symposium held in Tokyo three years ago, a large amount of work in the field of nuclear spectroscopy has been performed at the Laboratori Nazionali di Legnaro. Part of it will be reported to this Conference by Santo Lunardi and Andrès Gadea. I will report here some examples of experimental work done at GASP (the Great Array for nuclear SPectroscopy at L.N.L.) and related theoretical results.

The GASP array of L.N.L. has completed the first decade of successful operation in September 2002. A careful revision of the electronics and of the cryogenic system has been performed in the last few months and the system is now ready for a new cycle of operation. At the moment, after the end of the EUROBALL IV campaign in Strasbourg, GASP remains one of the most powerful arrays for gamma spectroscopy in Europe. Among the different lines of experimental and theoretical research performed by the GASP user community, I have selected the one concerning dynamical symmetries in nuclei, from the isospin symmetry to those related to the phase transitions in nuclear shape. I will discuss in particular three recent results, one concerning the Isospin symmetry, one the X(5) symmetry at the critical point of phase transition, and finally a new approach to the description of quadrupole + octupole deformation, with a possible example of phase transition in the octupole mode.

2. Isospin symmetry

The recent revival of interest for the consequence of isospin symmetry and for the effects of symmetry breaking interaction is related, on the experimental side, to the new investigations of nuclei with $Z \geq N$ up to the limit of the proton drip line and, on the theoretical side, to the progress in nuclear shell-model techniques which now permits to consider a model space as large as the entire $f-p$ shell. We will consider two sorts of observables: the *Coulomb energy differences* in isospin multiplets, which provide a sort of microscope to evidence tiny effects of nuclear structure, and the strengths of *mirror electromagnetic transitions* which can give a signature of the symmetry breaking. The amount of symmetry breaking, exemplified by the *isospin mixing* in the ground state (Fig. 1), is expected to increase with the nuclear mass A and, for a given A, to be maximum for $Z = N$.

It can be useful to recall now two *theoretical* results which have been obtained thanks to recent measurements of the Coulomb energy differences in the $f-p$ shell. The first concerns the collective behavior of the mirror pair

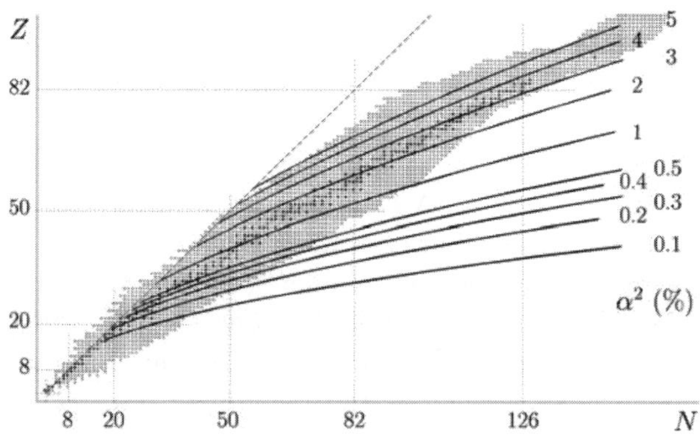

Figure 1. Contour plot for the isospin mixing in the ground state of even-even nuclei, due to the effect of the monopole Coulomb potential[1].

^{50}Cr – ^{50}Fe, predicted by large–scale shell model calculations. A rotational–like spectrum is predicted (and observed) at low spin, while at $J \geq 8$ the nucleon pairs start to break, and the nucleons align (as much as possible)

their individual angular momenta to the collective one. The experimental trend of the Coulomb Energy Differences (CED) reflects the decrease of Coulomb energy at the point where a proton pair breaks, and confirm that at $J = 8$ a proton pair breaks in ^{50}Fe and a neutron pair in ^{50}Cr, while the situation is reversed at $J = 10$ – as correctly predicted by the shell-model calculations[2].

The second result concerns the origin of the symmetry breaking. The calculations of CED in the $A = 50$ nuclei, reported in ref. [2], originally included only the Coulomb forces. More recently, it has been observed that additional isospin-violating terms must be introduced in the shell-model residual interactions, in order to reproduce the large body of experimental observables for the $f - p$ shell nuclei[3]. At the moment it is not clear, at least to me, whether this is due to the symmetry breaking at the level of fundamental interaction (which certainly exist, but at a rather low level) or to the necessary renormalization of residual interactions due to the finite size of the shell-model space.

The experimental work on the $f-p$ shell nuclei is still continuing. I will cite, one for all, the very recent results[4] on the excited states of ^{54}Ni. These results complete the information on the $(f_{7/2})^{-2}$ states of the $A = 54$ triplet which, in a model space limited to the $f_{7/2}$ shell, would be the particle-hole conjugates of those of the triplet $A = 42$.

Table 1. Experiments performed or under way at GASP and EUROBALL.

In the $f - p$ shell:		GASP	EUROBALL
^{47}V–^{47}Cr	D. Tonev *et al.*[5],		•
^{50}Cr–^{50}Fe	S.M. Lenzi *et al.*[2]		•
^{51}Mn–^{51}Fe	R. du Riez *et al.*[6], under way	•	
^{59}Cu–^{59}Zn	C. Andreoiu *et al.*[7]	•	
^{54}Ni	A. Gadea *et al.*[4]		•
^{62}Ga	D. Rudolph *et al.*[8]	•	
^{64}Ge	E. Farnea *et al.*[9]	•	•
In the $s - d$ shell:		**GASP**	**EUROBALL**
^{21}Ne–^{21}Na	S.Thummerer *et al.*[10]	•	
^{31}P–^{31}S	F. Della Vedova *et al.*[11], under way	•	
^{35}Cl–^{35}Ar	J.Ekman *et al.*[12], F. Della Vedova *et al.*[13].	•	

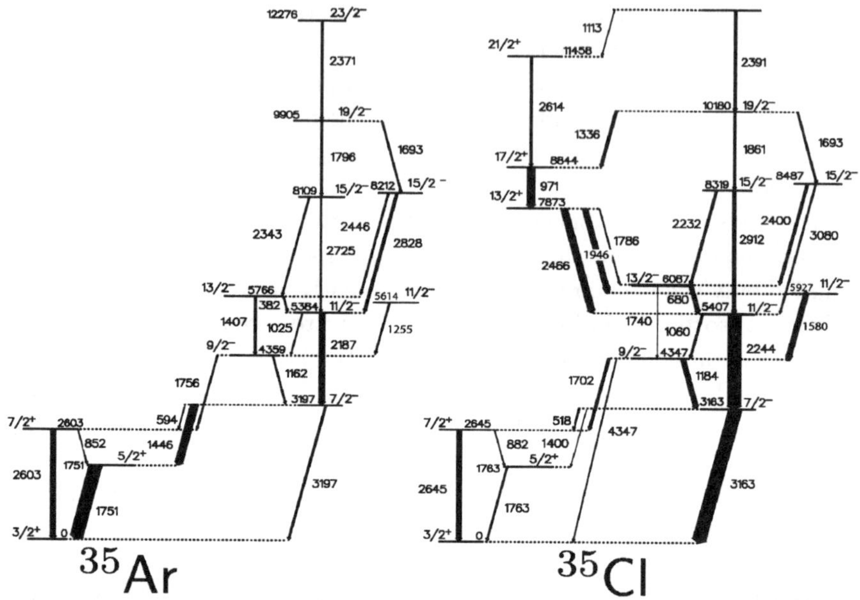

Figure 2. Level schemes and electromagnetic transitions in the mirror pair ^{35}Cl–^{35}Ar, as they result from recent measurements at GASP (by courtesy of F. Della Vedova[11]). In addition to the ^{35}Cl states shown in the figure, three more states of negative parity and seven more of positive parity have been identified in this work.

Recent measurements on $f_{7/2}$ nuclei, performed with GASP and EUROBALL (or both) are listed in the Table 1. More recently, the investigation has been extended to the $s-d$ shell nuclei, and this part is entirely carried out at GASP. I will discuss in some more detail the recent results on the $A = 35$ doublet, shown in the fig. 2. The measurements have been performed at GASP, with the reaction ^{16}O + ^{24}Mg at the beam energy of 70 MeV, and coincidences with charged particles (in the *ISIS Silicon ball*) and/or neutrons (in the forward *neutron ring*) have been used to select the very weak reaction channels leading to ^{35}Cl and ^{35}Ar.

It is impressive to observe how poor information was available, prior to this work, on such a light nuclei. In ^{35}Ar only the two lowest excited states were known, while in ^{35}Cl seven new positive–parity levels (up to $25/2^+$ and 16201 keV) and fourteen new negative–parity ones (up to $29/2^-$ and 22066 keV) have been identified in this measurement. At the moment, only the lowest part of the ^{35}Ar level scheme has been published on Phys. Rev. Lett.[12], while the rest has been reported at the Argonne Conference "Nuclei at the limits" last July[13]. These data will provide an excellent playground for the shell-model calculations, presently under way, which use a model space including the sub-shells $2s_{1/2}$, $1d_{3/2}$, $1f_{7/2}$, $2p_{3/2}$ (as in a former work[14] on ^{34}Cl). Here the situation is somewhat different from that of nuclei in the middle of the f-p shell. In fact, only the positive parity levels with $J \leq 9/2$ can be explained within the simple $2s_{1/2}-1d_{3/2}$ configuration. Negative-parity states, and higher-spin levels of both parities, require one or more particles to be promoted to the next major shell.

This shell-model space is perhaps not completely satisfactory, but is large enough to account for most of the level energies and transition probabilities, with the noticeable exception of E1 transitions. To account for the latter, one would need to include also the $1d_{5/2}$ subshell in order to complete the $s-d$ major shell and allow only a limited number of particle-hole excitations to the $f-p$ shell: a task which is probably not far from the present possibilities.

Actually, the comparison of mirror E1 strengths is somewhat surprising[12,13]. In fact, the branching ratios in the decay of the lowest $7/2^-$ state (two E1 and one M2 transitions, if $\delta = 0$) are very different in the two mirror nuclei ^{35}Cl, ^{35}Ar and this is quite unexpected if one considers the isospin symmetry rules. Namely, electromagnetic amplitudes (as all one-body amplitudes) include an isoscalar and an isovector part, which combine with opposite sign in the two mirror transitions:

$$\mathcal{M}(T_i \to T_f, T_3) = (-1)^{T_f - T_3}\, \mathcal{M}^{(0)}(T_i \to T_f) + (-1)^{T_f - T_3 + 1}\, \mathcal{M}^{(1)}(T_i \to T_f).$$

But, *in the long-wavelength limit* and *if the isospin symmetry holds*, the isoscalar part $\mathcal{M}^{(0)}$ vanishes in the E1 transition amplitudes, and mirror E1 transition should have *equal reduced strengths*. Similar, but weaker, symmetry rules (isoscalar part hindered by an order of magnitude compared to the isovector one) hold for M1 transitions and for those single-particle ML transitions[15] in which the change of single-particle angular momentum Δj equals the multipole order L.

The E1 symmetry rule can be to some extent violated due to the isospin–symmetry breaking in the parent and/or in the daughter state, and also to the *isoscalar terms* in the E1 amplitudes, which are small but not zero out of the limit $kR \ll 1$. In the part of the transition amplitude which depends on the charge density, the first isoscalar term comes from the second-order term in the series expansion

$$(1 + \partial/\partial r)\; j_1(kr) \approx 2\; kr\; [1 - (kr)^2/5 + ...]\;.$$

In our case, it is less than 10^{-4} in comparison to the leading one. Another isoscalar term comes from the coupling to magnetization densities, and can be estimated at the level of 10^{-3}. Therefore, these terms can be relevant when B(E1)$\leq 10^{-6}$W.u.

In ^{35}Cl, the mean life of the $7/2^-$ level is $\tau_m = 41.8$ ps and the reduced strengths of the γ decay transitions are B(E1)= 1.2 10^{-5} W.u., B(E1)= 2.0 10^{-8} W.u., and B(M2)= 0.277 W.u., for the transitions to the $7/2^+$, $5/2^+$ and $3/2^+$ state, respectively. The isoscalar part of the E1 amplitudes can be important at the level of 10^{-8} W.u., but not so much at the level of 10^{-5}, as in the case of the $7/2^- \rightarrow 7/2^+$ transition. Therefore, if the isospin symmetry holds, we can expect comparable values of B(E1) for the mirror transitions to the $7/2^+$ level. Taking into account the corresponding branching ratios, we can estimate for the $7/2^-$ level of ^{35}Ar a mean life $\tau_m \approx 40$ ps, and a reduced strength B(E1)$\approx$ 5.2 10^{-6} W.u. (B(M2)$\approx$ 0.042 W.u.) for its decay to the $5/2^+$ ($3/2^+$) level. There would be, in this case, a difference of two orders of magnitude between the B(E1) values of the mirror transitions to the $5/2^+$ state[a]. The strength B(M2) to the $3/2^+$ level, instead, would be six times smaller in ^{35}Ar than in ^{35}Cl. Alternatively, we could assume that the mirror B(M2) values are comparable (as expected for a single-particle $f_{7/2} \rightarrow d_{3/2}$ transition): then the mean life of the ^{35}Ar level should be seven times shorter and the B(E1) seven times larger than the above given values, and the difference between mirror E1 strengths would increase.

At this point, our conclusion is simple: one needs a measurement of the lifetime of the $7/2^-$ level of ^{35}Ar. In fact, such a measurement is in program at GASP[16], and will be performed soon. If this measurement will confirm a large asymmetry in the mirror strengths, at a level that cannot be explained by the small isoscalar amplitudes, does it imply a corresponding symmetry violation in the level wavefunctions? Yes, indeed, but not necessarily of the same amount. In fact, a large part of the isospin mixing in a low lying level is due to the *monopole* part of the Coulomb interaction, and involves states of the *Isovector Giant Monopole Resonance* (IGMR), two major shells above the level considered (let say, the final state of the E1 transition). The initial level has opposite parity, and is therefore located one major shell above (or below). The Giant Dipole Resonance (GDR) built upon it involves the next major shell, and partially overlaps with the IGMR built on the final state. It is possible to show[17,18] that the individual contributions from the mixing with levels of the IGMR add coherently in the forbidden E1 amplitude, so

---

[a] A significant contribution of M2 can be excluded for this transition in ^{35}Ar, as it should be at the level of 500 W.u.

that part of the E1 strength of the GDR is transferred to the forbidden transition. By using the closure approximation, the forbidden amplitude can be expressed in a form involving only the wavefunctions of the initial and final states, and can therefore be evaluated in a shell-model basis[b].

3. Phase transitions and the X(5) symmetry

From the isospin symmetry in light nuclei, let us now jump to much heavier nuclei and to the symmetries of collective models. It is well known that algebraical symmetries SU(3), U(5) and O(6) correspond, either exactly or in the limit of large particle numbers, to simple cases described by geometrical models as rigid rotation, harmonic vibration and gamma-unstable motion. An unified description of these limiting cases, as well as of all the intermediate ones situated inside or at the borders of the Casten triangle, is provided by the Arima – Iachello Interacting Boson Model. Obviously, this general model requires a number of adjustable parameters (at least 6, in the simplest case of IBM1), while the geometrical description of the limiting cases is essentially parameter free.

Recently, it has been shown by Iachello[19,20] that new symmetries (called E(5) and X(5)) exist at the critical point of phase transition between U(5) and O(6) or U(5) and SU(3). At these points, the geometrical model gives simple, parameter free predictions if one assumes that the potential energy remains practically constant when the deformation parameter β varies from zero to some upper limit β^0, and goes to $+\infty$ outside this interval. In the case of X(5) the asymmetry parameter γ is confined to very small values by a proper harmonic potential, while for E(5) it is not subject to restoring forces. We shall consider in particular the X(5) case. Evidence of phase transition is provided by a plot of the *order parameter* $E(4^+)/E(2^+)$ as a function of a proper *driving parameter*, such as the proton number Z (at constant N) or the neutron number N (at constant Z). The value predicted by the model for the order parameter at the critical point is 2.91. If it is so, also the excitation energies of all other levels are predicted by the model. Fig. 3 shows two examples of nuclei[21,22] whose level scheme is in a rather good agreement with the model predictions. The model, however, also predicts the ratios of reduced strengths B(E2) for the transitions in the ground-state band, and these predictions are in rough agreement with the experimental values for ^{152}Sm, but not[23] for ^{104}Mo. It is essential,

---

[b]I like to remember that this result was first presented at the Annual Meeting of the Physical Society of Japan in 1984.

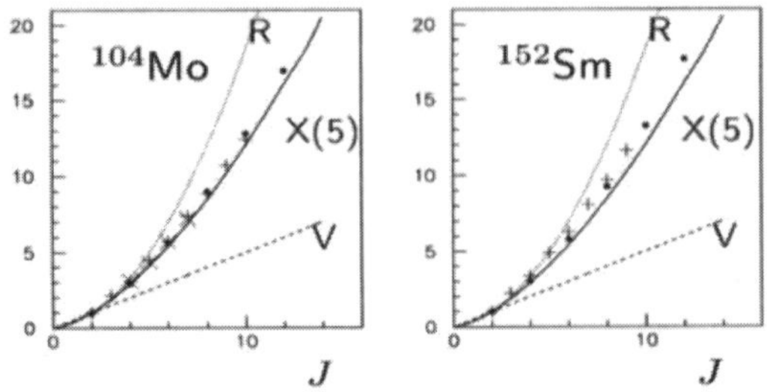

Figure 3. Excitation energy ratios $E(J)/E(2^+)$ for the levels of the ground-state band (•) and of excited γ bands (+, $K=2$; ×, $K=4$) in ^{104}Mo and ^{152}Sm.

therefore, that the B(E2) strength be measured in all other possible X(5) candidates. Several measurements of this kind have been performed or are planned at GASP[24,25,26]. I will report here some preliminary results of the measurements on ^{178}Os, the first example in a new region of possible X(5) candidates. Measurements have been performed by the recoil–distance method, using the Plunger device of the Cologne group.

Table 2. Experiments on X(5) nuclei performed or under way at GASP

^{156}Dy	P. Petkov *et al.*[24]
^{178}Os	A. Dewald *et al.*[25], under way
^{176}Os	A. Dewald *et al.*[26], in program

The preliminary results on the transition quadrupole moments Q_t, defined by the relation $B(\mathrm{E2}, J \to J-2) = (5/16\pi)(J,0,2,0|J-2,0)^2|Q_t|^2$, are shown in the fig. 4, and come out to be consistent with the model prediction. It is expected that smaller error bars will be reported when the data analysis will be completed.

4. Phase transitions in the octupole mode

The octupole mode of nuclear deformation has been the object of several measurements at GASP in the past years. Now, we are investigating – in analogy to the Iachello approach for the quadrupole case – the critical point of phase transition in the octupole mode. To this aim, a new theoretical frame has been developed[27] in order to describe the combined quadrupole and octupole motion, with the intrinsic reference frame referred

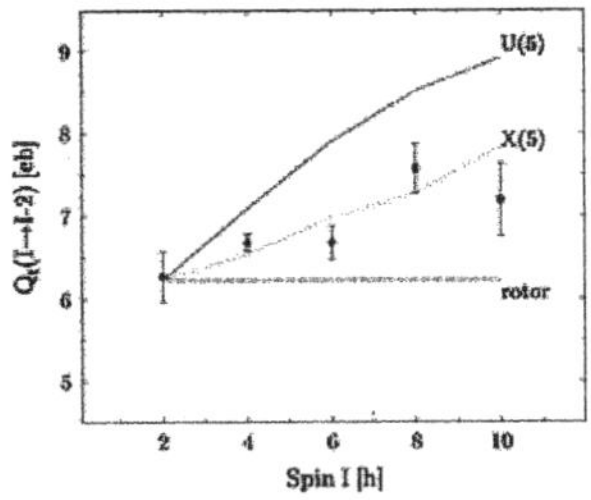

Figure 4. Comparison of the transition quadrupole moments Q_t derived from experimental values of the level mean life in ^{178}Os with the predicted values for harmonic vibration (U(5)), for rigid rotation, and at the critical point (X(5)). From ref. [25].

to the principal axes of the overall tensor of inertia. This treatment is valid when the non–axial amplitudes are small in comparison to the axial ones, but not necessarily frozen to zero. Detailed predictions[27,28] have been obtained for the phase transition between octupole oscillations and permanent octupole deformation in nuclei which already possess a stable quadrupole deformation $\bar{\beta}_2$. In this case, we have assumed for β_3 a square–well potential, extending symmetrically around zero, from $-\beta_3^0$ to $+\beta_3^0$, in order to respect fundamental symmetries (as time reversal). At variance with the X(5) model, our model has a free parameter, the ratio $b = \sqrt{2}\beta_3^0/\bar{\beta}_2$.

In the Thorium isotope chain, ^{226}Th turns out to be close to the model predictions for the critical point (Fig. 5). In the frame of the model, definite predictions can be obtained also for the ratios of transi-

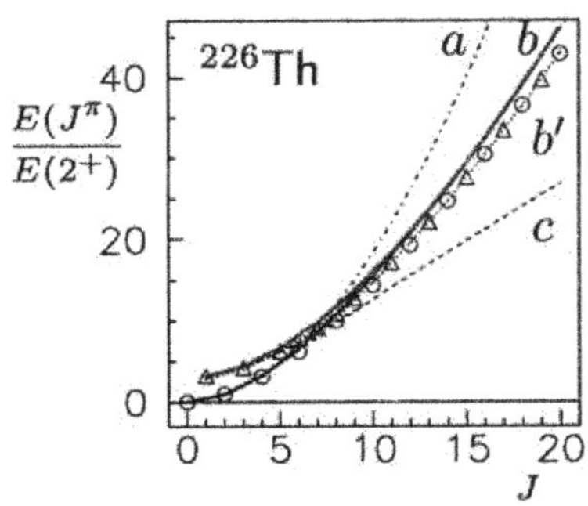

Figure 5. Values of the ratio $E(J^\pi)/E(2^+)$ for the positive- and negative-parity parts of the ground-state band of ^{226}Th, compared with different theoretical models: rigid rotator (a, positive parity only); present model with critical potential, fit on the 1^- level (b) or fit on the 20^+ level (b'); same model with harmonic potential, fit on the 1^- (c).

tion strengths. The known branching ratios are in rough agreement with the model predictions[27,28]. The measurement of the absolute E1 and E2 strengths, at least in the ground-state band, is an experimental problem that we are considering, but which will take some more time to be solved.

It is a pleasant duty to thank here all the members of the GASP group and of the GASP collaboration, and in particular Prof. A. Dewald, prof. S. Lenzi and Dr. F. Della Vedova for the permission to use their results.

References

1. G. Coló *et al.*, *Phys. Rev.* C **52**, 1175 (1995).
2. S.M. Lenzi *et al.*, *Phys. Rev. Lett.* **87**, 122501 (2001).
3. A.P. Zuker, S.M. Lenzi, G. Martinez-Pinedo and A. Poves, *Phys. Rev. Lett.* **89**, 142502 (2002).
4. A. Gadea *et al.*, *Int. Conf. "Nuclear Physics around the Coulomb barrier"* (Strasbourg, September 2004).
5. D. Tonev *et al.*, *Phys. Rev.* C **65**, 034314 (2002).
6. R. Du Riez *et al.*, *Annual Report L.N.L.* (2003) p. 5.
7. C. Andreoiu *et al.*, Europ. Phys. J, A **15**, 459 (2002).
8. D. Rudolph *et al.*, *Phys. Rev.* C **69**, 34309 (2004).
9. E. Farnea *et al.*, *Phys. Lett.* B **511**, 56 (2003).
10. S. Thummerer *et al.*, *J. Phys.* G **29**, 509 (2003).
11. F. Della Vedova *et al.* , *Annual Report L.N.L.* (2003) p. 3.
12. J. Ekman *et al.*, *Phys. Rev. Lett.* **92**, 132502 (2002).
13. F. Della Vedova, *Int. Conf. "Nuclei at the limits"* (Argonne National Lab., July 2004).
14. H. Mason *et al.*, *to be published.*
15. E.K. Warburton and J. Weneser, in *Isospin in Nuclear Physics* (Ed. D.H. Wilkinson, Amsterdam 1969), p. 173.
16. R. Du Riez *et al.*, *GASP proposal* 04/30 (approved).
17. M. Bini *et al.*, *Lett. Nuovo Cimento* **41**, 191 (1984).
18. P.G. Bizzeti, in *in Exotic nuclei at the proton drip line* (Ed.s C.M. Petrache and G. Lo Bianco, Camerino 2001), p.29.
19. F. Iachello, *Phys. Rev. Lett.* **85**, 3850 (2000).
20. F. Iachello, *Phys. Rev. Lett.* **87**, 052502 (2001).
21. R.F. Casten and N.V. Zamfir, *Phys. Rev. Lett.* **87**, 052503 (2001).
22. P.G. Bizzeti and A.M. Bizzeti-Sona, *Phys. Rev.* C **66**, 031301(R) (2002).
23. C. Hutter *et al.*, *Phys. Rev.* C **67**, 054315 (2003).
24. P. Petkov *et al.*, *Phys. Rev.* C **68**, 034328 (2003).
25. A. Dewald *et al.*, *Annual Report L.N.L.* (2003) p. 21.
26. A. Dewald *et al.*, *GASP proposal* 04/38 (approved).
27. P.G.B Bizzeti and A.M. Bizzeti-Sona, arXiv Nucl–Th/0409031; *Phys. Rev.* C **70** (in the press).
28. P.G. Bizzeti and A.M. Bizzeti-Sona, Europ. Phys. J, A **20**, 179 (2004).

REACTION MECHANISMS FOR STRUCTURE STUDIES OF HALO AND WEAKLY BOUND NUCLEI

ANGELA BONACCORSO *

INFN, Sez. di Pisa and Dipartimento di Fisica, Università di Pisa, Largo Pontecorvo 3, 56127 Pisa, Italy.
E-mail: bonac@df.unipi.it

In this paper we review two important aspects of the study of reaction mechanisms for structure studies of halo and weakly bound nuclei. The first is related to the importance of a proper interpretation of elastic scattering data on heavy nuclei, by including the Coulomb breakup effects on the microscopic optical potentials. The second has to do with understanding the special characteristics of two-neutron halo nuclei such as ^{14}Be, in which the two neutron pair is bound, while each single extra neutron is unbound in the field of the core. A model to describe projectile fragmentation, following which neutron-core coincidences are recorded and the neutron-core relative energy spectrum is reconstructed, is presented with an application to the study of ^{13}Be. We wish to see whether the neutron-^{12}Be relative energy spectra obtained from fragmentation of ^{14}Be or ^{14}B would show differences predictable in a theoretical model.

1. Introduction

A standard method to calculate elastic scattering angular distributions is to use DWBA. At high energies eikonal methods are also very accurate and much simpler to implement numerically in particular for very heavy systems. In both cases it is important to use reliable optical potentials. They are often just fitted to the data but there is also a large literature on microscopic optical potentials. Although they give sometimes less accurate fits, they are the only way to extract structure information from elastic scattering data.

Since the advent of Rare Isotope Beams (RIBs) [1] elastic nucleus-nucleus scattering with a radioactive projectile is a reaction which has been studied to a large extent in the attempt to find characteristics that would be typical

*In collaboration with G. Blanchon, A. Ibraheem, A. Garcia-Camacho, D. M. Brink, N. Vinh Mau.

for a weakly bound nucleus and would help understanding new phenomena such as the existence of the halos. It has been established that for halo projectiles breakup of the valence particles is responsible for a large angle damping in the elastic angular distribution. For weakly bound nuclei there are two mechanisms responsible for the breakup. The nuclear attraction of the neutron by the target and the core-target Coulomb repulsion which gives rise to an effective force on the neutron, causing it to separate from the core. These two mechanisms have recently been taken into account consistently in Refs. [2] and [3]. The corresponding optical potentials have been constructed in Refs.[4] and [5].

On the other hand light unbound nuclei have attracted much attention [1,9,12] in connection with two neutron halo nuclei. A precise understanding of unbound nuclei is essential to determine the position of the driplines in the nuclear mass chart. A fundamental question to answer, in order to understand the structure of matter, is indeed what makes a certain number of neutrons and protons to bound together, while adding an extra neutron would lead to an unbound nucleus. Sometimes adding instead two neutrons leads to bound nuclei. Those are two-neutron halo nuclei such as ^{6}He, ^{11}Li, ^{14}Be, in which the two neutron pair is bound, while each single extra neutron is unbound in the field of the core. In a three-body model these nuclei are described as a core plus two neutrons. The properties of core plus one neutron system are essential because structure models rely on the knowledge of angular momentum, parity, energies and spectroscopic strength for neutron resonances in the field of the core. An essential ingredient is therefore the neutron-core effective potential. The neutron elastic scattering at very low energies on the "core" nuclei is not feasible at the moment as many such cores, like ^{9}Li, ^{12}Be are themselves unstable and cannot be used as targets. Indirect methods, aiming at the determination of the energy and angular momentum of the neutron-core continuum states, have been used so far. One is projectile fragmentation, following which neutron-core coincidences are recorded and the neutron-core relative energy spectrum is reconstructed[1]. The other is neutron transfer from a deuteron or another weakly bound target[1,9] on a ^{9}Li or ^{12}Be beam.

2. Breakup Optical Potential Theory

The method we describe here is the same as in [4] and it is based on the extraction of an optical potential from the calculation of a phase shift. The elastic scattering probability is $P_{el} = |S_{NN}|^2 = e^{-4\delta_I(b)}$, given in terms of the nucleus-nucleus S-matrix. In a semiclassical approximation [4], the imaginary part of the nucleus-nucleus phase shift δ_I is related to the imaginary part of the optical potential by

$$\delta_I(b) = -\frac{1}{2\hbar}\int_{-\infty}^{+\infty} \left(W_V(\mathbf{r}(t)) + W_S(\mathbf{r}(t))\right) dt \tag{1}$$

where the volume potential is responsible for the usual inelastic core-target interaction, while the surface term takes care of the peripheral reactions like transfer and breakup. $\mathbf{r}(t) = \mathbf{b_c} + vt$ is the classical trajectory of relative motion for the nucleus-nucleus collision. The surface optical potential $W_S(\mathbf{r}(t))$ due to breakup can be related to the breakup probability by [4]

$$\int_{-\infty}^{+\infty} W_S(\mathbf{r}(t))dt = -\frac{\hbar}{2}P_{b_{up}}(b_c) \tag{2}$$

where $P_{b_{up}} = \sum_i p_i$ are the breakup probabilities in the various channels i. In order to obtain the surface imaginary potential Eq.(2) should be calculated as an identity in the distance of closest approach, which amounts to require that $W_S(r)$ be a local, angular momentum independent function. In semiclassical methods, the non locality, which is a characteristic of microscopic optical potentials, is usually transformed into an energy dependence [4].

On the basis of the results of Refs. [2,3,6], we argued in [5] that the imaginary part of the optical potential due to Coulomb breakup can also be calculated by Eq.(2) where now one of the p_i probabilities will be that of Coulomb breakup of the valence nucleon. Then, using Eq.(1) and (2), the nucleus-nucleus S-matrix, in the case of a halo projectile, can be written as $|S_{NN}|^2 = |S_{CT}|^2 e^{-P_{b_{up}}}$, where S_{CT} takes into account all core-target interactions while the term $e^{-P_{b_{up}}}$ depends only on the total halo neutron breakup probability, given by $P_{b_{up}}(b_c) = p^N_{b_{up}}(b_c) + p^C_{b_{up}}(b_c)$.

2.1. *Nuclear breakup*

The optical potential due to nuclear breakup was extensively discussed in Ref.[4]. We report here on some of the most important results which will help us also constructing the potential for the Coulomb breakup channel.

The nuclear breakup probability given by the eikonal model is

$$p^N_{b_{up}}(b_c) \approx \frac{C^2 S}{2\pi}\int \frac{d\varepsilon_f}{\hbar v}\int d\mathbf{b_n}(|1-S(\mathbf{b_n})|^2+1-|S(\mathbf{b_n})|^2)|\bar{\phi}_{l,m}(\rho,k_1)|^2 \tag{3}$$

where C^2S is the spectroscopic factor for the initial state and $\rho = |\mathbf{b_n} - \mathbf{b_c}|$. $\mathbf{b_c}$ and $\mathbf{b_n}$ are the core-target and the neutron-core impact parameters respectively. Eq.(3) is the neutron breakup probability from a definite single particle state of energy ε_i, momentum $\gamma = \sqrt{-2m\varepsilon_i}/\hbar$, and angular momentum l in thebehavior projectile to all possible final continuum state of energy ε_f with respect to the target. In our notation k_1 is the z components of the neutron momentum in the initial state, $\eta^2 = k_1^2 + \gamma^2$ is the modulus square of the transverse component of the neutron momentum.

The calculation of the nuclear breakup probability is done here in a reference frame with the center at the origin of the target. Neutron elastic breakup and absorption on the target are both included. The integrand in Eq.(3) is surface peaked and goes rapidly to zero at the interior of the projectile potential. The initial wave function one-dimensional Fourier transform has a behavior of the type $|\bar{\phi}_{l,m}(\rho, k_1)|^2 \approx C_i^2 \frac{e^{-2\eta|\mathbf{b_n}-\mathbf{b_c}|}}{2\eta|\mathbf{b_n}-\mathbf{b_c}|}$, which can be obtained analytically using asymptotic wave fuctions. A sum over m and average over l is intended. For the purpose of this paper the important thing is that in Eqs.(3) the main dependence on the core-target impact parameter b_c is contained in the exponential factor $e^{-2\eta|b_c-b_n|}$. After the b_n and ε_f integration, the breakup probability has still an exponential dependence on b_c for impact parameters larger than the strong absorption radius. Also Eq. (3) has a maximum in correspondence to the minimum value of $\eta = \gamma$. Therefore the b_c-dependence of the breakup probability $p^N_{b_{up}}(b_c)$ will be of the exponential form $p^N_{b_{up}}(b_c) \approx e^{-b_c/\alpha}$ with $a \approx (2\gamma)^{-1}$ where γ is the decay length of the neutron initial state wave function.

Assuming, as indicated earlier on, a straight line parameterization for the trajectory, then Eq.(2) reads $\int_{-\infty}^{+\infty} W_S^N(b_c, z)dz = -\frac{\hbar v}{2} p^N_{b_{up}}(b_c)$. At large distances, where no other reaction can occur, the absorptive potential due to nuclear breakup is expected to have the same exponential dependence as the probability $W_S^N(r) = W_0^N e^{-r/a}$. The LHS can be approximately evaluated as $\int_{-\infty}^{+\infty} W_S^N(b_c, z)dz = W_0^N \int_{-\infty}^{+\infty} e^{-(b_c + \frac{z^2}{2b_c})/a} dz = W_0^N \sqrt{2\pi b_c a} e^{-b_c/a}$, where we assumed $b_c >> z$ in the second step. Equating the RHS of the two expressions and renaming the distance b_c as r gives

$$W_S^N(r) = -\frac{\hbar v}{2} p^N_{b_{up}}(r) \frac{1}{\sqrt{2\pi a r}}. \tag{4}$$

This expression shows explicitly that the long range nature of the nuclear breakup potential originates from the large decay length of the initial state wave function. For a typical halo separation energy of 0.5MeV, $a = (2\gamma)^{-1} = 3.2fm$, while for a 'normal' binding energy of 10MeV,

$a = 0.7fm$ as expected. Therefore the parameter a will depend mainly on the projectile characteristics and not on the target.

2.2. *Coulomb breakup*

Coulomb breakup can be taken into account as well, following the formalism of [2] and using first order perturbation theory probability. The Coulomb breakup probability is best calculated in the projectile reference frame. Thus ε_k and k are now the neutron final energy and momentum with respect to the core. Obviously after integration over the final energy both nuclear and Coulomb probabilities are independent on the reference frame they were calculated in. If the initial state wave function is approximated by its asymptotic form, then after integration of the angular variables, the Coulomb breakup probability reads

$$p_{b_{up}}^{C}(b_c) \approx \frac{32}{3\pi} C^2 S \left(\frac{C_0 C_i}{b_c}\right)^2 \int d\varepsilon_k \frac{m_n}{\hbar^2 k} \left(\varpi^2 K_1^2(\varpi) + \varpi^2 K_0^2(\varpi)\right) \frac{k^4}{(k^2+\gamma^2)^4}. \tag{5}$$

The constant $C_0 = \beta_1 Z_P Z_T e^2/\hbar v$ is a dimensionless interaction strength and $\varpi = (\varepsilon_k - \varepsilon_i) b_c/\hbar v = \omega b_c/v$ is the adiabaticity parameter. The functions K_0 and K_1 are modified Bessel functions. We showed in the Appendix of [5] that inserting this result in Eq.(2) leads to an expression consistent with Eq. (12) of Andrés et al. [7]. Those authors have developed a semiclassical method to obtain a polarization potential for the Coulomb breakup channel at energies close to the barrier. Our formalism can then be viewed as a natural extension of the formalism of [7] in the high energy, straight line trajectory limit in which the eikonal model of the phase shift Eq.(1) is valid.

2.3. *Extraction of the imaginary potential*

To understand the characteristics of the optical potential and in order to use elastic scattering data for structure studies it is useful to deduce an analytical form of it. Details on the calculations and phenomenological inputs are given in [5]. We fitted the integrated probabilities to a sum of exponentials:

$$p^N(b_c) = \sum_n A_n \exp(-b_c/\alpha_n), \quad p^C(b_c) = \sum_n B_n \exp(-b_c/\beta_n), \tag{6}$$

obtaining the parameters given in Table 1. This allows a simple extraction of the potential from Eq.(4) and also a transparent physical interpretation of the large diffuseness. The parameters α_n and β_n show clearly that the exponential tails of the breakup probabilities have a very long range, in particular in the Coulomb breakup case. To derive forms of the potentials valid at all distances, we used the method outlined in Sec. 2.1, renaming again b_c with r such that

$$W_S^N(r) = -\frac{\hbar v}{2} P_0 \sum_n A_n \frac{e^{(-r/\alpha_n)}}{\sqrt{2\pi\alpha_n r}}, \quad W_S^C(r) = -\frac{\hbar v}{2} P_0 \sum_n B_n \frac{e^{(-r/\beta_n)}}{\sqrt{2\pi\beta_n r}}. \tag{7}$$

The core survival probability was parameterized as $P_0(b_c) = |S_{CT}|^2 = \exp(-\ln 2 e^{[(R_s - b_c)/a_0]})$ where the strong absorption radius is given by $R_s = 1.4(A_P^{1/3} + A_T^{1/3}) fm$, and $a_0 = 0.6 fm$.

Table 1. Best fit obtained for the nuclear and Coulomb halo breakup probabilities Eq.(6) for ^{11}Be incident on ^{208}Pb. The parameters α_n and β_n are in fm, while A_n and B_n are dimensionless. Incident energies in A.MeV

	Nuclear			**Coulomb**			
E_{inc}	20	40	80	20	40	80	
A_1	1860.38	303.23	339.45	9.672	3.922	1.734	B_1
A_2	19.73	7.91	1.91	1.558	0.6013	0.2549	B_2
-	-	-	-	0.1531	0.0587	0.0248	B_3
α_1	1.1924	1.5334	1.5356	2.7273	2.9438	3.1046	β_1
α_2	2.5403	2.7918	3.4955	6.5189	7.4850	8.2713	β_2
-	-	-	-	15.6863	19.9322	24.1546	β_3

3. Elastic Scattering Results

The potentials discussed in this work derive from the breakup of the $2s_{1/2}$ state of ^{11}Be, with separation energy $0.5MeV$, asymptotic normalization constant $C_i = 0.91 fm^{-1/2}$ and spectroscopic factor $C^2S = 0.77$. C_i was obtained from a wave function calculated in a Woods-Saxon potential with radius parameter $r = 1.25fm$, diffuseness $a = 0.8fm$ and depth $V_0 = -52.4MeV$, adjusted to fit the neutron separation energy.

The diffuseness of the nuclear breakup potential is a very important parameter. It depends weakly on the incident energy and its large value between 2 and 3 fm is due to the large decay constant of the neutron halo wave function and thus to the low separation energy. In the case of the Coulomb potential the diffuseness β_n is even larger, and furthermore increasing with energy. This is clearly a reflection of the long range, energy dependent effects of the Coulomb potential itself and of the core-recoil it causes. From the diffuseness parameters values shown in Table 1 we notice that at each incident energy the largest diffuseness values are $\beta_3 \approx \hbar v/(\varepsilon_k^{max} - \varepsilon_i)$ where

$\varepsilon_k^{max} = 2 \div 3 MeV$ is of the order of the largest continuum energy for which there is an appreciable breakup probability. This effect is clearly related to the behavior at very large distances of the Bessel functions $K_{0,1} \approx e^{-\varpi}$ in Eq.(5). Therefore in the case of the Coulomb breakup potential the diffuseness value is related to the adiabaticity parameter and it depends on the initial separation energy but also, strongly, on the incident energy. Thus this potential is much more dependent from the reaction dynamics than the nuclear breakup potential.

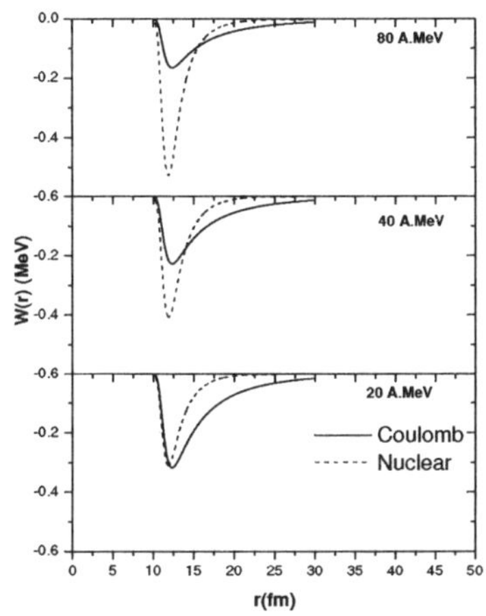

Figure 1. Nuclear and Coulomb breakup potentials

We show now in Fig. 1 the radial shapes of the potentials calculated for the breakup from the $2s$ state of ^{11}Be in the interaction with ^{208}Pb. Results are given for three laboratory incident energies: E_{inc}=20 A.MeV, 40 A.MeV and 80 A.MeV. The solid lines are the Coulomb breakup potentials while the dashed lines are the nuclear breakup potentials. Our results show that the potentials are energy dependent and the Coulomb breakup potential has a much longer tail than the nuclear breakup potential.

The optical potential has one of its most interesting application in the calculation of elastic scattering angular distributions. Since according to Eqs.(1) our phase shifts for the breakup channels are calculated in the eikonal model, it seems appropriate to calculate the angular distributions within the same model and we have used the code DWEIKO [10]. In Figs. 2 we show then the eikonal model angular distributions for ^{10}Be+^{208}Pb by the thin solid line. The choice of the optical model parameters for the volume parts of the bare potential is discussed in Ref.[5]. The thick solid line is for ^{11}Be+^{208}Pb with the bare volume potential of [5]. The dashed and dotted lines include the Coulomb breakup and the nuclear breakup potentials respectively, while the dot-dashed line includes both. As expected in most of the angular range the Coulomb breakup reduces the elastic scattering more than the nuclear breakup.

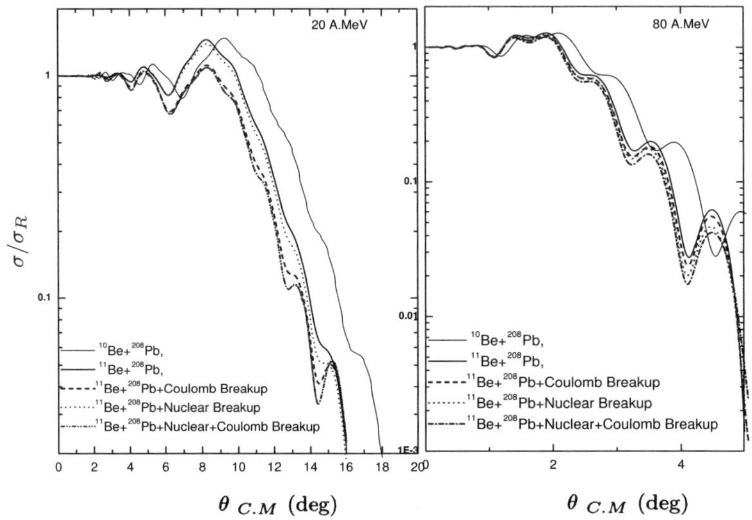

Figure 2. Angular distribution for the elastic scattering at 20 A.MeV and 80 A.MeV of ^{10}Be and ^{11}Be from ^{208}Pb.

Table 2. Values of the cross sections discussed in this paper. Cross sections in mb, energies in A.MeV.

σ	E_{inc}	20	40	80
$^{10}Be + ^{208}Pb$ total	σ_{CN}	2362	2897	2537
$^{11}Be + ^{208}Pb$ total with bare potential	σ_{NN}	2644	2969	2593
σ_{NN} including nuclear breakup	σ_{NN}^{N}	3185	3471	3060
σ_{NN} including Coulomb breakup	σ_{NN}^{C}	6685	5477	4152
σ_{NN} including Coulomb and nuclear breakup	σ_{NN}^{N+C}	7074	5900	4584
Nuclear halo breakup	$\sigma_{b_{up}}^{N}$	615	551	490
Coulomb halo breakup	$\sigma_{b_{up}}^{C}$	4280	2638	1570
n-^{208}Pb elastic free particle	σ^{el}	2554	2623	2467
n-^{208}Pb inelastic free particle	σ^{in}	2018	1902	2097
n-^{208}Pb total free particle	σ^{Tot}	4548	4437	4867
n-^{208}Pb experimental[8]	σ_{exp}	5888	4392	4817

Another significant effect of the imaginary surface potential is seen in the reaction cross sections obtained from the S_{NN} and given in Table 2. Also shown are a number of other cross section values necessary to justify the phenomenological inputs of our calculations and their consistency. A more detailed discussion is contained in Ref. [5]. As discussed in [4], the relation between the total reaction cross section and the breakup cross section, can be understood by expanding the exponential in σ_{NN} to first order in $P_{b_{up}}$ and integrating over the impact parameter b_c. One immediately finds $\sigma_{NN} = 2\pi \int b_c db_c \left(1 - |S_{NN}(b_c)|^2\right) \approx \sigma_{CT} + \sigma^N_{b_{up}} + \sigma^C_{b_{up}}$.

4. Projectile Fragmentation Model

We present now a model to describe the neutron breakup probability from a halo projectile due to the interaction with the target and including final state interaction with the original core nucleus. It is appropriate to describe coincidence measurements in which the neutron-core relative energy spectrum is reconstructed. For two-nucleon breakup the complete process including the second nucleon breakup will be discussed elsewhere [12]. According to Eq.(2.15) of [11] inelastic-like excitations can be described by a first order time dependent perturbation theory amplitude

$$A_{fi} = \frac{1}{i\hbar} \int_{-\infty}^{\infty} dt \langle \psi_f(\mathbf{r}, t) | V_2(\mathbf{r} - \mathbf{R}(t)) | \psi_i(\mathbf{r}, t) \rangle, \tag{8}$$

for a transition from a occupied nucleon bound state ψ_i to an unoccupied final state ψ_f. V_2 is the interaction responsible for the neutron transition. The wave function $\psi_i(\mathbf{r})$ for the initial state is calculated in a potential $V_1(\mathbf{r})$ which is fixed in space. The final state wave function $\psi_f(\mathbf{r})$ can be a bound state or a continuum state. The potential $V_2(\mathbf{r} - \mathbf{R}(t))$ moves past on a constant velocity path with velocity v in the z-direction with an impact parameter b_c in the x-direction in the plane $y = 0$. Eq.(8) can be put in a simple form by changing variables as $z' = z - vt$. Also $q = (\varepsilon_k - \varepsilon_i)/\hbar v$, and $V_2(x - b_c, y, q) = \int_{-\infty}^{\infty} dz V_2(x - b_c, y, z) e^{iqz}$. A simple analytical formula is obtained in the special case in which $V_2(r)$ is a delta function potential $V_2(r) = v_2 \delta(x)\delta(y)\delta(z)$, with $v_2 \equiv [MeV fm^3]$. Then the integrals over x and y can be calculated and

$$A_{fi} = \frac{v_2}{i\hbar v} \int_{-\infty}^{\infty} dz \; \psi_f^*(b_c, 0, z) \psi_i(b_c, 0, z) e^{iqz}. \tag{9}$$

The initial state is the numerical solution of the Schrödinger equation which fits the experimental neutron separation energy and the final continuum state is given by

$$\psi_f(b_c, 0, z) = C_f k \frac{i}{2} \left(h_{l_f}^{(+)}(kr) - S_{l_f} h_{l_f}^{(-)}(kr) \right) P_{l_f} \left(\frac{z}{r} \right). \quad (10)$$

$k = \sqrt{2m\varepsilon_k}/\hbar$ is the neutron momentum in the final states and C_f is the asymptotic normalization constant. S_{l_f} is the S-matrix representing the final state interaction of the neutron with the projectile core. The probability to excite a final continuum state of energy ε_k is an average over the initial state and a sum over the final states. Introducing the quantization condition and the density of final states [11] the probability spectrum reads

$$\frac{dP_{in}}{d\varepsilon_k} = \frac{2}{\pi} \frac{v_2^2}{\hbar^2 v^2} C_i^2 \frac{m}{\hbar^2 k} \Sigma_{l_f} (2l_f + 1) |1 - \bar{S}_{l_f}|^2 |I_{l_f}|^2. \quad (11)$$

Where $|I|$ and α are the modulus and phase respectively, of the integral in Eq.(9). Here $\bar{S} = e^{i2(\delta+\alpha)}$ does not represent the free neutron-core scattering, since there is an additional contribution α to the free neutron phase δ. An absorption term $1 - |\bar{S}_{l_f}|^2$ should be included if the neutron energy is higher than the core inelastic excitation energy threshold.

5. Application to ^{13}Be

One of the aims of this paper is to simulate the neutron-^{12}Be relative energy spectra obtained from fragmentation of ^{14}Be or ^{14}B on a ^{12}C target at 70 A. MeV and to see whether they would show differences predictable in a theoretical model. In the case of ^{14}B, which has a 2^- ground state, the valence neutron is either in a s-state or in a $d_{5/2}$-state, [13], coupled to the valence proton, which is in the $p_{3/2}$ state and we assume the binding energy $|\varepsilon_i| = 0.97 MeV$. In ^{14}Be a combination of s, p and d components can be supposed for the ground state and we take the binding energy $|\varepsilon_i| = 1.85 MeV$. Fig. 3 shows results obtained including the s and d bound initial states according to Eq.(11). The p-state is included only in the ^{14}Be case. In ^{12}Be the ground state wave function is a combination of such states with almost equal weight and none of them is fully occupied, then we assume that each of them has also components in the continuum. For each initial state a unit spectroscopic factor is taken. The result includes the transition bound to unbound from a s initial state to a s and d unbound states and from a bound d state to unbound s and d states and from a bound p state to an unbound p state. In our model transitions with change of parity give no contribution. Keeping fixed binding energies and the resonance energies of the p and d states, we have varied the scattering length of the final s state (given on the figure). The s-peak is dominant because of the well known threshold effect. It does not have a Lorentzian shape because it receives contribution from both the s and d bound components. In the case of a final bound s-state (dotted line) there is a very narrow peak close to threshold while for $a_s > 20fm$ there is no peak at all. In this case the p-state contribution appears as a little bump at 0.5 MeV but the peak takes less strength than that of the d-state around 2 MeV. This is due to the concentration at threshold of the s state, which has a less diffuse tail. When the s-state is unbound, the p-resonance peak disappears in the tail of the s- state, while the d-resonance peak can be clearly seen around 2 MeV because of the enhancement due to the $2l_f + 1$ factor in Eq.(11).

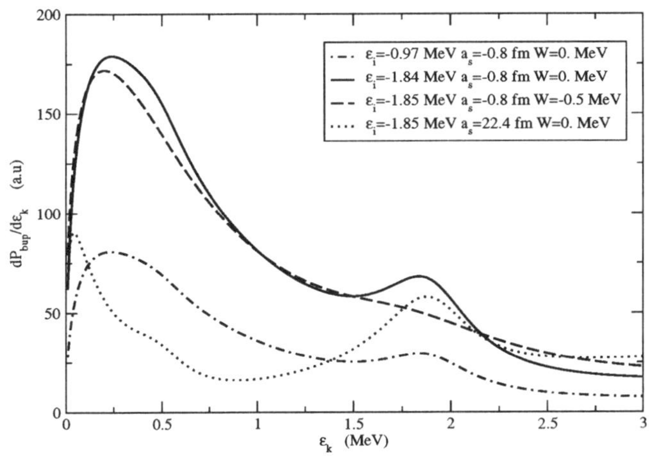

Figure 3. n-^{12}Be relative energy spectrum. Initial binding energy and final s-state scattering lengths are given on the figure.

Finally we wanted to address the issue of possible core excitation effects in ^{14}Be. They can be modeled in the present approach by considering a small imaginary part in the neutron-core optical potential. A potential of Woods-Saxon derivative form has been taken with a -0.5 MeV strength. The result is shown in Fig. 3 by the dashed line. The effect of the imaginary potential is to shift the s-state peak towards threshold and to wash out the d-resonance peak. It seems then that the spectrum of unbound nuclei would reflect the structure of the bound parent nucleus and that reaction mechanism models used to extract structure information should carefully include the effects discussed above. The model presented here seems to be promising in this respect. Details of potentials and physical parameters used will be given in a forthcoming publication[12].

6. Conclusions

In this paper we have discussed how reliable theoretical methods for reaction mechanisms involving halo or weakly bound nuclei can help extracting structure information. An explicit analytical form for the microscopic optical potential describing nuclear and Coulomb breakup has been obtained. We have shown that the large diffuseness of its exponential form is directly related to the decay length of the initial wave function for nuclear breakup and to the adiabaticity parameter for Coulomb breakup.

As a second example we have shown how one neutron-core relative energy spectra following projectile fragmentation of a two-neutron halo nucleus can be calculated. Information on the initial components of the bound state wave function and on the positions and widths of final resonance states can thus be obtained. This in turn helps determining the form of the two-body neutron-core potential.

References

1. B. Jonson, *Phys. Rep.* **389**, 1 (2004), and references therein.
2. J. Margueron, A. Bonaccorso and D. M. Brink, *Nucl. Phys.* **A720**, 337 (2003).
3. J. Margueron, A. Bonaccorso and D. M. Brink, *Nucl. Phys.* **A703**, 105 (2002).
4. A. Bonaccorso and F. Carstoiu, *Nucl. Phys.* **A706**, 322 (2002) and references therein.
5. A. A. Ibraheem, A. Bonaccorso, *Nucl. Phys.* **A748**, 414 (2005).
6. N. Fukuda et al., *Phys. Rev.* **C70**, 054606 (2004).
7. M.V. Andrés, J. Gómez-Camacho and M. A. Nagarajan, *Nucl. Phys.* **A579** 273 (1994).
8. R.W. Finlay et al., Phys.Rev. C 47 (1993) 237.
9. G. Blanchon, A. Bonaccorso and N. Vinh Mau, *Nucl. Phys.* **A739**, 259 (2004).
10. C. A. Bertulani, C. M. Campbell and T. Glasmacher, Comput. Phys. Commun. 152 (2003) 317
11. A. Bonaccorso and D.M. Brink, *Phys. Rev.* **C43** 299 (1991), *Phys. Rev.* **C38** 1776 (1988).

12. G. Blanchon, A. Bonaccorso, D.M. Brink and N. Vinh Mau, IFUP-TH 29/2004.
13. V. Guimarães et al., *Phys. Rev.* **C61** 064609 (2000).

FUSION REACTION OF DEFORMED NUCLEI AND EFFECT OF CLOSED SHELL STRUCTURE

H. IKEZOE, S. MITSUOKA, K. NISHIO AND K. TSURUTA

Japan Atomic Energy Research Institute,
Tokai, Ibaraki 319-1195, Japan
E-mail: ikezoe@popsvr.tokai.jaeri.go.jp

K. SATOU

University of Tsukuba,
Tsukuba, Ibaraki 305-8577, Japan

S. C. JEONG

Institute of Particle and Nuclear Studies, KEK,
Tsukuba, Ibaraki 305-0801, Japan

C. J. LIN

China Institute of Atomic Energy,
P.O. Box 275(10), 102413 Beijing, P. R. China

The effect of the nuclear deformation on heavy-ion fusion reactions were investigated for the reactions ^{60,64}Ni + ^{154}Sm and ^{76}Ge + ^{150}Nd, where ^{154}Sm and ^{150}Nd are well deformed nuclei. It is found that the fusion probability strongly depends on the orientation of these deformed nuclei. When projectiles collide at the tip of the deformed nuclei, the fusion probability is considerably reduced. On the other hand, when projectiles collide at the side of the deformed nuclei, the fusion occurs without any hindrance. This result suggests that the compact configuration at touching forms more easily a compound nucleus than an elongated configuration. The effect of the nuclear shell structure on heavy-ion fusion reactions was also investigated for the reaction systems ^{82}Se + ^{138}Ba, ^{82}Se + ^{134}Ba, ^{16}O + ^{204}Pb, ^{86}Kr + ^{134}Ba, ^{86}Kr + ^{138}Ba, and ^{82}Se + $^{\rm nat.}$Ce. The measured evaporation residue cross-sections for the reactions ^{82}Se + ^{138}Ba and ^{86}Kr + ^{138}Ba were two orders of magnitude and one order of magnitude larger than those for the reactions ^{82}Se + ^{134}Ba and ^{86}Kr + ^{134}Ba, respectively, at the excitation energy region of 10 $\sim$ 30 MeV. It was found that the evaporation residue cross-sections correlate strongly with the sum of the shell energies for both projectile and target nuclei, i.e., the evaporation residue cross-sections increase as the sum of the shell energy decreases.

1. Introduction

Fusion process between heavy nuclei to produce super-heavy elements has been extensively investigated so far. It is well known that the fusion probability between heavy nuclei is sensitive to the charge product $Z_p Z_t$ of the atomic numbers of projectile and target, i.e., the Coulomb repulsion at contact. When the charge product is less than ~1800, its fusion cross-section is well reproduced by the conventional one-dimensional barrier penetration model. When the charge product is larger than ~1800, its fusion cross-section is considerably reduced compared with the model calculation. This fact means that interacting nuclei in the fusion process not always fuse together to make a compound nucleus even if they have an enough kinetic energy to surmount the fusion barrier between two nuclei. This is because the saddle point of a heavy compound nucleus is more compact than a touching configuration of two nuclei. An extra kinetic energy is needed so that the reaction system can reach the saddle point after surmounting the fusion barrier. This extra kinetic energy begins to increase sharply at the charge product ~1800.

This situation can be changed in the case of the fusion reaction between deformed nucleus and spherical projectile. In this case, the distance between the mass centers at touching depends on the orientation of deformed nucleus, that is, the distance becomes short when projectile collides at the side of deformed nucleus and long when projectile collides at the tip of deformed nucleus. This suggests that the colliding angle with respect to the symmetric axis of deformed nucleus influences an effect on the fusion process. In order to investigate this effect, we measured evaporation residues (ERs) for the reactions ^{60,64}Ni + ^{154}Sm ($Z_p Z_t = 1736$)[1] and ^{76}Ge + ^{150}Nd ($Z_p Z_t = 1920$)[2], where the nuclei ^{154}Sm and ^{150}Nd have large nuclear deformations $\beta_2 = 0.321$ and 0.358, respectively.

A fusion probability between heavy nuclei at a low excitation energy is sensitive to not only the charge product $Z_p Z_t$ but also the nuclear structure of projectile and target. It has been reported that the number of a valence nucleon outside a major shell changes the fusion probability, i.e., the extra kinetic energy increases as the number of the valence nucleon.[3] Oganessian et al.[4] found a large difference between the fusion reactions ^{136}Xe + ^{86}Kr and ^{130}Xe + ^{86}Kr, where the nucleus ^{136}Xe has a closed neutron shell $N = 82$ and the neutron number of the nucleus ^{130}Xe is 76, six neutrons less than the closed neutron shell. The measured evaporation residue cross-sections for the fusion reaction ^{136}Xe + ^{86}Kr are almost two to three orders of magnitude larger than those for the fusion reaction ^{130}Xe + ^{86}Kr near the Coulomb barrier region. The enhancement of the evaporation residue cross-sections for the fusion reaction ^{48}Ca + ^{208}Pb is also pointed out in Ref. 5.

In order to investigate the effect of the nuclear shell structure on the heavy-ion fusion process, we measured the cross-sections of evaporation residues for the reactions ^{82}Se + ^{134,138}Ba[6], ^{86}Kr + ^{134,138}Ba[7], where the nucleus ^{138}Ba has the cosed neutron shell N=82 and the nucleus ^{134}Ba has 78 neutrons, only 4 neutrons less than the closed neutron shell.

2. Experimental Procedure and Discussions

The experiments were carried out at the tandem-booster facility of Japan Atomic Energy Research Institute (JAERI). The ERs were measured by the JAERI recoil mass separator (JAERI-RMS). The experimental details are described elsewhere[1,2].

2.1. *Effect of Deformation*

Evaporation residues cross-sections for the reactions ^{60}Ni + ^{154}Sm and ^{76}Ge + ^{150}Nd are shown in Fig. 1 and Fig. 2 as a function of the center-of-mass energy E_{cm}. The experimental data were compared with the statistical model calculations using the code HIVAP[13]. The details of the statistical model calculation are written in Ref. 1. Parameters in the calculation were adjusted so that the measured ERs for the reactions ^{32}S + ^{182}W and ^{28}Si + ^{198}Pt were well reproduced, which make the same compound nuclei ^{214}Th and ^{226}U as the reaction systems ^{60}Ni + ^{154}Sm and ^{76}Ge + ^{150}Nd, respectively.

The Coulomb barrier height varies as a function of the colliding angle θ_{coll} with respect to the orientation of deformed nucleus for the present two reaction systems. In the case of the system ^{60}Ni + ^{154}Sm, the calculated Coulomb barrier height is 172 MeV for the tip collision and 198 MeV for the side collision. The ER yields measured at E_{cm} = 180 MeV mainly come from the tip collision, because the *c.m.* energy 180 MeV is enough to surmount the Coulomb barrier at the tip collision, but not enough to surmount the Coulomb barrier at the side collision. As shown in Fig. 1, no ER yield was observed at E_{cm} = 175 and 182 MeV and the measured cross-sections of $2n$, $p2n$, αn, $\alpha 2n$ and αp channels below E_{cm} = 200 MeV are considerably smaller than the calculated results.

On the other hand, the measured cross-sections above E_{cm} = 200 MeV

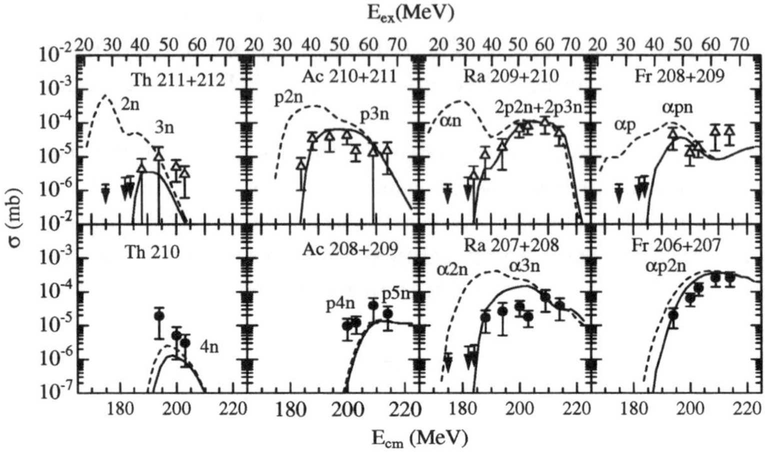

Figure 1. Measured ER cross-sections for the reactions $^{60}Ni + ^{154}Sm$ as a function of the *c.m.* energy. The upper abscissa indicates the excitation energy E_{ex} of the compound nucleus ^{214}Th. The dashed curves are the calculated results of HIVAP. The solid curves show the calculated results taking into account an extra-extra push energy.[1] The upper limits are shown as arrows.

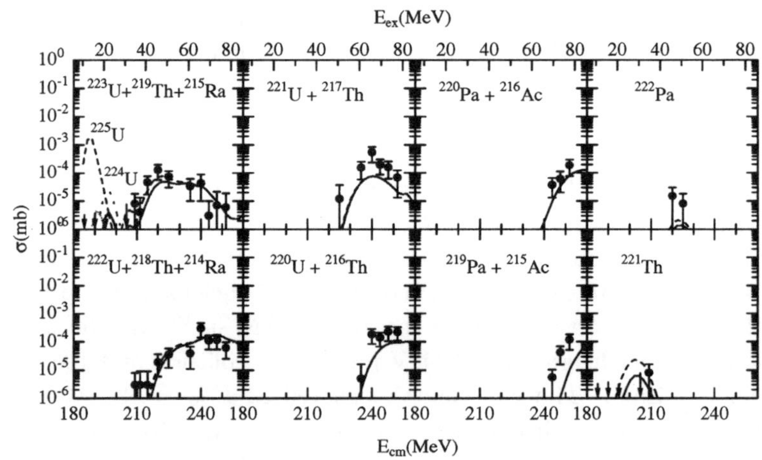

Figure 2. Same as Fig. 1 except for the reaction $^{76}Ge + ^{150}Nd$. The upper limits are shown as arrows at the lowest four energies.

are consistent with the present calculations. The fact that no ER was observed at the lowest two energies suggests that the fusion probability is significantly small for the collision in the vicinity of the tip of the deformed nucleus. Although every nuclear orientations come into play at the energy higher than $E_{cm} = 200$ MeV, the contribution of the side collision to the fusion cross-section large because of its larger solid angle than that of the tip collision.

The barrier height in the case of ^{76}Ge + ^{150}Nd also varies with the colliding angle from $E_{cm} = 184$ MeV at the tip collision to $E_{cm} = 214$ MeV at the side collision. No event was observed in $E_{cm} < 205$ MeV, whereas the measured cross-sections above $E_{cm} = 210$ MeV were consistent with the calculated results. Especially a large discrepancy between the calculation and the experimental data was found in the ^{225}U channel at $E_{cm} = 185$ - 190 MeV. The calculation predicted the cross-section of ~1 μb, but the experimental upper limit of this cross-section was about 3 - 4 nb. Here we found again the large fusion hindrance near tip collision ($E_{cm} < 185$ - 190 MeV) and no fusion hindrance at the energy higher than $E_{cm} = 210$ MeV where the side collision is dominant.[8]

Based on the present experimental results, it is concluded that the collision of projectile at the side ($\theta_{coll} > 50° - 60°$) of deformed nucleus results in a fusion without any hindrance, whereas the collision in the vicinity of the tip of deformed nucleus does not result in the formation of the compound nucleus even if the Coulomb barrier at that point is surmounted.

2.2. *Effect of Shell Structure*

The evaporation residue cross-sections for the reaction systems ^{82}Se + ^{138}Ba and ^{82}Se + ^{134}Ba have been reported in Ref. 6. The observed maximum cross-sections of the $2n$ and $3n$ channels in the reaction ^{82}Se + ^{138}Ba were about 100 μb and quite consistent with the calculated cross-sections for these channels. In addition, the measured cross-sections for the other evaporation channels were consistent with the calculated results, indicating that the fusion for the system ^{82}Se + ^{138}Ba was not hindered.

On the other hand, the measured evaporation residue cross-sections for the fusion reaction ^{82}Se + ^{134}Ba showed clear deviations from the calculated results mainly at the low excitation energy region $E_{ex} = 20 \sim 30$ MeV[6]. Large deficits of the evaporation residue cross-sections were seen, for instance, in the n channel (^{215}Th), $2n + 3n$ channel(^{213,214}Th), and also the p channel (^{215}Ac) at $E_{ex} = 20 \sim 30$ MeV. The observed cross-sections corresponding to the maximum of the $2n + 3n$ channel were $0.1 \sim 0.5\ \mu$b, which were small more than two orders of magnitude compared with the maximum cross-section of the $2n$ or $3n$ channel in the fusion reaction ^{82}Se + ^{138}Ba.

We plotted the reduced cross-sections $\sigma(\Sigma xn)k^2/\pi$ for the various reaction systems which make the compound nucleus ^{220}Th in Fig. 3(a) and the reaction systems which make the compound nucleus ^{216}Th in Fig. 3(b). Here, $\sigma(\Sigma xn)$ is the sum of the observed xn channels ($x = 1$ to 7) and k the wave number.

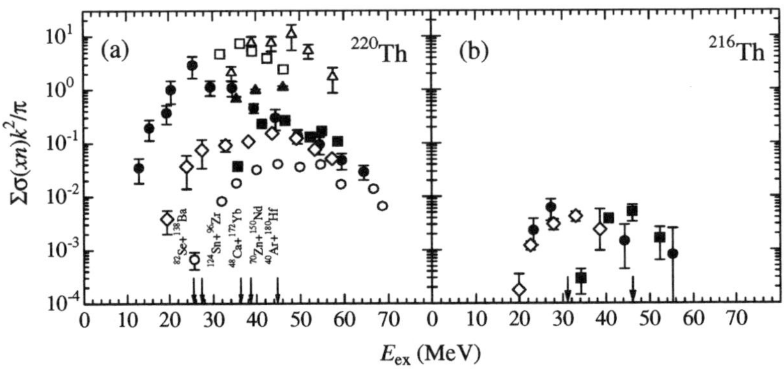

Figure 3. Reduced cross-sections of the sum of all xn channels measured in the several reaction systems, which make the compound nucleus ^{220}Th for the left figure (a) and ^{216}Th for the right figure (b). The solid circles and the open triangles show the present data. In the left figure (a), the data for the reactions ^{16}O + ^{208}Pb[9], ^{48}Ca + ^{172}Yb (solid triangles)[10], ^{40}Ar + ^{180}Hf (solid squares)[11], ^{124}Sn + ^{96}Zr (open diamonds)[10] and ^{70}Zn + ^{150}Nd (open circles)[12] are also plotted. The vertical arrows indicate the Bass barrier for each reaction system. In the right figure (b), the data for the reactions ^{40}Ar + ^{176}Hf (solid squares)[11], ^{124}Sn + ^{92}Zr (open diamonds)[10] are also plotted. The Bass barrier positions for these reaction systems are shown as arrows (E_{ex} = 31.1 MeV for ^{82}Se + ^{134}Ba and ^{124}Sn + ^{92}Zr, and 46.2 MeV for ^{40}Ar + ^{176}Hf).

We see that the reduced cross section for the reaction ^{82}Se + ^{138}Ba shows a peak exactly at the Bass barrier of this reaction system, that is, no fusion barrier shift and a narrow barrier distribution centered at the Bass barrier. On the other hand, the absolute value of the reduced cross-section for the reaction ^{82}Se + ^{138}Ba is one order of magnitude smaller than that for the fusion reaction ^{16}O + ^{204}Pb.

The reaction systems ^{124}Sn + ^{96}Zr and ^{70}Zn + ^{150}Nd show a broad barrier distribution extended to the energy by the amount of about 20 MeV larger than the Bass barriers. These broad barrier distributions are the typical indication of the fusion hindrance.

In Fig. 4(a) and 4(b), we plotted the reduced cross-sections at the Bass barrier energies for various reaction systems as a function of the sum of the shell energy of projectile and target nuclei. The effect of the excitation energy of the compound nucleus is shown as the solid lines in the figures. The reduced evaporation residue cross-sections are not totally correlated with the excitation energy of the compound nuclei and rather well correlated with the sum of the shell energy of projectile and target nuclei. It is clearly seen that the reduced cross-section increases as the sum of the shell energies decreases in negative values, i.e., the target and/or projectile having a closed shell structure makes the fusion probability high.

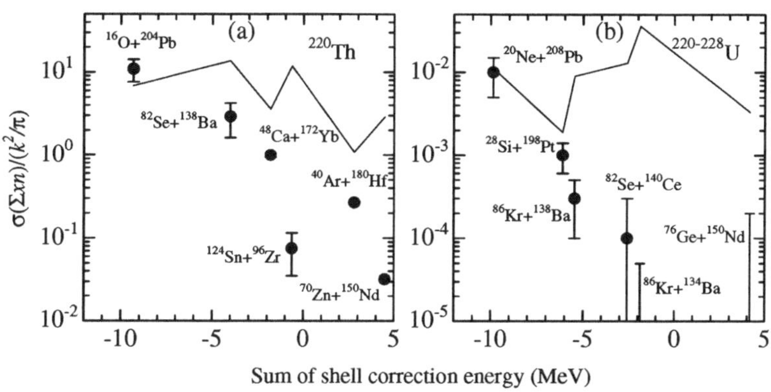

Figure 4. Reduced evaporation residue cross-sections at the Bass barrier energies as a function of the sum of the shell energy of projectile and target nuclei for various reaction systems which make the compound nuclei ^{220}Th for the left figure and $^{216,..226}$U for the right figure. The lines show the values calculated by HIVAP[13].

Myers and Swiatecki[14] pointed out that the shell energy resists neck growth at the time of contact between projectile and target, and then projectile nucleus can approaches closely to target nucleus with a small kinetic energy dissipation. They proposed the sum of the shell energy and the congruence energy for both projectile and target as a resistance factor against the neck growth. The observed trend seen in Fig. 4 is essentially unchanged even if the reduced evaporation residue cross-section is plotted as a function of the sum of the shell energy and the congruence energy. Therefore, the present data are consistent with this argument.

Möller and Sierk discussed the fusion process on the multi-dimensional microscopic potential energy surface.[15] When heavy nuclei first touch, mainly the Coulomb repulsive force push the system to deformations and eventually the system falls down the fission valley. If the system resists deformations away from the fusion path by the help of the shell energy, the system may proceed to a compound nucleus following the fusion path. This argument qualitatively explains the present data.

3. Conclusions

The present experimental results for the fusion between spherical projectile and deformed target nucleus indicate that the collision at the side (the equator) of a prolate nucleus favours the formation of compound nucleus compared with the collision in the vicinity of the tip of the prolate nucleus. This result is consistent with the theoretical predictions of the gentle fusion[16] and the hugging fusion[17] between two prolate nuclei. In these fusion models, it is predicted that when two well-deformed prolate nuclei approach under the condition of their symmetry axes orthogonal to each other, these two nuclei form the most compact configuration at the contact point and then tend to evolve into the compound nucleus formation with a high probability.

We also conclude that the fusion process for heavy reaction systems is strongly affected by the shell structure of interacting nuclei. It is important to realize theoretically the energy dissipation due to the friction after contact by taking into account both the shell structure of projectile and target nuclei.

Acknowledgments

We would like to thank the crew of the JAERI tandem booster facility for the beam operation.

References

1. S. Mitsuoka, H. Ikezoe, K. Nishio, K. Satou and J. Lu, *Phys. Rev.* C **65**, 054608 (2002).

2. K. Nishio, H. Ikezoe, S. Mitsuoka and J. Lu, *Phys. Rev.* C **62**, 014602 (2000).
3. A. B., Quint, W. Reisdorf, K. -H. Schmidt, P. Armbruster, F. P. Hessberger, S. Hofmann, J. Keller, G. Münzenberg, H. Stelzer, H. -G. Clerc, W. Morawek and C. -C. Sahn, *Z. Phys.* A **346**, 119 (1993).
4. Yu. Ts. Oganessian, " Heavy Elements and Related New Phenomena" , ed. by W. Greiner and R. K. Gupta (World Scientific, 1999) Vol.1, pp. 43-67.
5. Yu. Ts. Oganessian, V. K. Utyonkov, Yu. V. Lobanov, F. S. Abdullin, A. N. Polyakov, I. V. Shirokovsky, Yu. S. Tsyganov, A. N. Mezentsev, S. Iliev, V. G. Subbotin, A. M. Sukhov, K. Subotic, O. V. Ivanov, A. N. Voinov, V. I. Zagrebaev, K. J. Moody, J. F. Wild, N. J. Stoyer, M. A. Stoyer and R. W. Lougheed, *Phys. Rev.* C **64**, 054606 (2001).
6. K. Satou, H. Ikezoe, S. Mitsuoka, K. Nishio and S. C. Jeong, *Phys. Rev.* C **65**, 054602 (2002).
7. H. Ikezoe, K. Satou, S. Mitsuoka, K. Nishio and S. C. Jeong, *Phys. Atom. Nucl.* **66**, 1053 (2003).
8. K. Nishio, H. Ikezoe, S. Mitsuoka, K. Satou and S. C. Jeong, *Phys. Rev.* C **63**, 044610 (2001).
9. D. J. Hinde, M. Dasgupta and A. Mukherjee, *Phy. Rev. Lett.* **89**, 282701 (2002).
10. C.-C. Sahm, H.-G. Clerc, K.-H. Schmidt, W. Reisdorf, P. Arumbruster, F. P. Hessberger, J. G. Keller, G. Münzenberg and D. Vermeulen, Nucl. Phys. **A441**, 316 (1985).
11. H. -G. Clerc, J. G. Keller, C. -C. Sahm, K. -H. Schmidt, H. Schulte and D. Vermeulen, *Nucl. Phys.* **A419**, 571 (1984).
12. C. Stodel, PhD thesis, LPC Caen (1998).
13. W. Reisdorf and M. Schädel, *Z. Phys.* A **343**, 47 (1992)
14. W. D. Myers and W. J. Swiatecki, *Phys. Rev.* C **62**, 044610 (2000).
15. P. Möller and A. J. Sierk, Nature Vol. **422**, 485 (2003).
16. W. Nörenberg, " Fusion of Heavy Nuclei and a Novel fusion Path to Superheavies" , in the International Workshop on Heavy-Ion Fusion, edited by Stefanini et al., World Scientific, Padova, Italy, 1994, pp.248-253.
17. A. Iwamoto, P. Möller, J. R. Nix and H. Sagawa, *Nucl. Phys.* **A596**, 329 (1996)

INFLUENCE OF CHAOS ON THE FUSION ENHANCEMENT BY ELECTRON SCREENING

SACHIE KIMURA AND ALDO BONASERA

Laboratorio Nazionale del Sud, INFN, via Santa Sofia, 62, 95123 Catania, Italy
E-mail: kimura@lns.infn.it

We perform molecular dynamics simulations of screening by bound target electrons in low energy nuclear reactions. Quantum effects corresponding to the Pauli and Heisenberg principle are enforced by constraints. We show that the enhancement of the average cross section and of its variance is due to the perturbations induced by the electrons. This gives a correlation between the maximum amplitudes of the inter-nuclear oscillational motion and the enhancement factor. It suggests that the chaotic behavior of the electronic motion affects the magnitude of the enhancement factor.

1. Introduction

The relation between the tunneling process and dynamical chaos has been discussed with great interests in recent years.[1] Though the tunneling is completely quantum mechanical phenomenon, it is influenced by classical chaos. In the sense that the the chaos causes the fluctuation of the classical action which essentially determines the tunneling probability. We study the phenomenon by examining the screening effect by bound electrons in the low energy fusion reaction. In the low energy region the experimental cross sections with gas targets show an increasing enhancement with decreasing bombarding energy with respect to the values obtained by extrapolating from the data at high energies [2]. Many studies attempted to attribute the enhancement of the reaction rate to the screening effects by bound target electrons. In this context one often estimates the screening potential as a constant decrease of the barrier height in the tunneling region through a fit to the data. A puzzle has been that the screening potential obtained by this procedure exceeds the value of the so called adiabatic limit, which is given by the difference of the binding energies of the united atoms and of the target atom and it is theoretically thought to provide the maximum screening potential [3]. Over these several years, the redetermination of the bare cross sections has been proposed theoretically [4] and experimentally [5], using the Trojan Horse Method [6]. The comparison between newly obtained bare cross sections, i.e., astrophysical S-factors, and the cross sections by the direct measurements gives a variety of values for the screening potential. There are already some theoretical studies performed using the time-dependent Hartree-Fock(TDHF) scheme [7,8].

In this paper we examine the subject within the constrained molecular dynamics (CoMD) model [9], even in the very low incident energy region not reached experimentally yet. At such very low energies fluctuations are anticipated to play a substantial role. Such fluctuations are beyond the TDHF scheme. Not only TDHF calculations are, by construction, cylindrically symmetric around the beam axis. Such a limitation is not necessarily true in nature and the mean field dynamics could be not correct especially in presence of large fluctuations. Molecular dynamics contains all possible correlations and fluctuations due to the initial conditions(events). For the purpose of treating quantum-mechanical systems like target atoms and molecules, we use classical equations of motion with constraints to satisfy the Heisenberg uncertainty principle and the Pauli exclusion principle for each event [9]. In extending the study to the lower incident energies, we would like to stress the connection between the motion of bound electrons and chaos. In fact, depending on the dynamics, the behavior of the electron(s) is unstable and influences the relative motion of the projectile and the target. The feature is caused by the nonintegrablility of the 3-body system and it is well known that the tunneling probability can be modified by the existence of chaotic environment. We discuss the enhancement factor of the laboratory cross section in connection with the integrability of the system by looking the inter-nuclear and electronic oscillational motion. More specifically we analyze the frequency shift of the target electron due to the projectile and the small oscillational motion induced by the electron to the relative motion between the target and the projectile. We show that the increase of chaoticity in the electron motion decreases the fusion probability. We mention that the understanding of the fusion dynamics and fluctuations has a great potential for the enhancement of the fusion probability in plasmas for energy production.

The paper is organized as follows. In sect. 2 we determine the enhancement factor f_e and describe the essence of the Constrained molecular dynamics approach briefly. In sect. 3 we apply it to asses the effect of the bound electrons during the nuclear reactions. We discuss also the relation between the amplitudes of the inter-nuclear oscillational motion and the enhancement factor. We summarize the paper in sect. 4.

2. Formalism

2.1. *Enhancement Factor*

We denote the reaction cross section at incident energy in the center of mass E by $\sigma(E)$ and the cross section obtained in absence of electrons by $\sigma_0(E)$. The enhancement factor f_e is defined as

$$f_e \equiv \frac{\sigma(E)}{\sigma_0(E)}. \tag{1}$$

If the effect of the electrons is well represented by the constant shift U_e of the potential barrier, following [10,7], $(U_e \ll E)$:

$$f_e \sim \exp\left[\pi\eta(E)\frac{U_e}{E}\right], \tag{2}$$

where $\eta(E)$ is the Sommerfeld parameter [11].

2.2. *Constrained Molecular Dynamics*

We estimate the enhancement factor f_e numerically using molecular dynamics approach;

$$\frac{d\mathbf{r}_i}{dt} = \frac{\mathbf{p}_i c^2}{\mathcal{E}_i}, \qquad \frac{d\mathbf{p}_i}{dt} = -\nabla_{\mathbf{r}} U(\mathbf{r}_i), \tag{3}$$

where ($\mathbf{r}_i$,$\mathbf{p}_i$) are the position, momentum of the particle i at time t. $\mathcal{E}_i = \sqrt{\mathbf{p}_i^2 c^2 + m_i^2 c^4}$, $U(\mathbf{r}_i)$ and m_i are its energy, Coulomb potential and mass, respectively. We set the starting point of the reaction at 10Å inter-nuclear separation. In Eqs. (3) we do not take into account the quantum effect of Pauli exclusion principle and Heisenberg principle. As it is well known that these classical equations (3) can be derived by using the variational calculus of Lagrangian $\mathcal{L}$ of the classical system as well. So as to take the feature of the Pauli blocking into account in this framework, we use the Lagrange multiplier method for constraints.

Our constraints which correspond to the Pauli blocking is $\bar{f}_i \le 1$ in terms of phase space density, note that the phase space density can be directly related to the distance of two particles, i.e., $\mathbf{r}_{ij}\mathbf{p}_{ij}$, in the phase space. Here $\mathbf{r}_{ij} = |\mathbf{r}_i - \mathbf{r}_j|$ and $\mathbf{p}_{ij} = |\mathbf{p}_i - \mathbf{p}_j|$. The relation $\bar{f}_i \le 1$ is fulfilled, if $\mathbf{r}_{ij}\mathbf{p}_{ij} \ge \xi_P \hbar \delta_{S_i,S_j}$, where $\xi_P = 2\pi(3/4\pi)^{2/3}$. i, j refer only to electrons and $S_i, S_j (= \pm 1/2)$ are their spin projection. For the Heisenberg principle $\mathbf{r}_{ij}\mathbf{p}_{ij} \geq \xi_H \hbar$, where $\xi_H = 1$, i and j refer to not only electrons but the nucleus. It is determined to reproduce the correct energy of hydrogenic atoms. Obviously the conditions $\mathbf{r}_{ij}\mathbf{p}_{ij} = \xi_{H(P)}\hbar$ must be fulfilled in the ground state configuration rather than $\mathbf{r}_{ij}\mathbf{p}_{ij} > \xi_{H(P)}\hbar$.

Using these constraints, the Lagrangian of the system can be written down as

$$\mathcal{L} = \sum_i \frac{\mathbf{p}_i^2 c^2}{\mathcal{E}_i} - \sum_{i,j(\neq i)} U(\mathbf{r}_{ij})$$
$$+ \sum_{i,j(\neq i)} \lambda_i^H \left(\frac{\mathbf{r}_{ij}\mathbf{p}_{ij}}{\hbar} - 1\right) + \sum_{i,j(\neq i)} \lambda_i^P \left(\frac{\mathbf{r}_{ij}\mathbf{p}_{ij}}{\xi_P \hbar}\delta_{S_i,S_j} - 1\right), \tag{4}$$

where λ_i^P and λ_i^H are Lagrange multipliers. The variational calculus leads

$$\frac{d\mathbf{r}_i}{dt} = \frac{\mathbf{p}_i c^2}{\mathcal{E}_i} + \frac{1}{\hbar}\sum_{j(\neq i)} \left(\frac{\lambda_i^H}{\xi_H} + \frac{\lambda_i^P}{\xi_P}\delta_{S_i,S_j}\right) \mathbf{r}_{ij}\frac{\partial \mathbf{p}_{ij}}{\partial \mathbf{p}_i}, \tag{5}$$

$$\frac{d\mathbf{p}_i}{dt} = -\nabla_{\mathbf{r}} U(\mathbf{r}_i) - \frac{1}{\hbar}\sum_{j(\neq i)} \left(\frac{\lambda_i^H}{\xi_H} + \frac{\lambda_i^P}{\xi_P}\delta_{S_i,S_j}\right) \mathbf{p}_{ij}\frac{\partial \mathbf{r}_{ij}}{\partial \mathbf{r}_i}. \tag{6}$$

In order to obtain the atomic ground-state configuration, We perform the time integration of the eqs. (5) and (6). The value of λ_i^H and λ_i^P are determined depending on the magnitude of $\mathbf{r}_{ij}\mathbf{p}_{ij}$. If $\mathbf{r}_{ij}\mathbf{p}_{ij}$ is (smaller)larger than $\xi_{H(P)}\hbar$, λ has positive(negative) sign. Thus we change the phase space occupancy of the system. The constraints restrict us to variations $\Delta\mathcal{L} = 0$ that keep the constraints always true. The approach has been successfully applied to treat fermionic properties of the nucleons in nuclei and the quark system [9]. It can be extended easily in the case of the Heisenberg principle, as stated above. In this way we obtain many initial conditions which occupy different points in the phase space microscopically. Notice that in the D+d case that we investigate here, since one electron is involved, only the Heisenberg principle is enforced for each event. We obtain -13.56 eV as the binding energy of deuterium atom and 0.5327 Å as its mean square radius. These values can be compared with the experimental value of -13.59811 eV [12] and Bohr radius $R_B = 0.529$ Å, respectively [13].

In order to treat the tunneling process, we define the collective coordinates $\mathbf{R}^{coll}$ and the collective momentum $\mathbf{P}^{coll}$ as

$$\mathbf{R}^{coll} \equiv \mathbf{r}_P - \mathbf{r}_T; \qquad \mathbf{P}^{coll} \equiv \mathbf{p}_P - \mathbf{p}_T, \tag{7}$$

where $\mathbf{r}_T, \mathbf{r}_P$ ($\mathbf{p}_T, \mathbf{p}_P$) are the coordinates(momenta) of the target and the projectile nuclei, respectively. When the collective momentum becomes zero, we switch on the collective force, which is determined by $\mathbf{F}_P^{coll} \equiv \dot{\mathbf{P}}^{coll}$ and $\mathbf{F}_T^{coll} \equiv -\dot{\mathbf{P}}^{coll}$, to enter into imaginary time [14]. We follow the time evolution in the tunneling region using the equations,

$$\frac{d\mathbf{r}_{T(P)}^{\Im}}{d\tau} = \frac{\mathbf{p}_{T(P)}^{\Im}}{\mathcal{E}_{T(P)}}; \qquad \frac{d\mathbf{p}_{T(P)}^{\Im}}{d\tau} = -\nabla_{\mathbf{r}} U(\mathbf{r}_{T(P)}^{\Im}) - 2\mathbf{F}_{T(P)}^{coll}, \tag{8}$$

where τ is used for imaginary time to be distinguished from real time t. $\mathbf{r}_{T(P)}^{\Im}$ and $\mathbf{p}_{T(P)}^{\Im}$ are position and momentum of the target (the projectile) during the tunneling process respectively. Adding the collective force corresponds to inverting the potential barrier which becomes attractive in the imaginary times. The penetrability of the barrier is given by [14]

$$\Pi(E) = \left(1 + \exp\left(2\mathcal{A}(E)/\hbar\right)\right)^{-1}, \tag{9}$$

where the action integral $\mathcal{A}(E)$ is

$$\mathcal{A}(E) = \int_{r_b}^{r_a} \mathbf{P}^{coll}\, d\mathbf{R}^{coll}, \tag{10}$$

r_a and r_b are the classical turning points. The internal classical turning point r_b is determined using the sum of the radii of the target and projectile nuclei. Similarly from the simulation without electron, we obtain the penetrability of the bare Coulomb barrier $\Pi_0(E)$.

Since nuclear reaction occurs with small impact parameters on the atomic scale, we consider only head on collisions. The enhancement factor is thus given by eq.(1),

$$f_e = \Pi(E)/\Pi_0(E) \tag{11}$$

for each event in our simulation. Thus we have an ensemble of f_e values at each incident energy.

3. Application to the Electron Screening Problem

Fig. 1 shows the incident energy dependence of the enhancement factor for the D+d reaction. The averaged enhancement factors $\bar{f}_e$ over events in our simulation are shown with stars and its variance $\Sigma = \left(\bar{f_e^2} - (\bar{f}_e)^2\right)^{1/2}$ with error bars. In the figure we show also several estimations of the enhancement factor by the latest analysis of the experimental data using quadratic(dotted) and cubic(dot-dashed) polynomial fitting [4] with the screening potentials $U_e = 8.7$ eV and 7.3 eV respectively. The dashed curve shows the enhancement factor in the adiabatic limit $f_e^{(AD)}$ for an atomic deuterium target and it is obtained by assuming equally weighted linear

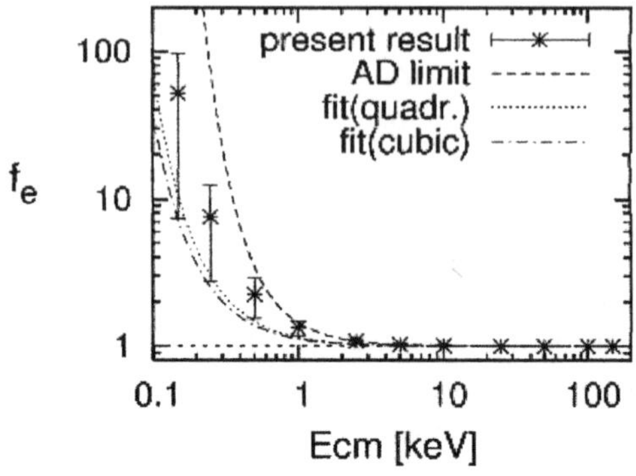

Figure 1. Enhancement factor as a function of incident center-of-mass energy for the D+d reaction.

combination of the lowest-energy gerade and ungerade wave function for the electron, reflecting the symmetry in the D+d, i.e., $f_e^{(AD)} = \frac{1}{2}\left(\exp(\pi\eta(E)\frac{U_e^{(g)}}{E}) + \exp(\pi\eta(E)\frac{U_e^{(u)}}{E})\right)$, where $U_e^{(g)} = 40.7$ eV and $U_e^{(u)} = 0.0$ eV [8,7]. In the low energy region the enhancement factor is very large and exceeds 50. However the averaged enhancement factor does not exceed the adiabatic limit. We performed also a fit of our data using eq. (2) including very low energy region and obtained $U_e = 15.9 \pm 2.0$ eV. This value, between the sudden and the adiabatic limit, is in good agreement with TDHF calculations[7,8].

Now we look at the oscillational motions of the particle's coordinates as the projection on the z-axis (the reaction axis). We denote the z-component of $\mathbf{r}_T, \mathbf{r}_P$ and $\mathbf{r}_e$ as z_T, z_P and z_e, respectively. Practically, we examine the oscillational motion of the electron around the target $z_{Te} = z_e - z_T$ and the oscillational motion of the inter-nuclear motion, i.e., the motion between the target and the projectile, $z_s = z_T + z_P$, which essentially would be zero due to the symmetry of the system in the absence of the perturbation. In Fig. 2 these two values are shown for 2 events, which have the enhancement factor $f_e = 170.8$ (ev. A), and $f_e = 6.5$ (ev. B), at the incident energy $E_{cm} = 0.15$ keV. The panels show the z_s, z_{Te} as a function of time. The stars indicate the time at which the system reaches the classical turning point. It is clear that in the case of event B the orbit of the electron is much distorted from the unperturbed one than in event A. Characteristics of z_s are that (1) its value often becomes zero, as it is expected in the un-

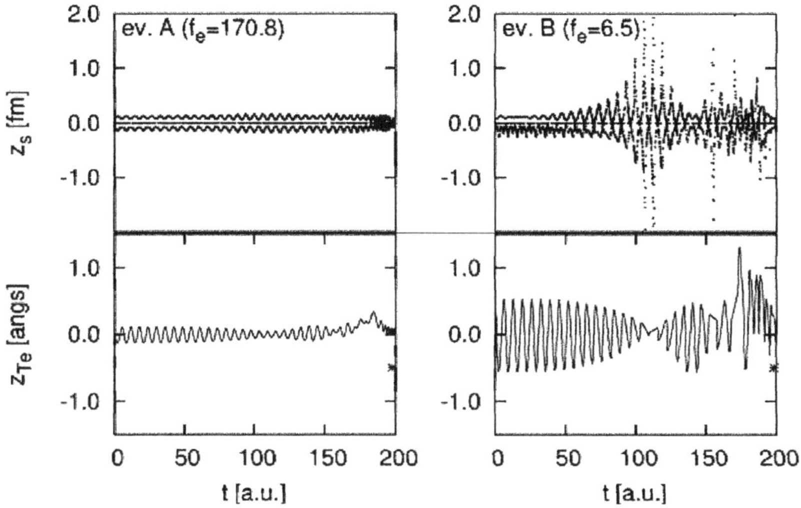

Figure 2. The oscillational motion of the electron around the target (lower panels) and the inter-nuclear motion (upper panels) as a function of time, in atomic unit, for two events, with large f_e(ev. A) and small f_e(ev. B), for the D+d reaction at the incident energy 0.15keV. The inter-nuclear separation is 10Å at $t = 0$.

perturbed system, and (2) the component of the deviation from zero shows periodical behavior. It is remarkable that the amplitude of the deviation becomes quite large at some points in the case of event B which shows the small enhancement factor. Note that in event B one observes clear beats, i.e., resonances. Thus for two events, with the same macroscopic initial conditions, we have a completely different outcome, which is a definite proof of chaos in our 3-body system. We can understand these results in first approximation by considering the motion of the ions to be much slower than the rapidly oscillating motion of the electrons. [15] From the Fig.2 we can deduce the following important fact. If the motion of the electron is initially in the plane perpendicular to the reaction axis, the enhancement factor is large, event A(notice $|z_{Te}| \ll R_B$, i.e., the Bohr radius, at $t \sim 0$). On the other hand if there is a substantial projection of the electron motion, as in event B(the amplitude of $|z_{Te}| \sim R_B$ at $t \sim 0$), on the reaction axis the enhancement factor is relatively small because of the increase of chaoticity. The fact suggests that if one performs experiments at very low bombarding energies with *polarized targets*, the enhancement factor can be controlled by changing the *polarization*. The largest enhancement would be gained with targets polarized perpendicularly to the beam axis.

In order to test this estimation, we prepared ensembles of target atoms which are polarized perpendicular and parallel to the beam axis, numerically. Fig. 3 shows the initial configuration dependence of the enhancement factor at the incident energy E_{cm} =1.25keV. The open circles are the enhancement factors for perpendicular polarized targets and the closed squares are ones for parallel polarized targets. It is clear that the perpendicular targets give always large enhancement factors and small variance, as is shown with the error bars in the figure. Instead the parallel targets gives relatively small enhancement factors and large variance. This is the case which corresponds to the event B in Fig. 2 and chaotic behavior of the system affects the determination of the enhancement factor. Remarkable thing is that with the parallel targets the enhancement factor often becomes less than 1. It means that in this case the bound electron gives the effect of hindrance to the tunneling probability.

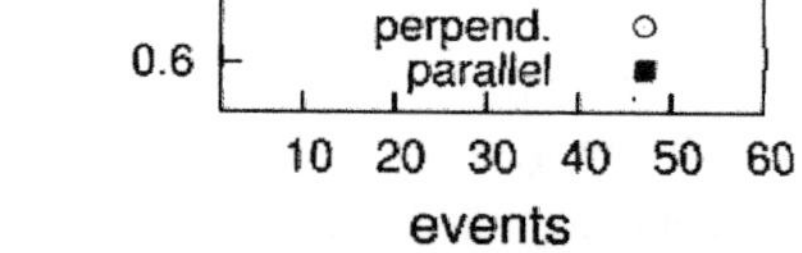

Figure 3. The initial configuration dependence of the enhancement factor at the incident energy E_{cm} =1.25keV. The enhancement factors for perpendicular polarized targets(open circles) and for parallel polarized targets(closed squares) are shown.

We notice in passing that event A is a case where cylindrically symmetry is approximately satisfied, since the electronic motion remains practically on the $xy-$plane in the target reference frame. This case gives a screening potential U_e =19.5eV closest to the adiabatic limit and to the TDHF result[7,8].

4. Summary

We discussed the penetrability of the Coulomb barrier by using molecular dynamics simulations with constraints and imaginary time. We have shown that both the enhancement factor and its variance increase as the incident energy becomes lower. However we obtained the averaged screening potential smaller than the value in the adiabatic limit. The chaoticity of the electron motion affects the enhancement factor of the cross section. We suggest to perform experiments on fusion at very low energies with polarized targets in order to obtain the large enhancement by the bound electrons.

References

1. W.A.Lin and L.E. Ballentine, Phys. Rev. Lett. **65**, 2927(1990); O. Bohigas, S. Tomosvic, and D. Ullumo. Phys. Rev. Lett., **65**, 5(1990); A. Shudo, and K.S.Ikeda, Phys. Rev. Lett., **74**, 682(1995).
2. A. Krauss, *et al.*, Nucl. Phys. A **467**, 273(1987); S. Engstler *et al.*, Phys. Lett. B **202**, 179 (1988).
3. C. Rolfs, and E. Somorjai, Nucl. Instrum. Meth. B **99**, 297 (1995).
4. F. C. Barker, Nucl. Phys. A **707**, 277 (2002).
5. M. Junker, *et al.* Phys. Rev. C **57**, 2700 (1998).
6. A. Musumarra, *et al.* Phys. Rev. C **64**, 068801 (2001).
7. T. D. Shoppa, S. E. Koonin, K. Langanke, and R. Seki, Phys. Rev. C **48**, 837 (1993).
8. S. Kimura, N. Takigawa, M. Abe, and D.M. Brink, Phys. Rev. C **67**, 022801(R) (2003).
9. M. Papa, T. Maruyama, and A. Bonasera, Phys. Rev. C **64**, 024612 (2001); S. Terranova, and A. Bonasera, Phys. Rev. C **70**, 024906 (2004).
10. H. J. Assenbaum, and K. Langanke and C. Rolfs, Z. Phys. A **327**, 461(1987).
11. D. D. Clayton, *Principles of Stellar Evolution and Nucleosynthesis* (University of Chicago Press, 1983) Chap. 4.
12. J.A. Bearden, and A.F. Burr, Rev. Mod. Phys, **39**, 125(1967).
13. S. Kimura, and A. Bonasera, physics/0409008.
14. A. Bonasera, and V. N. Kondratyev, Phys. Lett. B **339**, 207(1994); T. Maruyama, A. Bonasera, and S. Chiba, Phys. Rev. C **63**, 057601(2001).
15. S. Kimura, and A. Bonasera, Phys. Rev. Lett. **93**, 262502 (2004).

THE MAGNETIC SPECTROMETER PRISMA AT LNL

E. FIORETTO, L. CORRADI, A.M. STEFANINI, S. SZILNER*, B.R. BEHERA, G. DE ANGELIS, A. GADEA, A. LATINA, G. MARON, D.R. NAPOLI, A. PISENT, Y.W. WU

INFN - Laboratori Nazionali di Legnaro, Viale dell'Università, 2 Legnaro (PD), 35020, Italy

S. BEGHINI, G. MONTAGNOLI, F. SCARLASSARA

Dipartimento di Fisica dell'Università and INFN Sezione di Padova, Via Marzolo 4 Padova, 35131, Italy

M. TROTTA, A. DE ROSA, G. INGLIMA, M. LA COMMARA, D. PIERROUTSAKOU, M. ROMOLI, M. SANDOLI

Dipartimento di Fisica dell'Università and INFN Sezione di Napoli, Via Cintia Napoli, 80126, Italy

G. POLLAROLO

Dipartimento di Fisica dell'Università and INFN Sezione di Torino, Via P. Giuria 1 Torino, 10125, Italy

The new magnetic spectrometer PRISMA for heavy-ions has been installed at the end of 2002 and is now operating at LNL. In-beam tests evidenced experimental resolutions consistent with the design characteristics of the setup. A first experimental campaign has been completed at the end of 2004.

1. Introduction

The magnetic spectrometer PRISMA [1,2,3] has been recently installed at LNL. It has been designed to be used with heavy-ion beams accelerated at energies up to E = 5-10 AMeV by means of the Tandem-ALPI accelerator complex of the Laboratori Nazionali di Legnaro. The next operation of the new superconductive injector PIAVE and its coupling with the ALPI Linac will allow to extend the nuclear physics studies in the A=100-200 mass region. Moreover, PRISMA is

* Present address: R. Boskovic Institute, Bijenicka c. 54, HR-10001 Zagreb, Croatia.

the natural candidate for use in future experiments with radioactive ion beams which will be produced by the proposed SPES facility at LNL.

The spectrometer is located in the East target area of the Tandem building and its commissioning has been completed at the beginning of 2003. In July of 2003 the installation of the γ-array CLARA [4] around the target point of PRISMA (in the backward hemisphere) was started. Such array is composed of 25 CLOVER detectors of the EUROBALL setup. The first experimental campaign started in April 2004.

Few- and multi-nucleon transfer reactions [5,6] (from grazing to deep-inelastic collisions), elastic and inelastic scattering will be studied with specific experiments by using PRISMA in order to extract the properties of the nuclear residual interaction and to understand the evolution of the single-particle levels. Moreover, binary reactions will be used as a powerful tool to populate moderately neutron-rich isotopes and to study the nuclear structure in interesting mass regions (near to ^{78}Ni and ^{132}Sn). These experiments will be performed in conjunction with the CLARA setup. The operation of the spectrometer in Gas-Filled Mode will allow to extend the nuclear dynamics studies to the fusion-evaporation reactions.

A detailed description of the spectrometer and its entrance and focal plane detectors will be given. The results of the in-beam tests and first experiments will be also reported.

2. The PRISMA setup

2.1. *Spectrometer*

The optical design of the spectrometer consists of a magnetic quadrupole (30 cm diameter, 50 cm long) followed at 60 cm by a 60° bending angle magnetic dipole (20 cm gap, 100 cm usable width, 120 cm curvature radius). A picture of the spectrometer is shown in Fig. 1. It is mounted on a rotating platform that allows rotations around the target point in the $-20° \leq \theta \leq +130°$ angular range with respect to the beam direction (an important requirement for the study of two-body reactions such as multi-nucleon transfer). A small sliding seal scattering chamber houses a couple of small area Si detectors (~ 50 mm^2) working as beam monitors and/or as normalization devices. The spectrometer is equipped with entrance and focal plane detectors that allow the complete identification and trajectory reconstruction of the reaction products by using the Time of Flight (TOF) and ΔE-E techniques together with the position information at the entrance and the focal plane, respectively.

The most interesting features of the spectrometer are:

- a solid angle of 80 msr;
- angular acceptances of $\Delta\theta=\pm 6°$ and $\Delta\phi=\pm 11°$;
- a momentum acceptance of ±10%;
- a maximum rigidity around 70 AMeV;
- a mass resolution up to 1/300 (via TOF);
- an energy resolution up to 1/1000 (via TOF).

Figure 1. The PRISMA spectrometer placed at 90° with respect to the beam direction. The magnets are clearly seen on the rotating platform.

These performances are made possible by the software reconstruction of the ion tracks.

2.2. *Entrance Detector*

The entrance detector is composed of a double layer of large-area (80x100 mm^2) Micro-Channel Plates (MCP) mounted in the Chevron configuration [7]. Its high counting rate capability allowed to mount the detector close to the target (about 25 cm from it) where, usually, a very high background due to δ-electrons, X-rays and light charged particles is present. In this location it matches the whole solid angle covered by the spectrometer. The ejectiles coming from the target enter the active volume of the detector through a grid made of 20 μm gold-plated tungsten wires biased at the same high voltage of the MCP pair (typically around -2200 V). Then the ions impinge and pass through a thin self-supporting carbon foil (~20 $\mu g/cm^2$), tilted by 45° with respect to the central trajectory and biased at about -2500 V, and enter into the PRISMA spectrometer. The backwards emitted electrons are accelerated towards the MCP by means of the electric field produced by a second grid placed 4 mm apart from the carbon foil and biased at about -2200 V. In order to preserve the ion position information, a magnetic field of about 100-120 Gauss, parallel to the accelerating electric field, is applied using an external coil. The accelerated and focused electrons reach the MCP pair surface and undergo multiplication. The produced charge is collected by a position sensitive anode. It is composed of two orthogonal delay lines made of 70 μm Cu-Be wires wrapped around rounded plexiglass frames of two different diameters. The outer delay line is 3 mm apart from the MCP output face. The whole anode structure is directly fixed to the printed circuit board housing the home-made electronics for the signal processing. All electronics is installed in vacuum.

A picture of the detector is shown in Fig. 2.

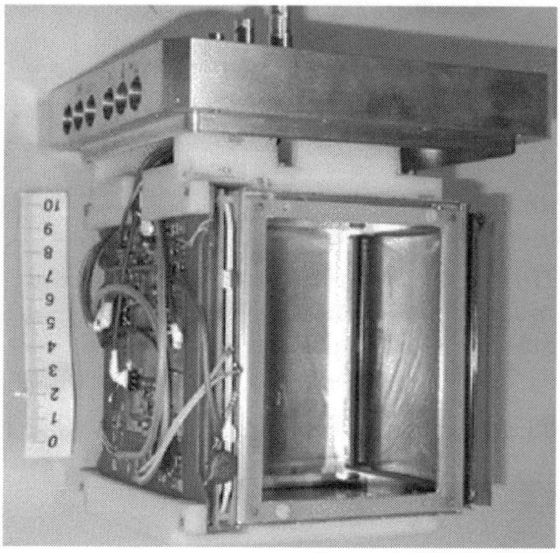

Figure 2. View of the MCP detector. The printed circuit board with the home-made electronics and the carbon foil are clearly visible on the left and on the right, respectively.

The detector provides a timing signal to be used as start reference for TOF measurements and for measuring the (X,Y) entrance position of the reaction products in the spectrometer. The position information is obtained from the difference in arrival time of the signal at one end of the delay line with respect to the reference time signal. The latter is extracted from the output face of MCP through a 100-240 pF capacitor.

2.3. *Focal Plane Detector*

The focal plane detector is composed of a large area Multi-Wire Parallel Plate Avalanche Counter (MWPPAC) followed by a transverse field multi-anode Ionization Chamber (IC) at about 50 cm of distance [8].

In order to match the large acceptance of the spectrometer, the MWPPAC covers a total active area of 100x13 cm^2 and has a three electrode structure: a central cathode and two anodic wire planes (X and Y) orthogonally oriented to each other and symmetrically placed with respect to the cathode at a distance of 2.4 mm from it. The cathode is made of 20 μm diameter gold-plated tungsten wires with a spacing of 0.3 mm (3300 wires) and electrically divided in ten equal and independent sections. Each of them has an active area of 10x13 cm^2 and provides a timing signal for both TOF measurements and Data Acquisition trigger. The X-plane is equally segmented as the cathode and each section consists of 100 wires of the same type with a spacing of 1 mm. The cathode and X-plane wires are soldered to a printed-circuit on G-10 frames. The Y-plane is common to all cathode sections with 130 wires (100 cm long, with the same diameter and 1 mm step as for the X wires) soldered on the other side of the same frame of the cathode. The cathode-anode gap is 2.4 mm. The position information is extracted from the X and Y anodes by using a delay-line readout. The cathode is biased at a negative high voltage (500-600 V with isobutane at working pressures of about 7-8 mbar) while the anodes are electrically grounded through 100 kΩ resistors. The detector is inserted in a stainless steel vessel closed on both sides by two large mylar windows (100x13 cm^2) facing the anodes wire planes at a distance of 3 mm. Windows are made of 1.5 μm thick mylar foils glued on a proper metallic frame and supported by 100 μm stainless steel wires spaced 3.5 mm. A picture of the MWPPAC is shown in Fig. 3.

Figure 3. View of the MWPPAC inserted in its vacuum vessel.

The IC dimensions in the dispersion plane are 110x17.6 cm^2 while the active depth is 120 cm. Its cathode and anode are divided in ten sections along the focal plane. Each of them is segmented in four pads (265.0 mm long by 99.0 mm wide) providing independent ΔE signals. This choice is an optimum compromise for preserving good Z and E resolution over the whole focal plane, where the event rate may overcome several kHz/section, keeping at a reasonable level the number of pads and consequently the electronic equipment. The cathode and anode are obtained by a photoengraving process of Cu layer deposited onto a 2 mm thick G-10 plate. The Frisch grid (common to all sections) is made of 1000 golden tungsten wires with a 100 μm diameter spaced 1 mm apart and 1 m long. The wires are soldered on both sides of a G-10 frame after proper stretching (100 g). The anode-Frisch grid and cathode-Frisch grid distances are 1.6 and 16 cm, respectively. The 1.5 μm thick mylar window is glued on a stainless steel nose and is supported by 1000 stainless steel wires (100 μm thick, 1 mm step). The cathode (all pads) is held at ground potential while the Frisch grid and the anode are biased in such a way as to maximize the electron drift velocity in the active region and to minimize the shielding inefficiency of the Frish grid in the collection region. The detector is routinely used with methane (CH_4, 99% purity), mainly due to its high electron drift velocity. However for energetic heavy-ions, CF_4 has to be used because of its high stopping power and electron drift velocity.

3. In-beam tests and first experiments

The performances of the PRISMA spectrometer have been tested with several heavy-ion beams (^{32}S, ^{40}Ca, ^{56}Fe, ^{58}Ni, ^{82}Se, ^{90}Zr) at energies ranging from 4 to 6 AMeV impinging onto different targets.

A typical X (Y) position resolution of 1 mm (2 mm) has been obtained for the MWPPAC. Position spectra obtained for the 230 MeV ^{40}Ca+^{208}Pb reaction are shown in Fig. 4. The different shapes of X and Y spectra are due to the PRISMA optics, i.e. dispersing in X and focusing in Y. The detector efficiency is close to 100% for a wide range of heavy-ions and energies when compared with the reference number of events collected in the IC placed downstream.

In Fig. 5 the ΔE-E scatter plot obtained with the IC for the 505 MeV ^{82}Se+^{238}U reaction is shown for one of the central sections. Its multi-anode structure allows to optimize the partition between the ΔE and residual E in order to identify at the same time reaction products coming from very different nuclear processes. In fact, for this reaction the magnetic fields of the spectrometer select both transfer and fission-like fragments. A good identification of the different proton stripping (up to 8-10 proton transfer) and pick-up channels has been evidenced. The fission fragments are well identified and separated from the transfer products because of their lower range in the IC.

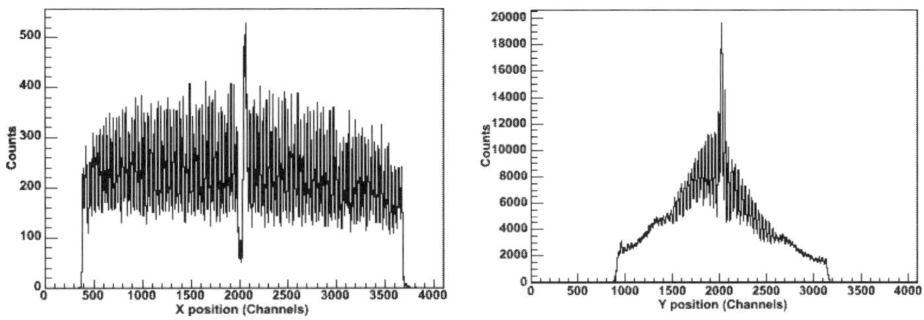

Figure 4. X (only one section) and Y position spectra obtained for the 230 MeV $^{40}Ca+^{208}Pb$ reaction.

The Z resolving power is Z/ΔZ~60 evaluated for Z=34. The measured energy resolution is about 2%.

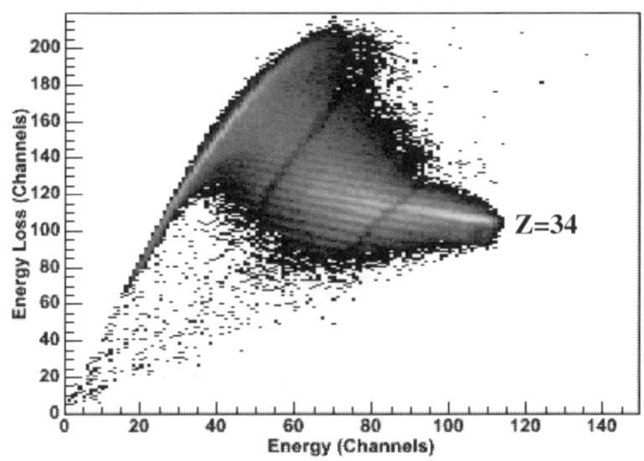

Figure 5. ΔE-E matrix obtained for the central section of the IC in the 505 MeV $^{82}Se+^{238}U$ reaction.

The 235 MeV ^{40}Ca + ^{208}Pb (50 $\mu g/cm^2$) reaction was used to check the energy (momentum) resolution of PRISMA. Elastically and inelastically scattered ^{40}Ca ions were analyzed by means of PRISMA (placed at 90°) and identified at the focal plane. Fig. 6 shows the X position spectrum at the focal plane obtained by gating on A=40 in the mass spectrum and selecting events corresponding to Z=20 in the ΔE-E matrix of the IC with the condition Δθ~1.5° (gate on the position spectrum of the MCP). The elastic peak is clearly separated from the inelastic octupole excitations of both projectile and target. The deduced energy resolution is roughly 700 keV, corresponding to about 1/300, mainly depending on the target thickness and the reaction kinematics.

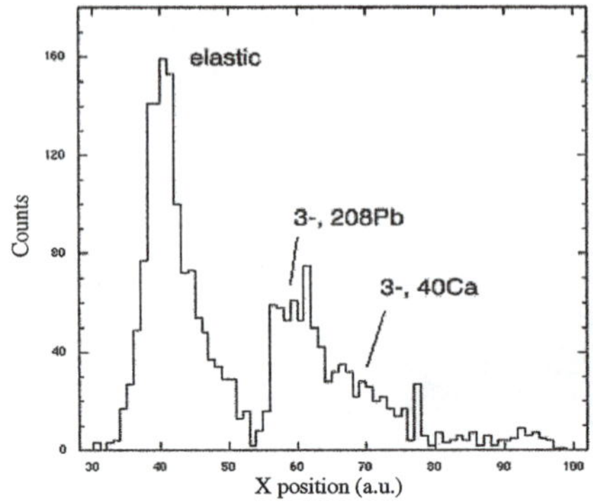

Figure 6. X position spectrum along the focal plane of PRISMA.

A significant test of the mass resolution was performed by studying the 263 MeV ^{58}Ni + ^{124}Sn reaction. The ejectiles were analyzed by PRISMA placed at $\theta_{lab} = 68°$ (near the grazing angle) and several mass spectra have been obtained by selecting different Z in the ΔE-E matrix of the IC. Multi-proton stripping channels were mainly observed besides strong neutron pickups as expected from the Q-value systematics for the various reaction channels.

A typical X-TOF scatter plot obtained for one of the central section of the MWPPAC gated by a 10x40 mm^2 window in the X-Y position matrix of the MCP is shown in Fig. 7.

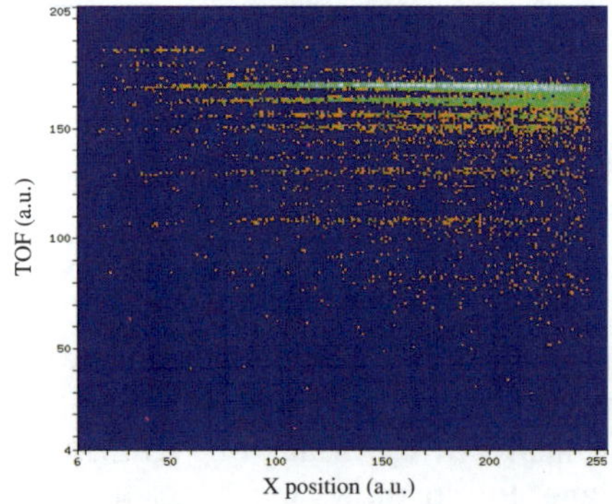

Figure 7. X-TOF scatter plot obtained for the 263 MeV ^{58}Ni+^{124}Sn reaction.

The different bands correspond to different mass number (A) to atomic charge state (q) ratios for the reaction products analyzed by PRISMA. A mass resolution up to ΔA/A=1/280 was measured. This is also routinely obtained during the experiments presently being performed using the PRISMA-CLARA combined facility.

An example of mass spectrum is reported in Fig. 8. It has been obtained in the 505 MeV ^{90}Zr+^{208}Pb reaction after ion-track reconstruction and corresponds to the mass distribution of Zirconium isotopes for a part of the total accumulated statistics.

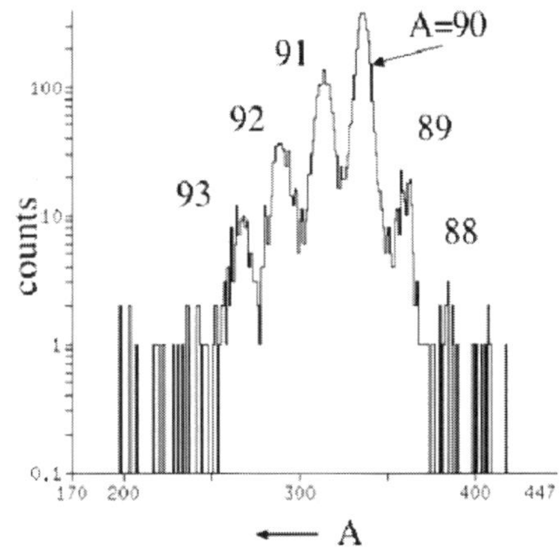

Figure 8. Zirconium isotopes populated by neutron transfer in the 505 MeV ^{90}Zr on ^{208}Pb.

The scattered Zr-like ions were analyzed by the magnetic spectrometer centered around the grazing angle ($\theta_{lab} \sim 62°$). One sees that one- and multi-neutron pick-up channels are favoured with respect to the stripping ones, as expected from Q-value systematics.

The first experimental campaign was started in April 2004. Up to now nine experiments have been performed with the PRISMA-CLARA setup. Many of these experiments have been devoted to nuclear structure studies and some of their main goals were:

- studying the shell closure in neutron-rich nuclei at N=20 and N=28;
- studying the shell closure evolution at the magic number N=50;
- the spectroscopy of neutron-rich nuclei in the A=60 mass region;
- studying the shell model in the ^{48}Ca region;
- studying the multi-nucleon transfer channels for the ^{90}Zr+^{208}Pb reaction.

The data analysis of the experiments is in progress.

4. Next developments

The use of PRISMA in a "gas filled mode" (GFM) constitutes an important extension of its capabilities to the measurements of evaporation residues recoiling at or near to 0°. The GFM is characterized by high transmission efficiency (a very important feature in the case of low production cross sections) but the main drawback is the loss of mass and energy resolution. Then, PRISMA in GFM will be used as a separator rather than a spectrometer. In this configuration evaporation residues can be efficiently focused onto a reasonably small area at the exit of the magnetic dipole, where they can be implanted for decay studies (recoil-α tagging, α-decay studies, etc.).

Simulations for GFM operation of the PRISMA spectrometer are in progress. The first results seem to indicate that a suitable detection system for implanting the recoil nuclei could be composed of a matrix of 2x5 Si detectors with a thickness of 300 μm and an active area of 5x5 cm^2. To get the position information Si strips will be used. A first prototype of this detection system is being assembled. The drift region existing downstream of the magnetic dipole was introduced mainly in order to increase the base for time of flight and ultimately to optimize the mass resolution in vacuum, but it is mostly useless in gas. Indeed the implantation system can be placed at only 60 cm of distance from the exit of the magnetic dipole.

5. Summary

The commissioning of the PRISMA+CLARA setup has been completed at the end of 2003. The in-beam tests and first experiments confirmed the performances foreseen for PRISMA and its detectors. Nine experiments have been already completed during 2004. The construction of the new detection system for the GFM operation of the PRISMA spectrometer is in progress.

References

1. A.M. Stefanini et al., *Nucl. Phys.* **A701**, 217c (2002).
2. F. Scarlassara et al., *Nucl. Phys.* **A746**, 195c (2004).
3. A. Latina et al., *Nucl. Phys.* **A734**, E1 (2004).
4. A.Gadea et al., *Eur. Phys. J.* **A20**, 193 (2004).
5. L. Corradi et al., *Phys. Rev.* **C66**, 024606 (2002).
6. S. Szilner et al., *Eur. Phys. J.* **A21**, 87 (2004).
7. G. Montagnoli et al., *INFN-LNL(REP)* - **202/2004**, 149 (2004).
8. E. Fioretto et al., *INFN-LNL(REP)* - **198/2003**, 148 (2003).

DESCRIPTION AND FIRST RESULTS OF THE CLARA-PRISMA SETUP

A.GADEA,* N.MARGINEAN, G.DE ANGELIS, D.R.NAPOLI, L.CORRADI, E.FIORETTO, A.M.STEFANINI, J.J. VALIENTE-DOBÓN, S.SZILNER, L.BERTI, P.COCCONI, D.ROSSO, N.TONIOLO, M.AXIOTIS, B.R.BEHERA, A.LATINA, I.V.POKROVSKIY, C.RUSU, W.ZHIMIN

INFN, Laboratori Nazionali di Legnaro Padova, Italy

E.FARNEA, S.M.LENZI, C.A.UR, R.ISOCRATE, D.BAZZACCO, S.BEGHINI, S.LUNARDI, G.MONTAGNOLI, R.MENEGAZZO, F.SCARLASSARA, F.DELLA VEDOVA, M.NESPOLO, R.MARGINEAN,

Dipartimento di Fisica, Universitá and INFN Sez. di Padova, Italy

A.BRACCO, F.CAMERA, S.LEONI, B.MILLION, M.PIGNANELLI, G.BENZONI, O.WIELAND

Dipartimento di Fisica Universitá and INFN Sez. di Milano, Italy

P.G.BIZZETI, A.M.BIZZETI-SONA

Dipartimento di Fisica, Universitá and INFN Sez. di Firenze, Italy

G.POLLAROLO

Dipartimento di Fisica, Universitá and INFN Sez. di Torino, Italy

M.TROTTA

Dipartimento di Fisica, Universitá and INFN Sez. di Napoli, Italy

AND THE CLARA AND PRISMA2 COLLABORATIONS.

*corresponding author (andres.gadea@lnl.infn.it)

CLARA is an array of 25 Clover (EUROBALL type) Ge detectors, placed at the target position of the large acceptance magnetic spectrometer PRISMA. Due to the granularity of the CLARA array (100 crystals), the photopeak efficiency (≈3%) and the PRISMA large acceptance, the setup is an excellent tool to investigate the structure of neutron rich nuclei, populated in multinucleon transfer reactions and deep inelastic collisions with stable beams. The setup is now fully operational, and the preliminary outcome of the firsts experiments will be discussed.

1. Introduction

Multinucleon-transfer reactions and deep-inelastic collisions have been used successfully in the last two decades to study the structure of nuclei far from stability in the neutron-rich side of the nuclear chart. Already in the 80s, Guidry and collaborators [1] suggested the possibility to populate high spin states in transfer reactions induced by heavy projectiles. Since then the use of these reactions in nuclear spectroscopy studies has increased, following the evolution of the $\gamma-$multidetector arrays, in some cases competing successfully with results from first generation radioactive beam facilities. A good example are the neutron-rich nuclei around ^{68}Ni, the structure of this nucleus has revealed the quasi-doubly-magic character of N=40 Z=28 [2]. Nuclei in this region has been investigated both with fragmentation and deep inelastic collision techniques [2,3,4].

Ancillary devices capable of identifying the reaction products or at least one of them, were already used in early works: PPAC counters in kinematic coincidences [5,6,7] or Si telescopes to identify the light fragment [8].
Increasing the $\gamma-$ray efficiency in Compton-suppressed arrays allowed selection techniques purely based on the detection of $\gamma-\gamma$ coincidences between unknown transitions from the neutron-rich nucleus and known ones from the reaction partner. The method was first used by Broda and coworkers [9] and since then, the use of these kind of reactions in combination with modern $\gamma-$ray arrays (GASP, Gammasphere, Euroball, etc.) has increased substantially the amount of spectroscopic information of nuclei in the neutron-rich side of the nuclide chart.
Recent cross section measurements, for selected multinucleon transfer reactions, with neutron rich stable targets have shown the potentiality of this reaction mechanism to populate neutron rich nuclei with sizeable cross section values [10].

The interest in studying phenomena only present in nuclei very far from stability, especially in neutron-rich medium mass or heavy nuclei, has led to the necessity of new techniques to assign the $\gamma-$transitions to the corresponding reaction product. Recently, a collaboration working at ANL and MSU, have used the information obtained from beta-decay to select $\gamma-$rays produced in deep-inelastic collisions and detected by the Gammasphere array [11]. This technique is limited to nuclei where some states are populated both in the parent $\beta-$decay and in in-beam experiments with deep-inelastic collisions. The assignment of any other transition to the nucleus is done exclusively on the basis of $\gamma-$coincidences.

In a joint effort, $\gamma-$spectroscopy and reaction mechanisms groups belonging to INFN, in collaboration with several European institutes, have developed a new setup by coupling the array of Euroball Clover detectors CLARA [12] to the large acceptance magnetic spectrometer PRISMA [13]. This setup is a step forward in the use of the multinucleon transfer and deep inelastic collisions in γ spectroscopy, and aims at measuring in-beam prompt coincidences of $\gamma-$rays detected with CLARA and the reaction product seen by the PRISMA detectors. The setup allows in most cases to assign unambiguously the transitions to the emitting nucleus by identifying the mass (A) and atomic (Z) numbers of the product going into PRISMA. Therefore, it will lower the sensitivity limit in the measurements and consequently allow to study excited states of nuclei away from stability produced with low cross sections.

2. Description of the CLARA-PRISMA setup

PRISMA is a large acceptance magnetic spectrometer for heavy ions [13], The optical design of PRISMA, consists of a quadrupole singlet and a dipole separated by 60 cm, and its most interesting features are: its large solid angle, approximately 80 msr; momentum acceptance $\pm10\%$; mass resolution up to $\Delta A/A \approx 1/300$ achieved via TOF; Z resolution $\Delta Z/Z \approx 1/60$ using the segmented focal plane ionization chamber; energy resolution up to 1/1000 and rotation around the target in a large angular range ($-20^o \leq \theta \leq 130^o$). Some of these performance figures are achieved by software reconstruction of the ion tracks using the energy measured at the focal plane and the position and time of flight build with the MCP entrance [14] and focal-plane multi wire parallel plate detector [15]. Within the mentioned Z and A resolutions, PRISMA allows for the unambiguous identification of the reaction products. A more detailed description of PRISMA is present in this volume in the contribution of Fioretto [16].

Figure 1. Photographic view of the CLARA-PRISMA setup. At the left of the picture is the CLARA array of Clover detectors followed by the quadrupole and dipole magnets of the PRISMA magnetic spectrometer. The picture ends at the right with the flight chamber and focal plane detector of the spectrometer.

At the target position of PRISMA is installed CLARA, an array of 25 composite EUROBALL Clover detectors [17]. These detectors are composed of four HP-Ge crystals mounted in a single cryostat and, to make use of the escape-suppression technique, the detectors are surrounded by a BGO anti-Compton shield.

CLARA is placed on a mobile platform which rotates together with the spectrometer, in such a way that the reaction products detected in the start and focal plane detectors of the spectrometer, in coincidence with the $\gamma-$rays, will have a forward trajectory with respect to the array. The PRISMA start detector (MCP) [14] allows to determine the outcoming angle of the products with an angular resolution $\Delta\theta < 1^o$. The software reconstruction of the trajectory allows to determine accurately the velocity of the products. Therefore, the final Doppler broadening is due only to the angular aperture of the Ge crystals.

The main performance figures of the array are: efficiency $\approx$3% for E_γ=1.3 MeV, Peak/Total ratio $\approx$45% and energy resolution $\approx$1% for v/c=10%. CLARA is installed at the target position of PRISMA.

Clover composite detectors have an excellent sensitivity for measuring the degree of linear polarization of the $\gamma-$rays emitted by the products [18]. In nuclear reactions where the outgoing products have a preferential direction for the alignment of the angular momentum, the polarization coincidence measurements in oriented nuclei (PCO), together with the angular distributions or correlation (DCO) information, allows one to determine the character and multipolarity of the electromagnetic transitions and thus to obtain fundamental information on the angular momentum and parity of excited states.

The combination of the CLARA Clover array with the PRISMA spectrometer allows to perform lifetime measurements with several techniques. For short lifetimes, below $\approx$1 ps, a technique based on the lineshape analysis can be used. This technique, developed at the EUROBALL Recoil Filter Detector [19], is based on the straggling of the ions in the target. Since the PRISMA start MCP detector can give an accurate trajectory of the outcoming ion and therefore a precise Doppler correction, it is possible to distinguish between the fraction of events emitted in the target where the ion can still change the trajectory (imprecise Doppler correction) and the fraction emitted outside the target for which the Doppler correction is right. The lifetime range covered by this technique, strongly dependent on the product velocity, goes from $\approx$ 50 fs to $\approx$ 0.5 ps using different target thicknesses.

A lifetime technique covering the range from $\approx$ 1 ps to $\approx$ 1 ns is based on the differential recoil distance method. The idea is to place a thin metallic foil at certain distances after the production target. This foil will act as energy degrader, and the velocity will be changed for the projectile-like reaction products going towards PRISMA. Depending on the velocity distribution of the projectile-like nuclei and on the desired lifetime range, the target-degrader distance will vary from few microns to hundreds of microns. This technique will be tested during the next semester with differential plunger targets consisting on a thin production target and an energy degrader foil placed at fixed distance, in a single mechanical setup. Several plunger targets, corresponding to different fixed distances, will be placed in the normal target holder inside the reaction chamber. The angular straggling of 2-3 degrees due to the degrader does not influence significantly the geometrical acceptance of the spectrometer and the velocity distribution shift can be compensated with proper magnetic field settings.

Finally, for long lifetimes, the Recoil Shadow Method, developed at EUROBALL with the Clover detectors [20], can measure up to several ns, again depending on the product velocity. This technique is based on the shadowing of the Clover detector crystals by the heavy metal collimator. It allows to measure the average distance from the target to the point where the radiation has been emitted.

3. Experimental activity

As it has been mentioned before the CLARA-PRISMA setup is an ideal tool to study the structure of moderately neutron-rich nuclei using multi-nucleon transfer and deep-inelastic collisions. In this sense, the spectroscopic information provided by the setup is complementary to that provided by first generation radioactive beam facilities. Therefore, several topics are common to the mentioned facilities and our setup, i.e. shell evolution at N=20, 28 and 50 and the onset of collectivity for the Cr and Fe isotopes towards N=40.

The experimental program of the CLARA-PRISMA setup started on March 2004. During this experimental campaign only 22 Clover detectors were used and therefore the efficiency of CLARA was $\approx$2.6%.
A consistent fraction of the experimental activity is connected to the study of the magic numbers in neutron-rich nuclei. Regarding this topic, nuclei in the vicinity of N=50 have been studied and some preliminary results will be described.
The appearance of unexpected magic numbers and the onset of collectivity in nuclei beyond this new magicity, in particular in neutron-rich nuclei with A$\approx$60 and N$\approx$34 have also concentrated experimental efforts and some preliminary results obtained with CLARA-PRISMA in this region will be also shown.

3.1. *The N=50 shell closure in the vicinity of ^{78}Ni*

The large N/Z ratio of the neutron-rich doubly magic nucleus ^{78}Ni qualifies the region around this nucleus for searching for shell effects connected with nuclei with large neutron excess.

Recent works extensively discussed the changes of the nuclear potential due to the difference between the proton and neutron root mean square radius in neutron rich nuclei [21,22,23,24]. The vicinity to the drip-line is not

the only effect that can modify the shell structure in neutron-rich nuclei. It has been suggested recently that the attractive tensor interaction between spin-flip orbitals (repulsive between non spin-flip) may contribute to the weakening of the shell gaps in neutron-rich nuclei [25].
The spectroscopic information provided by experiments in the ^{78}Ni region can be compared with shell model calculations, and from the comparison it is expected to infer possible changes in the N=50 shell gap. This comparison will as well provide fundamental information to extract the contribution of the tensor interaction in the evolution of the shell gaps.
The calculations performed in the Infinite Nuclear Matter framework by Nayak and collaborators [26] predicted that changes in the shell gap at N=50 start in ^{82}Ge (Z = 34). On the contrary, the relativistic mean field calculations performed by Geng and collaborators [27] predict observable effects on the N=50 shell gap only for nuclei bellow Z=28-26.

The experimental activity of CLARA-PRISMA in this region [28] has been performed using a ^{82}Se beam at 505 MeV, delivered by the Tandem-ALPI complex, impinging on a $^{238}UO_2$ 400 μgr/cm^2 target. In order to select mainly the quasi-elastic projectile-like reaction products from the multi-nucleon transfer process, the spectrometer was placed at the grazing angle ($\theta_g = 64^o$). Spectra from more than 50 nuclear species, from Kr to Cr isotopes, were obtained in a 4-days experiment, with a beam intensity of 5 to 6 pnA.

The limited CLARA and PRISMA efficiencies and the short beamtime, prevented the measurement of $\gamma-\gamma-$product coincidences, for many of the measured nuclei. Therefore, to build the level scheme it was necessary to resort to a previous GASP experiment [29] performed again with ^{82}Se beam at an energy of 460 MeV bombarding a thick ^{192}Os target.

For the more exotic N=50 isotones, due to the low population cross-section and the limited duration of the experiment, it was only possible to identify a candidate for the 4^+ state in ^{82}Ge [28]. In Fig. 3.1 the confirmed 4^+ in ^{84}Se and the candidate for the 4^+ in ^{82}Ge are shown together with the systematics and shell model calculation performed by Lisetskiy and collaborators [30]. These calculations are done with a new effective interaction based in a G-matrix Bonn-C Hamiltonian fitted to the experimental data available in the region, and takes into account the changes in the effective single particle energies due to evolution of the monopole interactions, between ^{56}Ni and ^{78}Ni (Z=28 isotopes) and between ^{78}Ni and ^{100}Sn (N=50 isotones), for the proton and neutron orbitals respectively.

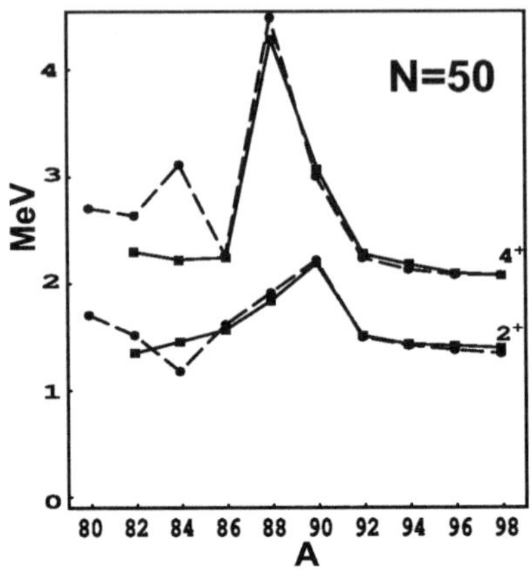

Figure 2. Calculated [30](circles) and experimental (squares) 2^+ and 4^+ excitation energies for the even-even N=50 isotones from ^{98}Cd to ^{82}Ge. The excitation energy of the 4^+ in ^{84}Se has been confirmed and a preliminary value for ^{82}Ge is also reported.

3.2. *The onset of deformation in neutron rich A=60 nuclei*

The excitation energy of the first 2^+ states in even-even Cr isotopes [31] is decreasing very fast when going from the N=32 to the N=40 Cr isotope. The 2^+ states in ^{58}Cr, ^{60}Cr and ^{62}Cr are placed at 880, 646 and 446 keV respectively, suggesting the possible onset of deformation towards N=40.

Neutron-rich nuclei in the A=60 mass region, where protons are partially filling the $1f_{7/2}$ shell and the neutrons are filling the other pf orbitals, are predicted to develop deformation. In particular, recent shell-model calculations [32] predict the decrease of the excitation energy of the first 2^+ state and large $E2$ transition probabilities for its decay to the ground state, for even mass Cr and Fe isotopes with neutron number approaching N=40, a magic number at stability, due to the $g_{9/2}$ and $d_{5/2}$ orbitals entering into play.

An experiment aimed at studying the structure of Cr and Fe isotopes in this region [33], has been performed at CLARA-PRISMA setup. In this experiment it has been identified for the first time the ground state band in the nucleus ^{58}Cr (Z = 24;N = 34). The nucleus ^{58}Cr has been populated in the reaction ^{64}Ni + ^{238}U. The ^{64}Ni beam, with the energy of 400 MeV was delivered by the Tandem-ALPI accelerator complex. The thickness of the

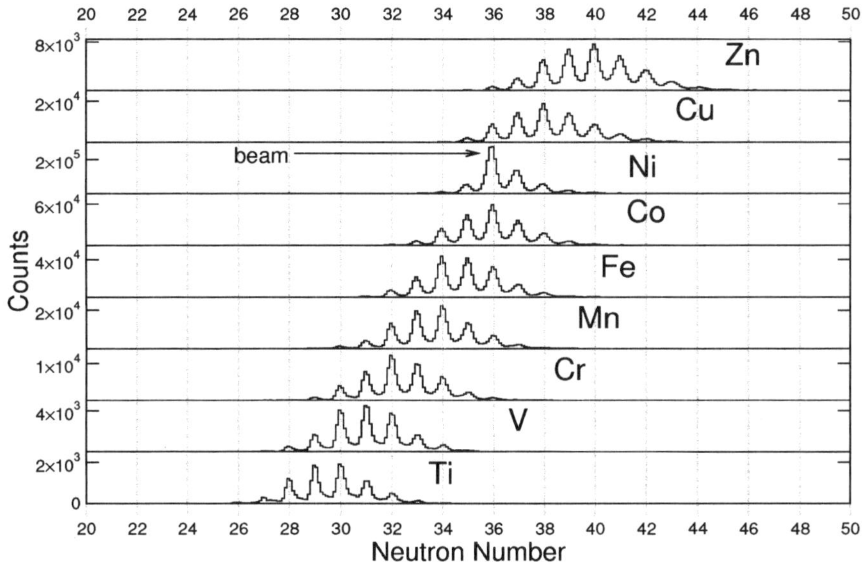

Figure 3. Mass distribution for isotopes from Zn to Ti, measured with the PRISMA spectrometer following the ^{64}Ni 400 MeV + ^{238}U reaction. In the axis the neutron number has been plotted.

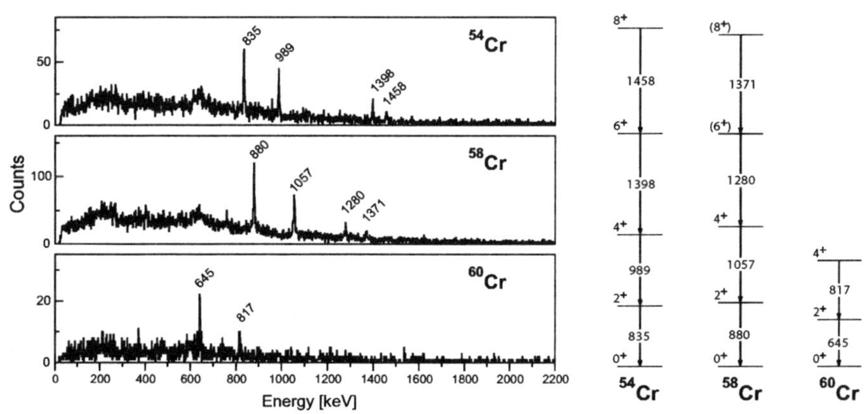

Figure 4. CLARA $\gamma-$ray spectrum and level scheme obtained for few Cr even-even isotopes in the ^{64}Ni beam experiment.

Uranium target was 400 $\mu g/cm^2$. Projectile-like nuclei produced following multinucleon transfer were detected with the PRISMA spectrometer placed at 64°, which corresponds to the grazing angle of the quasi-elastic reaction.

The mass spectrum obtained in the analysis of for Zn to Ti isotopes is shown in Fig. 3. A CLARA $\gamma-$ray spectrum is obtained for this nucleus setting a condition on the ^{58}Cr in the PRISMA data. The spectrum for ^{58}Cr together with close by Cr isotopes, are shown in Fig. 4 [33]. Only the 2^+ level at 880 keV, was previously known from $\beta-$decay data [34]. The tentative location of the (4^+) state at 1937 keV excitation energy, ratio $E(4^+)/E(2^+) = 2.2$, characterizes the ^{58}Cr as a transitional nucleus. The excitation energy of the (6^+) state is preliminary assigned to 3217 keV and therefore, the ratio between the excitation energy of the 6^+ and 2^+ states is equal to 3.65. The two aforementioned ratios are very closed to the expected values for a nucleus described by the E(5) critical point symmetry [35], i.e. a nucleus at the U(5)-0(6) shape phase transition.

4. Conclusions

This contribution describes the CLARA-PRISMA setup and shows the preliminary results of two of the seven experiments already performed at the setup. CLARA-PRISMA is now fully operational and the results obtained show the high potential of the multinucleon-transfer and deep-inelastic reactions with stable beams in populating neutron rich nuclei. The upgrades realized for the LNL accelerators (low beta cavities for the ALPI linac and the new PIAVE injector) open new perspectives concerning the variety of nuclear species that will be available for the users in 2005.

5. Acknowledgments

The author acknowledge the technical groups, involved in the construction and running of the CLARA-PRISMA setup and the LNL accelerators, for the excellent work done.

References

1. M.W.Guidry et al., *Phys. Lett.* **B163**, 79(1985).
2. R.Broda et al., *Phys. Rev. Lett.* **74**, 868 (1995).
3. R. Grzywacz et al., *Phys. Rev. Lett.* **81**, 766 (1998).
4. T.Ishii et al., *Phys. Rev. Lett.* **81**, 4100 (1998).
5. A.O.Machiavelli et al., *Nucl. Phys.* **A432**, 436 (1985).

6. C.Y.Wuet al., *Phys. Lett.* **B188**, 25 (1987).
7. S.Juutinen et al., *Phys. Lett.* **B192**, 307 (1987).
8. H.Takai et al., *Phys. Rev.* **C38**, 1247 (1988).
9. R.Broda et al., *Phys. Lett.* **B251**, 245 (1990).
10. L.Corradi et al., *Phys. Rev.* **C63**, 021601 (2001) and refs. therein.
11. R.V.F.Janssens et al., *Phys. Lett.* **B546**, 55 (2002).
12. A.Gadea, *Eur. Phys. J.* **A** in press
13. A.M.Stefanini et al., *Nucl. Phys.* **A701**, 109c (2002), A.Latina et al., *Nucl. Phys.* **A734**, E1 (2004).
14. G.Montagnoli et al., *Nucl. Instr. & Meth.* **A** in press.
15. S.Beghini et al., *Nucl. Instr. & Meth.* **A** in press.
16. E.Fioretto et al., in this proceedings
17. G.Duchêne et al., Nucl. Instr. & Meth. A 432, (1999) 90
18. L. Garcia-Rafi et al., *Nucl. Instr. & Meth.* **A391**, 461 (1997).
19. P.Bednarczyk et al., private communication
20. E.Gueorguieva et al., *Nucl. Instr. & Meth.* **A474**, 132 (2001).
21. G.A.Lalazissis et al., *Phys. Rev.* **C57**, 2294 (1998).
22. D.Vretenar et al., *Phys. Rev.* **C57**, 3071 (1998).
23. J.Meng et al., *Nucl. Phys.* **A650**, 176 (1999).
24. M.Del Estal et al., *Phys. Rev.* **C63**, 044321 (2001).
25. T.Otsuka, proceedings of the "XXXIX Zakopane School of Physics", *Acta Phys. Pol.* in press.
26. R.C.Nayak et al., *Phys. Rev.* **C60**, 064305 (1999).
27. L.S.Geng et al., *nucl-th/0402083.*
28. G.de Angelis, G.Duchêne, N.Marginean et al., private communication.
29. Y.H.Zhang et al., *Phys. Rev.* **C70**, 024301 (2004).
30. A.F.Lisetskiy et al., *Phys. Rev.* **C70**, 044314 (2004).
31. O.Sorlin et al., *Eur. Phys. J.* **A16**, 55 (2003).
32. E.Caurier et al., *Eur. Phys. J.* **A15**, 145 (2002).
33. S.M.Lenzi, S.J.Freeman, N.Marginean et al., private communication.
34. J.I.Prisciandaro et al., *Phys. Lett.* **B510**, 17 (2001).
35. F.Iachello, *Phys. Rev. Lett.* **85**, 3580 (2000).

CONSTRUCTION OF POLARIZED PROTON TARGET AND FIRST EXPERIMENT WITH ^{6}HE

H. SAKAI, M. HATANO, H. IWASAKI, H. KUBOKI, T.K. ONISHI, T. SAITO, M. SASANO, AND K. YAKO

Department of Physics, University of Tokyo,
Hongo 7-3-1, Bunkyo, Tokyo 113-0033, Japan
E-mail: sakai@phys.s.u-tokyo.ac.jp

T. WAKUI, T. UESAKA, T. KAWABATA, AND K. SUDA

Center for Nuclear Study, University of Tokyo,
Hirosawa 2-1, Saitama 351-0198, Japan

N. AOI, N. SAKAMOTO, K. SEKIGUCHI, AND Y. YANAGISAWA

RIKEN, Hirosawa 2-1,
Wako, Saitama 351-0198, Japan

T. IKEDA, AND K. ITOH

Department of Physics, Saitama University,
Shimo-Ohkubo 255, Saitama 338-8570, Japan

A. TAMII

Research Center for Nuclear Physics,
Osaka University, Mihogaoka 10-1, Ibaraki,
Osaka 567-0047, Japan

A polarization asymmetry measurement with an unstable beam of ^{6}He with an energy of 71 MeV/u has, for the first time, been successfully carried out. For this purpose a new type of the spin polarized solid proton target was constructed which can be operated under a low magnetic field of 0.08 T and a high temperature of 100 K.

1. Introduction

Nuclear reactions such as $(\vec{p},p)$, $(\vec{p},d)$, $(\vec{p},pp)$, etc. with spin polarized beam have provided rich information on nuclear structures as well as on reaction mechanisms. In order to apply such polarization measurements to the radio-isotope (RI) beam experiments, a polarized proton target is needed. A polarized '*gas*' proton target system has been established technically. However the low density of the gas target coupled with low intensity of RI beam, i.e. low luminosity, makes the RI beam experiment very difficult. Consequently a polarized '*solid*' proton target (PSPT) is a natural choice. Here one must take account of the fact that the RI beam experiments utilize the inverse kinematics and therefore it involves detection of low energy protons or deuterons which are emitted to large angles ($> 50°$ in the lab. system). In this respect a conventional PSPT system is inconvenient for the RI beam experiments since it requires a high magnetic field (≥ 2.5 T) as well as a very low temperature (≤ 1 K). This severe experimental environment makes the detection of low energy particles difficult. We have constructed a PSPT which overcomes those drawbacks and allows us to pursue various spin polarized experiments with RI beams.

2. Construction of PSPT

Our target material is a crystal of naphthalene ($C_{10}H_8$) (host) doped with pentacene ($C_{22}H_{14}$) (guest). It was first reported by Henstra that a proton in this aromatic material can be polarized under a low magnetic field and at a high temperature [1].

The basic principle to polarize a proton is a pulsed dynamic nuclear polarization (DNP) method. First, an alignment of electron population which appears in the lowest triplet state of pentacene is produced by means of optical pumping by a laser. This electron alignment is subsequently transferred to a proton polarization by using the integrated solid effect (ISE). The ideal proton polarization is 72.8%, if the polarization transfer is perfect and relaxation of polarization is negligible.

We first constructed a PSPT system for the off-line test which consisted of a C-type magnet, the laser system with a Ar-ion laser operated at 514 nm and the cylindrical cavity with a microwave generator with 9.1 GHz.

We have established a method to fabricate a single crystal of naphthalene doped with pentacene (0.01 mol%) by the Bridgman method [2]. The highly purified naphthalene must be prepared beforehand by the zone-melting refinement. Figure 1 is a photograph of the naphthalene single crystal produced by us. The pentacene has a light violet color.

By using this system we have achieved the proton polarization of about 37% under the magnetic field of 0.3 T and the temperature of 100 K with

Figure 1. Crystallized naphthalene doped with pentacene by the Bridgman technique.

the naphthalene crystal size of 4×4×3 mm^3 [3]. The relaxation time was measured to be about 22 hours. The proton polarization was measured by the weak pulse NMR method and calibrated by using the 90° pulse NMR signal at a thermal equilibrium. Uncertainty of the proton polarization is 11%.

Figure 2 shows the polarization buildup as a function of time.

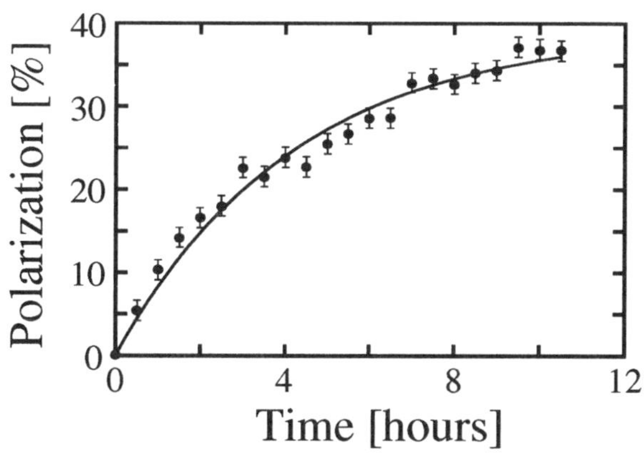

Figure 2. Polarization buildup as a function of time.

Based on this experience gained in off-line tests we designed a PSPT system which could be operated under the experimental conditions required by the RI beam [3,2]. Firstly, we needed to produce a large and thin target crystal to cope with a large size of the RI beam due to the secondary beam. Secondly, a completely new design was needed for the resonator system of microwave, to keep minimum material along the recoil proton trajectory to avoid energy loss. We introduced a loop-gap resonator which is made of a Teflon thin film of 25 μm on both sides of which copper stripes with a thickness of 4.4 μm are printed [4]. Finally, special care was taken for the target chamber design to cool down the target. The target chamber has another small chamber nesting inside it. The small chamber in which the target crystal and the devices for polarizing the target such as the loop-gap resonator are equipped is cooled by cold nitrogen gas. The volume between the two chambers is evacuated and connected to the beam pipe. Each chamber has a glass window for the laser light input and three exit windows with thin Kapton foils (50 μm), one for the beam and two for the recoil particles at left and right sides of the chambers with respect to the beam direction.

3. Polarization Measurement

The first polarization asymmetry measurement [2] was performed with the unstable beam of ^{6}He with 71 MeV/u produced by the RIKEN Projectile-fragment Separator (RIPS). We chose elastic scattering since the event identification is relatively easy. Moreover, the cross section data existed for the scattering angle of $\theta_{\rm cm} = 20° - 50°$ [5]. The cross section and the polarization asymmetry (=target polarization$\times$analyzing power) were measured for $\theta_{\rm cm} = 40° - 80°$ which covers the second diffraction peak of cross sections. This angle region is very interesting since all optical model predictions based on a microscopic folding model [6,7] or a phenomenological model [8] show a large *positive* polarization asymmetry, i.e. positive analyzing powers.

The PSPT system was operated under a magnetic field of 0.08 T and a temperature of 100 K. The target crystal had a thickness of 1 mm with a diameter of 14 mm. The spot size of the ^{6}He beam was about 10 mm in diameter (FWHM) and its intensity was on the average 1.7×10^5 particles/sec. The proton polarization during the measurement was monitored by a pulsed NMR method. The NMR amplitude as a function of time is shown in Fig. 3. The direction of the polarization was reversed during the experiment to minimize the systematic uncertainties associated with geometry.

The scattered ^{6}He were detected by a multi-wire drift chamber (MWDC) and three planes of plastic scintillator hodscopes. The recoil protons were detected by the multi-stripped position sensitive silicon de-

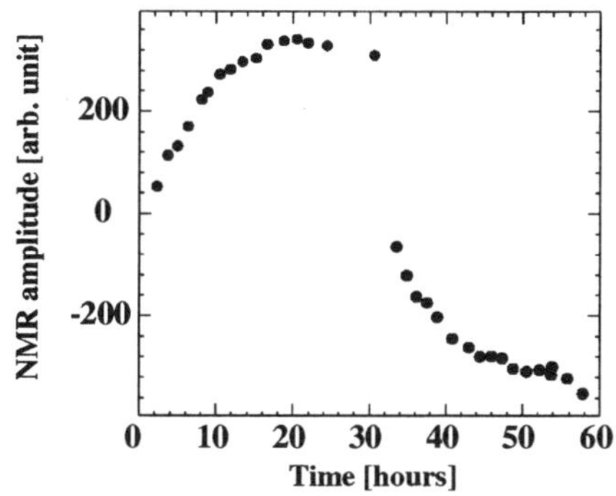

Figure 3. NMR amplitudes during the measurement

tectors backed by the plastic scintillation counters on left and right sides with respect to the beam.

The polarization asymmetry ϵ can be derived from the measurement by

$$\epsilon = \frac{\sqrt{Y}-1}{\sqrt{Y}+1}, \tag{1}$$

$$Y = \frac{N_L^{\uparrow} \cdot N_R^{\downarrow}}{N_L^{\downarrow} \cdot N_R^{\uparrow}}, \tag{2}$$

where $N_{L(R)}^{\uparrow(\downarrow)}$ is the counts detected by the left(right) $L(R)$ detector with the target polarization state 'up'('down') $\uparrow$ ($\downarrow$).

ϵ is related to the target polarization P_t and the analyzing power A_y for the elastic scattering as

$$\epsilon = P_t \times A_y. \tag{3}$$

To derive A_y we need to know P_t. Unfortunately we could not perform the calibration of P_t during the experiment and we thus made our best guess on the P_t value. We assumed P_t =0.21. Systematic uncertainty of P_t could be as large as 50%. Note that this brings about large ambiguity in the magnitude of A_y but never changes its sign.

Figure 4 shows the cross sections and the analyzing powers for the $\vec{p}$+^{6}He scattering at 71 MeV/u as a function of c.m. scattering angles. Present results are plotted by the solid circle and previous results by Korsheninnikov [5] are plotted by the open circle. Only statistical errors are shown in this figure. Errors for the cross section are smaller than the size of the symbol in most cases. Systematic error for the cross section is esti-

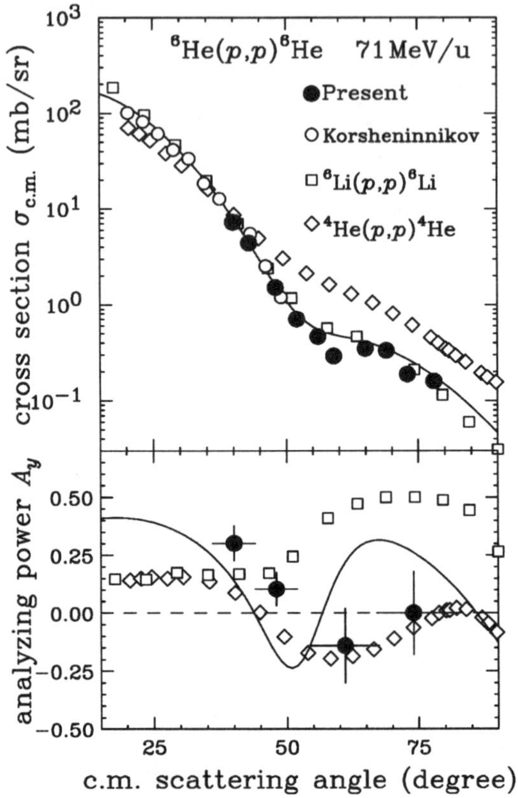

Figure 4. Cross sections and analyzing powers. See text for detail.

mated to be ±9 %. For the analyzing powers, the horizontal bar indicates the bin width of the angle integrated.

It might be very interesting to compare our $\vec{p}$+^{6}He results to those of $\vec{p}$+^{6}Li at the similar bombarding energy by Henneck [9]. The $\vec{p}$+^{6}Li data are plotted by the open square symbols in Fig. 4. It is surprising to find that the magnitude and its angular dependence of cross sections are almost identical. This indicates that both nuclei have a similar potential radius. However, it is even more surprising to find that the observed polarization asymmetry shows a remarkable difference between the $\vec{p}$+^{6}He and $\vec{p}$+^{6}Li scatterings. The polarization asymmetry changes the sign from positive to negative values for $\vec{p}$+^{6}He, while it increases rapidly from the small positive value to the large positive value for the same angular range in case of $\vec{p}+{}^6$Li. This feature of the *negative* values contradicts largely with the optical model predictions [5,6,7]. It should be noted that the similar sign change in A_y is also seen in the $\vec{p}+{}^4$He scattering at the same bombarding energy by Burzynski [10]. Data are plotted in Fig.4 by the diamond symbols after modifying the scattering angle so as to give the same momentum transfer.

4. Analysis and Discussion

We have carried out the optical model analysis by using the cross section data. The standard Wood-Saxon shape is assumed for real and imaginary central potentials (V_R and W_R). The cross sections are easily reproduced by the calculation as shown in the upper panel of Fig.4.

Obtained parameters are shown in Table 1. For a comparison purpose, those by Henneck [9] for the $\vec{p}+{}^6$Li reaction at 72 MeV/u are included in Table 1 and plotted in Fig. 4 with dashed curve.

Nucleus	V_R	r_R	a_R	W_v	r_{wv}	a_{wv}
^{6}He	−20.2	1.27	0.57	−19.2	0.92	0.64
^{6}Li	−31.7	1.10	0.75	−14.1	1.15	0.56
	$V_{\ell s}$	$r_{\ell s}$	$a_{\ell s}$			
^{6}He	−3.36	1.30	0.94			
^{6}Li	−3.36	0.90	0.94			

The peculiar behavior of the sign change of A_y can be better reproduced when $r_{\ell s}$ is set to be larger compared to the standard value of this mass region. Roughly speaking, the spin-orbit potential locates outside further by about 0.8 fm compared to that for ^{6}Li. A detail analysis is in progress.

5. Summary

A polarized solid proton target which works under the condition of a low magnetic field of 0.8 T and a high temperature of 100 K has been built and used, for the first time, for the polarization asymmetry measurement with the ^{6}He beam at 71 MeV/u. The asymmetry shows an unexpected behavior which disagrees with the optical model predictions which were made before this experiment.

Acknowledgments

We would like to thank I. Tanihata and T. Suda for their continuous support on this work. We are indebted to M. Iinuma and K. Takeda for helpful suggestions in the early stage of this work. We also wish to thank the RIKEN accelerator group for their excellent work. This work has been supported in part by the Grant-in-Aid for Scientific Research No. 15740139 of the Ministry of Education, Culture, Sports, Science, and Technology of Japan.

References

1. A. Henstra, T. -S. Lin, J. Schmidt, and W. Th. Wenckebach, *Chem. Phys. Lett.* **165**, 6 (1990).
2. M. Hatano, PhD Thesis, University of Tokyo, unpublished, 2004.
3. T. Wakui, M. Hatano, H. Sakai, A. Tamii, and T. Uesaka, *AIP Conf. Proc.* **675**, 911 (2003) and T. Wakui, *et al.*, *Nucl. Instr. Meth.* **A526**, 182 (2004).
4. T. Uesaka *et al.*, *Nucl. Instr. Meth.* **A526**, 186 (2004).
5. A.A. Korsheninnikov *et al.*, *Nucl. Phys.* **A616**, 45 (1997).
6. S.P. Weppner, O. Garcia and Ch. Elster, *Phys. Rev.* **C61**, 044601 (2000).
7. K. Amos, private communication.
8. D. Gupta, C. Samanta and R. Kanungo, *Nucl. Phys.* **A674**,77 (2000).
9. R. Henneck *et al.*, *Nucl. Phys.* **A571**, 541 (1994).
10. S. Burzynski, *et al.*, *Phys. Rev.* **C39**, 56 (1989).

THE GAMMA DECAY OF THE GIANT DIPOLE RESONANCE AT FINITE TEMPERATURE : PROBING SHAPES AT VERY HIGH SPINS

A. BRACCO
Dipartimento di Fisica Università di Milano
via Celoria 16,
20133 Milano, Italy
E-mail: Bracco@mi.infn.it

A. MAJ
The Niewodniczański Institute of Nuclear Physics, Polish Academy of Sciences
E-mail: Adam.Maj@ifj.edu.pl

The study of the gamma decay from the giant dipole resonance (GDR) built on highly excited states continues to be a useful tool to explore the basic nuclear nuclear properties at finite temperature and angular momenta. In particular the dependence of the GDR width as a function of temperature and angular momentum provides information on the evolution of the nuclear shapes and of the damping mechanisms of this collective state. Here we focus on results of new exclusive measurements concerning a region at the low temperature at around 1-1.5 MeV and at high angular momenta close to the fission limit. Two different behaviours of the nuclear shape are found for the mass region A=40 and A= 220. In the first case some evidence for the Jacobi transition (oblate to prolate shape transition) is found while for the heavy mass the data show the persistence of nearly spherical shapes up to the fission limit.

1. Introduction

During the last years considerable efforts have been made to investigate the properties of the giant dipole resonance built on excited states. The field has progressively expanded resulting in significant nuclear structure and nuclear reaction information. Among the structure and reaction effects explored with the GDR there are, for example, the determination of the nuclear shape, the coupling to low lying states and to quadrupole deformation, damping due to collisions and thermal shape fluctuations, fission time scales, loss of collectivity at high temperatures, entrance channel effects and pre-equilibrium giant dipole vibrations. In this conference these last two topics have been discussed by D. Santonocito who has presented new interesting results obtained with measurements made at Laboratori Nazionali del Sud and reported in this volume.

In contrast, this paper focuses on the problem of the nuclear shape and of the damping mechanisms [1] as addressed by the measurements of the GDR width. In this connection it is important to stress that the existing investigations made in different mass regions in the interval of temperature 1-2 MeV give results for the GDR widths which are larger than the T=0 values and that are in general rather well reproduced by the model predictions. The latter include the effects of thermal shape fluctuations and assume that the collisional damping does not change with temperature. Nevertheless, a much more stringent test to the theory in the regions at the limit of temperature and angular momentum is needed and this requires exclusime measurements to be made.

In the case of high temperatures, namely at T>2.5- 3 MeV it is not yet clear whether or not the width becomes roughly constant with temperature as expected from the fact that thermal shape fluctuation effects should saturate. The experimental approach to address this question requires exclusive measurements of particles and gamma-rays to determine well the temperature of the compound nucleus on which the GDR is built. Recent measurements have been made at LNL using the GARFIELD apparatus equipped with gamma-ray detectors [2] (the analysis is presently in progress) which show the importance of choosing the proper heavy ion reaction to thermalise the nucleus at high temperature and consequently to limit the contribution of pre-equilibrium emission.

In the present paper we will discuss exclusive measurements addressing the problem of shapes at very high angular momenta near to the fission limit.

The measurements concern two regions of the mass table, namely A≈40 and A≈220 which are expected to behave differently at high spins and finite temperature. In particular, a search for the Jacobi shape transition (the change from oblate to prolate shape induced by rotation in a liquid drop rotating synchronously) has been made in connection with the study of the A≈40 region [3]. The heavy mass nuclei are not expected to show this shape transition at angular momenta lower than the fission limit as the corresponding rotational frequency is to low for causing it. To probe this behaviour the GDR in the ^{216}Rn nucleus has been measured.

2. Search for the Jacobi shape transition in the hot ^{46}Ti

Light nuclei, with mass ranging from A=30 to A=70, open new horizons for studying the mechanisms driven by rotation and producing nuclear excitations. On one hand, the number of nucleons is here large enough to align and to form high spins. However, because of the relatively light mass and consequently of the small values of the moment of inertia, the angular velocities associated with high spins are extremely large. One would, therefore, expect to observe much easier effects which are related to the rapid rotation, e.g. change in deformation, Coriolis effects, etc. On the other hand, the number of nucleons is small enough so that the band termination is easily reached, and the related single-particle

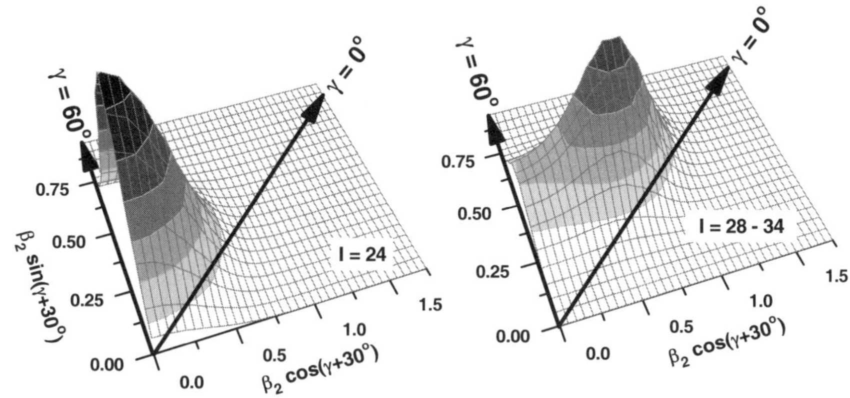

Fig. 1. Boltzmann factor distributions for ^{46}Ti, obtained from the LSD model for two spin regions. The maximum of the distribution for I = 24 $\hbar$ is centered on the oblate shape ($\beta_2 \approx$ 0.3-0.4), while the distribution for the 28 $\hbar < I <$ 34 $\hbar$ shows importance of elongated triaxial shapes ($\beta_2 \approx$ 0.6-0.7). The width of these probability distributions reflects the extent of the thermal shape fluctuations on the β-γ plane. At $I \approx$ 30 $\hbar$ the rotational frequency becomes $\hbar\omega_{rot} \approx$ 2.8 - 3 MeV

effects are expected to prevail at highest spins. In addition, the shape fluctuations with such low number of nucleons are predicted to be sizeable.

At finite temperature the nucleus behaves as a liquid drop whose shape and deformation depend on the rotational frequency which in this case can reach the highest possible values. In particular, the Jacobi shape transition has been predicted to appear in many nuclei at angular momenta close to the fission limit. This consists in an abrupt change of the nuclear shape from an oblate ellipsoid rotating non-collectively around its symmetry axis to an elongated prolate or triaxial shape, rotating collectively around the shortest axis.

Calculations based on the recently developed LSD (Lublin-Strasbourg Drop) model [4,5] have been made for the Jacobi transition mechanism in ^{46}Ti nucleus. The results of these calculation [3] have shown, that the equilibrium shape of the nucleus (in this liquid drop approximation) is spherical at I=0 and nearly spherical for I<10 hbar and becomes oblate at larger spin value. The size of the oblate deformation becomes larger when the spin increases. At around I=28 hbar the Jacobi shape transition sets in: the nucleus becomes unstable towards triaxial shape and then (for I>34 hbar) it follows prolate shape configurations, with rapidly increasing size of the deformation up to the fission limit (around I=40 hbar). This is shown in figure 1 where the Boltzmann factor at T = 1.5 MeV is shown for two different spin regions.

In order to test these predictions, an experiment was performed at the VIVITRON accelerator, using the EUROBALL phase IV Ge-array coupled to the HECTOR array [6] and the charged particle detector EUCLIDES.

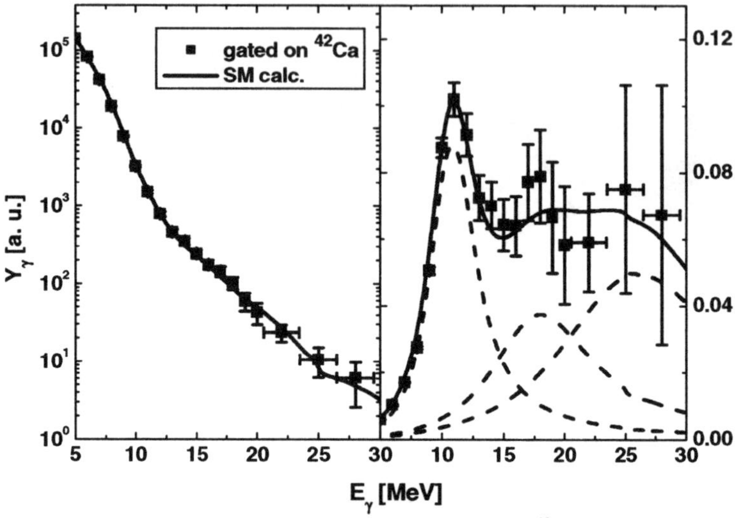

Fig. 2. Left panel: The high-energy γ-ray spectrum gated by the ^{42}Ca transitions and by high fold region, in comparison with the best fitting Montecarlo statistical model calculations (full drawn line) assuming a 3-Lorentzian GDR line shape with energies: E_{GDR} = 10.8(1), 18(1) and 26(2) MeV; widths: Γ_{GDR} = 4.0(5), 10(1) and 16(3) MeV; and strengths: σ_{GDR} = 0.35(5), 0.38(10) and 0.7(2). Right panel: The deduced experimental GDR strength function (full drawn line) together with best fitting 3-Lorentzian function and its individual components.

The ^{46}Ti compound nucleus was populated with the ^{18}O + ^{28}Si reaction at E_B = 105 MeV. The excitation energy of the ^{46}Ti was 86 MeV and the maximum angular momentum $l_{max} \approx 34\ \hbar$. To obtain the clean high-energy γ-ray spectrum, associated with the high spin part of the ^{46}Ti compound nucleus, the gates were set on known, well resolved low energy γ-ray transitions of ^{42}Ca measured in the Ge-detectors of the EUROBALL array. This particular residual nucleus was populated only in the $2p2n$- or α emission from the highest spin region of the compound nucleus. In addition, one should note that the low spin transitions in ^{42}Ca, used as gates, were found (see [7] and references therein) to be strongly fed from the highly deformed bands in the same nucleus. The left panel of Fig. 2 shows the measured high-energy γ-ray spectrum obtained with such condition together with statistical model predictions obtained using the Monte Carlo Cascade code. The quality of the fit can be judged more clearly by inspecting the right panel of Fig. 2, where the GDR line shape, i.e. the extracted absorption cross-section using the method described in e.g. Ref. [8] is shown. One feature of the obtained GDR line shape is a broad high-energy component centered at around 25 MeV. Much more pronounced, however, is a narrow low-energy component at 10.5 MeV. This fit shows also that the average GDR line shape has to be approximated with at least 3 components.

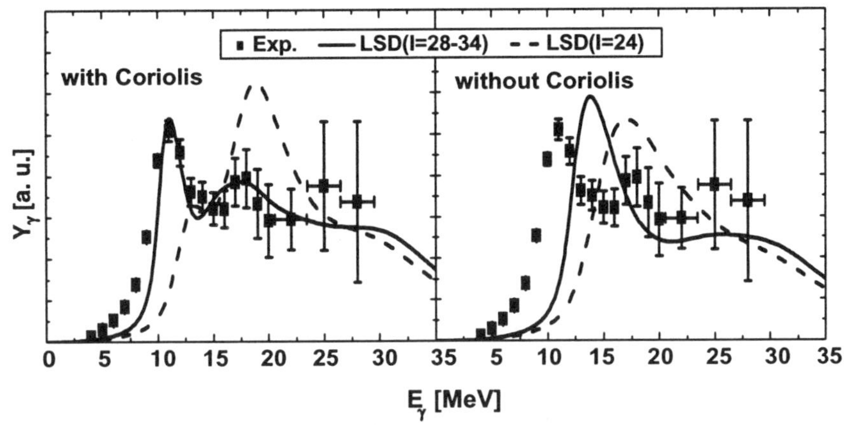

Fig. 3. Left-panel: The full drawn line shows the theoretical prediction (at $\langle T \rangle \approx 2$ MeV and in the spin region $28 \div 34\ \hbar$) of the GDR line shape in ^{46}Ti obtained from the thermal shape fluctuation model (including the Coriolis splitting of the GDR components) based on free energies from the LSD model calculations. The dashed line shows similar prediction for $I = 24\ \hbar$. The filled squares are the experimental data shown also in the right panel of Figure 2. Right panel: the same, as in the left panel, but in the calculations the Coriolis splitting was not taken into account.

To interpret this result for the GDR line shape, we used the same approach as that adopted in many studies concerning the GDR in hot and rotating nuclei, namely we performed calculations within the thermal shape fluctuation model (see Ref.[1] and references therein) which assumes that the average GDR line shape is the weighted sum of individual (i.e. at given deformation) GDR line shapes. The weighting factors (the probability of finding the nucleus at a given deformation value) are calculated by using the macroscopic deformation-dependent LSD liquid drop energies. In the calculations, we included also the possibility of Coriolis splitting of the GDR strength function for a given spin and deformation values using the rotating harmonic oscillator model [9]. Thus, for each deformation point the GDR line shape consists in general of a 5-Lorentzian parametrization. The results for the spin region I=28-34 are presented together with the experimental GDR strength function in the left panel of Fig. 3.

A remarkable good agreement between the theoretical predictions and the present experimental results can be observed. For comparison, the calculated averaged GDR line shape for I=24, i.e. in the oblate regime, is shown with dashed line in the same figure. This agreement might be an evidence of observation of the predicted Jacobi shape transition. Additional signature for the Jacobi transition may come from the presence of two other broad components in the strength function at higher energies both in the experiment and calculations.

One component, at around 17 MeV, is clearly seen in the present data, and the other which is very broad (20-30 MeV) can also be identified.

To demonstrate the importance of the Coriolis effect, the right hand side panel of Fig. 3 shows the same calculations of the average GDR line shape when the Coriolis splitting is neglected. As can it can be seen, in this case the low-energy component has at higher value than the experimental one, and also the entire GDR line shape does not reproduce the experimental data. The predictions for the oblate regime (I=24) are very weakly sensitive to the Coriolis effect.

In addition, the GDR feeding of the normal deformed bands and that of the well-deformed (possibly superdeformed) collective band in ^{42}Ca [10] were investigated. It is found that the ratio of the intensity of transitions from well-deformed ("sd") and normal deformed ("nd") states is not 1 but rather 1.5 when gating with high-energy γ-rays in the region 8-10 MeV. This might indicate that the low energy component of the GDR in the compound nucleus ^{46}Ti feeds preferentially the "sd" (well-deformed) band in the ^{42}Ca evaporation residue.

3. Probing the nuclear shape around the fission limit in ^{216}Rn

An interesting and still open question concerning the mechanism of nuclear fission induced by fast rotation, is the nuclear shape evolution on the way to scission. This problem can also be addressed with the study of the GDR which, as discussed before, is sensitive to the average nuclear shape at finite temperature [1,11,12]. Nuclear deformations corresponding to the saddle point configurations are often expected to be very large. The related experimental information is particularly interesting as it provides a sensitive test of both the macroscopic methods of calculating the total nuclear energy at high temperatures as well as the microscopic ones taking into account thermal excitations.

The term 'fission limit' used throughout this paper applies by definition to the spin value, say I_f , for which the fission-barrier penetrability from the static equilibrium deformation equals 50%. Within a model, such a spin can be relatively easily determined in the calculations but much less so through measurement. Calculation results and their comparison with experiment suggest that in the present context $I_f \sim 40\ \hbar$, within a few $\hbar$ inaccuracy [13].

In order to obtain information about the nuclear shape evolution along the fission path, it is possible to use at least two different approaches. One of them consists in studying the GDR in fissioning nuclei, by measuring the GDR decay during the fission process. From the high-energy γ-ray spectra in coincidence with the fission products, in principle one may expect to obtain the information about the shape of fissioning nuclei and thus about their shape evolution along the fission path. However, the existing data show that this is not at all an easy task because of the presence in the spectra of the γ-rays from the fission products [14,15].

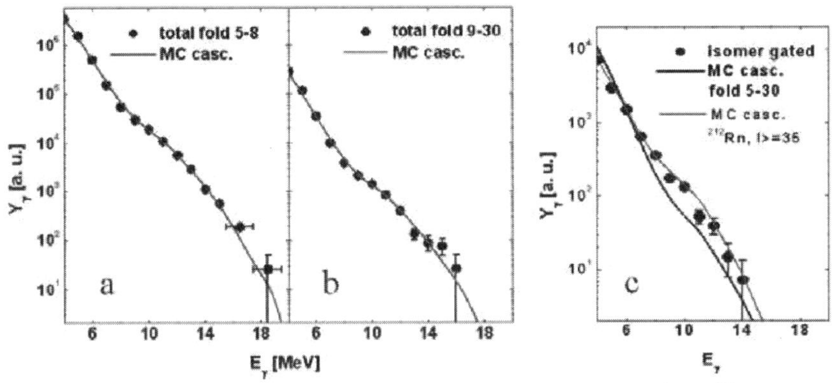

Fig. 4. The high energy spectra measured in the compound fusion reaction ^{18}O + ^{198}Pt at 96MeV. Panels (a) and (b) show the spectra measured for low energy coincidence folds 5–8 and 9–30 corresponding to an average spin of 23 and 29 , respectively. Panel (c) shows the high energy γ-ray spectrum measured in coincidence with the isomer stopped in the catcher. In all panels, the thin light continuous lines show the results of the statistical model calculations. In panel (c) the calculations have been done requiring ^{212}Rn as a residue and compound nucleus with a spin higher than 35 . The thick line shows the results of the calculations where such conditions are lifted. In all three cases the GDR centroid value is 13.2 MeV and the GDR width is 7.0,7.0 and 7.3 MeV respectively.

Another approach consists in measuring the GDR decay in the fusion-evaporation reaction by selecting very high spins around the critical spin-value at which fission process is beginning to dominate. Such measurements are also in general very difficult, because one has to select a very narrow spin window and at the same time to reject the fission contaminations. This can be achieved by tagging either on the fusion-evaporation products or on the γ-rays from specific residual nuclei. So far, investigations of this type have provided high-energy γ-ray spectra through gating on different residual nuclei (see e.g. the case of ^{194}Hg [16]. However, the existing measurements did not allow to make a spin selection in a region very close to the fission limit.

More recently measurements were made using the fusion-evaporation reaction ^{18}O+^{198}Pt at the bombarding energy of 96~MeV, populating with a rather high cross-section residual nuclei characterised by the presence of long-lived high-spin isomeric states. The decaying ^{216}Rn compound nucleus, strongly feeds the isomeric states of I=30 in ^{212}Rn and I=63/2 in ^{211}Rn. In addition, since only two feeding transitions to these isomeric states were found [17], one expects that the fission limit in ^{216}Rn is below or around 40 ħ. Therefore, the selection of the GDR γ-decay in coincidence with delayed γ-ray transitions is expected to probe mainly the compound nuclei which survive fission, yet, with angular momenta close to the fission limit.

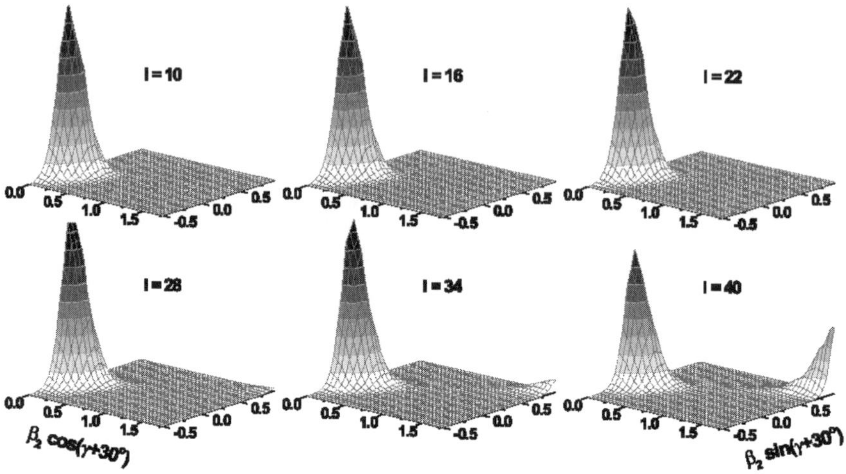

Fig.5 The probability distributions calculated using the Boltzmann factor. Note the stability of the position of the maximum of the displayed distributions with increasing spin values. However, the distributions spread out as function of spin, their maxima get lower and, at I ~40 hbar the trace of the fission path become visible.

The multiplicity of low energy γ-rays measured around the target in coincidence with the isomeric-states is of the order of only 5 (the average value measured in the reaction is of the order of 13–15) indicating that the isomers are populated in the reaction and stopped in the catcher which collects the rest of the decay. Fig. 4 shows the measured high energy γ-rays spectra. Panels (a) and (b) show the spectra in coincidence with the fold intervals 5–8 and 9–30, respectively. Panel (c) shows the -ray spectra in coincidence with the isomer. In all plots, the continuous grey line shows the results of a statistical model calculation performed using the Monte Carlo version of the statistical model code CASCADE. As it can be seen, the statistical model calculations reproduce well the measured spectra in all cases. The GDR centroids and widths are found to be very similar in the three cases (see figure caption of Fig.4). In panel (c) the thick dark line shows the calculation which reproduces the total un-gated spectrum while the thin one shows the calculation where the condition of populating the ^{212}Rn residues and of having initial spin larger than 35 ħ are required. It should be noted that the measured values of the GDR width are found to be : (i) larger than in the ground state nuclei (almost by factor of two); (ii) independent of the angular momentum of the compound system. The observed discrepancy between the T=0 value and those at finite temperature is

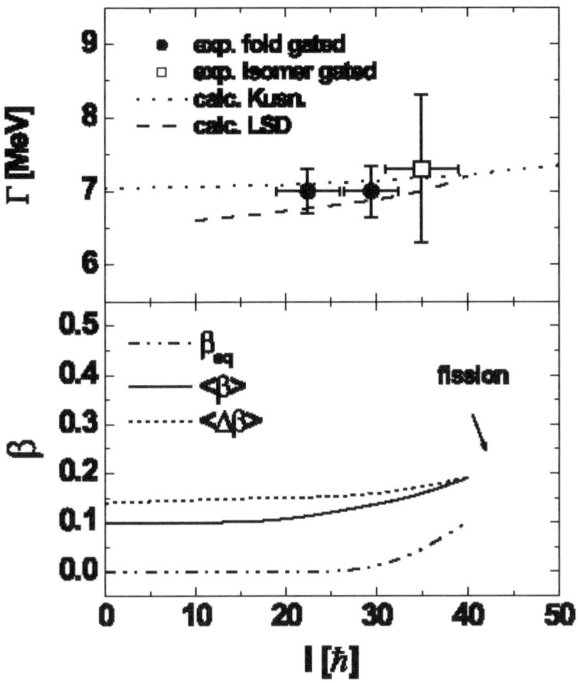

Fig. 6 ***Top Panel:*** The GDR width as a function of spin obtained for experimental GDR strength functions is shown with points in the top panel in comparison with theoretical predictions (the horizontal error bars correspond to the half-width of the spin distribution selected by the gating conditions). The dotted line corresponds to results of calculations performed according to the Kusnezov formula [13] The GDR width calculated using thermal shape fluctuations based on potential energies obtained with LSD model is drawn with the dashed line.
Bottom panel: The LSD model predictions for equilibrium deformation β_{eq} is presented in the bottom panel with dash-dotted line. The dashed line $\langle\Delta\beta\rangle$ (square root of the variance of the β distribution) indicates the extent of the possible deformations and $\langle\beta\rangle$ (solid line) is the average deformation.

of the same order of magnitude of that found in lighter nuclei which was interpreted in terms of thermal shape fluctuations of the hot nuclei.

In addition, the fact that the experimental results are independent of the angular momentum indicates that the average deformation of the nucleus is not changing significantly by increasing the spin. Such a scenario is theoretically confirmed as it can be seen in Fig.5 where the potential energy maps of ^{216}Rn calculated using the LSD model [13] are shown. In such calculations, the ^{216}Rn equilibrium and average shape remains almost spherical up to 40 ħ , behaviour which is in contrast to that found in the very light nuclei (see Fig. 1). For the highest spins 40> ħ, the fission channel dominates. Fig. 6 shows the measured GDR width compared to the prediction of the model of thermal shape fluctuations in which the free energy is calculated using the LSD model [4,5] (dashed line). The dotted line represents the prediction of the scaling law of Ref.

[11]. In both cases, the predictions agree significantly well with the experimental data indicating that the thermal shape fluctuation approach can be successfully extended up to mass A = 220 (where fusion-fission competition is remarkably strong) for all angular momenta.

4. Conclusion

The study of the GDR built on excited states at very high spins around the fission barrier has provided interesting results which are different in the light and heavy mass regions.

In the case of the ^{46}Ti nucleus a highly fragmented GDR strength function has been found with a broad structure at 15-25 MeV and a narrow low energy component at around 10.5MeV. This can be interpreted as the result of Jacobi shape transition and strong Coriolis effects. The Jacobi shape transition in the compound nucleus is also very interesting because it could constitute a kind of a gateway to very elongated, rapidly rotating cold nuclear shapes.

The picture emerged from the study of ^{216}Rn is very different. The present experimental results for the GDR width indicate that nucleus is nearly spherical with a small oblate deformation which does not change with increasing spin up to the spin value at which it fissions.

The two examples here discussed show that the progress in the field of nuclear structure at finite temperature requires new exclusive measurements of the GDR on excited nuclei with good selection of spin, temperature but also of beam and target combination. In connection with this last point it is important to stress that the availability radioactive beam facility of second generation such as SPES and SPIRAL2 will open new perspectives for investigating nuclear shapes at finite temperature for nuclei farther away from the stability line.

Acknowledgments

The work presented here could not be possible without the precious contributions and continuous efforts of our collaborators. They are : F. Camera, G. Benzoni, S. Leoni, B. Millio, O. Wieland, S. Brambilla, M. Pignanelli and N. Blasi from Milano, J. Styczen'n, M. Kmiecik, P. Bednarczyk, M. Brekiesz, J. Grcebosz, M. Lach, W. Mceczynski, M. Ziceblinski, K. Zuber, from Cracow, B. Herskind from NBI Copenhagen, E. Farnea, G. de Angelis and D. Napoli from LNL Legnaro; M. Kici\'nska-Habior from Warsaw; J. Nyberg from Uppsala; C.M. Petrache from Camerino. D. Curien, J. Dudek and N. Dubray from Strasbourg and K. Pomorski from Lublin.

A financial support from the Polish State Committee for Scientific Research (KBN Grant No. 2 P03B 118 22), the European Commission contract EUROVIV and the Italian INFN is acknowledged.

References

1. P. F. Bortignon, A. Bracco and R.A. Broglia, Giant Resonances: Nuclear Structure at Finite Temperature, New York: Gordon Breach (1998).
2. O. Wieland et al., J. Phys. G. in print and F.Gramegna et al., Acta Physica Polonica B36(2005)1155
3. A. Maj et al., Nucl. Phys. **A731**,319 (2004).
4. K. Pomorski and J. Dudek, Phys. Rev. C **67**, 044316 (2003).
5. J. Dudek and K. Pomorski, Eur. Phys. J. **A20**,15 (2004)
6. A. Maj et al., Nucl. Phys. **A571**, 185 (1994).
7. M. Lach et al., Eur. Phys. J. **A12**, 381(2001)
8. M. Kici'nska-Habior et al., Phys. Rev. C **41**, 2075 (1990).
9. K. Neergard, Phys. Lett. **B110,** 7 (1982)
10. M. Kmiecik et al., Acta Physica Polonica B36(2005)1169
11. D. Kusnezov et al. Phys. Rev. Lett. **81**,542(1998)
12. D. Kusnezov and W. E. Ormand, Phys. Rev. Lett. **90**, 042501 (2003).
13. M. Kmiecik et al., Phys. Rev. C 70 (2004)064317
14. R. Butsch et al., Phys. Rev. Lett. **41**, 1530 (1990).
15. T. Tveter et al., Phys. Rev. Lett. **76**, 1035 (1996).
16. F. Camera et al., Phys. Rev. C **60**, 014306 (1999).
17. G. D. Dracoulis et al., Phys. Lett. B **246**, 31 (1990).

EVOLUTION OF THE PROMPT DIPOLE γ-RAY EMISSION WITH INCIDENT ENERGY IN FUSION REACTIONS

D. PIERROUTSAKOU, B. MARTIN,* G. INGLIMA, A. BOIANO, A. DE ROSA, M. DI PIETRO, M. LA COMMARA, R. MORDENTE, M. ROMOLI, M. SANDOLI, M. TROTTA and E. VARDACI

Dipartimento di Scienze Fisiche, Università di Napoli "Federico II" and INFN, Sezione di Napoli, Napoli, Italy

T. GLODARIU,† M. MAZZOCCO and C. SIGNORINI

Dipartimento di Fisica, Università di Padova and INFN, Sezione di Padova, Padova, Italy

L. STROE‡

INFN Laboratori Nazionali di Legnaro, Legnaro, Italy

C. AGODI, R. ALBA, M. COLONNA, R. CONIGLIONE, A. DEL ZOPPO, M. DI TORO, C. MAIOLINO, N. PELLEGRITI, P. PIATTELLI, D. SANTONOCITO and P. SAPIENZA

Dipartimento di Fisica e Astronomia, Università di Catania and INFN, Laboratori Nazionali del Sud, Catania, Italy

G. CARDELLA, E. DE FILIPPO, A. PAGANO and S. PIRRONE

INFN, Sezione di Catania, Italy

V. BARAN

NIPNE-HH and Bucarest University, Romania

*Electronic address: martin@na.infn.it
†On leave from: NIPNE-HH, Romania
‡Present address: NIPNE-HH, Romania

We studied the prompt dipole γ-ray emission, associated with entrance channel charge asymmetry effects, as a function of incident energy in the $^{32,36}S + ^{100,96}Mo$ (E_{lab}=6 MeV/nucleon) and $^{36,40}Ar + ^{96,92}Zr$ fusion reactions (E_{lab}= 16 and 15 MeV/nucleon, respectively). With the above reaction pairs the ^{132}Ce compound nucleus was formed, from entrance channels having different charge asymmetries, at an excitation energy of 117 and 304 MeV with identical spin distribution. By studying the differential γ-ray multiplicity spectra related to the above fusion reactions, it was shown that at the higher compound nucleus excitation energy the Giant Dipole Resonance γ-ray intensity increases by ~14% for the more charge asymmetric system while at the lower one no difference between the data was seen within the experimental uncertainties. Calculations based on a collective bremsstrahlung analysis of the reaction dynamics are presented and compared with the experimental findings.

1. INTRODUCTION

By studying the differential γ-ray multiplicity spectra related to heavy-ion reactions where the same composite system is formed from entrance channels having different charge asymmetry, it was evidenced an excess of Giant Dipole Resonance (GDR) γ-rays emitted from the more charge asymmetric system for deep inelastic[1,2] and fusionlike events[2,3,4,5]. This excess was ascribed to the prompt dipole γ-ray emission which occurs during the charge equilibration process in addition to the statistical one originating in the thermal excitation of the hot compound nuclei dipole vibration.

In recent theoretical works[6,7] it was suggested that this kind of pre-equilibrium emission depends on the incident energy, taking a maximum value in an appropriate energy region situated between the low incident energies near the Coulomb barrier and the higher ones near the Fermi energy domain, where it diminishes. A systematic study of its dependence on beam energy is of great interest due to the fact that it can be of aid in the formation of superheavy elements. The emission of pre-equilibrium dipole photons could represent an efficient new cooling mechanism of the composite system in charge asymmetric "hot" fusion reactions, increasing thus its survival probability against fission.

To maximize this emission one has to choose the suitable reactions having large entrance channel charge asymmetries and to find the appropriate energy range where it becomes larger.

The first attempt towards this direction was undertaken in the present work by investigating the evolution of the prompt dipole γ-ray emission with beam energy in the fusion reaction pairs: $^{32,36}S + ^{100,96}Mo$ at E_{lab} = 196 and 214.2 MeV, respectively[8], and $^{36,40}Ar + ^{96,92}Zr$ at E_{lab} = 576.9 and

605.8 MeV, respectively. In each reaction pair the ^{132}Ce compound nucleus was formed with identical spin distribution at the same excitation energy, through entrance channels having different charge asymmetry and thus, different dipole moment. The dipole moment changes by 16.6 fm from the ^{40}Ar + ^{92}Zr system to the more N/Z asymmetric one, ^{36}Ar + ^{96}Zr, while it changes by 16.5 fm from the ^{36}S + ^{96}Mo to the ^{32}S + ^{100}Mo system. With the chosen incident energies the ^{32,36}S + ^{100,96}Mo reaction pair populate the ^{132}Ce compound nucleus at an excitation energy of 117 MeV while the ^{36,40}Ar + ^{96,92}Zr one at 304 MeV (by taking into account the energy lost due to pre-equilibrium particle emission). The fusion critical angular momentum for all of the above reactions was $L_{max} = 83\hbar$ according to PACE2[9] calculations. There was only a small difference in the entrance channel mass asymmetry between the two systems for each reaction pair. However, as all of the studied systems were located above the critical curve in the fissility-mass asymmetry plane[10] the effects associated with the mass asymmetry were negligible[11]. Therefore, since all the parameters were kept identical in each reaction pair, any difference in the γ-ray emission between the two reactions can be safely ascribed to the difference in the entrance channel charge asymmetry.

The results of the present work can be directly compared with those related to the ^{32,36}S + ^{100,96}Mo fusion reactions performed at beam energy of ~9 MeV/nucleon where the ^{132}Ce compound nucleus was formed at an excitation energy of 173.5 MeV[5]. In that case an increase of the GDR γ-ray intensity of ~25% was found for the more charge asymmetric system in the bremsstrahlung-subtracted linearized spectra.

This paper is organized as follows: in section **2** the experimental techniques are discussed, in section **3** the differential γ-ray multiplicities in coincidence with the evaporation residues are shown. In section **4** a theoretical analysis, based on a collective bremsstrahlung approach, is compared with the experimental findings. Section **5** is devoted to the conclusions.

2. EXPERIMENT

2.1. $^{32,36}S + {}^{100,96}Mo$

The reactions ^{32,36}S + ^{100,96}Mo were performed by using the ^{32}S and ^{36}S pulsed beams provided by the heavy-ion 15 MV Tandem-XTU accelerator of the Laboratori Nazionali di Legnaro (Italy), impinging on 550 μg/cm^2 thick ^{100}Mo and ^{96}Mo self-supporting targets, isotopically enriched to 97.42% and 95.9%, respectively. The beam consisted of ~2 ns wide bunches with a

400 ns separation and it was measured in a Faraday cup shielded with lead and paraffin to reduce the background due to γ-rays and neutrons. Beam current was about 10 nA.

The γ-rays were detected by using six seven-pack clusters of BaF_2 scintillators situated at 28 cm from the target and at the following θ angles with respect to the beam direction: 70°, 90°, 110°, and 135°. The total solid angle covered by the BaF_2 detectors was 1.6 sr. The BaF_2 clusters were surrounded by a 3 mm thick lead shield which reduced the counting rate due to the low energy γ-rays ($E_\gamma \leq 1$ MeV) to 50% and stopped the charged particles. The fusion-evaporation residues were detected by four position sensitive Parallel Plate Avalanche Counters (PPAC's) located symmetrically around the beam direction at 70 cm from the target. The PPAC's were centered at $\theta = 7°$ with respect to the beam direction, subtending 7° in θ. The total solid angle covered by the PPAC's was 0.089 sr. They provided the time of flight (TOF) with respect to the radiofrequency signal of the accelerator, the energy loss (ΔE) and the position of the reaction products.

The energy calibration of the γ-ray detectors was obtained by using the sources ^{60}Co, ^{88}Y, the composite source of ^{241}Am+^{9}Be and the 15.1 MeV γ-rays from the p+^{12}C reaction.

Down-scaled single events together with coincidence events between a PPAC and at least one fired BaF_2 scintillator were collected during the experiment. A coincidence event was accepted if the deposited energy in a BaF_2 cluster was greater than $\sim$ 5 MeV. The threshold of each BaF_2 scintillator was set at $\sim$ 100 keV. The coincidence request eliminated any cosmic ray contamination of the γ-ray spectra.

2.2. $^{36,40}Ar + {}^{96,92}Zr$

The reactions ^{36,40}Ar + ^{96,92}Zr were performed by using the ^{36}Ar and ^{40}Ar pulsed beams provided by the Superconducting Cyclotron of the Laboratori Nazionali del Sud (Italy), impinging on a 450 μg/cm^2 thick ^{96}ZrO$_2$ target (enriched to 95.63% in ^{96}Zr) and on a 600 μg/cm^2 thick ^{92}ZrO$_2$ target (enriched to 95.36% in ^{92}Zr). The targets were evaporated on carbon layers 90 and 60 μg/cm^2 thick, respectively. The beam consisted of $\sim$1 ns wide bunches with a 150 ns separation. Beam current was about 2 nA.

The γ-rays and the light charged particles were detected by using the 180 BaF_2 modules of the MEDEA experimental apparatus[12] that covers the polar angular range between θ=30° and θ=170° and the full range in the azimuthal angle ϕ. The total solid angle covered by the MEDEA apparatus was 3.7π. The fusionlike residues were detected by the same PPAC's described previously for the ^{32,36}S + ^{100,96}Mo reaction pair.

The discrimination between γ-rays, light charged particles and high-energy neutrons (E>20 MeV) was performed by means of a shape analysis of the BaF_2 signal; the discrimination between γ-rays and low energy neutrons was achieved with a TOF measurement with respect to the radiofrequency signal of the Cyclotron. The energy calibration of the γ-rays was done in the same way as for the ^{32,36}S + ^{100,96}Mo reaction pair while the calibration of the charged particles was performed as described elsewhere[13]. Down-scaled single events together with coincidence events between a PPAC and at least one fired BaF_2 scintillator were collected during the experiment. The threshold of the BaF_2 modules was set at ~6 MeV for the detection of γ-rays.

3. DATA ANALYSIS AND RESULTS

3.1. $^{32,36}S + ^{100,96}Mo$

The fusion events were selected off-line in the bidimensional plot of the ΔE versus the TOF of the reaction products detected in each PPAC. At the present beam energies, the incomplete fusion cross section should be small, representing no more than 15% of the total fusion cross section[14]. This kind of events cannot be discarded in the TOF spectrum of the reaction products because they have overlapping velocity distributions with those of the complete fusion reactions[15]. It has been seen, that the pre-equilibrium particle emission is insensitive to the details of target and projectile nuclei and that a scaling of the pre-equilibrium particle multiplicity with the bombarding energy above the Coulomb barrier can be assumed[15]. In this hypothesis, the average excitation energy loss due to pre-equilibrium particle emission is found to be approximately 0 for both reactions, by using the empirical relation given by Kelly et al.[16]

The differential γ-ray multiplicity obtained with all the BaF_2 clusters in coincidence with evaporation residues, in the center of mass reference frame, is plotted in Figure 1. ϵ_{det} appearing in the Oy axis is the energy dependent efficiency of the experimental apparatus. The squares and the circles correspond to the ^{32}S +^{100}Mo and the ^{36}S +^{96}Mo reactions, respectively. The spectra were integrated in 4π by assuming an isotropic emission in the center of mass reference frame.

Different mechanisms may feed the γ-ray spectra at E_γ greater than 20

MeV, the nucleon-nucleon bremsstrahlung mechanism being the dominant one[17]. At the present incident energies this contribution to the spectra can be neglected.

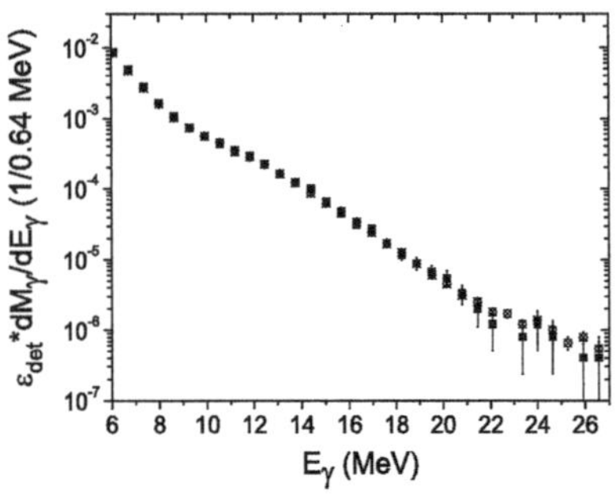

Figure 1. Center of mass experimental differential γ-ray multiplicity of the ^{32}S + ^{100}Mo (squares) and of the ^{36}S + ^{96}Mo (circles) reaction for fusion events with all of the BaF_2 scintillators.

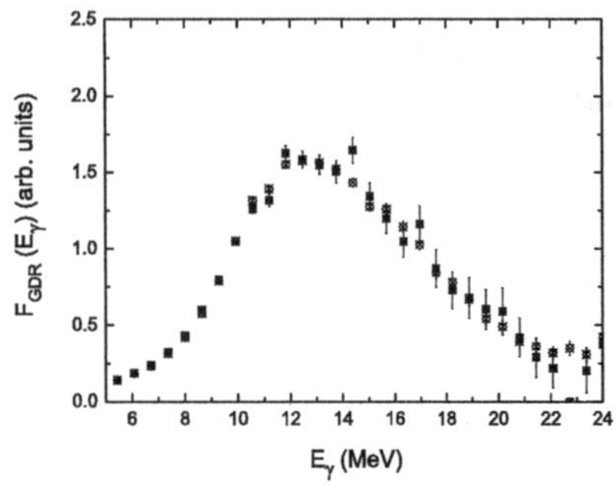

Figure 2. Linearized experimental γ -ray spectrum of the ^{32}S + ^{100}Mo (squares) and of the ^{36}S + ^{96}Mo (circles) reaction for fusion events with all of the BaF_2 scintillators.

From Figure 1 one can see that the γ-ray multiplicity is identical for both reactions within the experimental uncertainties in the whole energy range. In order to emphasize the details in the GDR energy region, a linearization procedure of the experimental γ-ray spectra was performed. In this way, the exponential behaviour due to the nuclear level density is eliminated from the data. The γ-ray spectra associated with the two reactions (Figure 1) were linearized by dividing them by the same theoretical γ-ray spectrum, calculated with the CASCADE code[18] where a constant dipole strength and a level density parameter a=$A/8$ MeV^{-1} were considered. Moreover, the constant strength theoretical γ-ray spectrum was folded by the experimental set up response function by using the GEANT code[19]. The linearized data of the ^{32}S + ^{100}Mo (^{36}S + ^{96}Mo) reaction are displayed by the squares (circles) in Figure 2. From Figure 2 it can also be seen that there is no difference between the linearized data of the two reactions within the error bars.

3.2. $^{36,40}Ar + ^{96,92}Zr$

As the analysis of the data is not yet completed, here preliminary results are reported. The fusionlike events were selected off-line in the bidimensional plot of the ΔE versus the TOF of the reaction products detected in each PPAC. At the present incident energies, the incomplete fusion cross section represents approximately the 90% of the total fusion cross section[14]. The average excitation energy lost due to pre-equilibrium particle emission for the considered fusionlike events was evaluated by means of the empirical relation given by Kelly et al.[16] and was taken into account when calculating the appropriate beam energy for each reaction to form the compound nucleus at the same excitation energy. It was found to be approximately 81 and 76 MeV for the ^{36}Ar + ^{96}Zr and ^{40}Ar + ^{92}Zr reaction, respectively. Subsequently, the quality of such an evaluation was checked by looking at the proton multiplicity spectra obtained with the BaF_2 scintillators in coincidence with fusionlike residues for the two systems. These spectra were found to be identical with each other within the experimental uncertainties at backward angles where the emission comes mainly from the hot compound nucleus. Thus, we can affirm that the ^{132}Ce compound nucleus was indeed formed at the same average excitation energy in both reactions.

In Figure 3 the differential γ-ray multiplicity spectrum for the ^{40}Ar + ^{92}Zr (solid line) and the ^{36}Ar + ^{96}Zr (dotted line) for fusionlike reactions at $\theta_{lab} = 90°$ is plotted. ϵ_{det} appearing in the Oy axis is the energy dependent efficiency of the experimental apparatus. The spectra were integrated in 4π by assuming isotropic emission in the center of mass reference frame. The dashed line is the difference between the bremsstrahlung component of the ^{36}Ar + ^{96}Zr and the ^{40}Ar + ^{92}Zr system evaluated by fitting the plotted data for energies larger than 32 MeV with an exponential function[17]. From this figure it's clear that the data present a definite difference (squares) in the compound nucleus GDR energy region coming from the different initial dipole moment in the two entrance channels.

The γ-ray spectra associated with the two reactions (Figure 3) were linearized by dividing them by the same theoretical γ-ray spectrum, calculated with the CASCADE code[18] where a constant dipole strength and a level density parameter $a=A/10$ MeV^{-1} were considered. Moreover, the constant strength theoretical γ-ray spectrum was folded by the response function of the MEDEA experimental apparatus[20]. The linearized γ-ray spectrum of the ^{40}Ar + ^{92}Zr (^{36}Ar + ^{96}Zr) reaction is displayed by the triangles (circles) in Figure 4. If the linearized spectra of Figure 4 are integrated between 8 and 21 MeV, an increase of the GDR γ-ray intensity of $\sim$ 14% is found by going towards the more charge asymmetric system.

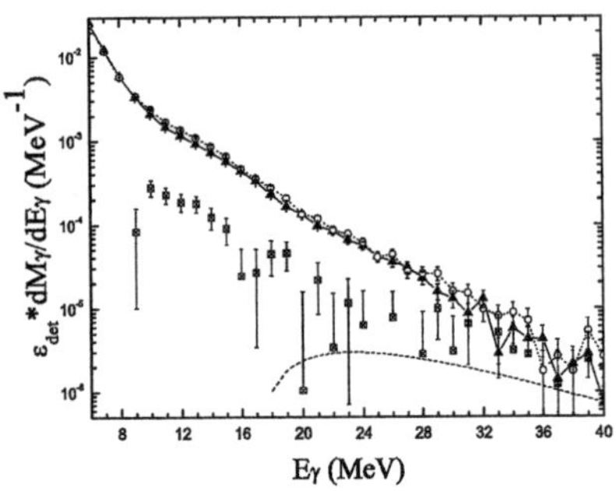

Figure 3. Experimental differential γ-ray multiplicity of the ^{40}Ar + ^{92}Zr (solid line) and of the ^{36}Ar + ^{96}Zr (dotted line) reaction for fusionlike events at θ=90° and the corresponding difference (squares). The dashed line is the difference between the corresponding Bremsstrahlung components.

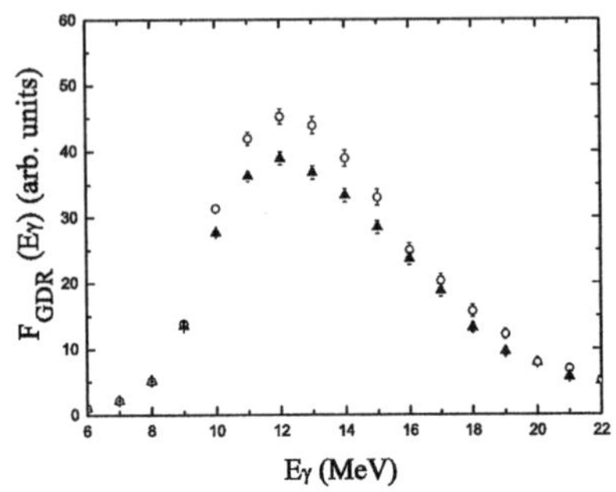

Figure 4. Linearized experimental γ -ray spectrum of the ^{40}Ar + ^{92}Zr (triangles) and of the ^{36}Ar + ^{96}Zr (circles) reaction for fusionlike events at θ= 90°.

4. DISCUSSION

The present results can be compared directly with those related to the ^{32,36}S + ^{100,96}Mo reactions at $E_{lab}\sim$ 9 MeV/nucleon where the ^{132}Ce compound nucleus was formed at an excitation energy of 173.5 MeV[5]. As mentioned previously, in that case an increase of the GDR γ-ray intensity of $\sim$25% was found for the more charge asymmetric system in the bremsstrahlung-subtracted linearized spectra. From all these experimental findings one can see that the prompt dipole γ-ray emission presents a maximum at incident energy of $\sim$ 9 MeV/nucleon decreasing towards lower and higher incident energies.

In order to understand the origin of the observed experimental differences for the $^{32,36}S$+$^{100,96}Mo$ reaction pair we compared the dynamical dipole mode in the entrance channel for the considered systems at incident energy of 6 and 9 MeV/nucleon in the framework of a BNV transport model[6]. Moreover, by using a bremsstrahlung approach [7] we calculated the photon emission probability associated with this dipole mode. Details of these calculations are presented elsewhere[8]. In our simulations the pre-equilibrium dipole photon emission probability for the almost charge symmetric system $^{36}S +^{96} Mo$ is so small that it can be neglected. Therefore, here we will focus our study on the results obtained for the N/Z asymmetric reaction $^{32}S +^{100} Mo$.

In Figure 5 one can see the corresponding bremsstrahlung spectra related to the dynamical dipole mode for the $^{32}S +^{100} Mo$ system obtained at the two energies for impact parameters b=0, 2 and 4 fm (solid lines). In all cases, as a reference, the first step statistical spectrum (dashed line) is also included. The latter is just the product of the statistical γ decay rate, times the mean life time of the equilibrated source. From this figure it is evident that the prompt dipole γ-ray emission presents an increasing trend with incident energy from 6 to 9 MeV/nucleon.

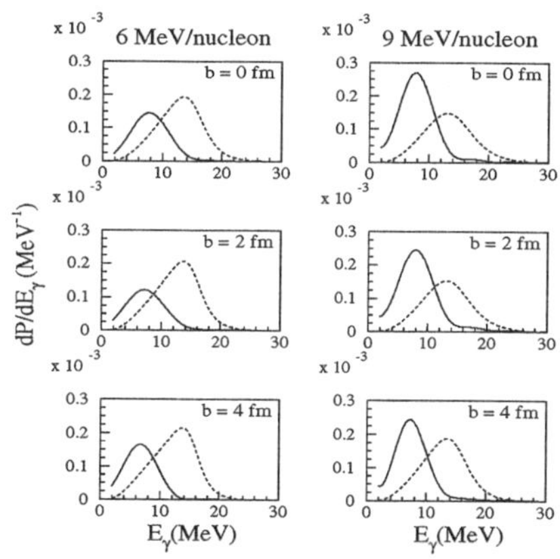

Figure 5. The bremsstrahlung spectra for the ^{32}S + ^{100}Mo system at incident energy of 6 and 9 MeV/nucleon (solid line) and the first step statistical spectrum (dashed line) for three impact parameters.

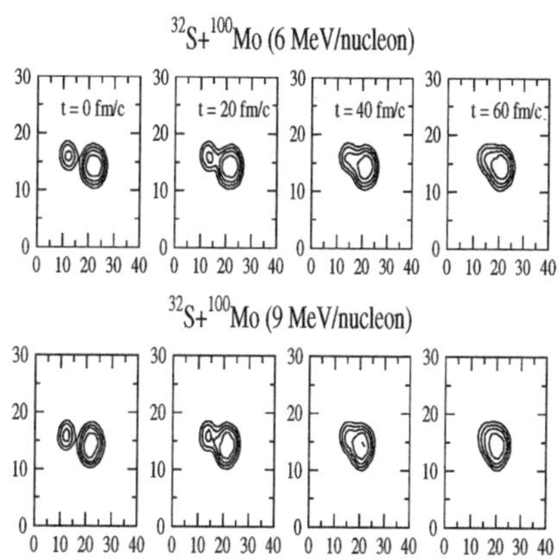

Figure 6. Density plots of the neck dynamics for the ^{32}S + ^{100}Mo system at incident energy of 6 and 9 MeV/nucleon.

A possible explanation of this effect is related to a slower transition from a dinuclear configuration to a mononucleus which hinders the isovector collective response for the lower incident energy. From the density plots of the neck dynamics, reported in Figure 6, a slower dynamics of the neck is evidenced at 6 MeV/nucleon. On the contrary, at 9 MeV/nucleon a faster disappearence of the neck leads to physical conditions which favour an early and complete collective isovector response. The time evolution of the dipole acceleration at the two energies supports this scenario. At 9 MeV/nucleon this quantity presents a larger amplitude in the early stage of the reaction, which will also lead to more persistent dipole oscillations. Both effects are contributing to the enhancement of the total bremsstrahlung yield for the higher incident energy.

The theoretical ratio C_{th} of total pre-equilibrium to total statistical GDR γ-ray multiplicity, this latter being calculated by means of the CASCADE code and integrated between 8 and 21 MeV, can be compared to the corresponding experimental quantity, $C_{exp} = \frac{(c-d)}{d}$ with c and d equal to the integral: $\int_8^{21} \left(\epsilon_{\rm det} * \frac{dM_\gamma}{dE_\gamma}\right) dE_\gamma$, for the ^{32}S $+^{100}$Mo and ^{36}S $+^{96}$Mo reaction, respectively. The theoretical value C_{th} is equal to 0.04 and 0.08 for incident energy of 6 and 9 MeV/nucleon, respectively, in reasonable agreement with the experimental one, C_{exp}, which varies from (0.015 ± 0.020) at 6 MeV/nucleon to 0.13 at 9 MeV/nucleon (from previous data[5]).

At high beam energies, above (15-20) MeV/nucleon, the extra dipole radiation is expected to be hindered by a faster damping of the dinuclear collective dipole mode and by a larger pre-equilibrium neutron emission from the most neutron-rich partner that will reduce the initial dipole amplitude of the dinuclear system. However, calculations for the ^{36,40}Ar + ^{96,92}Zr reaction pair at about 16 MeV/nucleon are not yet available.

5. CONCLUSIONS

In the present work we investigated the evolution of the prompt dipole γ-ray emission with beam energy by studying two reaction pairs ^{32,36}S + ^{100,96}Mo (^{36,40}Ar + ^{96,92}Zr) at E_{lab} = 196 MeV and 214.2 MeV (E_{lab}=576.9 MeV and 605.8 MeV), respectively, in order to form the ^{132}Ce compound nucleus at an excitation energy of 117 (304) MeV with identical spin distribution. The only difference between the two systems of each reaction pair concerns the initial dipole moment.

The linearized γ-ray spectra, associated with the ^{32,36}S + ^{100,96}Mo reactions, are identical with each other within the experimental uncertainties, while a GDR γ-ray intensity increase of $\sim$14% was evidenced for the more charge asymmetric system of the ^{36,40}Ar + ^{96,92}Zr reaction pair. The present result can be directly compared with that reported for the ^{32,36}S + ^{100,96}Mo fusion reactions performed at $\sim$ 9 MeV/nucleon. It can be seen that the prompt dipole γ-ray emission presents a maximum at incident energy of ~9 MeV/nucleon, decreasing towards lower and higher beam energies.

Calculations performed within the BNV transport model framework and based on a collective bremsstrahlung approach at beam energy of 6 and 9 MeV/nucleon predict a lower dynamical dipole yield at the lower incident energy related to a slower neck dynamics which obstructs a full collective response.

In future, the investigation of the prompt dipole γ-ray emission should be pursued in association with the use of radioactive beams which allow the maximization of this kind of emission. Such studies could be of aid to form superheavy elements by taking advantage of the prompt dipole γ-ray emission taking place in "hot" fusion reactions between charge asymmetric colliding ions.

References

1. D. Pierroutsakou *et al.*, *EPJ* **A16**, 423 (2003) and *NP* **A687**, 245c (2001)
2. F. Amorini *et al.*, *Phys. Rev.* **C69**, 014608(2004)
3. S. Flibotte *et al.*, *Phys. Rev. Lett.* **77**, 1448 (1996)
4. M. Cinausero *et al.*, *Il Nuovo Cimento* **111**, 613 (1998)
5. D. Pierroutsakou *et al.*, *Eur. Phys. J.* **A17**, 71 (2003)
6. V. Baran *et al.*, *Nucl. Phys.* **A679**, 373 (2001)
7. V. Baran *et al.*, *Phys.Rev.Lett.* **87**, 182501 (2001)
8. D. Pierroutsakou *et al.*, *Phys. Rev.* **C71**, 054605 (2005)
9. A. Gavron, *Phys. Rev.* **C21**, 230 (1980)
10. W.J. Swiatecki, *Physica Scripta* **24**, 113 (1981)
11. M. Thoennessen *et al.*, *Phys. Rev.* **C51**, 3148 (1995)
12. E. Migneco *et al.*, *Nucl. Instr. Meth.* **A314**, 31(1992)
13. A. Del Zoppo *et al.*, *Nucl. Instr. Meth. Phys. Res.* **A327**, 363(1993)
14. H. Morgenstern *et al.*, *Phys. Rev. Lett.* **52**, 1104 (1984)
15. M.P. Kelly, *et al.*, *Phys. Rev.* **C56**, 3201 (1997)
16. M.P. Kelly *et al.*, *Phys. Rev. Lett.* **82**, 3404 (1999)
17. H. Nifenecker and J.A. Pinston, Ann. rev. nucl. part. sci. **44**, 113 (1990)
18. F. Puhlhofer, *Nucl. Phys.* **A280**, 267 (1977) and M.N. Harakeh extended version (private communication)
19. R. Brun *et al.*, *CERN Report No.CERN-DD/EE/84-1* (unpublished, 1986)
20. G. Bellia *et al.*, *Nucl. Instr. Meth. Phys. Res.* **A329**, 173(1993)

SELECTION RULES ON K-QUANTUM NUMBER IN WARM ROTATING NUCLEI

G.BENZONI

INFN sezione di Milano and Università degli Studi di Milano,
Via Celoria 16,
20133- Milano, Italy
E-mail: Giovanna.Benzoni@mi.infn.it

To study the order-to-chaos transition in nuclei we investigate the validity of the K-quantum number in the excited rapidly rotating ^{163}Er nucleus, analysing the variance and covariance of the fluctuations of counts of spectra γ-cascades feeding into low-K and high-K bands. The data are compared to simulated spectra obtained using microscopic cranked shell model calculations. K-selection rules are found to be obeyed in the decay along excited unresolved rotational bands of internal excitation energy up to around 1.2 MeV. At higher internal energy, from about 1.2 to 2.5 MeV, the selection rules are found to be only partially valid.

1. Introduction

In the present contribution I report on a work addressing the problem of the persistence of selection rules on the K quantum number in deformed nuclei at excitation energies where the transition between the ordered regime (characterising the discrete rotational bands) and the chaotic regime (compound system) is expected to take place. The tool is the γ-decay associated to the intermediate region corresponding to transitions governed by a distribution of the rotational quadrupole strength which acquires a width that is usually referred to as *rotational damping width.*

The fact that, at low excitation energies, γ-transitions corresponding to large changes in K are hindered is well established and deduced by the presence of high-K isomeric states in several nuclei[1]. However, one expects at higher excitation energiesand spins the Coriolis force and the residual interaction to cause a mixing of these states, with a consequent weakening of the selection rules.

The issue of the validity of the K-quantum number has been previously addressed by studying the γ-decay from neutron resonances at excitation energy U $\approx$ 8 MeV[2,3]. Our analysis of the rotational quasi-continuum instead allowes to focus on the region where the order-to-chaos transition is expected to take place, namely above 1 MeV of internal energy, where the contribution from the residual interaction begins to be relevant also due to the high level density.

The present work is based on new data on the nucleus ^{163}Er, from a EUROBALL experiment this nucleus was already studied in a previous experiment[4]. Two novelties are presented. First of all, a detailed experimental investigation is carried out in the region of high internal energy up to U $\approx$ 2.5 MeV. The experimental analysis is based on a statistical analysis of quasi-continuum $\gamma - \gamma$ spectra. Secondly, and most important, a direct and realistic comparison between experiment and theory is made for the first time.

2. Experimental details and data analysis

The experiment was performed using the EUROBALL[5] array at the IReS Laboratory (France), employing the reaction ^{18}O + ^{150}Nd, at E_{beam}= 87 and 93 MeV. The ^{150}Nd target was made of a stack of two thin foils for a total thickness of 740 $\mu g/cm^2$. The maximum angular momentum reached in the reaction at the two different energies was calculated to be 40 and 45 $\hbar$ respectively. This reaction produces ^{162,163}Er nuclei as main evaporation residua. A total of $\approx 3\times10^9$ of triple and higher Ge-folds events were finally obtained.

The data have been sorted into a number of two-dimensional (2D) matrices in coincidence with specific γ-transitions of ^{163}Er[6]. In particular, three different classes of 2D-matrices were created: the first one, later referred to as *Total*, collects the events populating the nucleus ^{163}Er and is created by gating on low-lying transitions belonging to the specific nucleus; the second class, *Low-K*, is created by gating on transitions belonging to the low-K (K=5/2) signature and parity configurations labelled A=(1/2,+), B=(-1/2,+), E=(1/2,-) and F=(-1/2,-). Finally a third class of matrices, *High-K*, is created gating on transitions feeding the high-K (K=19/2) bands labelled K1 (negative parity), K2 and K4 (positive parity). The corresponding background matrices were also sorted setting gates close to the transitions used to select the different configurations. In fig.1 we show examples of gated spectra of the mentioned configurations: panel a) shows band E, panel b) band F, panel c) band K1 and panel d) K2+K4 bands. All these spectra are obtained summing single gates on transitions belonging to the band and then subtracting the gate-related background. In order to perform an analysis in terms of statistical fluctuations[7], all known peak-peak and peak-background coincidences have been subtracted from each 2D spectrum using the Radware software[8]. Finally, the individual gated matrices have been summed into one *low-K* (A+B+E+F) and one *high-K* (K1+K2+K4) matrix.

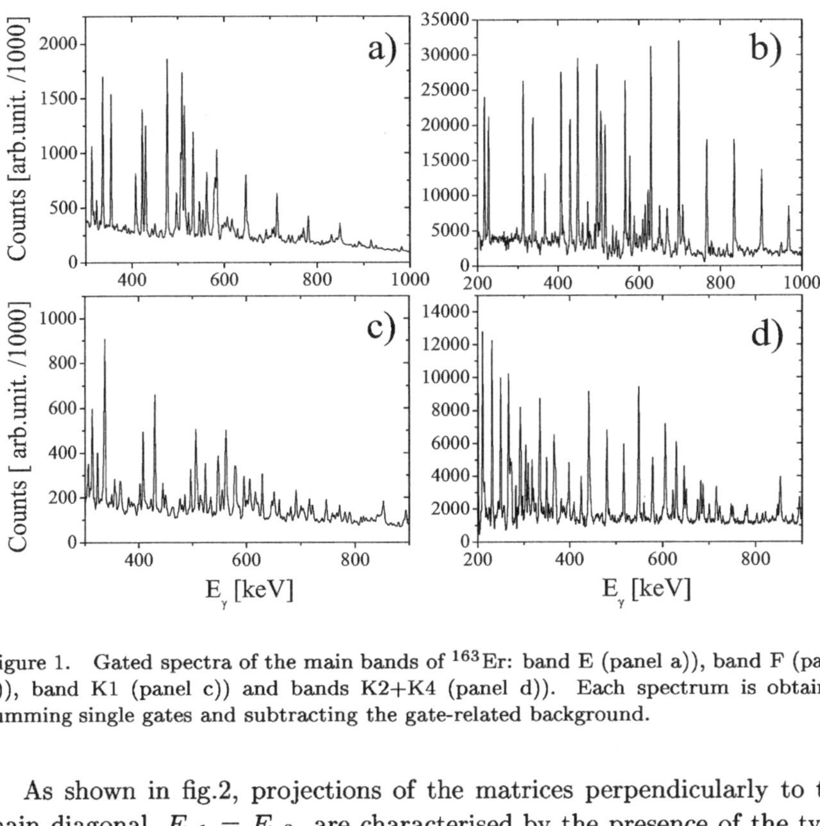

Figure 1. Gated spectra of the main bands of ^{163}Er: band E (panel a)), band F (panel b)), band K1 (panel c)) and bands K2+K4 (panel d)). Each spectrum is obtained summing single gates and subtracting the gate-related background.

As shown in fig.2, projections of the matrices perpendicularly to the main diagonal, $E_{\gamma1} = E_{\gamma2}$, are characterised by the presence of the typical ridge-valley structure, where ridges are fed by discrete but unresolved bands, while the valley is formed by transitions in the rotational damping regime.

Two experimental observables are used to test the validity of the K-quantum number at increasing internal energy, namely the number $N^{(2)}_{path}$ of decay paths measured in coincidence with low-K (high-K) discrete bands and the correlation coefficient r[7,9]. While $N^{(2)}_{path}$ is related both to the level

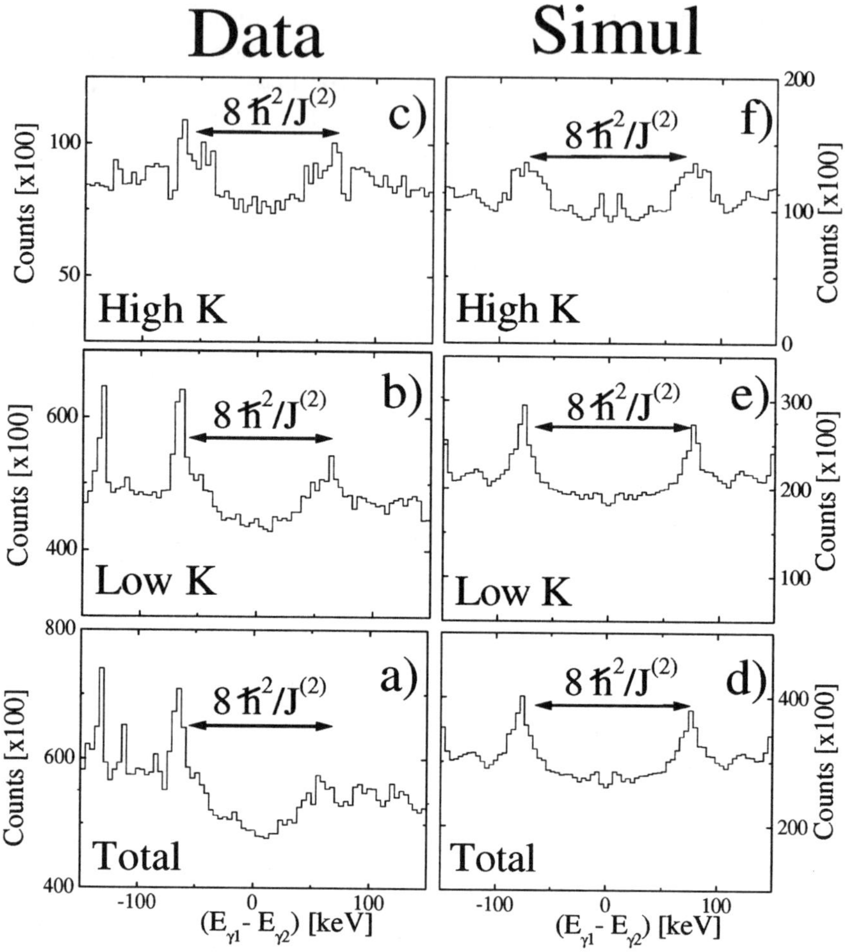

Figure 2. 60 keV wide projections perpendicular to the $E_{\gamma_1} = E_{\gamma_2}$ diagonal of experimental and simulated 2D spectra of ^{163}Er (left and right panels, respectively), at the average transition energy $< E_\gamma >$=900 keV. The spectra collect either the total γ-decay flow (panel a) and d)) or the γ-decay in coincidence with low-K (panel b) and e)) or high-K (panel c) and f)) configurations. The ridge-valley structure typical of rotational nuclei is seen in all spectra, with a separation between the two most inner ridges equal to $8\hbar^2/J^{(2)}$, as indicated by the arrows. The reduced intensity observed in the $E_{\gamma_1} \geq E_{\gamma_2}$ region of the spectra is due to the subtraction of all discrete lines known from the level scheme, in the case of the experimental data, and of the yrast and first excited bands, in the case of the simulation.

density and to the rotational damping width, r measures the similarity of the cascades recorded in coincidence with low-K or high-K bands: a large value of r ($\approx$1) indicates that many decay paths can lead both to low-K and high-K configurations, thus suggesting a weakening of the selection rules.

The experimental results, discussed in the next section, are compared to the simulation calculations,also described in the following section.

3. The simulation

To obtain predictions on the validity of the K-quantum number in excited states, we have performed simulations based on band mixing calculations including both the residual interaction and a term that takes into account the angular momentum carried by the K-quantum number[10]. The calculations are made diagonalizing the Hamiltonian in a basis of np-nh excitations in a cranked Nilsson potential. The lowest eigenstates are retained, covering an energy interval above yrast of $\approx$ 2.5 MeV. Inspecting the states resulting from the band mixing calculations, one finds that every state is characterized by a Gaussian distribution of K with a FWHM which is found to increase with U. Below U $\approx$ 1.5 MeV the FWHM of the Gaussian distribution is found to be smaller than the typical spreading in the average value of K, so that it is possible to define two different sets of states (i.e. low-K with K $\leq$ 8 and high-K with K $>$ 8), which turn out to correspond to level densities differing by a factor of $\approx$ 3. For U $\geq$ 2-2.5 MeV the average value of K converges towards $\langle K \rangle \approx 7$ and the intrinsic FWHM is larger than the spreading in the average values. The distinction between low-K and high-K configurations is therefore blurred, corresponding to the statistical limit of strong K-mixing. For energies around U $\approx$ 1.5 MeV one finds an intermediate regime associated to the onset of K-mixing.

Starting from the microscopically calculated bands, we have generated simulated $\gamma - \gamma$ spectra by means of a MonteCarlo code, MONTESTELLA[11], describing the competition between collective E2 and statistical E1 transitions.

While the E2 transitions are calculated microscopically, the statistical E1 transitions are obtained using the calculated level density and a GDR strength function corresponding to a prolate nucleus with quadrupole deformation $\beta = 0.25$ and rotating collectively.

Each γ-cascade is started from initial values of internal energy U and spin I randomly chosen from a two-dimensional entry distribution of Gaussian shape, with centroids and widths reproducing the experimental condi-

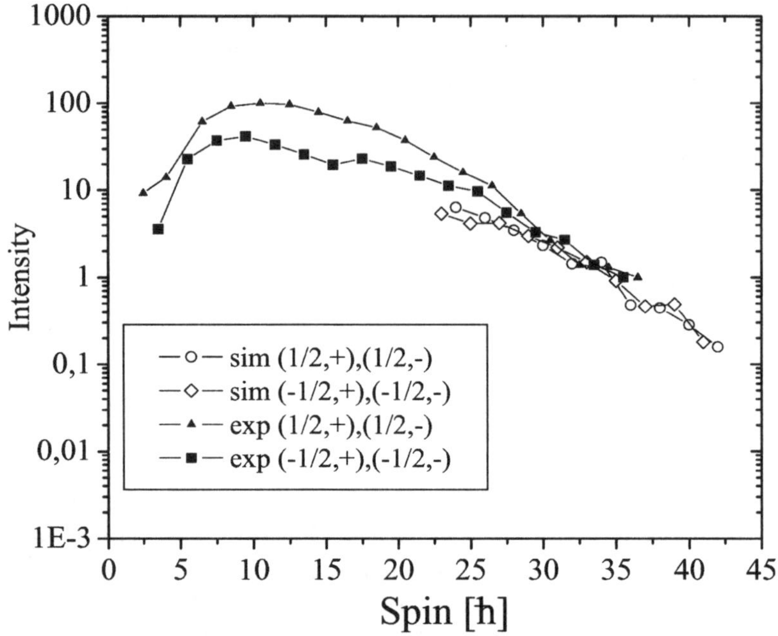

Figure 3. Comparison between experimental and calculated intensities of the yrast transitions of signature 1/2 (full triangles and open circles for experimental data and simulation respectively) and of signature -1/2 (full squares and open diamonds).

tions (i.e. $<U> = 4$ MeV, $FWHM_U = 4$ MeV, $<I> = 44\ \hbar$, $FWHM_I = 20\ \hbar$). The resulting simulated intensities well reproduce the experimental values both for the ridge structures and low spin yrast transitions, as shown in fig.3. One has to note that, since pairing effects are not included in the model, the simulation can only proceed down to spin $\approx 20\hbar$.

In order to reproduce the selection rules on the K-quantum number the E1 transition probability has been modified to include an exponential quenching factor depending on the difference between the values of K of the initial and final states (K_i and K_f):

$$P_K(K_i, K_f) = \iint e^{\frac{(K_i-k_1)^2}{2\sigma_i^2}} * e^{\frac{(K_f-k_2)^2}{2\sigma_f^2}} * e^{\frac{k_1-k_2}{\sigma}}\, dk_1\, dk_2 \qquad (1)$$

Here σ_i and σ_f are the FWHM of the K distribution associated to the initial and final states, while σ is the hindrance on K used to mimic the selection rules. A good reproduction of such selection rules is obtained using $\sigma = 0.3$, which is analogous to the one employed in the analysis of the E1 decay-out from isomeric states[12,13].

The MONTESTELLA code creates both 1D and 2D spectra with requirements similar to the ones used to update experimental matrices. Examples of projections of gated and total 2D matrices are shown in fig.2 in the right panels. These spectra can be analysed in terms of the fluctuations and the covariances of the spectral counts, to get a direct comparison with experimental data.

4. Results and discussion

4.1. *Number of paths*

The number of paths obtained from the analysis of the first ridge of the 2D matrices gated by individual bands is found, in average, to be ≈ 10 for each of the four low-K and of the three high-K configurations. Adding together the number of paths relative to specific configurations, taking also into account their relative intensities and correlations[9], a total number of ≈ 20 paths is found both for low-K and high-K states, as shown in panel a) of fig.4 by open circles and squares, respectively. From the analysis of the *Total* matrix one finds that a total of ≈ 45 discrete rotational bands exist at internal energies below the onset of damping, as shown by filled triangles in fig.4a).

The experimental results are well reproduced by the theory, represented in fig.4a) by lines: the dashed line is the number of paths for high-K states, the dotted one for low-K and the full line for the total ^{163}Er nucleus.

In contrast to the results of the ridge analysis, the number of paths obtained from the valley region is found to depend significantly on the nuclear configuration for $E_\gamma \leq 1$ MeV. This result is shown in panel b) of fig.4 together with the number of paths deduced from the *Total* spectrum. As the valley is probing the region in which the rotational bands are strongly mixed, this result suggests that the mixing process is indeed different for high-K and low-K states. Such clear difference in the number of paths between low-K and high-K configurations is also seen by the simulation up to $E_\gamma \approx 1$ MeV ($U \approx 2$ MeV), while the two quantities tend to converge at higher transition energies. This can be understood in terms of the onset of K-mixing: in the lower energy region, not only the level density for high-K states is ≈ 3 times lower than for the low-K ones[10], but also the rotational damping width has been measured to be $\approx 30\%$ lower for high-K states[14].

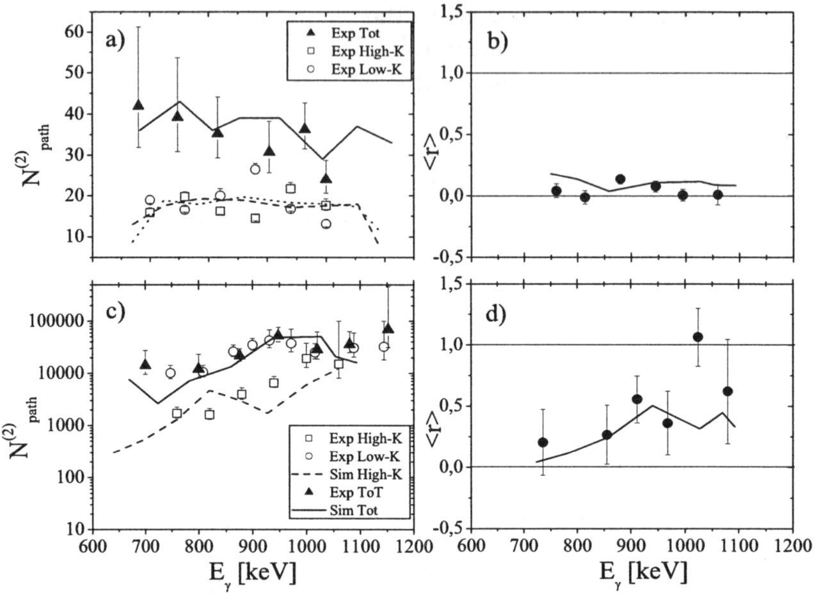

Figure 4. Panel a): Number of decay paths extracted from the fluctuation analysis of the ridge structures of ^{163}Er. The open circles (squares) refer to the number of unresolved rotational bands populating the ridges of γ-γ matrices gated by low-K (high-K) configurations, while the filled triangles give the number of discrete paths obtained from the total matrix. The corresponding values for simulated spectra are shown by dotted, dashed and full lines (low-K, high-K and total, respectively). Panel c): The effective number of transitions among mixed bands obtained from the fluctuation analysis of the valley region is shown by open circles (squares) for low-K (high-K) gated spectra, while the full triangles show the results obtained from the total matrix. The dashed and solid lines give the theoretical expectations for total and high-K cascades obtained from simulated spectra. Panels b) and d): Results of the covariance analysis on ridge and valley structures of ^{163}Er, respectively. Full circles in both panels show the correlation coefficient r obtained from the experimental analysis of the *low-K* versus the *high-K* matrices. The theoretical values, as obtained from the covariance analysis of simulated spectra, are represented by full lines.

4.2. *Covariance*

Figures 4b) and d) show the average values of the correlation coefficient r extracted from the covariance analysis of the ridge and valley structures of γ-γ coincidence spectra of ^{163}Er. In each panel the solid horizontal lines indicate the two opposite limits expected in the case of a complete conservation of selection rules ($r = 0$) and of a compound nucleus regime ($r = 1$).

In the case of the ridge analysis (fig. 4b)), r is found to be approximately zero for combinations of low-K and high-K spectra. This proves that there are basically no cross-transitions between the $\approx$ 20 bands feeding high-K states and the $\approx$ 20 bands feeding low-K states.

In the valley (fig.4d)) the r coefficient between low-K and high-K configurations, instead, increases from 0.2, at $E_\gamma \approx 700$ keV, up to $r \approx 0.5$ for $E_\gamma \approx 1$ MeV. Together with the fact that in the valley region also the number of paths of the low-K and of the high-K configurations approaches each other, this represents an experimental indication of the weakening of selection rules associated with the K-quantum number with increasing rotational frequency and internal excitation energy.

A reasonable agreement between simulations and experimental data is generally found both in the ridge and in the valley (full lines in panels b) and d)). In particular, in the valley, the simulation confirms the increasing trend of the r coefficient, suggesting a progressive weakening of K-selection rules due to the mixing of K-states. The flattening observed for $E_\gamma \geq 1$ MeV in the simulated coefficient between low-K and high-K configurations is associated with the fact that the excitation energy of the experimental γ-cascades at high spins extends up to values of internal energy larger than the range covered by the simulated bands ($U \leq 2.5$ MeV).

5. Conclusions

In conclusion, the onset of K-mixing in excited rapidly rotating nuclei has been discussed by comparing the results from high statistics experimental data on ^{163}Er with recently developed band mixing calculations. For the lower interval of internal energy up to approximately 1.2 MeV, a rather strict conservation of the K-quantum number is found, as deduced by the analysis of the ridge structure in $\gamma - \gamma$ matrices. At higher internal energy ($1.2 \leq U \leq 2.5$ MeV), probed by the weaker and more numerous transitions populating the valley, a partial conservation of the K-quantum number is found, as deduced both from the fluctuation and covariance analysis.

Acknowledgments

The author would like to thank all collaborators for this work. In particular I hereby acknowledge the active partecipation of A. Bracco, S. Leoni, N. Blasi, F. Camera, B. Million, A.Paleni, M. Pignanelli, E. Vigezzi, O. Wieland (INFN sez. Milano and Università degli Studi di Milano); M. Matsuo (Graduate School of Science and Technology, Niigata University);

T. Døssing, B. Herskind, G.B. Hagemann, J. Wilson (The Niels Bohr Institute Copenhagen); A. Maj, M. Kmiecik (The Niewodniczanski Institute of Nuclear Physics, Krakow); G. Lo Bianco, C. M. Petrache (Università di Camerino); M. Castoldi, A. Zucchiati (INFN sez. Genova); G. De Angelis, D. Napoli (Laboratori Nazionali di Legnaro); P. Bednarczyk, D. Curien (Istitute de Recherches Subatomic,Strasbourg).

The work has been supported by the Italian Institute of Nuclear Physics, by the Danish Natural Science Foundation Research Council, by the Polish Committee for Scientific Research (KBN Grant No. 2 P03B 118 22), by EU Transnational Access to Major Research Infrastructures (Contract No. HPRI-CT-1999-00078) and by the EU TMR project (Contract No. ERBFMRXCT970123).

References

1. P. Walker and G. Dracoulis, Nature **399**, (1999) 35.
2. I. Huseby et al., Phys. Rev. **C55** (1997) 1805.
3. V.G. Soloviev, Phys. Lett. **B317**(1993) 501.
4. P. Bosetti et al., Phys. Rev. Lett. **76** (1996) 1204.
5. J.Simpson, APH N.S., *Heavy Ion Physics* **6** (1997) 253.
6. G.B. Hagemann et al., *Nucl. Phys.* **A618** (1997) 199.
7. T. Døssing et al., Phys. Rep. **268** (1996) 1.
8. D.C. Radford, Nucl. Inst. Meth. **A361**, (1995) 297.
9. S. Leoni et al., Nucl. Phys. **A671** (2000) 71.
10. M. Matsuo et al., Nucl. Phys. **A736** (2004) 241.
11. A. Bracco et al., Nuc. Phys. **A673**, (2000) 64.
12. P. Walker et al., Phys. Lett. **B408** (1997) 242.
13. F.G. Kondev et al., Nuc. Phys. **A632** (1998) 473.
14. S. Leoni, Phys. Rev. Lett.**93** (2004) 022501-1.

SECTION II

NUCLEAR DYNAMICS AND HOT NUCLEI

LOW-ENERGY NUCLEAR REACTIONS WITH UNSTABLE BEAMS

CARLOS H. DASSO

Departamento de Física Atómica, Molecular y Nuclear, Universidad de Sevilla, Spain

ANDREA VITTURI

Dipartimento di Fisica Galileo Galilei and Istituto Nazionale di Fisica Nucleare, Padova

Coupling to reaction channels (including continuum break-up channels) seems to predict enhancement of fusion below the barrier (for total fusion and, to a lesser degree, also for complete fusion), when the reaction process is treated within the conventional coupled-channel approach, with proper modifications due to weak-binding nature of halo systems (as for example the long range of formfactors and potentials)

Our knowledge of the nuclear chart has significantly expanded over the last few years. Thanks to the new dedicated facilities for radioactive beams and improved large-scale detection systems the spectroscopy of nuclear systems has moved farther away from the valley of stability and, in some cases (mainly for light systems or on the proton-rich side), has reached the drip lines. New features characterizing the structure of these nuclei have been found. For instance, the occurrence of skins and haloes, novel excitation modes of collective or not collective character and modifications of the usual single-particle sequence of nucleon orbitals. As we know, nuclear structure properties strongly influence the diverse mechanisms associated with nuclear reactions. For heavy-ion induced processes this is especially valid at energies around the Coulomb barrier. It is therefore natural to expect that the specific features of systems far from stability will also lead to characteristic signatures in nuclear reactions that involve such nuclei.

This has been the case, for example, for the description of subbarrier fusion reactions. Since the early eighties a lot of activity – both in the theoretical and experimental fronts – has focused on the understanding of the fusion process between heavy ions at energies close or below the Coulomb barrier. The motivation for these studies was provided by the fact that the measured cross sections far exceeded (i.e. beyond repair) the theoretical predictions based on the barrier-penetration models that had successfully accounted for the experimental results obtained with lighter systems.

Although there is continuous progress in this field the picture that gradually emerged from all these investigations required a revision of the standard one-dimensional tunneling mechanism. It was gradually understood that the process is fundamentally affected when the system in question is able to exploit a multitude of other available degrees of freedom to improve its chance of going through a classically forbidden area. Or, in other words, that the mechanism of quantum tunneling should be carefully considered whenever it takes place in the presence of couplings to other reaction channels.

One popular way to visualize this situation is by thinking that everything happens as if the total incident flux of projectiles that impinge on the target actually can be divided in many different components that face not just the original potential barrier but, rather, a *collection* of barriers with splittings and relative weights that can be constructed following simple and well-defined prescriptions. This intuitive way of understanding the multidimensional barrier-penetration problem has sufficient appeal to have caused that a significant fraction of the data collected in the laboratories is presented or discussed in terms of experimental "barrier distributions".

This established picture may need to be revised in the case of fusion process involving weakly-bound nuclei. In fact, many aspects of the problem could be modified due to the particular nature of nuclear systems in the vicinity of the drip lines. We will breafly discuss here a couple of aspects of the problem that acquire particular relevance in the case of weakly-bound systems.

The first topic concerns the impact of the long-range of the radial wavefunctions associated with the valence orbitals characteristic of these nuclei on the effective height of the potential barriers. A dependence of the fusion cross sections with the number of neutrons involved has been known from the earliest stages of the game. This is observable even with nuclei close to the stability valley and is sometimes referred to as the "isotopic effect". In fact, as the number of neutrons in the projectile increases and one moves into heavier and heavier isotopes of a given element, the potential barrier $V_B(A)$ is systematically expected to decrease. We should note that this is essentially an $A^{1/3}$-effect and should become independent of the binding energies as soon as their values are ≈2 MeV or more. At the very end of the chain, however, the reduction in barrier height should become more prominent due to the long-range of the relevant weakly-bound orbital on the nuclear ion-ion potential. Calculations based on a microscopic description of the nuclear density (as given for example by mean-field calculations) predict that the halo contribution is capable of reducing the barrier height by several MeV, i.e. *more* than what is expected from the genuine isotopic effect.

Besides the purely static effects discussed above one also expects others of a dynamical nature, associated with the multidimensional character of our problem. These should be investigated in detail using the interacting potentials and formfactors within a coupled-channel formalism. The crucial point is the choice of the proper model space and the construction of the corresponding non-diagonal coupling matrix elements. In the case of weakly-bound systems an essential role is played by the break-up channels, which therefore must be included in the coupled-channel scheme. This is an interesting topic because there has existed for quite a while some disagreement insofar as whether the coupling to these channels enhances or de-enhances the observable cross sections.

The most straighforward way of incorporating break-up channels in the coupled-channel formalism is to interpret the projectile break-up in terms of inelastic excitation of single-particle states (or cluster-states) in the continuum. If one neglects the coupling in-between continuum states is it straightforward to slice the continuum and construct representative formfactors to each of the continuum bins. These are obtained by folding the Coulomb and nuclear field induced by the target with the microscopic transition density given, within our model, by the product of single-particle bound (initial) and continuum (final) states. The distribution in energy exhibits a strong peak at low excitation energy, reflecting the analogous behaviour of the corresponding low-lying (non-resonant) strength distribution. The outstanding feature is, however, that the weakly-bound nature of the initial orbital leads to nuclear formfactors that extend to much larger distances than in the case of standard, well-bound systems.

Within this formalism a complete coupled-channel analysis of the break-up and fusion processes has been recently performed by Hagino et.al. for the case of $^{11}Be+^{208}Pb$ [1]. The study lead to the conclusion that the cross sections of the complete-fusion (CF) and incomplete-fusion (IF) processes at energies well below the Coulomb barrier are indeed enhanced as a consequence of the couplings to the continuum. These results are supported by an even more recent state-of-the-art effort by Diaz-Torres et al. [2] within the coupled discretized continuum channel (CDCC) approach. In this work the coupling in-between continuum channels has been taken into account, something that had been neglected in earlier calculations. We should note, however, that opposite results have been obtained by Yabana et al. [3] with a different approach based on a time-dependent wave-packet formalism.

The problem therefore still deserves further consideration. Most of the coupled-channel calculations suffer, however, from technical limitations associated with the large number of continuum channels involved. There are also uncertainties and difficulties in discerning the contribution of the different ingredients entering in the calculation. For these reasons we have once more recurred to a simple, schematic, multi-dimensional barrier penetration model. This is implemented within a space confined to a reduced number of "continuum" channels which are coupled through real, gaussian potentials and formfactors. To retain the relevant features of the physical problem, one can play with the parameters of the couplings to recreate, realistically, the prevalent conditions of a weakly-bound nucleus. The results of this study confirm [4] that couplings to continuum reaction channels that mock up the weak-binding nature of halo systems still predict an enhancement of the total fusion cross section below the barrier. (The same conclusion remains valid – although to a somewhat lesser extent – in the complete-fusion regime.)

References

1. K. Hagino, A. Vitturi, C.H. Dasso and S.M. Lenzi, Phys.Rev. **61**, Art. 037602 (2000)
2. A. Diaz-Torres and I.J. Thompson, Phys.Rev. C **65**, 024606 (2002)
3. K.Yabana, M.Ito, M.Kobayashi, M.Ueda, T.Nakatsukasa, Nucl.Phys. A**738**, 303 (2004)
4. C.H. Dasso and A. Vitturi, to be published

REACTION MECHANISMS OF WEAKLY BOUND AND EXOTIC NUCLEI AT NEAR-BARRIER ENERGIES

D. PIERROUTSAKOU,* A. DE ROSA, M. DI PIETRO, G. INGLIMA, M. LA COMMARA, B. MARTIN, M. ROMOLI, M. SANDOLI, and E. VARDACI

INFN, Sezione di Napoli and Dipartimento di Scienze Fisiche, Università di Napoli "Federico II", Napoli, Italy

T. GLODARIU,† M. MAZZOCCO, C. SIGNORINI

Dipartimento di Fisica, Università di Padova and INFN, Sezione di Padova, Padova, Italy

F. SORAMEL

Dipartimento di Fisica, Università di Udine and INFN, Sezione di Udine, Udine, Italy

R. BONETTI, A. GUGLIELMETTI

Dipartimento di Fisica, Università di Milano and INFN, Sezione di Milano, Milano, Italy

L. STROE

NIPNE-HH, Romania

The experimental results concerning the reaction mechanisms of light weakly bound stable as well as radioactive nuclei, on various targets at near-barrier energies are reviewed and discussed. In order to study the reaction mechanisms of radioactive nuclei at near-barrier energies our collaboration developed a new line (EXOTIC) for the production of light exotic nuclei, by inverse kinematics nuclear reactions using the beams from the Tandem XTU accelerator of the Laboratori Nazionali di Legnaro impinging on H_2 or CH_4 gas target.

*Electronic address:pierroutsakou@na.infn.it
†On leave from: NIPNE-HH, Romania

1. INTRODUCTION

In the first part of this contribution the experimental results on the reaction mechanisms of light weakly bound stable and unstable nuclei, on various targets at near-barrier energies are discussed. In view of the similarity of the weakly bound stable systems with their associate (isobar and/or isotope) radioactive ones their study is relevant for the understanding of the radioactive nuclei structure and reaction mechanisms.

When going towards weakly bound stable and exotic nuclei the picture concerning the elastic scattering and the fusion process is expected to be different from that of well bound species because of the appearance of new phenomena. The loose binding of the weakly bound nuclei yields in a strong breakup channel which influences both the elastic scattering and the fusion cross section at the barrier. Its effect on the fusion cross section is strongly disputed: if the breakup is considered as any other channel, it should increase the tunnelling probability and therefore it can cause an enhancement of the fusion cross section. On the other hand the breakup can cause an inhibition of the fusion cross section due to the fact that it prevents the capture of the whole projectile by the target nucleus. Furthermore, the halo distribution of the outer nucleons, if present, could also influence the fusion cross section at the barrier, as the extended matter distribution can cause strong force begin acting at large distances yielding an enhancement of the fusion cross section. The theories developed in order to predict the behaviour of the fusion cross sections at near-barrier energies, have controversial results as the sub-barrier fusion cross sections have been found to be either reduced[1,2] or enhanced[3,4].

For what concerns the elastic scattering of weakly bound nuclei, it was suggested[5] that the threshold anomaly (for a review see [6]), which appears as a localized peak in the strength of the real potential associated with a sharp decrease in the strength of the imaginary potential at the barrier, may disappear because of the coupling to breakup channels. From an experimental point of view, an anomalous behaviour of the weakly bound stable ^{6,7}Li and ^{9}Be nuclei elastic scattering was observed in the vicinity of the barrier which contradicts the conventional threshold anomaly met in well bound nuclei.

Therefore, many questions arise about the behaviour of the weakly bound stable and unstable nuclei elastic scattering and fusion cross section around the barrier. It is actually clear, that the key issue to disentangle them is the knowledge of all the reaction channel cross section in this energy region.

In the second part of this contribution the new facility EXOTIC developed by our collaboration for the production of light exotic nuclei at near-barrier energies, by inverse kinematics nuclear reactions using the beams from the Tandem XTU accelerator of the Laboratori Nazionali di Legnaro (LNL), is presented. This line is dedicated to the study of light exotic nuclei reaction mechanisms at near-barrier energies. The EXOTIC line is associated with the EXODET experimental apparatus[7] for the detection of the reaction products and is connected with the large acceptance magnetic spectrometer PRISMA[8]. The characteristics of this facility and the possible beams which can be produced are reported.

2. REACTION MECHANISMS AT THE BARRIER

2.1. *Elastic scattering and breakup*

By studying the elastic scattering of weakly bound stable nuclei like ^{6,7}Li on different targets at near-barrier energies, it was evidenced that the obtained picture contradicts the one met with tightly bound nuclei and expressed with the threshold anomaly[9]. By using the optical model to analyze the elastic scattering data, it was seen in that work that the real potential remains constant down to and below the barrier, while the imaginary potential increases (decreases), in the case of ^{6}Li (^{7}Li), approaching the barrier from higher to lower energies. It was suggested by the authors that the influence of strong breakup and/or transfer channel, developed around the barrier, produces a more repulsive polarization potential which compensates the attractive part of the real potential. Moreover, in these experiments a strong inclusive breakup channel giving α particles was observed and it was evidenced that the α particle cross section follows a universal behaviour independently of the target[10]. By approaching the Coulomb barrier, the α production yield either due to breakup or to transfer almost exhausts the total reaction cross section. By studying the α-particle angular distributions in the ^{7}Li+ ^{28}Si scattering it was deduced that, while the compound nucleus mechanism is substantial, d-transfer and t-transfer are the dominant mechanisms at near-barrier energies[11]. Therefore, it was suggested by the authors that the unusual behaviour observed in the elastic scattering of ^{6,7}Li on various targets is directly related to the behaviour of the α production.

By studying the elastic scattering of the weakly bound stable ^{9}Be projectile on a ^{209}Bi target at energies around the barrier, an increasing imaginary potential was observed[12] along with a large incomplete fusion cross section. This latter originated in the partial fusion with the target of He fragments coming from the ^{9}Be breakup. In the cited work, the sum of the fusion and of the α production cross section was found to exhaust the total reaction cross section[13,14].

Furthermore, by studying the ^{6}Li+^{208}Pb inclusive and exclusive breakup at energies around the barrier, it was seen that the exclusive breakup cross section (α+d, α+p) is much smaller than the inclusive one (α only) which is comparable with the fusion cross section[15]. Also in that work, the reaction cross section was found to be equal to the sum of fusion and α inclusive breakup cross section.

By studying the inclusive breakup in p+^{16}O of the unstable proton rich ^{17}F on a ^{208}Pb target at beam energies of 98 and 120 MeV[16] it was found that the angular distributions of ^{16}O are dominated by stripping breakup. The angle integrated stripping cross section was found to be around 30% of the fusion cross section near the barrier. While the inclusive breakup cross section of the ^{17}F is large at near barrier energies, the exclusive breakup cross section measured near the grazing angle at beam energy of 170 MeV[17] suggests that the breakup channel seems too small to influence fusion in a significant way near the barrier. This result is in a good agreement with that reported from the measurement of the exclusive breakup of the ^{17}F on a ^{208}Pb target at backward angles and at a beam energy of 90.4 MeV, which shows that the two-body breakup is rather weak[18]. In the latter experiment, the ^{17}F scattering angular distributions were measured at backward angles and the data were analyzed in the optical model framework. It was found that the ^{17}F reaction cross section is very similar to those of the well bound nuclei ^{16,17}O on a ^{208}Pb target and it is three times smaller than that of the ^{19}F. From this observation the authors suggest that the probability to excite collective modes in ^{19}F is larger than the breakup probability in ^{17}F.

The elastic scattering angular distributions of the halo nucleus ^{6}He on a ^{64}Zn target were measured[19] and the total reaction cross section extracted by optical model analysis was found to be a factor two larger than the one of the ^{4}He +^{64}Zn at the same center of mass energies. The largest fraction of the total reaction cross section for the ^{6}He-induced reaction is due to direct processes like breakup and transfer, in a good agreement with the findings for the ^{6}He+^{209}Bi scattering[20].

In conclusion, from the above mentioned data it seems that at near-barrier energies a strong inclusive breakup and/or transfer channel is open for weakly bound stable and unstable nuclei while the exclusive breakup, when measured, is found to be much smaller than the inclusive one. The question which arises is the following: how does such a strong inclusive breakup influence the fusion cross sections at near-barrier energies? And which is the role of the halo structure?

2.2. *Fusion cross section*

A first answer to the previous questions can be given by the experiments where the fusion cross section of weakly bound stable and unstable light systems on different targets was investigated so far. Complete fusion excitation function data of ^{6,7}Li+ ^{209}Bi and ^{9}Be+^{208}Pb show a suppression of the complete fusion cross section with respect to that predicted by fusion models by 25-30% depending on the breakup threshold[21,22]. The large cross sections observed for incomplete fusion products support the interpretation that the observed fusion suppression is caused by the projectile breakup before reaching the fusion barrier. However, data of fusion excitation functions concerning the ^{9}Be on light-medium targets do not evidence any suppression of the complete fusion cross section due to breakup[23,24], showing thus a dependence on the mass of the target. According to the authors, the short range nuclear breakup dominates the breakup process and does not inhibit the fusion. Data on ^{9,10,11}Be+^{209}Bi show very similar fusion cross sections above and below the barrier while, from a theoretical point of view, one should expect a larger sub-barrier fusion cross section for the ^{11}Be because of its small binding energy and its halo structure[25]. Data on ^{4,6}He+^{238}U show that the sub-barrier complete fusion cross sections are very similar with each other, whereas the fission cross section in the ^{6}He-induced reaction is much larger than the one in the reaction induced by the ^{4}He projectile. This larger fission cross section is due to a large two neutron transfer cross section from the ^{6}He projectile to the target[26]. Similar results were obtained in the study of the ^{4,6}He+^{64}Zn fusion reactions[19] where no enhancement of the sub-barrier fusion cross section was seen in the ^{6}He-induced reaction, while the larger reaction cross section observed was ascribed to breakup or transfer processes as mentioned in the previous section. The fusion cross section (complete+incomplete) of ^{17}F and ^{19}F projectiles on a ^{208}Pb target are very similar with each other at energies below and above the barrier, with a slight hindrance for the ^{17}F-induced reaction at the lowest energy point[28]. The only experiment evidencing a larger sub-barrier fusion cross section for a halo nucleus is the ^{6}He+^{209}Bi fusion [29]. Following the authors, this enhancement may arise from the coupling to positive Q value neutron transfer channels, resulting in a "neutron flow" between the projectile and the target.

Therefore, from a large part of the existing data we can deduce that the halo nuclei do not present an enhancement of the sub-barrier fusion cross section as it could be expected from a theoretical point of view. On the other hand, the effect of a strong breakup channel is to reduce the complete fusion cross section of weakly bound stable nuclei like ^{9}Be and ^{6,7}Li on heavy targets. However, it is clear that the weakly bound nuclei reaction mechanisms at the barrier constitute a very complicated subject and more data, especially for medium mass and light targets, are needed to go further in our understanding.

3. EXOTIC BEAM LINE

The Exotic beam line[30,31,32] is a facility installed at the INFN National Laboratories of Legnaro (LNL) and it is devoted to the experimental research on radioactive light nuclei at incident energies around the Coulomb barrier. This facility was planned with the aim to cover the time gap between the design and the operation of the new large international advanced facilities for the production of radioactive beams. The facility design takes advantage of the already existing devices, in particular the use in connection with the PRISMA large acceptance magnetic spectrometer is planned.

The EXOTIC beam line facility will produce light radioactive nuclei through inverse kinematics reaction on light targets. The first beam is the ^{17}F through the reaction p(^{17}O,^{17}F)n induced on a H_2 or CH_4 gas target. The final goal is to reach a RIB current intensity in the range of $\sim 10^6$ pps with very low contaminants background. This nucleus is a very interesting candidate to study, since it is very loosely bound, it has only one bound state below the breakup threshold and its first excited state has a halo structure. It breaks into two charged fragments (^{16}O and p) easy to detect. The same technique can be also used for the production of other exotic beams exploiting suitable inverse kinematics reactions. Some possible cases to be considered are the following: ^{3}He(^{6}Li,^{8}B)n, ^{7}Li(^{9}Be,^{6}He)^{10}B, p(^{7}Li, ^{7}Be)n, p(^{14}N,^{14}O)n, ^{3}He(^{16}O,^{18}Ne)n, ^{3}He(^{20}Ne,^{22}Mg)n, ^{3}He(^{36}Ar,^{38}Ca)n.

The designed beam facility consists of: i) a gas target which is devoted to the production of the exotic nuclei. It consists of a small cylindrical cell with metal (Havar) entrance and exit windows which are strong enough to sustain the internal pressure of the gas target and thin enough to reduce to an acceptable level the uncertainties due to the energy loss of both the incident primary beam and the secondary particles crossing the windows. To keep constant the effective thickness of the target, the internal pressure and temperature of the cell is continuously monitored. A proper security system prevents the collapse of the accelerator vacuum in case of window breaking. The gas cell can be operated at room or at liquid N_2 temperatures with an operational pressure up to 1 atm. ii) a beam selection and transport system consisting of several magnetic and electric devices having the task to separate and focus the most part of the exotic ions leaving the gas cell onto the target where the reaction under study will be performed. The scattering chamber can be connected to the PRISMA magnetic spectrometer. An important feature of this system is its ability to suppress the residual primary beam exiting the production target. iii) a detection system for both light and heavy reaction products. Light charged products are detected and charge identified by means of an array of position sensitive silicon detector telescopes (EXODET apparatus). This array is housed in the scattering chamber which can be connected to the entrance window of the PRISMA magnetic spectrometer which can detect the heavy charged reaction products. The EXODET apparatus has been designed to be used also without the spectrometer.

3.1. *The beam selection and transport system*

To define the features of the system, the p(^{17}O,^{17}F)n reaction has been used as a reference case. The basic characteristics of the ion production beam line are the large secondary beam acceptance of the optics elements and a maximum suppression capability of the unwanted scattered beams (primary beam, its halos and spurious beams with the same magnetic rigidity as the selected RIB). The final configuration of the beam selection and transport system is the following. A triplet of large diameter quadrupole lenses (ϕ=160 mm) collects the ejectiles exiting the production target and focuses them at the entrance of a bending magnet. The bending magnet is followed by a second triplet of quadrupole lenses. Due to the different magnetic rigidity this system separates the fully stripped ions of the secondary exotic beam from the residues of the primary one. However, some tails of such a beam, having the same magnetic rigidity as the secondary beam, will pass through the full system. To eliminate these tails the second quadrupole triplet is followed by a Wien filter which is sensitive to the energy to charge ratio of the ions. The main features of the components of the beam, selection and transport system are the following:

Solid angle: $\Delta\Omega$=20 msr
Energy acceptance: ΔE/E=$\pm$10%
Momentum acceptance: Δp/p=$\pm$5%
Horizontal acceptance: $\Delta\theta$=$\pm$3.4°
Vertical acceptance: $\Delta\phi$=$\pm$5.2°
Magnetic rigidity: Bρ=0.98 Tm

In order to stop the beam contaminants, two sets of slits have been placed along the beam line: S1 after the first quadrupole triplet, S2 after the bending magnet. A diaphragm is also placed inside the second quadrupole triplet.

3.2. *Final tests*

The performed tests pointed out the importance of the primary beam injection to reduce the background due to scattering on the gas target holder and to maximize the transmission of the secondary exotic beam through the separator. To improve the primary beam positioning and focusing a magnetic quadrupole of the standard LNL type, a four sector slit system and two steerer dipole magnets were mounted on the beam line before the gas target.

To monitor the secondary beam a two stage telescope detector was mounted on the beam line at the exit of the intermediate scattering chamber. The first stage of the detector consisted of a gas ionization chamber while the second stage was a Silicon strip detector. The 16 strips, 3 mm wide each, were orthogonal to the deflection plane of the separator dipole magnet to measure the dispersion of the secondary beam at the reaction target point. The ionization chamber was operated at a gas pressure of 40 mbar, suitable to the charge identification of the ejectiles by means of the standard ΔE-E technique. The target was filled by CH_4 gas at a pressure of 0.8 atm which at a temperature of 300 K determines a thickness of 2.57 mg/cm^2. To prevent the beam monitor damage the tests were performed with a current of 0.5 nA ^{17}O beam from the tandem. Taking into account the energy loss in the HAVAR windows and in the gas target, to get a secondary ^{17}F beam having an energy distribution centered at 90 MeV, a ^{17}O primary beam of 7^+ charge state and 119 MeV energy is required. After the measurement of the secondary beam parameters at low intensity of the primary beam, the ^{17}O current was raised up to 300 nA and the secondary beam intensity was determined, by measuring the elastic scattering cross section on a 2 μg/cm^2 thick ^{197}Au self-supporting target at 40°. At this angle the scattering can be considered of a pure Rutherford nature and the secondary beam intensity can be deduced.

The left-hand side panel of Fig. 1 shows the energy spectrum of the ions transmitted through the separator when the Wien filter is switched off. Each peak is characterized by a different charge of the ions. The energy calibration was performed by standard α sources. The field of the dipole magnet was tuned at B=0.885 T to optimize the transmission of the ^{17}F ions. Then, all the ions having the same magnetic rigidity as the ^{17}F ones are transmitted. The suppression of the unwanted residues of the primary beam is performed by the Wien filter. Such an effect is visible in the right-hand side panel of Fig. 1 where the Wien filter was tuned at 90 kV, which is about 75% of its maximum operation voltage. From this figure one can see that only the ^{16}O ions with 8+ charge were not completely suppressed because they have a velocity very similar to that of ^{17}F ions and consequently the Wien filter voltage should be further increased in order to completely suppress them. The deduced intensity of the ^{17}F secondary beam was 7.5*10^5 pps on target at 300 nA of primary beam ^{17}O. Furthermore, the secondary beam has a FWHM in the x direction of 9 mm as it was evidenced with the beam monitor.

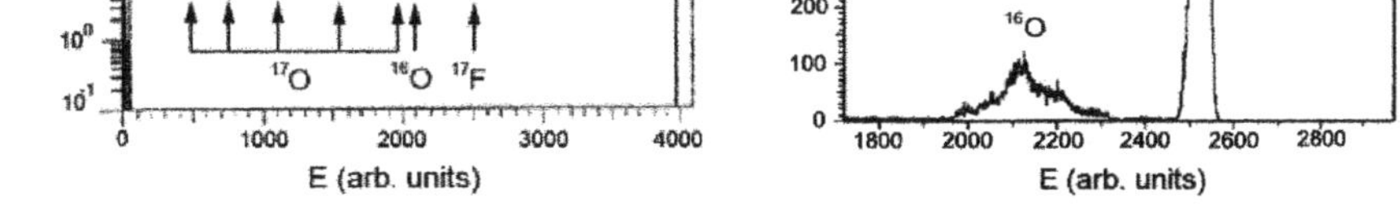

Figure 1. Energy spectrum of the ^{17}F(^{17}O) + ^{197}Au elastic scattering with the Wien filter switched off (left-hand side panel) and at 75% of its maximum operation voltage (right-hand side panel).

3.3. *Experiments to be performed with the EXOTIC line*

The experiments designed to be performed by using this facility concern measurements of elastic scattering, fusion, breakup and transfer cross section of the produced radioactive beams on heavy and medium mass targets. In this way, a complete study of all the reaction mechanisms of the produced radioactive nuclei at near-barrier energies will be carried out. The detection of the reaction products will be done in the large solid angle EXODET apparatus which allows a complete reconstruction of the event kinematics. If necessary, the PRISMA spectrometer can be used along with EXODET for the detection of the heavy ejectiles.

References

1. M. S. Hussein *et al.*, *Phys. Rev.* **C46**, 377 (1992)
2. N. Takigawa *et al.*, *Phys. Rev.* **C47**, R2430 (1993)
3. C.H. Dasso *et al.*, *Phys. Rev.* **C50**, R12 (1994)
4. K. Yabana *et al.*, *Prog. Theor. Phys.* **97**, 437 (1997)
5. C. Mahaux *et al.*, *Nucl. Phys.* **A 449**, 354 (1986)
6. G.R. Satchler *et al.*, *Phys. Rep.* **199**, 147 (1991)
7. M. Romoli *et al.*, *EPJ direct* **A**, (2005), in press and M. Romoli, this issue
8. F. Scarlassara *et al.*, *Nucl. Phys.* **A746**, 195c (2004) and E. Fioretto, this issue
9. A. Pakou *et al.*, *Phys. Lett.* **B556**, 21 (2003) and *Phys. Rev.* **C69**, 054602 (2004)
10. A. Pakou *et al.*, *Phys. Rev. Lett.* **90**, 202701 (2003)
11. A. Pakou *et al.*, *Phys. Rev.* **C**, (2005), in press
12. C. Signorini *et al.*, *Phys. Rev.* **C61**, 061603(R) (2000)
13. C. Signorini *et al.*, *EPJA* **5**, 7 (1999)
14. C. Signorini *et al.*, *Prog. of Th. Phys. Sup* **154**, 272 (2004)
15. C. Signorini *et al.*, *Phys. Rev.* **C67**, 044607(2003)
16. J.F. Liang *et al.*, *Phys. Rev.* **C67**, 044603 (2003)
17. J.F. Liang *et al.*, *Phys. Lett.* **B491**, 23 (2000)
18. M. Romoli *et al.*, *Phys. Rev.* **C69**, 064614 (2004)
19. A. Di Pietro *et al.*, *Phys. Rev.* **C69**, 044613 (2004)
20. E. F. Aguilera *et al.*, *Phys. Rev. lett.* **84**, 5058 (2000)
21. M. Dasgupta *et al.*, *Phys. Rev.* **C66**, 041602(R) (2002)
22. M. Dasgupta *et al.*, *Phys. Rev. Lett.* **82**, 1395 (1999)
23. R.M. Anjos *et al.*, *Phys. Lett.* **B534**, 45 (2002)
24. S.M. Moraes *et al.*, *Phys. Rev.* **C61**, 064608 (2000)
25. C. Signorini *et al.*, *Nucl. Phys.* **A735**, 329 (2004)
26. R. Raabe *et al.*, *Nature* , 02984 (2004)
27. K. E. Rehm *et al.*, *Phys. Rev. Lett.* **81**, 3341 (1998)
28. K.E. Rehm *et al.*, *Phys. Rev. Lett.* **81**, 3341 (1998)
29. J.J. Kolata *et al.*, *Phys. Rev. Lett.* **81**, 4580 (1998)

30. M. Sandoli *et al.*, *LNL Annual report*, 192 (2001)
31. T. Glodariu *et al.*, *LNL Annual report*, 153 (2003)
32. V.Z. Maidikov *et al.*, *Nucl. Phys.* **A746**, 389c (2004)

SCATTERING PROCESS FOR THE SYSTEM $^{11}Be + ^{209}Bi$ AT COULOMB BARRIER ENERGIES WITH THE EXODET APPARATUS

M. MAZZOCCO, C. SIGNORINI, T. GLODARIU* AND L. STROE*

Dipartimento di Fisica and INFN, via Marzolo 8, I-35131, Padova, Italy
E-mail: marco.mazzocco@pd.infn.it

M. ROMOLI, A. DE FRANCESCO, M. DI PIETRO, E. VARDACI, A. DE ROSA, G. INGLIMA, M. LA COMMARA, B. MARTIN, D. PIERROUTSAKOU AND M. SANDOLI

Dipartimento di Scienze Fisiche and INFN,
Complesso Universitario di Monte S. Angelo,
via Cinthia, I-80125, Napoli, Italy

R. BONETTI AND A. GUGLIELMETTI

Istituto di Fisica Generale ed Applicata and INFN,
via Celoria 16, I-20133, Milano, Italy

F. SORAMEL

Dipartimento di Fisica and INFN, via delle Scienze 208, I-33100, Udine, Italy

K. YOSHIDA, A. YOSHIDA, R. KANUNGO, N. KHAI AND T. MOTOBAYASHI

The Institute of Physical and Chemical Research (RIKEN),
2-1 Hirosawa, Wako-shi, Saitama, 351-0198, Japan

T. NOMURA, T. ISHIKAWA, H. ISHIYAMA, S. JEONG, H. MIYATAKE, M. H. TANAKA, I. SUGAI AND Y. WATANABE

Institute of Particle and Nuclear Studies (KEK),
1-1 Oho, Tsukuba-shi, Ibaraki, 305-0801, Japan

*and INFN - Laboratori Nazionali di Legnaro. On leave from NIPNE, Magurele-Ilfov, Romania

The scattering process for the system ^{11}Be + ^{209}Bi has been measured in the energy range around the Coulomb barrier at the RIKEN RIPS facility. The scattered ^{11}Be ions have been detected with a new type of large solid angle pixel-structure detector array, named EXODET. The experimental data evaluated at 44-MeV ^{11}Be beam energy have been compared with those obtained for the stable system ^{9}Be + ^{209}Bi. The comparison allowed to investigate for the first time the effects due to the ^{11}Be halo structure and low binding energy on the scattering process.

1. Introduction

Over the last decades, the study of nuclear reactions induced by light weakly bound nuclei has attracted the interest of many scientific communities around the world. Recent developments together with a larger availability of radioactive ion beams (RIB's) might give the possibility to observe new phenomena. In fact exotic nuclei have in many cases extremely different properties from those reported for nuclei along the valley of stability and, therefore, they are expected to behave differently from stable well-bound nuclei, especially at Coulomb barrier energies.

The new degrees of freedom that come into the game in reactions involving loosely bound nuclei, are the following:

(i) the **halo** structure. The wave functions of the last loosely bound nucleons extend well outside the nuclear core originating an r.m.s. radius larger than that deduced from the systematics of stable nuclei.
(ii) the **neutron skin** structure. Last neutrons are mostly orbiting in the outer part of the nucleus in a skin-like structure.
(iii) the **small binding energy** of the last nucleons (around one order of magnitude smaller than for nuclei close to the valley of stability).

All these phenomena are usually not independent from each other and generate opposite and sometimes conflicting effects.

The interest for the nuclear reactions induced from light RIB's motivated the study of the scattering process for the system ^{11}Be + ^{209}Bi in the energy range around the Coulomb barrier. In fact the ^{11}Be ion ($T_{1/2}$ = 13.8 s) has some interesting features: (i) a well-known neutron halo structure [1], (ii) a very small binding energy (S_n = 0.504 MeV) and (iii) only one excited level below the particle threshold. Moreover, in this energy range the elastic scattering for the stable system ^{9}Be + ^{209}Bi (S_n = 1.573 MeV) has been measured with high statistical accuracy [2] and could represent a good reference for comparison with our system.

2. Experimental Set-Up

2.1. ^{11}Be *secondary beam production*

In this experiment, the ^{11}Be radioactive beam was produced via (In-Flight) fragmentation of a $^{13}C^{6+}$ primary beam at 101.23 MeV/nucleon and $\sim$ 3 eμA intensity. The production target was a 12-mm-thick Beryllium foil, corresponding to a thickness of 2.22 g/cm^2. The produced 66.12 MeV/nucleon ^{11}Be secondary beam was selected among all the projectile fragments with the RIKEN Projectile-Fragment Separator (RIPS) [3] of the RIKEN Accelerator Research Facility (RARF). The high kinetic energy of the resulting secondary beam was reduced to the required range by the experiment using two thick Aluminium degraders located in suitable positions along the beam line [3].

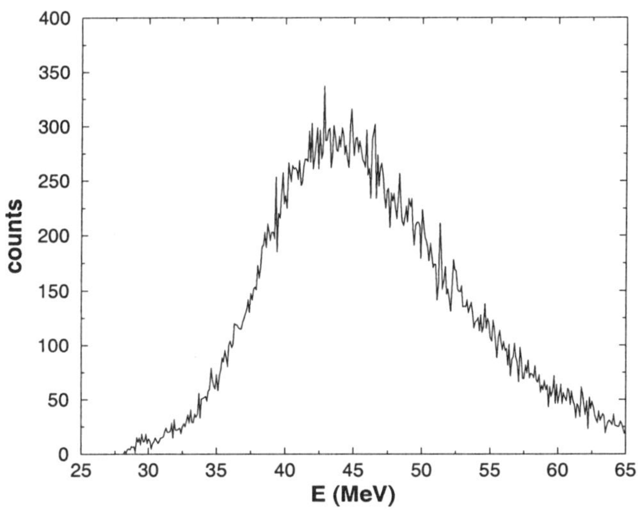

Figure 1. *Spectrum of the ^{11}Be incoming beam energy.*

The heavy degradation produced a ^{11}Be secondary beam with a bell-shaped energy spectrum centered at around 43 MeV (see Fig. 1) and a large beam spot at the target position: 26.0 mm (horizontal) $\times$ 19.0 mm (vertical), as shown in Fig. 2. The total secondary beam intensity integrated over the energy range 38-50 MeV was $\sim$ (1.0×10^6) particle/s.

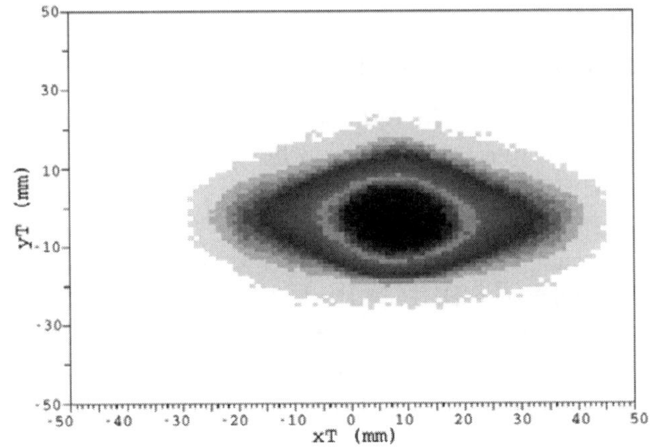

Figure 2. *Reconstructed beam distribution at the target plane for the heavy degraded* ^{11}Be *secondary beam. The full width at half maximum of the beam was: 26.0 mm (horizontal)* $\times$ *19.0 mm (vertical).*

2.2. *Secondary beam tagging system*

Since the secondary beam had a large energy spread and emittance, it was necessary to provide an event-by-event tracking in order to measure the incident energy of the incoming particles and to reconstruct their scattering angles. To do this, two sets of position sensitive Parallel Plate Avalanche Counters (PPAC) [4] were used.

Each PPAC had a total sensitivity area of $\sim (100.0 \times 100.0)$ mm^2 and allowed a position resolution of $\sim (1.0 \times 1.0)$ mm^2. The two detectors were used at the same time to determine the target position hit by each incident particle (see Fig. 2) and together with a plastic scintillator to provide a beam energy reconstruction via Time-of-Flight (ToF) measurement. The flight-path between the plastic scintillator and the position of the first (second) PPAC was 4.768 m (5.072 m) long, corresponding to a ToF of 184.0 ns (195.7 ns) for a ^{11}Be energy of 43 MeV. The target was located 682.5 mm downstream the second PPAC.

2.3. *Detection equipment: the EXODET array*

The ^{11}Be scattered ions were detected with the EXODET apparatus, placed around the target in a secondary reaction chamber.

The EXODET array consists of 16 Silicon detectors arranged into 8 two-stage telescopes with a total solid angle coverage of around 72% of 4π. The front side of each detector is segmented in 100 strips with a pitch size of 0.5 mm. The strips of the ΔE layer are mounted orthogonally to those of the E stage, allowing a pixel resolution of $\sim$ (0.5 $\times$ 0.5) mm^2 for particles passing through the first layer. More details about this detector set-up as well as its electronics are described in Ref. [5].

The readout of the multi-strip detectors was provided with an ASIC chip [6]. The chip digital output data stream contains the number of the detector hit strip, the time correlation between the signal and the trigger and a rough information on the energy lost of the recorded event. The thickness of the used detectors was: 40 μm (70 μm) for six (two) ΔE layers and 500 μm for all the E stages. The target was 3.03-mg/cm^2-thick ^{209}Bi evaporated on a 1-μm-thick Mylar foil.

3. ^{11}Be beam energy reconstruction

The heavy degradation needed to decrease the secondary beam energy to the required range by our experiment produced a ^{11}Be beam with a large energy spread (see Fig. 1). In order to provide an event-by-event energy reconstruction of the incoming particles, we performed a Time-of-Flight measurement between a plastic scintillator and the two PPACs placed at a distance of about 5 m.

The ToF had to be preliminarily energy calibrated using either primary or secondary beams with very sharp energy spreads. For this purpose several beams were opportunely selected changing the magnetic fields of RIPS spectrometer. Unfortunately the ^{11}Be energy at the exit of the second PPAC could not be directly derived from the ToF measurements, since the low-energy ^{11}Be secondary beam had a quite large energy lost inside the two PPACs, that had to be properly taken into account in the energy reconstruction. We used the code SRIM [7] to preliminarily evaluate this energy loss: at 40-MeV (60-MeV) beam energy the ^{11}Be ions lost about 20% (10%) of their kinetic energy.

This energy reduction could systematically affect the energy reconstruction provided by the second PPAC, since the distance between the two PPACs (304 mm) was covered by the particles with velocities smaller than those between the plastic scintillator and the first PPAC. The relationship between the final ^{11}Be beam energy and the ToF-value was obtained placing an energy calibrated Silicon Solid State Detector (Si SSD) at the end of the flight-path, just in front of the EXODET apparatus. In such a configuration, the energy lost in both PPACs was experimentally taken into account by comparing the energy reconstructed from the ToF measurement with the energy signal collected by Si SSD detector (see Fig. 3). Since the energy lost increases as the incident energy decreases, the relationship between these two quantities could not be fitted using a simple linear regression. However we found that a good description could be achieved with a quadratic fit of the curve shown in Fig. 3. The second order correction, although much smaller than the linear coefficient, was able to reproduce quite well the experimental deviations from the linearity.

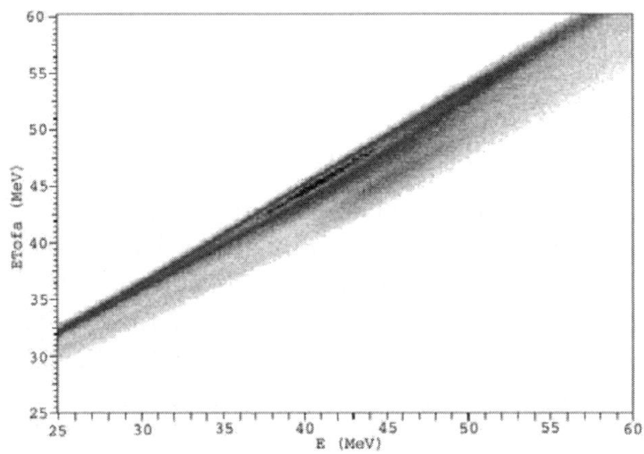

Figure 3. *Matrix of the energy signal collected by the Si SSD and the incident energy reconstructed via ToF measurement between the plastic scintillator and the first PPAC, 4.768 m downstream.*

Up to now we have performed a preliminary evaluation of the systematic errors introduced by this calibration procedure. However a qualitative analysis of Fig. 3 shows that, for example, a ToF-reconstructed energy $E_{ToF} = 45$ MeV corresponds to an average energy signal in the Si SSD of $E_{SSD} \sim 42$ MeV, with a FWHM of ~ 1.5 MeV. Therefore, the accuracy of the ^{11}Be energy reconstruction was assumed to be $\sim 4\%$.

4. Data analysis

In order to evaluate the ^{11}Be scattering angular distributions and to compare them with those measured for the system ^{9}Be + ^{209}Bi, the continuum spectrum of the incoming energy shown in Fig. 1 was subdivided into 2-MeV bins. This value was found to be a good compromise between statistics, energy resolution of the resulting angular distributions and accuracy of the incoming beam energy reconstruction via ToF measurements. The reduction of the effective beam energy interaction due to the target thickness was not taken into account, since it was rather small in comparison with both energy binning and incident energy uncertainty. In fact the energy loss for 42-MeV ^{209}Bi ions into a 1.5-mg/cm^2-thick ^{209}Bi foil (half of the target thickness) results to be $\sim$ 0.8 MeV.

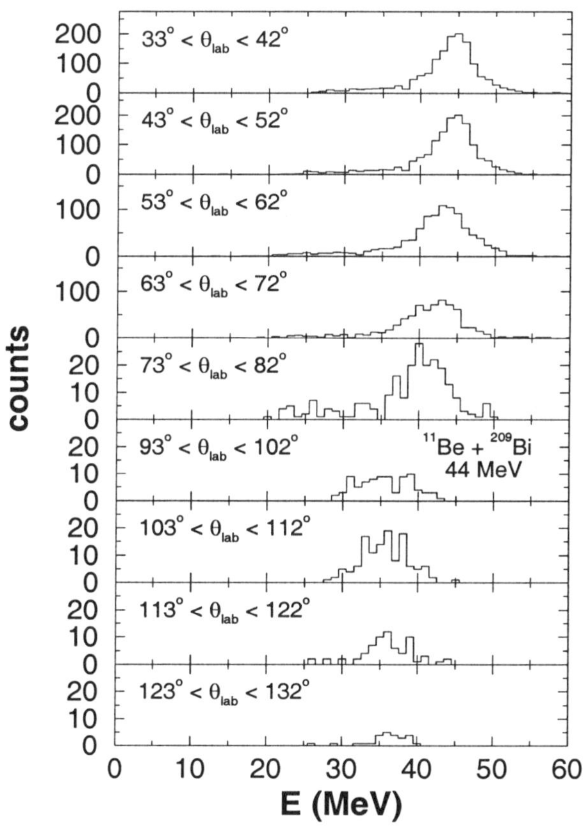

Figure 4. *Total energy spectrum (sum of the energy released in the ΔE and E telescope stages) for the nine angular bins used to evaluate the the scattering differential cross section for the system ^{11}Be + ^{209}Bi at 44-MeV beam energy.*

The range of the polar angles θ has been divided into $10°$-bins. Fig. 4 shows the ^{11}Be total energy spectrum (sum of the energy released in the ΔE and E telescope stages) for the nine angular bins used to evaluate the scattering angular distribution at 44-MeV beam energy. All these spectra have a quite Gaussian shape, in particular those at forward angles, with very large widths ($\sim$ 7-10 MeV). The origins of these broad structures are manifold: the energy resolution of the incoming beam due to the 2-MeV binning ($\sim$ 5%), the accuracy in the reconstruction of the incident energy via ToF measurements ($\sim$ 4%), the energy resolutions of the detectors ($\sim$ 4% for the ΔE layers and $\sim$ 2% for the E ones), the energy lost into the target ($\sim$ 2%) and finally the quite large angular range ($10°$) necessary to collect enough statistics for each bin.

The experimental energy resolution did not allow to distinguish "pure" elastic scattering events from quasi elastic events (with either ^{11}Be first excited state at 0.320 MeV or ^{209}Bi first excited states at 0.896 MeV and 1.609 MeV being populated in the interaction). These contributions to the quasi-elastic cross section were estimated within the Distorted Wave Born Approximation (DWBA) formalism. Target excitations were found to be negligible in a first approximation in comparison with the pure elastic scattering differential cross section, whereas those arising from the projectile excitation resulted to be up to 30% of the elastic scattering angular distribution.

The experimental data were normalized using the ^{11}Be low energy events, under the assumption that the measured cross section at 38-MeV beam energy was purely Rutherford. DWBA and Coupled-Channel calculations ensured that this hypothesis is realistic up to $\theta \sim 120/130°$, whereas it introduces a systematic error ($\sim +$ 10-20%) for larger scattering angles where the resulting angular distribution could be slightly overestimated.

Fig. 5 shows the ^{11}Be scattering differential cross section at 44-MeV beam energy. We evaluated five points at forward angles and only four in the backward direction since the last one could not be measured because of the the limited statistics. The plotted errors along the y axis take into account only the statistical uncertainties. On the other side, since the polar angles have been grouped into $10°$-bins, an average indetermination of $5°$ was assumed for the plotted θ angles.

Although rapid variations in the differential cross section could not be observed due to the large angular binning and a better energy resolution of the secondary beam could improve the quality of the resulting differential cross section, present data are sufficient to a primary investigation of the

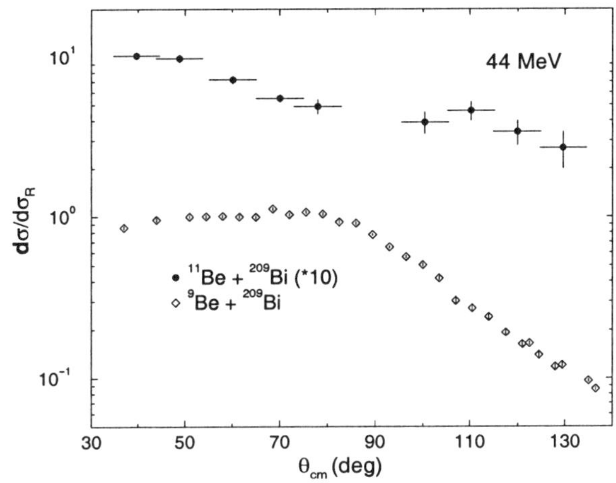

Figure 5. *Experimental angular distributions for the scattering process in the systems* $^{9,11}Be + ^{209}Bi$ *at 44-MeV beam energy.*

effects of the ^{11}Be halo structure and low binding energy on the elastic scattering process.

5. Discussion

Fig. 5 shows the comparison between the systems ^{9,11}Be + ^{209}Bi at 44-MeV beam energy. We clearly see that, contrary to a first expectation, the two scattering angular distributions resulted to be rather similar at backward angles, within the experimental errors. This is quite surprising since the ^{11}Be binding energy (S_n = 0.504 MeV) is about one third of the ^{9}Be one (S_n = 1.573 MeV) and, considering the ^{11}Be well-known halo structure. Therefore one would have expect a much larger flux into non-elastic channels for the reaction induced by the unstable isotope. This reduction has not been experimentally observed. On the contrary the differential cross section for the ^{9}Be scattering process from a ^{209}Bi target at 44-MeV beam energy resulted to be even slightly larger than that measured for the system ^{11}Be + ^{209}Bi at the same colliding energy [2]. Nevertheless we have to keep in mind that, as already mentioned in the previous section, the data for the system ^{11}Be + ^{209}Bi contains also a non-negligible contribution coming from inelastic excitations.

We may underline that in this energy range a similar behavior has been observed also for the fusion cross sections of these two colliding systems [8].

In the forward direction the influence of non-elastic channels seems to begin at smaller scattering angles for the ^{11}Be projectile than for the stable isotope. However, because of some uncertainties in the evaluation of the detector efficiency in the region $53° \leq \theta \leq 82°$, the corresponding values need to be further investigated.

6. Conclusions

From the performed analysis, we observed that at 44-MeV beam energy the scattering angular distributions of the systems ^{9,11}Be + ^{209}Bi resulted to be rather similar at backward angles. Therefore, since both Beryllium isotopes are very loosely bound whereas only ^{11}Be has a well-known halo structure, we can conclude for the moment that the effects due to the low binding energy (*i.e.* mainly the possibility that a projectile breakup process occurs) should be larger than those arising from the ^{11}Be halo structure (namely a reduced Coulomb barrier and a consequent higher probability for ^{11}Be to fuse with respect to ^{9}Be).

In addition, since the ^{9}Be binding energy is around three times larger than the ^{11}Be one, the similarity of the scattering differential cross section suggests that, within the experimental accuracy, the probability to excite the ^{9}Be rotational band based on a $K = 3/2^-$ g.s.[9,10] could compensate the effects due to the ^{11}Be much lower binding energy.

All these conclusions have been derived only for one beam energy and need to be verified in a much wider energy range, both below and above the Coulomb barriers of the two systems. Data collected during this experiment are still under analysis and should permit to clarify the whole scenario.

References

1. I. Tanihata *et al.*, *Phys. Rev. Lett.* **55**, 2676 (1985).
2. C. Signorini *et al.*, *Phys. Rev.* **C61**, 061603(R) (2000).
3. T. Kubo *et al.*, *Nucl. Instrum. Methods Phys. Res.* **B70**, 309 (1992).
4. H. Kumagai *et al.*, *Nucl. Instrum. Methods Phys. Res.* **A470**, 562 (2001).
5. M. Romoli *et al.*, *Phys. Rev.* **C69**, 064614 (2004).
6. A. Perazzo *et al.*, BABAR note no. 501, 1999.
7. SRIM - The Stopping and Range of Ions in Matter. Website: *http://www.srim.org.*
8. C. Signorini *et al.*, *Nucl. Phys.* **A735**, 329 (2004).
9. C. Signorini *et al.*, *Nucleus-Nucleus Collisions*, Proceedings of the International Conference BO2000, Bologna, Italy, 2000, edited by G.C. Bonsignori *et al.* (World Scientific, Singapore, 2001) p. 413.
10. D. Hinde *et al.*, *Phys. Rev. Lett.* **82**, 1395 (1999).

FISSION DYNAMICS IN SYSTEMS OF INTERMEDIATE FISSILITY WITH 8πLP APPARATUS

E. VARDACI, A. BRONDI, G. LA RANA, R. MORO, M. TROTTA, A. ORDINE, A. BOIANO, R. MORDENTE

Dipartimento di Scienze Fisiche, Universita' di Napoli "Federico II" and INFN, Sezione di Napoli, 80126 Napoli, Italy

M. CINAUSERO, E. FIORETTO, G. PRETE, V. RIZZI, D. SHETTY

Laboratori Nazionali di Legnaro, Viale dell'Universita', Legnaro, Padova, Italy

M. BARBUI, D. FABRIS, M. LUNARDON, S. MORETTO, G.VIESTI

Dipartimento di Fisica and INFN, Padova, Italy

N. GELLI, F. LUCARELLI

Dipartimento di Fisica and INFN, Firenze, Italy

V.A. RUBCHENYA

Department of Physics, University of Jyväskylä, Finland, and Khlopin Radium Institute, St. Petersburg, Russia

P.N. NADTOCHY

Department of Theoretical Physics, Omsk State University, Omsk, Russia

We have started a research program at LNL with the apparatus 8πLP aimed at studying the fission dynamics in systems of intermediate fissility using light charged particles as probes. These systems are characterized by an evaporation residues (ER) cross section comparable or larger than the fission cross section, and by a relatively higher probability for charged particle emission in the pre-scission channel. In a theoretical framework in which time scale estimates rely on statistical model calculations, the additional analysis of particle emission in the ER channel allows to put more constraints on the input parameters of statistical and dynamical models. The presentation will report about an overall view of the systems studied so far and will provide clues for interesting physical cases. Recent advancement will also be discussed on the interplay between the fission and the evaporation residues channels as implied by a dynamical approach to the description of the fission process.

1. Introduction

Fission dynamics has been the subject of a large variety of studies in the last two decades [1-14]. Most of the studies have shown that the pre-scission multiplicities of neutrons and charged particles increase monotonically with the bombarding energy in contrast with the calculations of the standard statistical model (SM). Since the fission process, as well as the fission time scale, is thought to be affected by nuclear dissipation [15], this result is considered as the evidence that fission is a slow process with respect to the lifetime for the emission of light particles. With increasing excitation energy, the particle decay lifetime decreases and becomes smaller than the time necessary for the build up of the collective motion of the nuclear matter toward the saddle point. Consequently, fission does not compete as effectively as predicted by the SM in the early stages of the decay, and particles and GDR γ-ray emissions can occur more favorably. The overall cause of the establishment of these transient effects is believed to be associated with the nuclear matter viscosity which slows down the collective flow of mass from equilibrium to scission and does not allow the fission decay lifetime to be reduced with increasing excitation energy as in the case of light particles. An energy domain has further been identified [17] above which the SM predictions begin to deviate from the data.

Several variants of the statistical model have been proposed in the literature to take explicitly into account time scales as well as viscosity [1,3,4,6,11,18,19]. Following the initial idea of the "neutron clock" [1,3], the common trend is to split the path from the equilibrium to the scission point configuration into two regions, the pre- and the post-saddle. The total fission time is defined as $\tau_f = \tau_d + \tau_{ssc}$, where τ_d is the pre-saddle delay, namely, the characteristic time which the composite system spends inside the barrier, and τ_{ssc} is the time necessary to travel the path from saddle to scission. The relevant observables are computed using τ_d and τ_{ssc} as free parameters, along with the other input parameters relative to the specific ingredients of the model, and fitted to the experimental data. In spite of the extensive work, estimates of the fission time scales are however quite controversial, ranging from ~5 to ~400×10^{-21} s, depending on the system and on the experimental probe. Furthermore, such estimates are weakened by the fact that different sets of input parameters can result in equally good fits within the same model [7,11,12].

Modified statistical models as well as dynamical models based on the Euler-Lagrange, Fokker Planck and Langevin equations [20-28] have been used in order to gain insight on the nature of dissipation. Following Kramer's work [15], the inclusion of dissipative effects results in an effective fission decay width Γ_{eff}

which is smaller than the standard Bohr-Wheeler decay width by a hindrance factor,

$$\Gamma_{eff} = \Gamma_{BW}\left(\sqrt{1+\gamma^2}+\gamma\right)$$

where γ is the nuclear friction coefficient and Γ_{BW} is the Bohr-Wheeler fission decay width. γ is usually associated to the fission delay time [11,16,18]. The values of γ estimated from the fits to the particle multiplicities, both from statistical and dynamical models, provides a contradictory picture on the values of γ, which range over an order of magnitude, and on the one-body or two-body origin of the nuclear dissipation. More dramatic is the situation from the pure theoretical point of view, where the predicted values of γ, on the basis of microscopic models, are spread over three orders of magnitude [29]. It must be pointed out that only neutron multiplicities have been measured in most of the studies and mostly for heavy systems ($A \geq 200$), and the lack of a sufficient number of constraints to the models could be, in several cases, the source of discrepancies. In order to withdraw a more consistent picture of nuclear dissipation it seems crucial the take into account a larger number of observables which can be expected to be sensitive to the nuclear dissipation and to try to reproduce the variety of observables with a unique set of input parameters.

In this framework, the 8πLP collaboration has started a research program at the Laboratori Nazionali di Legnaro aimed at studying the fission dynamics in systems of intermediate fissility (χ=0.5-0.6). These systems are very little studied although they offer several advantages.

They are characterized by an evaporation residue cross-section comparable or larger than the fission cross-section, and by a shorter path in the deformation space from the saddle-to-scission point [30]. Consequently: 1) the input parameters of the models can be further constrained by the energy spectra and multiplicities of the light particles in the evaporation residue channel; 2) the effect of the fission delay over the fission and ER cross section is much more pronounced with respect to heavier systems because the emission of a charged particle in the pre-saddle region strongly enhances the probability of producing an evaporation residue as consequence of both a reduction of the fissility and the large value of the angular momentum necessary to ignite fission.

The fact that the potential energy surface is characterized by a shorter path from the saddle to scission [30] means that the role of the pre-saddle dynamics relative to the saddle to scission dynamics is enhanced and, therefore, some of the ambiguities on the not-well identified separation and interplay between pre- and post-saddle might be reduced in the interpretation of the data.

In conclusion, we expect that the measurements of neutron and charged particle multiplicities and energy spectra in the two channels as well as the measurements of the cross sections of the channels themselves will allow putting more severe constraints onto the models. This should provide more reliable values of fission delay and of the friction parameter.

This presentation will review some of the measurements performed recently with the 8πLP apparatus in Legnaro and will provide some recent results based on a new version of the SM code PACE2 which includes the Kramer's factor plus a more consistent treatment of the deformation effects. A dynamical model approach is finally used in the attempt to overcome the limits of the picture of the SM. The paper is thus organized as follows. First, a short review of the 8πLP apparatus is given. Second, we present the study of the fission time scale in the system ^{32}S + ^{100}Mo at 200 and 240 MeV. Third, we give some preliminary results about the application of a dynamical model to the system ^{32}S + ^{109}Ag at 180 MeV.

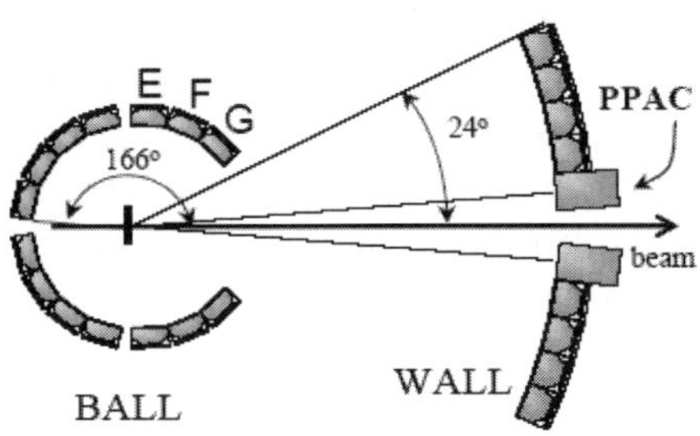

Figure 1. Schematic layout of the 8πLP apparatus.

2. The 8pLP apparatus

The 8πLP apparatus [31], shown schematically in Figure 1, is a light charged particle detector assembly which covers about 80% of 4π. It consists of two detector subsystems, each made out of two-stage telescopes: the WALL and the BALL. The WALL contains 116 telescopes and is placed at 60 cm from the target. Each of the WALL telescopes consists of a 300μm Si detector backed by a 15 mm CsI(Tl) crystal and has an active area of 25 cm^2 which corresponds to an angular opening of about 4°. The WALL covers the angular range from 2° to 24°. The BALL has a diameter of 30 cm and consists of 7 rings placed coaxially around the beam axis. Each ring contains 18 telescopes and covers an angular opening of about 17°. The telescopes of the BALL are made out of a 300 μm Si detector mounted in the *flipped* configuration (particle entering from the Ohmic side) backed by a 5 mm CsI(Tl) crystal. The BALL has a total of 126 telescopes and covers the angular range from 34° to 166°. The rings are labeled from A to G going from backward to forward angles.

Particle identification is carried out by the ΔE-E method for the ions that are not stopped in the ΔE stage. The particles stopping in the first stage are identified by the TOF method in the case of the WALL telescopes, and by the Pulse Shape Discrimination (PSD) technique in the case of the BALL telescopes. In this configuration we are able to measure energies up to 64 AMeV in the WALL and 34 AMeV in the BALL with energy thresholds of 0.5 MeV for protons and 2 MeV for alpha particles.

Heavy fragments can be detected in the telescopes of the BALL. The PSD technique allows the separation between heavy fragments and light particles stopping in the same detector but does not provide any information about the mass or charge of the fragments. The selection between symmetric and asymmetric binary mass splitting can nevertheless be achieved on a kinematics ground as shown later. A more selective trigger detector based on a time-of-flight spectrometer is presently under study.

In the 8πLP setup it is also possible to detect Evaporation Residues (ER). The WALL detectors between 2.5° and 7.5° around the beam axis are in fact replaced by 4 Parallel Plate Avalanche Counter (PPAC) modules, each one subtending a solid angle of about 0.3 msr. Each module consists of two coaxial PPACs mounted and operating in the same gas volume at a distance of 15 cm from each other. By adjusting the gas pressure, it is possible to stop the ER between the two PPACs, and let the lighter ions to impinge on the second PPAC. Consequently, ERs are sorted out from the first PPAC signals using the signals from the second PPAC as a veto.

3. Dynamical effects in the system $^{32}S + ^{100}Mo$ at 200 an 240 MeV

In both reactions energy spectra of protons and alpha particles have been measured in coincidence with fission fragments and ER with an almost 4π coverage. Heavy fragments were detected in the telescopes of the ring F and G of the BALL. The selection between symmetric and asymmetric binary mass splitting has been achieved on a kinematical ground. An example of such a selection for the reaction at 200 MeV is in Figure 2 where we show the fragment-

fragment energy correlation corresponding to two fragments detected by two different telescopes in the same ring F (a), in the same ring G (b) and in the rings F-G (c), respectively. The plots show clearly the transition between the symmetric and asymmetric mass splitting by a proper choice of the detecting geometry, which means, a variable coverage of the folding angle. The angular correlations are such that when the two fragments are detected in two different telescopes of the ring F, which corresponds to a folding angle $\Delta\theta$ between 105° and 138°, only the symmetric component of the fragment mass distribution is detected; in the case of ring G instead, only the most asymmetric component is selected. The case of rings F and G corresponds to an intermediate condition, namely to an angular range centered on the most probable folding angle for symmetric fission.

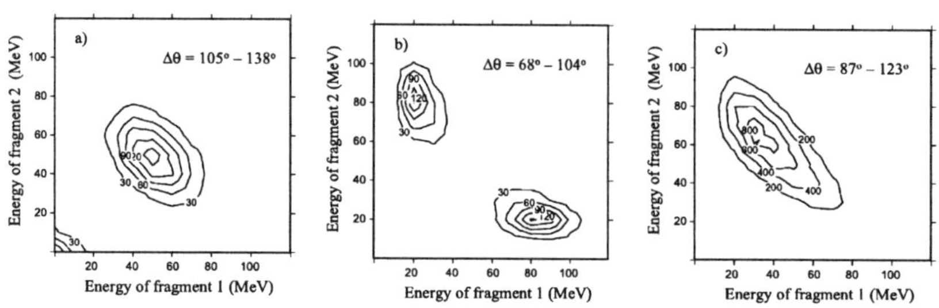

Figure 2. Energy-energy correlation matrix of the measured fragments: a) both fragments in ring F, b) both fragments in ring G, c) one fragment in ring F and the other in ring G.

Laboratory energy spectra of the protons alpha particles in triple coincidence (fragment-fragment-particle) were extracted for all the correlation angles allowed by the geometry (12 in plane and 56 out of plane). Examples of multiplicity spectra are shown as histograms in Figures 3 for the reaction at 240 MeV, and in Figure 4 for the reaction at 200 MeV.

The position of each particle detector with respect to the beam has been translated to a position relative to a trigger plane (defined by the position of the two fired fragment detectors) using the in-plane α ($0° < \alpha < 360°$) and the out-of-plane β ($-90° < \beta < 90°$) angles. The values of α and β are shown in Figures 3 and 4. Each particle spectrum has been obtained as the sum of the alpha particle spectra corresponding to the same in-plane and out-of-plane angles, and normalized to the number of its corresponding trigger fragment-fragment events.

To extract the pre- and post-scission integrated multiplicities, particle spectra have been analyzed considering three evaporative sources: the composite nucleus prior to scission (CE, Composite Emission) and the two fully accelerated fission fragments F1 and F2 (FE, Fragment Emission).

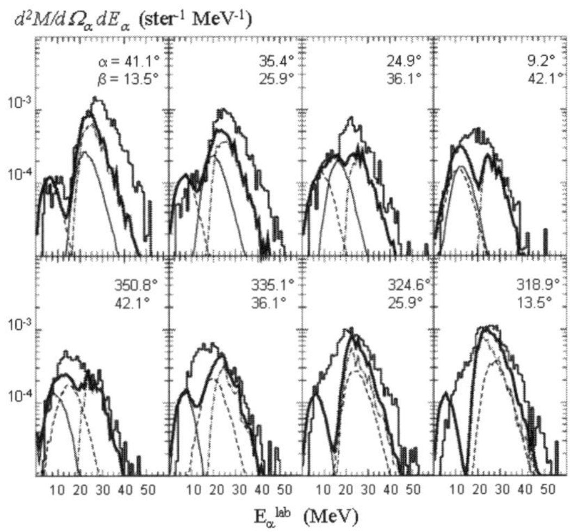

Figure 3. Out-of-Plane multiplicity spectra of the alpha particles in the ring G in coincidence with fission fragments detected in the rings F-G in the reaction $^{32}S + ^{100}Mo$ at 240 MeV.

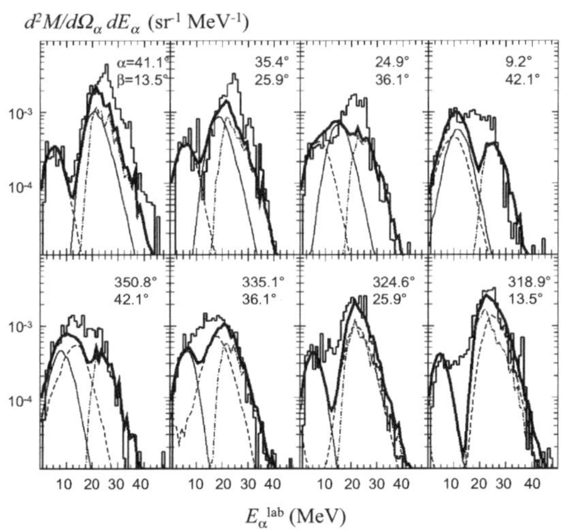

Figure 4. Same as Figure 3 but for the same reaction at 200 MeV.

We have used a well established procedure which employs the Monte Carlo Statistical code GANES [2,32,33]. Particle evaporative spectra are computed separately for each source of emission in the trigger configuration defined in the experiment, taking into account the detection geometry. Afterwards, the calculated spectra are normalized to the experimental ones, and the integrated multiplicities are evaluated for each emitting source.

The curves superimposed on the histograms in the Figures 3 and 4 represent calculated multiplicity spectra for CE (dot-dashed line), F1 (thin solid line) and F2 (dashed line) components, along with their sum (thick solid line). A large deformation of the composite system prior to scission was necessary to fit simultaneously the energy spectra and the angular distributions. The deformation of the emitter affects both the mean energy of the evaporated charged particles (because of the change in the evaporation barriers) and the out-of-plane angular distribution (because of the increase in the moment of inertia). In our data the best fit to the energy spectra provides for the CE component a prolate shape with axis ratio b/a = 3, both at 200 and 240 MeV. This emitter deformation results into mean energies of the alpha particles which are $\approx$ 2 MeV lower than those expected for the case of a spherical emitter.

Although the bulk of the experimental spectra is very well reproduced at both energies, also considering the wide angular coverage of the detecting array, there are some important deviations which indicate contributions not accounted for by the CE and FE components which are mainly of two kinds: an excess of high energy alpha particles at most forward angles, and a surplus of particles with energies intermediate between those corresponding to CE and FE components. The same is observed for the protons. These two types of contributions have already been observed in other experiments of the same kind as presented here [2] and have been ascribed to pre-equilibrium and near-scission emission [34-36], respectively.

From the fit to experimental spectra, we have extracted the pre and post scission particle multiplicities which we have compared with the prediction of a modified version of the code PACE2 [37]. In particular, a fission delay parameter τ_d is included into the code so that the fission width is zero up to a time τ_d and has the full statistical value subsequently. Since we expect that the particle multiplicities are sensitive to the value of the ratio a_f/a_v and to the delay time τ_d, we performed a grid of calculations for $1.00 < a_f/a_v < 1.10$ and $0 < \tau_d < 50 \times 10^{-21}$s. The result relevant for the present presentation is that the SM is able to reproduce the particle prescission multiplicities in the reaction at 240 MeV whereas at 200 MeV a delay of about 25 x 10^{-21}s is necessary. This result is at variance with what is expected with increasing excitation energy and at variance with the prediction of the systematics in ref [17], where a sizable deviation from the statistical description is expected at both energies.

In our analysis [14] with the statistical model as implemented in the code PACE2, the lack of a fission delay at 240 MeV was ascribed to the opening of the fast-fission channel [38] at the higher energy. Fast-fission is a decay mode which is open when the angular momentum of the composite system is such that the fission barrier height is reduced to zero. Similar findings are reported in [39] where it is shown that in systems of mass around 150 the fission delay reduces coherently in correspondence of the progressive opening of the fast fission channel.

From calculations in the system ^{32}S + ^{100}Mo at 200 MeV the fast-fission cross section is about 13% of the total fission cross-section, whereas at 240 MeV the fraction of fast-fission grows up to 61%. Since in both experiments it was not possible to separate experimentally the typology of binary fragmentation, the delay time extracted at 240 MeV is affected by the mixing of two processes (and eventually others) which are characterized by different time scales, and this may justify the lack of a fission delay. However, even though this conjecture might seem reasonable to justify the lack of a fission delay at the higher energy, it cannot stand as an explanation of an unexpected result we have found for this system: the α particle pre-scission multiplicity remains approximately constant with increasing excitation energy. It is well documented indeed for systems with $A \geq 150$ that the pre-scission light particles multiplicities increase with the excitation energy, even in the cases where the fast fission decay channel is open [3]. Such data on systems with $A < 150$ are really missing and no conclusion can be withdrawn on this apparent anomaly. In fact this is a region of masses poorly investigated for which no data exists and for which the fission process may show a different dynamics. The system ^{32}S + ^{100}Mo produces the compound nucleus ^{132}Ce. This nucleus happens to be in the range of masses where the barrier height is in the region of the absolute maximum [40]. Moreover, the calculated saddle and scission points are almost indistinguishable. What is the behavior of such systems with respect to the fission as known for heavier systems is a matter not investigated and may well reveal a different manifestation of dynamical effects. In this regards, for the systems of intermediate mass we consider reasonable the more general concept of binary fragmentation rather than fission decay.

In order to clarify the above physical scenario, in our opinion it is necessary to use as many physical probes as possible in a larger range of excitation energy. It is in fact already in our program to perform an experiment on the same system, in a larger range of excitation energy, in which also neutron multiplicities are to be measured along with the fission cross sections. In the next section we will discuss an example of the effect of the use of a more complete set of observables.

4. Dynamical effects in the system $^{32}S + {}^{107}Ag$ at 180 MeV

Protons and alpha particle multiplicities have also been measured for the S + Ag system in the fission and evaporation residues (ER) channels. The same analysis as before has been applied to the fission channel to extract the pre- and post-scission proton and alpha particle multiplicities. In order to extract the particle multiplicities in the ER channel we have integrated the in-plane and out-of-plane angular distributions by using the statistical code LILITA_N97 [41] which includes the detector geometry.

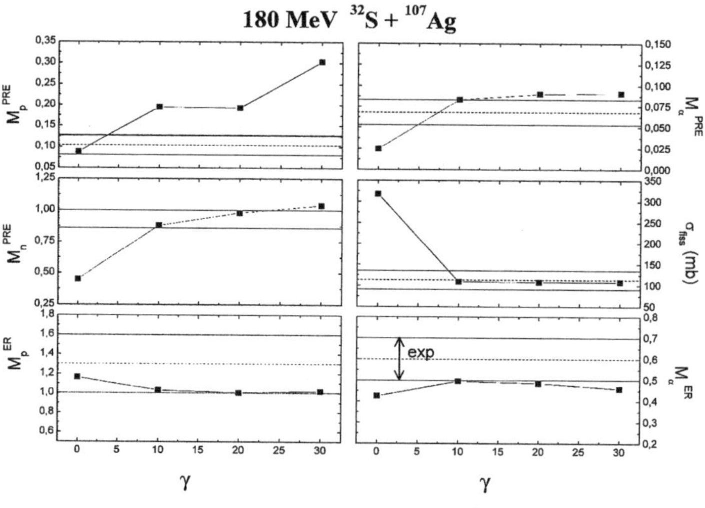

Figure 5. Measured particle prescission multiplicities (PRE), proton and alpha multiplicities in the ER channel, and fission cross sections (horizontal lines) compared to the predictions of the model (squares) for different values of the friction coefficient γ.

The experimental data are shown in Figure 5 by dashed lines along with the range of error by solid lines. M_p^{PRE}, M_α^{PRE} and M_n^{PRE} are the prescission multiplicities, respectively, for protons, alpha particles and neutrons. The experimental value of M_n^{PRE} has been extrapolated from the data of a similar system. M_p^{ER} and M_α^{ER} are the protons and alpha particles multiplicities in the ER channel, and σ_{fiss} is the fission cross section. Figure 5 also shows the results of the calculation performed with a new version of PACE2 [37] for a range of values of the friction coefficient. This version of PACE2 includes a new consistent approach to the evaporation of charged particles from a deformed nucleus. Dynamical effects have also been included in the code through the Kramer factor [15] where the friction coefficient is related to the fission delay according to the well known prescription of Grangé et Weidemuller [16].

For this system, the statistical model modified to include only the delay is able to reproduce only the particle multiplicities in the fission channel and the fission cross section with a fission delay of $\approx 27 \times 10^{-21}$s but is unable to reproduce at the same time the particle multiplicities in the ER channel. The inclusion of the Kramer factor and a more detailed treatment of the deformation of the composite system [37] produce the pattern shown in Figure 5. The value of $\gamma=10$, which is consistent with one body dissipation, allows to reproduces most of the observables. The proton pre-scission multiplicity M_p^{PRE} seems to require a smaller value of the friction parameter ($\gamma=2$).

In order to gain more insight on the presence of dynamical effects we have also used the dynamical model of the fission implemented as a three dimensional Langevin stochastic approach the detail of which can be found in ref. [24,27]. The code has recently been coupled to the statistical code LILITA_N97 [41] in order to allow the evaporation of light particles from the composite system during the evolution along trajectories in the phase space. We have performed several sets of calculations using the one-body dissipation prescription, by changing the strength of such dissipation by a factor k_s and the maximum angular momentum J_{max} of the rotating system. From this preliminary calculations it turns out that by modulating the values of k_s in the range $0.25 \div 1$ and J_{max} in the range $60 \div 64\ \hbar$ the dynamical model may reach a reasonable overall agreement with whole set of observable shown in Figure 5. This result supports the conclusion that the dynamical approach to fission decay is very promising in describing both fission and evaporation residues channel within the same model.

References

1. A. Gavron et al., *Phys. Rev.* **C35**, 579 (1987)
2. R. Lacey el at., *Phys. Rev.* **C37**, 2540 (1988)
3. D. J. Hinde et al., Phys. Rev. C 39 (1989) 2268; D.J. Hinde, D. Hilscher, H. Rossner, et al., *Phys. Rev.* **C45**, 1229 (1992)
4. J.P. Lestone et al., *Nucl. Phys.* **A559**, 277 (1993)
5. H. Ikezoe, Y. Nagame, I. Nishinaka, et al., *Phys. Rev.* **C49**, 968 (1994)

6. A. Saxena, A. Chatterjee, R. Choudhury et al., *Phys. Rev.* **C49**, 932 (1994)
7. A. Chatterjee, A. Navin, S. Kailas et al., *Phys. Rev.* **C52,** 3167 (1995)
8. L. Fiore, G. D'Erasmo, D. Di Santo, et al., Nucl. Phys. A**620**, 71 (1997)
9. D.J. Hofman, B.B. Back, and P. Paul, *Phys. Rev.* **C51**, 2597 (1995)
10. V.A. Rubchenya et al., *Phys. Rev.* **C58**, 1587 (1998)
11. I.Diószegi, N.P. Shaw, I. Mazumdar et al., *Phys. Rev.* **C61**, 24613 (2000)
12. N.P. Shaw, I.Diószegi, I. Mazumdar et al., *Phys. Rev.* **C61**, 44612 (2000)
13. A. Saxena, D. Fabris, G. Prete, et al., *Phys. Rev.* **C65**, 64601 (2002)
14. G. La Rana et al., *Eur. Phys. J.* **A16**, 199 (2003)
15. A. Kramer, Physica (Amsterdam) **7**, 284 (1940)
16. P. Grangé et H. A. Weidenmuller, *Phys. Lett.* **B96** 26 (1980)
17. M. Thoennessen and G.F. Bertsch, *Phys. Rev. Lett.* **71**, 4303 (1993)
18. V.A. Rubchenya, *Proceedings of Int. Conf. Large-Scale Collective Motion of Atomic Nuclei*, Brolo (Messina, Italy),1996, Eds. G. Giardina, G. Fazio and M. Lattuada, World Sci., Singapore, 1997) p.534.
19. G. Giardina, *Proceedings of 6th Int. School-Seminar on Heavy Ion Physics, Dubna (Russia), 1997*, Eds.Yu.Ts. Oganessian and R. Kalpakchieva, (World Sci., Singapore, 1998) p.628
20. T. Wada et al., Phys. *Rev. Lett.* **70**, 3538 (1993)
21. C. Bhattacharya et al., *Phys. Rev.* **C53**, 1012 (1996)
22. P. Frobrich and I.I. Gontchar, *Phys. Rep.* **292**, 131 (1998)
23. A.K. Dhara et al., *Phys. Rev.* **C57**, 2453 (1998)
24. A.V. Karpov et al, *Phys. Rev.* **C63,** 54610 (2001)
25. G. Chaudhuri and S. Pal, *Phys. Rev.* **C65** 54612 (2002)
26. P. Frobrich and I.I. Gontchar, *Europhys. Lett.* **57** , 355 (2002)
27. P.N. Nadtochy et al., *Phys. Rev.* **C65,** 64615 (2002)
28. W. Ye, Progr. *Theor. Phys.* **109**, no. 6, 933 (2003)
29. C. Badimon, Ph.D. Thesis – CENBG, Bordeaux, 2001
30. M.G. Itkis, A. Ya. Rusanov, *Phys. Part. Nucl.* **29**, (1998) 160
31. E. Fioretto et al., *IEEE Trans. Nucl. Scie.* **44**, (1997) 1017
32. N.N. Ajitanand et al., *Nucl. Instr. Meth. Phys. Res.* **A243**, 111 (1986)
33. N.N. Ajitanand, G. La Rana, R. Lacey, et al., *Phys. Rev.* **C34,** 877 (1986).
34. E. Duek et al., *Phys. Lett.* **131B,** 297 (1983)
35. L. Schad et. al., *Z. Phys.* **A318,** 179 (1984)
36. E. Vardaci, M. Kaplan et al., *Phys. Lett.* **480B,** 239 (2000)
37. M.A. DiMeo, *Ph.D. Thesis*, Universita' di Napoli "Federico II" (2002).
38. C. Ngô et.al, *Nucl. Phys.* **A400**, (1983) 259c
39. G. La Rana et al., Proc. Int. Conf. on Nuclear Reaction Mechanisms, June 5-9, 2000, Varenna, Italy
40. A. Sierk, *Phys. Rev.* **C33** 2039 (1986)

41. LILITA_N97 is an extensively modified version of the original LILITA program made by J. Gomez del Campo and R. G. Stockstad, Oak Ridge National Laboratory, (Rep. No TM7295, 1981 unpublished)

SEARCH FOR ISOSPIN EFFECTS ON THE LEVEL DENSITY OF NUCLEI INVOLVED IN THE DECAY OF ^{139}Eu

N. GELLI AND F. LUCARELLI

Dipartimento di Fisica and INFN, Firenze
E-mail: gelli@fi.infn.it

A. BRONDI, G. LA RANA, R. MORO, M. TROTTA, E. VARDACI, A. ORDINE, A. BOIANO AND R. MORDENTE

Dipartimento di Scienze Fisiche and INFN, Napoli

M. CINAUSERO, E. FIORETTO, G. PRETE

INFN, Laboratori Nazionali di Legnaro

M. BARBUI, D. FABRIS, M. LUNARDON, V. RIZZI

Dipartimento di Fisica and INFN, Padova

I. GOVIL

Panjab University, Chandigarh, India

Evaporative proton and alpha particle energy spectra and angular distributions have been measured in the decay of the compound nucleus ^{139}Eu produced at an excitation energy of 90 MeV by the ^{32}S + ^{107}Ag reaction. In-plane and out-of-plane angular distributions and energy spectra of alpha particles have been compared with theoretical calculations performed using the statistical Montecarlo code Lilita. Three expressions for the level density parameter have been used: *a)* without isospin dependence *b)* with N-Z isospin dependence and *c)* with Z-Z_0 isospin dependence, based on the distance of the nucleus from the valley of stability. The predictions by the Z-Z_0 prescription fail to reproduce the data. Isospin independent and N-Z dependence reasonably well reproduce the data.

1. Introduction

Nuclear level density is a fundamental quantity for the static and dynamic properties of nuclei and in particular, in nuclear astrophysics. Much work has been devoted to study this quantity whose most commonly used expression is based on the Fermi-gas model. Starting from the Bethe work[1], based on an energy independent level density of single particle states, angular momentum as well as pairing and shell effects have been included to get more realistic expressions[9] . At present, the level density has only been studied in nuclei close to the valley of stability and mainly on the neutron-deficient side. In all these studies, isospin effects have been usually neglected, as they are expected to be relatively small for nuclei close to the stability valley, as those produced by stable beams. In the framework of the Fermi-gas model the isospin effects on the level density can be taken into account including a factor related to the isospin component $T_3=(N-Z)/2$ as well as to the density of single-particle states at the Fermi energy and to the excitation energy of the system. According to this, the level density is expected to decrease with increasing $|T_3|$.

Recently, the analysis of the level density data for excitation energies below 8 MeV for the nuclei $20 \leq A \leq 70$ listed in the ENSDF data file[3] and three nuclei with A=100 and two nuclei with A=140 has been carried out comparing different expressions of the level density[4]. It has been shown that data support the need for an isospin dependence. Compared to the prescription based on the N-Z dependence, a reduction factor of the level density based on the distance from the valley of stability $Z-Z_0$, where Z_0 is the Z of the beta stable isotope with the same mass, provides a better description of data. These prescriptions, in particular the $Z-Z_0$ dependence, would have strong implications for nuclei far from the stability line. Such effects could be relevant also in nuclei relatively close to the valley of stability.

Therefore, it is of interest to look for isospin effects in nuclei at higher excitation energies as those involved in nuclear reactions. In particular, the inclusion of isospin dependence in the level density expression could account for the failure of the statistical model in reproducing experimental evaporative particle multiplicity in some reactions[5].

In this framework, we have undertaken the study of the nuclear level density of nuclei produced in the particle decay of the highly excited ^{139}Eu nucleus. For this system, statistical model calculations predict large differences in the light particle multiplicities and shape of spectra, when the $Z-Z_0$ dependence is used with respect to the results obtained without isospin. We will report here preliminary results regarding the measured evaporative alpha particle emission from the system ^{139}Eu at an excitation energy of 90 MeV.

2. Experimental details

The ^{139}Eu nucleus has been produced by the 180 MeV ^{32}S + ^{107}Ag reaction. A ^{32}S beam delivered by the XTU-Tandem accelerator of Laboratori Nazionali di Legnaro was impinging on a target of 300 μg/cm^2 thickness.

We detected light charged particles emitted in the evaporation channel by the telescopes of the 8πLP apparatus [6]. The detector system is formed by 250 telescopes made by a first stage Si detectors followed by a second stage of CsI detectors geometrically arranged in two sections: the Ball and the Wall. The Ball and Wall detectors cover angular ranges from 34° to 165° and from 2° to 24°, respectively. Ball and Wall detectors have angular openings of 20° and 5°, respectively.

The evaporation residues were measured by four detectors placed symmetrically around the beam axis : two of them (up, and down) in the vertical plane and the other two (right and left) in the horizontal plane at an average angle of about 4°, 60 cm away from the target, each subtending a solid angle of about 0.8 msr. Each detector is composed by a sequence of two parallel plate avalanche counters (PPAC's) separated by an absorber. Evaporation residues are identified by the Time Of Flight (TOF) technique taking the stop signal by the first PPAC. The second PPAC was used as a veto to the elastic scattering events.

We have analysed the coincidences between particles, detected in the Ball and Wall sections, and evaporation residues detected in the PPAC. From these data multiplicity spectra of protons and alpha particles evaporated from the compound nucleus were extracted. By integrating spectra, differential multiplicity distributions were obtained. Absolute angular multiplicity distributions of alpha particles were obtained by the particles detected in 17 telescopes placed in the vertical plane in coincidence with evaporation residues detected by the up and down PPACs for the in-plane one and by the left PPAC for the out-of-plane one, respectively. These multiplicities were obtained by the number of coincident events for each telescope divided by the solid angle of the detector and by the number of the evaporation residues detected in the PPAC in single mode. The out-of-plane distribution is plotted as a function of the detector angle, taken anti-clockwise, with respect to the beam direction. The in-plane angular distribution is plotted as a function of the correlation angle between the particle detector and the up or down PPAC's.

In order to fix the value of the maximum angular momentum for the evaporative decay, to be entered in the statistical code, the corresponding evaporation residue cross section was measured in a dedicated experiment carried at LNL. For this purpose the incoming beam was deflected out by an electrostatic separator[7]. Evaporation residues focused at small angles were detected and further discriminated by residual beamlike particles using an E - TOF telescope based on microchannel plates and a silicon detector. The measured cross section resulted to be 794 $\pm$ 15 mb corresponding to a sharp cut-off maximum angular momentum for the residue production $\ell_{ER} = 58\ \hbar$.

3. Statistical model analysis and results

Theoretical calculations have been carried out using the MonteCarlo code LILITA_N97 [8] implemented with the inclusion of the N-Z or Z - Z_0 isospin dependence, as proposed in ref.[4]. The code uses as standard prescription the Fermi-gas level density formula with spin dependence. Transmission coefficients for spherical nuclei derived by *a)* Optical Model (OM)[9] and *b)* Ingoing Wave Boundary Condition Model (IWBCM)[10] were used. To compare calculations with the experimental results, simulated events were filtered by a program which takes into account the response function of the 8πLP apparatus.

Calculations were performed with No Isospin, N-Z and Z-Z_0 prescriptions using the parameter values extracted by a best fit to experimental data, performed by Al-Quraishi et al.[4]. In particular, the used expressions are:

$$a = 0.1097\ A\,, \tag{1}$$

$$a = 0.1110\ A\ exp[-0.000641(N-Z)^2]\,, \tag{2}$$

$$a = 0.1138\ A\ exp[-0.0493(Z-Z_0)^2]\,, \tag{3}$$

for No Isospin, N-Z and Z-Z_0 prescriptions, respectively. Z_0 is calculated by the expression derived in ref.[4].

From the simulation we extracted alpha particle multiplicities, angular distributions and spectra. Calculations performed using OM transmission coefficients gave much better agreement with experimental data with respect to the results obtained using the IWBCM transmission coefficients. Therefore, we will present here only the results obtained by Optical Model. Values of alpha particle multiplicities of 0.81, 0.88 and 1.07 were obtained for the No Isospin, N-Z and Z-Z_0 prescriptions,respectively.

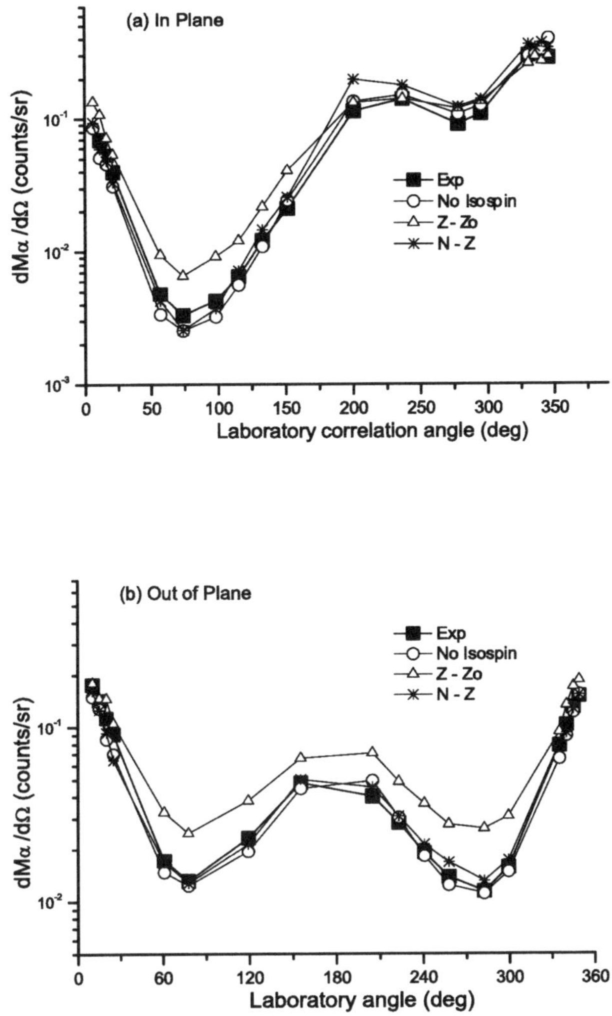

Figure 1. Comparison between experimental and theorethical angular multiplicity distributions for alpha particles: (a) in-plane (b) out-of-plane.

In the figures 1a and 1b we present the alpha in-plane and out-of-plane angular multiplicity distributions.

In figure 2 the multiplicity spectra of alpha particles for a detector placed at backward angle in the out-of-plane angle distribution, namely at

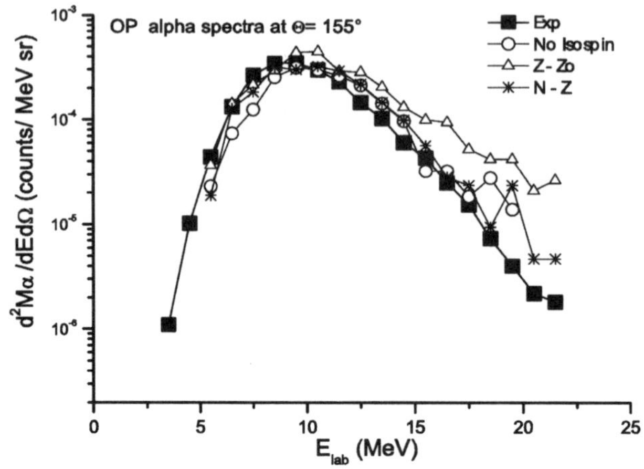

Figure 2. Experimental and theorethical multiplicity alpha particle spectra at a forward angle.

155°, is shown as an example.

As it can be seen, the prediction by the Z-Z_0 prescription fails to reproduce the data. In particular, the out of plane distribution is more significant different from the experimental data. Isospin independent and N-Z dependence reasonably well reproduce the data. The N-Z dependence does not produce a significant difference with respect to the No Isospin recipe.

Based on the analysis of the evaporative alpha decay, we can draw the preliminary conclusion that isospin effects are not present or are negligible in this system at this excitation energy. Analysis of proton data is under way to confirm this finding.

4. Perspectives

The effect of the isospin in the level density should be further tested in the evaporative charged particle emission on other systems especially at lower excitation energy and involving more exotic nuclei as those produced by radioactive beams available today. In the latter case the isospin effects would be larger. In addition the selection of specific decay channels could enhance the sensitivity.

References

1. H.A. Bethe, *Phys. Rev.* **50**,332 (1936).
2. J.R. Huizenga and L.G. Moretto, *Ann. Rev. Nucl. Sci.* **22**,427(1972).
3. M.R. Bhat, in *Evaluated Nuclear Structure Data File, Nuclear Data For Science and Technology* edited by S. M. Qaim, Springer Verlag, Berlin 1992 p.817
4. S.I. Al-Quraishi, S.M. Grimes, T.N. Massey and D.A. Resler, *Phys. Rev.* **C63**, 065803 (2001) 56.
5. J. M. Alexander et al., *Symposium on Nuclear Dynamics and Nuclear Disassembly*; American Chemical Society; Dallas, Texas, April 10-14, 1989,p.211, J. B. Natowitz, ed., World Scientific (1990)
6. E. Fioretto, M. Cinausero, M. Giacchini, M. Lollo, G. Prete, R. Burch, A. Brondi, G. La Rana, R. Moro, E. Vardaci, A. Ordine, A. Boiano, A. Zaghi, N. Gelli and F. Lucarelli, *IEEE Trans. Nuc. Sci.* **44**, 1017 (1997).
7. S. Beghini, C. Signorini, S. Lunardi, M. Morando, G. Fortuna, A.M. Stefanini, W. Meczynski and R. Pengo, *Nucl. Inst. Meth. Phys. Res.* **A239**, 585 (1985).
8. Lilita_N97 is an extensively modified version of the original Lilita program made by J. Gomez del Campo and R.G. Stockstad, Oak Ridge National Laboratory, (Rep. No TM7295, 1981 unpublished)
9. J.R. Huizenga and G. Igo, *Nucl. Phys.* **29**, 462 (1961).
10. S. Landowne and S.C. Piper, *Phys. Rev.* **C29**, 1352 (1984).

ISOSPIN IN REACTION DYNAMICS. THE CASE OF DISSIPATIVE COLLISIONS AT FERMI ENERGIES

M. DI TORO

Laboratori Nazionali del Sud INFN, Phys.Astron. Dept. Catania University
Via S. Sofia 62,
I-95123 Catania, Italy
E-mail: ditoro@lns.infn.it

A key question in the physics of unstable nuclei is the knowledge of the EOS for asymmetric nuclear matter (ANM) away from normal conditions. We recall that the symmetry energy at low densities has important effects on the neutron skin structure, while the knowledge in high densities region is crucial for supernovae dynamics and neutron star properties. The *only* way to probe such region of the isovector EOS in terrestrial laboratories is through very dissipative collisions of asymmetric (up to exotic) heavy ions from low to relativistic energies. A general introduction to the topic is firstly presented. We pass then to a detailed discussion on the $neck - fragmentation$ process as the main dissipative mechanism at the Fermi energies and to the related isospin dynamics. From Stochastic Mean Field simulations the isospin effects on all the phases of the reaction dynamics are thoroughly analysed, from the fast nucleon emission to the mid-rapidity fragment formation up to the dynamical fission of the *spectator* residues. Simulations have been performed with an increasing stiffness of the symmetry term of the EOS. Some differences have been noticed, especially for the fragment charge asymmetry. New isospin effects have been revealed from the correlation of fragment asymmetry with dynamical quantities at the freeze-out time. A series of isospin sensitive observables to be further measured are finally listed.

1. Introduction

While we are planning second and third generation facilities for radioactive beams our basic knowledge of the symmetry term of the EOS is still extremely poor. Effective interactions are obviously tuned to symmetry properties around normal conditions and any extrapolation can be quite dangerous. Microscopic approaches based on realistic NN interactions, Brueckner or variational schemes, or on effective field theories show a rather large variety of predictions. While all the potential symmetry energies obviously cross at normal density ρ_0, quite large differences are present for values, slopes and curvatures in low density and particularly in high density regions. Moreover even at the relatively well known "crossing point" at normal density the various effective forces are presenting controversial predictions for the momentum dependence of the fields acting on the nucleons and consequently for the splitting of the neutron/proton effective masses, of large interest for nuclear structure and dynamics.

In the recent years under the stimulating perspectives offered from nuclear astrophysics and from the new Radioactive Ion Beam (RIB) facilities a relevant activity has started in the field of the isospin degree of freedom in heavy ion reactions, see [1,2,3]. Here we quickly review the field trying to pin down the most interesting theory questions and eventually the related key observables.

The slope and curvature of the density dependence of the symmetry term around saturation is related to physics properties of exotic nuclei, bulk densities, neutron distributions and monopole frequencies. The momentum dependence of the interactions in the isovector channel can be analysed discussing the effects on the energy-slope of the Lane Potential at normal density and on the symmetry field seen by high momentum nucleons [4,5].

The study of symmetry properties at low density is important for the Liquid-Gas Phase Transition in Asymmetric Matter. A dynamical approach of great relevance since it leads to the space-time properties of the unstable normal modes which give rise to fragments inside the extended spinodal boundary. Characteristic coupled dispersion relations must be solved. We remind that the coupling of the isoscalar and isovector collective response is well known for stable modes [6,7] in neutron rich exotic nuclei, with transition densities that show a mixed nature. We can extend the same results to the unstable responses of interest in fragmentation reactions. The mixed nature of the unstable density oscillations will naturally lead to the *isospin distillation* effect, i.e. a different *concentration* (N/Z) between the liquid and the gas phase [8,9,10,11]. We can expect that quantitatively this effect will be dependent on properties of the symmetry energy in very dilute matter, well below the saturation point.

The early emission of particles in heavy ion collisions is relevant for understanding both the reaction dynamics and the mechanism of particle production. The density behavior of the symmetry potential largely influences the value of the pressure that is governing the early reaction dynamics [12]. Therefore we can predict important symmetry effects on transport properties (fast particle emission, collective flows) in the asymmetric NM that will be probed by Radioactive Beam collisions at Fermi and intermediate energies. Different parametrizations in the momentum dependence of the isovector channel can be tested, in particular the transport effect of the sign of the n/p effective mass splitting $m_n^* - m_p^*$ in asymmetric matter at high baryon and isospin density [13,5].

The Fermi energy domain represents the transition region between a dynamics mainly driven by the mean-field, below $15 - 20\ AMeV$, and the one where the nucleon-nucleon collisions play a central role, above $100\ AMeV$. We can have then a very rich variety of dissipative reaction mechanisms, with related different isospin dynamics. One of the new distinctive features is the enhanced production of Intermediate Mass Fragments ($IMF, 3 \leq Z \leq 20$). We analyse then with great detail the isovector channel effect on the onset of the fragmentation mechanism and on the isotopic content of the produced fragments. In the reaction simulations it is essential the use of a *Stochastic Transport Approach* since fluctuations, instabilities and dynamical branchings are very important in this energy range [14,15]. With the centrality of the collision we can have different scenarios for fragment production, from the growing instabilities of dilute matter in central reactions to the cluster formation at the interface between low and normal density regions in semicentral collisions. We can then expect a large variety of isospin effects, probing different regions of the symmetry energy below saturation.

The traditional approach to nuclear physics starts from non-relativistic formalisms in which non-nucleonic degrees of freedom are integrated out giving nucleon-nucleon potentials. Then nuclear matter is described as a collection of quantum nonrelativistic nucleons interacting through an istantaneous effective potential. Although this approach has had a great success, a more appropriate set of degrees of freedom consists of strongly interacting effective hadron fields, mesons and baryons. These variables are the most efficient in a wide range of densities and temperatures and they are the degrees of freedom actually observed in experiments, in particular in Heavy Ion Collisions at intermediate energies. Moreover this framework appears in any case a fundamental "Doorway Step" towards a more microscopic understanding of the nuclear matter. Relativistic contributions to the isospin physics for static properties and reaction dynamics deserve an appropriate discussion.

The QHD (Quantum-Hadro-Dynamics) effective field model represents a very successful attempt to describe, in a fully consistent relativistic picture, equilibrium and dynamical properties of nuclear systems at the hadronic level [16,17]. The same isospin physics described in detail at the non-relativistic level can be naturally reproduced. Moreover some new genuine relativistic effects can be revealed, due to the covariant structure of the effective interactions. One of the main points of the discussion is the relevance of the coupling to a scalar isovector channel, the effective $\delta[a_0(980)]$ meson, not considered in the usual nuclear structure studies [18,19]. Particular attention is devoted to the expectations for the splitting of the nucleon effective masses in asymmetric matter, with a detailed analysis of the relationship between the Dirac masses of the effective field approach and the Schrödinger masses of the non-relativistic models. A relativistic Dirac-Lane potential is deduced, which shows a structure very similar to the Lane potential of the non-relativistic optical model, but now in terms of isovector self-energies and coupling constants [5].

The Relativistic Mean Field (RMF) approximation is allowing more physics transparent results, often even analytical. We must always keep a close connection to the more microscopic Dirac-Brueckner-Hartree-Fock ($DBHF$) approaches, in their extension to asymmetric matter, [20,21,22,23] It is well known that correlations are naturally leading to a density dependence of the coupling constants, see ref.[21] for the $DBHF$ calculations and ref.[24] just for the basic Fock correlations. Within the RMF model we can get a clear qualitative estimation of the contribution of the various fields to the nuclear dynamics. The price to pay is that when we try to get quantitative effects we are forced to use different sets of couplings in different baryon density regions. In this case we can use the $DBHF$ results as guidelines [25,26,27].

Results for relativistic Heavy Ion Collisions (HIC) in the $AGeV$ beam energy region, for collective flows, charged pion production and isospin stopping, are presented in refs. [25,26,27]. We recall that intermediate energy HIC's represent the only way to probe in terrestrial laboratories the in-medium effective interactions far from saturation, at high densities as well as at high momenta. Within a relativistic transport model it is shown that the isovector-scalar δ-meson, which affects the high density behavior of the symmetry term and the nucleon effective mass splitting, influences the isospin dynamics. The effect is largely enhanced by a relativistic mechanism related to the covariant nature of the fields contributing to the isovector channel. The possibility is emerging of a direct measurement of the Lorentz structure of the effective nuclear interaction in the isovector channel. Very sensitive quantities appear to be the elliptic flow, related to the time scale of the particle emissions, and the isospin transparency in central collisions.

2. Neck Dynamics at the Fermi Energies: Theory Survey

The possibility of observation of new effects, beyond the Deep-Inelastic binary picture, in fragment formation for semicentral collisions with increasing energy was advanced on the basis of the reaction dynamics studied with transport models [28,29,30]. The presence of a time matching between the instability growth in the dilute overlap zone and the expansion-separation time scale was suggesting the observation of mean field instabilities first at the level of anomalous widths in the mass/charge/... distributions of Projectile-Like or Target-Like (PLF/TLF) residues in binary events, then through a direct formation of fragments in the $neck - region$, [31,4]. It is clear that in the transport simulations stochastic terms should be consistently built in the kinetic equations in order to have a correct description of instability effects. Stochastic Mean Field approaches have been introduced, that well reproduce the presently available data and reveal a large predictive power, [32,33,34,5].

In conclusion at the Fermi energies we expect an interplay between binary and $neck - fragmentation$ events, where Intermediate Mass Fragments (IMF, in the range $3 \leq Z \leq 10$) are directly formed in the overlapping region, roughly at mid-rapidity in semicentral reactions. The competition between the two mechanisms is expected to be rather sensitive to the nuclear equation of state, in particular to its compressibility that will influence the interaction time as well as the density oscillation in the $neck-region$. In the case of charge asymmetric colliding systems the poorly known stiffness of the symmetry term will also largely influence the reaction dynamics. An observable sensitive to the stiffness of the symmetry term can be just the relative yield of incomplete fusion vs. deep-inelastic vs neck-fragmentation events [35]. Moreover for n-rich systems in the asy-soft case we expect more interaction time available for charge equilibration. This means that even the binary events will show a sensitivity through a larger isospin diffusion. At variance, in the asy-stiff case the two final fragments will keep more memory of the initial conditions.

Systematic transport studies of isospin effects in the neck dynamics have been performed so far for collisions of Sn-Sn isotopes at 50AMeV [32,33], Sn-Ni isotopes at 35AMeV [34,5] and finally Fe-Fe and Ni-Ni, mass 58, at 30 and 47 AMeV [36].

We show in Fig. 1 the density contour plots of a neck fragmetation event at $b = 6fm$, for the reaction ^{124}Sn+^{124}Sn at $50AMeV$, obtained from the numerical calculations based on the stochastic transport approach described before, [33].

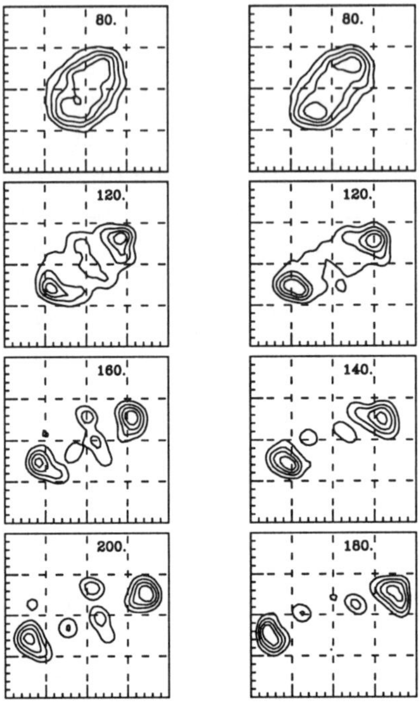

Figure 1. $^{124}Sn +^{124} Sn$ collision at $50AMeV$: time evolution of the nucleon density projected on the reaction plane. Left column: $b = 4fm$. Right column: $b = 6fm$.

We clearly see neck-fragmentation events even at lower energies, now mostly ternary. For the system ^{124}Sn+^{64}Ni at $35AMeV$ ($b = 6fm$), neck instabilities favour the appearance of $NIMF$'s, after $150fm/c$, in a variety of places and ways as can be seen by looking at Fig.2, [34]. For four events, we illustrate two characteristic stages, the early phase of the fragment formation process and the configuration close to freeze-out.

3. Neck Dynamics at the Fermi Energies: Experimental Survey

It is now quite well established that the largest part of the reaction cross section for dissipative collisions at Fermi energies goes through the *Neck Fragmentation* channel, with IMFs directly produced in the interacting zone in semiperipheral collisions on very short time scales. We can expect different isospin effects for this new fragment formation mechanism since cluster are formed still in a dilute asymmetric matter but always in contact

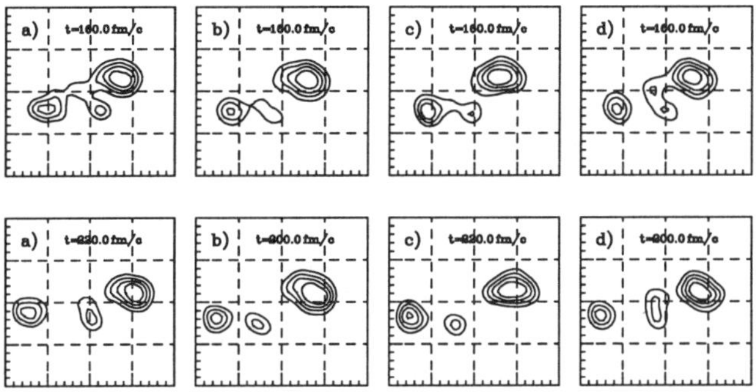

Figure 2. Density contour plots for $^{124}Sn+^{64}Ni$ at $35AMeV$ $(b = 6fm)$. Early stage of fragment formation (top) and same events close to the freeze-out (bottom) for four ternary cases a), b), c), d) in neck fragmentation.

with the regions of the Projectile-Like and Target-Like remnants almost at normal densities.

A first evidence of this new dissipative mechanism was suggested at quite low energies, around $19AMeV$, in semicentral $^{100}Mo+^{100}Mo$, $^{120}Sn+^{120}Sn$ reactions [37,38]. A transition from binary, deep-inelastic, to ternary events was observed, with a fragment formed dynamically that influences the fission-like decay of the primary projectile-like (PLF) and target-like (TLF) partners. Similar conclusions were reached in [39]. Consistent with the dynamical scenario was the anisotropic azimuthal distribution of IMF's. In fact the IMF *alignement* with respect to the (PLF^*) velocity direction has been one clear property of the "neck-fragments" first noticed by Montoya et al.[40] for $^{129}Xe+^{63,65}Cu$ at $50AMeV$.

A rise and fall of the neck mechanism for mid-rapidity fragments with the centrality, with a maximum for intermediate impact parameters $b \simeq \frac{1}{2}b_{max}$, as observed in [41], suggests the special physical conditions required. The size of the participant zone is of course important but it appears that a good time matching between the reaction and the neck instabilities time-scales is also needed, as suggested in refs. [28,31]. In fact a simultaneous presence, in noncentral collisions, of different IMF production mechanisms at midrapidity was inferred in several experiments [42,43,44,45,46,47,48,49].

We can immediately predict an important isospin dependence of the neck dynamics, from the presence of large density gradients and from the possibility of selecting various time-scales for the fragment formation. The first evidences of isospin effects in neck fragmentation were suggested by Dempsey et al. in [50] from semiperipheral collisions of the systems

$^{124,136}Xe+ {}^{112,124}Sn$ at $55AMeV$, where correlations between the average number of IMF's, N_{IMF}, and neutron and charged particle multiplicities were measured. The variation of the relative yields of $^{6}He/^{3,4}He$, $^{6}He/Li$ with v_{par} for several Z_{PLF} gates shows that the fragments produced in the midvelocity region are more neutron rich than are the fragments emitted by the PLF. Enhanced *triton* production an midrapidity was considered in ref.[41], and more recently in [51], as an indication of a neutron neck enrichement.

P.M. Milazzo et al. [52,53] analyzed the the IMF parallel velocity distribution for $^{58}Ni+^{58}Ni$ semiperipheral collision at $30AMeV$. The two-bumps structure for IMFs with $5 \leq Z \leq 12$, located around the center of mass velocity and close to the quasiprojectile (PLF^{*}) source respectively, was explained assuming the simultaneous presence of two production mechanisms: the statistical disassembly of an equilibrated PLF^{*} and the dynamical fragmentation of the participant region. The average elemental event multiplicity $N(Z)$ exhibits a different trend for the two processes: more n-rich in the mid-rapidity source.. This experiment has a particular importance since isospin effects were clearly observed, *in spite of the very low inital asymmetry.* The same reactions have been recently studied at the Cyclotron Inst. of Texas $A\&M$ at various beam energies with measurements of the correlations of fragment charge/mass vs. dynamical observables (emission angles and velocities) [54,55].

Isospin Diffusion

The isospin equilibration appears of large interest even for more peripheral collisions, where we have shorter interaction times, less overlap and a competition between binary and neck-fragmentation processes. The low density neck formation and the preequilibrium emission are adding essential differences with respect to what is happening in the lower energies regime. Tsang et al. [56] have probed the isospin diffusion mechanism for the systems $^{124}Sn+ {}^{112}Sn$ at $E = 50AMeV$ in a peripheral impact parameter range $b/b_{max} > 0.8$, observing the isoscaling features of the light isotopes $Z = 3-8$ emitted around the projectile rapidity. An incomplete equilibration has been deduced. The value of the isoscaling parameter $\alpha = 0.42 \pm 0.02$ for $^{124}Sn+ {}^{112}Sn$ differs substantialy from $\alpha = 0.16 \pm 0.02$ for $^{112}Sn+ {}^{124}Sn$. The isospin imbalance ratio [57], defined as

$$R_i(x) = \frac{2x - x^{124+124} - x^{112+112}}{x^{124+124} - x^{112+112}} \tag{1}$$

($i = P, T$ refers to the projectile/target rapidity measurement, and x is an isospin dependent observable, here the isoscaling α parameter) was estimated to be around $R_P(\alpha) = 0.5$ (vs.$R_P(\alpha) = 0.0$ in full equilibration). This quantity can be sensitive to the density dependence of symmetry energy term since the isospin transfer takes place through the lower density neck region.

4. Neck Observables

Properties of Neck-Fragments, mid-rapidity IMF produced in semicentral collisions: correlations between N/Z, alignement and size

The alignement between $PLF - IMF$ and $PLF - TLF$ directions represents a very convincing evidence of the dynamical origin of the mid-rapidity fragments produced on short time scales [34]. In fact a very selective correlation is provided by the anisotropy in the fragment emission, measured by the in-plane azimuthal angle, Φ_{plane}, defined as the angle between the projection of the $PL - IMF$ scission axis onto the reaction plane and the separation axis between PLF and TLF, [38]. A $|\Phi_{plane}| \simeq 0$ collects events corresponding to asymmetric "fissions" of the PL-system very aligned along the outgoing $PL - TL$ separation axis. At variance a statistical fission dominance would correspond to a flat Φ_{plane} behavior. The form of the Φ_{plane} distributions (centroid and width) can give a direct information on the fragmentation mechanism [58].

Time-scale measurements

The estimation of time scales for fragment formation from velocity correlations appears to be a very exciting possibility [49,59]. With a good event by event detection of the TL/PL residues we can measure the violations of the Viola systematics for the $TLF - IMF$ and $PLF - IMF$ systems which tell us how much the $IMFs$ are uncorrelated to the *spectator* remnants [61].

For each Neck-IMF we can evaluate the ratios $r = v_{rel}(PLF)/v_{viola}(PLF)$, ($r1 = v_{rel}(TLF)/v_{viola}(TLF)$). In Figure 3 we plot $r1$ against r for each $NIMF$ for the $Sn + Ni$ reaction [34]. We call such a representation a $Wilczynski - 2$ plot [61]. The solid lines represent the loci of the PL-($r = 1$) and TL-($r1 = 1$) fission events respectively. The values ($r, r1$) appear simultaneously larger than 1 suggesting a weak $NIMF$ correlation with *both* PLF and TLF, in contrast to a statistical fission mechanism. The process has some similarities with the participant-spectator scenario. However the dynamics appear much richer than in the simple sudden abrasion model, where the locus of the $r - r1$ correlation should be on mainly the bisectrix. Here the wide distributions of Fig.3 reveal a broad range of fragment velocities, typical of the instability evolution in the neck region that will lead to large dynamical fluctuations on $NIMF$ properties.

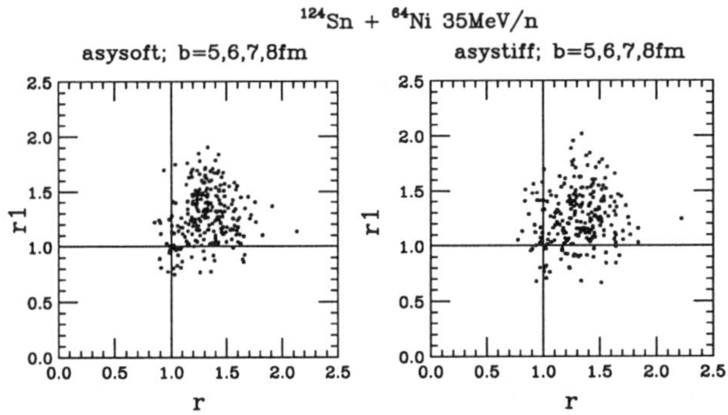

Figure 3. Wilczynski-2 Plot: correlation between deviations from Viola systematics, see text. Results are shown for two $asy - EOS$.

The Φ_{plane} distributions, corresponding to the same $NIMF$ events analysed before, are covering a quite limited angular window, close to the full alignment configuration, $\Phi_{plane} = 0°$. The distribution becomes wider when we approach the two $r, r1 = 1$ lines of the Fig.3.

With appropriate cuts in the velocity correlation plots and Φ_{plane} distributions we can follow the properties of clusters produced from sources with a "controlled" different degree of equilibration. We can figure out a continous transition from fast produced fragments via neck instabilities to clusters formed in a dynamical fission of the Projectile(Target) residues up to the evaporated ones (statistical fission). Along this line it would be even possible to disentangle the effects of volume and shape instabilities. The isospin dynamics will look different in the various scenarios and rather dependent on the symmetry term of the EOS.

A more promising observable seems to be the isotopic content of the Neck-IMF. For the three $asy - EOSs$ introduced before we plot in Fig.4 the average isotopic composition I of the $NIMF$'s as a function of the PLF r-deviation from Viola systematics. At all impact parameters clear

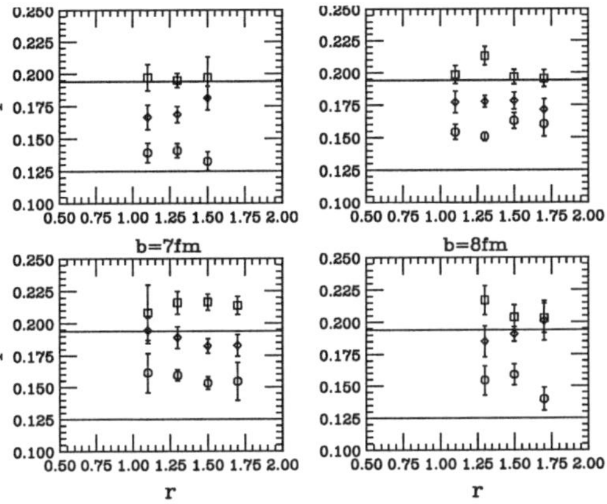

Figure 4. $124Sn + 64Ni$ at $35AMeV$. $NIMF$ isospin content for $asysoft$ (circles), $asystiff$ (rombs) and $superasystiff$ (squares) EOS as a function of r-deviation from Viola systematics, at impact parameters from $b = 5fm$ to $8fm$. The two solid lines represent the mean asymmetries of the projectile (top) and target (bottom).

differences are evident. The average asymmetry does not depend strongly on r. We notice however that it increases with the stiffness of the symmetry potential around and below saturation. The *superasystiff* parametrization, i.e. with an almost parabolic increasing behavior around ρ_0, produces systematically more neutron rich $NIMF$s. This effect is clearly due to a different neutron/proton migration at the interface between PL/TL "spectator" zone of normal density and the dilute neck region where the $NIMF$s are formed. The Wilczynski-2 plot will help to make the selections. We have to remind that all the results presented here refer to properties of primary (excited) fragments. The neutron excess signal appears likely to be washed out from later evaporation decays. A reconstruction of the primary fragments with neutron coincidence measurements would be very important.

Isospin Dynamics

Isospin effects on the reaction dynamics and $Isospin\ Migration$: an interesting neutron enrichment of the overlap ("neck") region is expected, due to the neutron migration from higher (spectator) to lower (neck) density regions. This effect is also nicely connected to the slope of the symmetry energy. Neutron and/or light isobar measurements in different rapidity regions appear important. Moreover, moving from mid- to "spectator"-rapidities an increasing hierarchy in the mass and N/Z of the fragments is

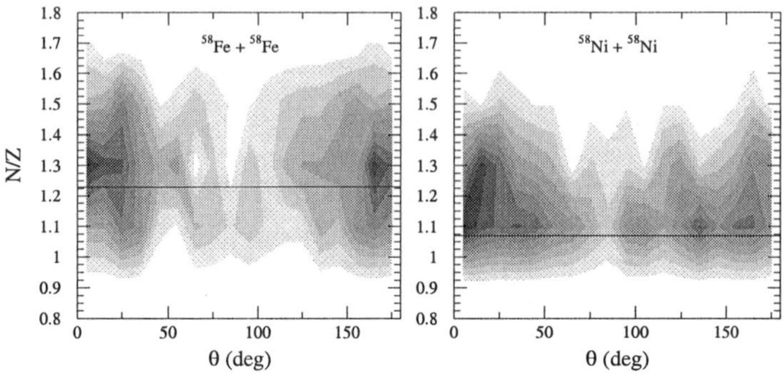

Figure 5. Fe, Fe vs. Ni, Ni reactions at $47AMev$ ($b = 4fm$): Contour plots of the N/Z distribution versus the emission angle.

expected [36]. Some experimental evidences are in ref.[48]. An interesting related observable is the corresponding angular correlation due to the driving force of the Projectile(Target)-like partners [36]. The analysis of the N/Z distribution versus the emission angle reveals that larger fluctuations are present close to forward and backward angles, suggesting that IMF's that are more correlated to the spectator matter may become more neutron-rich, see Fig.5 for the $Fe, Fe/Ni, Ni$ systems at $47AMeV$ [36].

We stress again that the $neck - IMF$ always present a neutron enrichment, *even in the case of a n-poor system.* The latter paradox is due to the pre-equilibrium isospin dynamics. In fact, as anticipated above, due to pre-equilibrium, the system will loose some protons and acquire a N/Z larger than the initial one. Then, before the di-nuclear system reseparates, the neutron excess is transferred to the neck region that is at low density.

Isospin Diffusion

Measure of charge equilibration in the "spectator" region in semicentral collisions: *Imbalance Ratios* for different isospin properties. Test of the interplay between concentration and density gradients in the isospin dynamics [5,62,63]. For the reasons noted before we expect to see a clear difference in the isospin diffusion between binary (deep-inelastic like) and neck-fragmentation events.

We report its dependence on the interaction time (impact parameter) for asysoft (squares) and asysuperstiff (circles) EOS in Fig.6, for the Sn, Sn case [5,62]. A good degree of isospin equilibration corresponds to a "convergence" to 0 of both R_P (upper curves) and R_T (lower curves). From our results we conclude that an asystiff-like EOS provides a better agreement to the experimental observations, shown as arrows in Fig.6, [56]. In fact in the MSU experiment there is no particular selection on binary events. We expect that the presence of events with production of Neck-$IMFs$ will induce an apparent larger isospin equilibration since more neutrons will migrate to the neck region (and viceversa for protons), as discussed before.

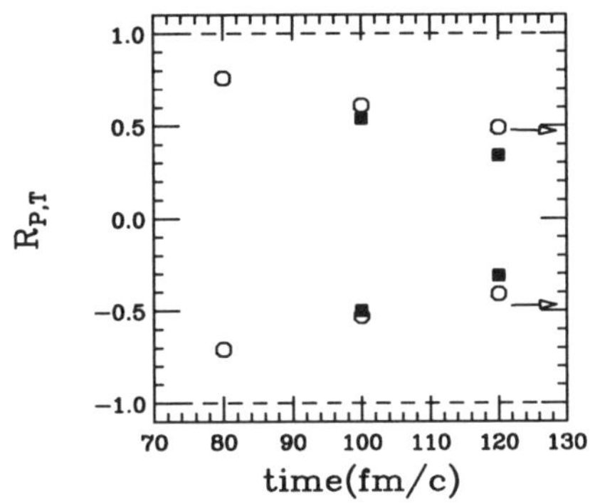

Figure 6. $^{124}Sn+^{112}Sn$ 50 $AMeV$ collision: interaction time evolution of the projectile (upper) and target (lower) isospin imbalance ratios. Squares: *Asysoft EOS*. Circles: *Superasystiff*. The arrows correspond to the data of ref.[56].

A clear difference between the two equations of state is evident especially for the longer interacting time, $b = 8fm$ case. Smaller values of the isospin imbalance ratios for asysoft EOS point towards a faster equilibration rate. In ref. [56] a possible explanation was proposed related to the observation that below normal density the asysoft EOS has a larger value of symmetry energy. Therefore an enhanced isospin equilibration has to be triggered if the diffusion takes place at lower density. In fact the mechanism of charge equilibration is more complicated at these energies due to reaction dynamics (fast particle emissions, density gradients, etc.), with interesting compensation effects.

"Pre-equilibrium" emissions

As already remarked, the isospin content of the fast particle emission can largely influence the subsequent reaction dynamics, in particular the isospin transport properties (charge equilibration, isospin diffusion). We can reach the paradox of a detection of isospin dynamics effects in charge symmetric systems.

Finally the simultaneous measurements of properties of fast nucleon emissions and of the neck dynamics can even shed lights on the very controversial problem of the isospin momentum dependence [5,63].

Outlook

We stress the richness of the phenomenology and nice opportunities of getting several cross-checks from completely different experiments. Apart the interest of this new dissipative mechanism and the amazing possibility of studying properties of fragments produced on an almost continous range of time scales, we remark the expected dependence on the isovector part of the nuclear EOS. From transport simulations we presently get some indications of "asy-stiff" behaviors, i.e. increasing repulsive density dependence of the symmetry term, but not more fundamental details. Moreover, all the available data are obtained with stable beams, i.e. within low asymmetries.

Acknowledgements

This report is deeply related to ideas and results reached in very pleasant and fruitful collaborations with extremely nice people: V.Baran, M.Colonna, G.Fabbri, G.Ferini, Th.Gaitanos, V.Greco, R.Lionti, B.Liu, S.Maccarone, F.Matera, M.Zielinska-Pfabe', J.Rizzo, L.Scalone and H.H.Wolter. I have learnt a lot from all of them in physics as well as in human relationships.

References

1. B.-A.Li, C.M.Ko, W.Bauer, Int.J.Mod.Phys. E7 (1998) 147.
2. *Isospin Physics in Heavy-ion Collisions at Intermediate Energies*, Eds. Bao-An Li and W. Udo Schröder, Nova Science Publishers (2001, New York).
3. M.Di Toro, V.Baran, M.Colonna, V.Greco, S.Maccarone, M.Cabibbo, Eur.Phys.J. A13 (2002) 155
4. M.Di Toro et al., Progr.Part.Nucl.Phys. 42 (1999) 125-136.
5. V.Baran, M.Colonna, V.Greco, M.Di Toro, *Reaction Dynamics with Exotic Nuclei*, arXiv:nucl-th/0412060, Physics Reports (2005) in press
6. Contributions of P.G.Reinhard, H.Sagawa, M.Di Toro, G.Colo', E.G.Lanza, in *GR98*, Nucl.Phys. A649 (1999) 1c-446c.
7. I.Hamamoto, Phys.Rev. C60 (1999) 031303.
8. H.Müller, B.D.Serot, Phys.Rev. C52 (1995) 2072.
9. V.Baran, M.Colonna, M.Di Toro, V.Greco, Phys.Rev.Lett. 86 (2001) 4492.
10. M.Colonna, Ph.Chomaz, S.Ayik, Phys.Rev.Lett. 88 (2002) 122701.
11. J.Margueron, P.Chomaz, Phys. Rev. C67 (2003) 041602.
12. P.Danielewicz, Nucl.Phys. A685 (2001) 368c.
13. J.Rizzo, M.Colonna, M.Di Toro, V.Greco, Nucl.Phys. A732 (2004) 202.
14. M.Colonna et al., Nucl.Phys. A642 (1998) 449.
15. M.Colonna et al., Nucl.Phys. A742 (2004) 337
16. J.D.Walecka, Ann.Phys.(N.Y.) 83 (1974) 491.
17. B.D.Serot, J.D.Walecka, Int.J.Mod.Phys. E6 (1997) 515.
18. B.Liu, V.Greco, V.Baran, M.Colonna, M.Di Toro, Phys.Rev. C65 (2002) 045201.
19. V.Greco, M.Colonna, M.Di Toro, F.Matera, Phys.Rev. C67 (2003) 015203.
20. C.H.Lee, T.T.S.Kuo, G.Q.Li, G.E.Brown, Phys.Rev. C57 (1998) 3488.
21. F.de Jong, H.Lenske, Phys.Rev. C57 (1998) 3099.
22. F.Hofmann, C.M.Keil, H.Lenske, Phys.Rev. C64 (2001) 034314 .
23. E.N.E.van Dalen, C.Fuchs, A.Faessler, Nucl.Phys. A744 (2004) 227.
24. V.Greco, M.Colonna, M.Di Toro, G.Fabbri, F.Matera, Phys.Rev. C64 (2001) 045203.
25. V.Greco, M.Colonna, M.Di Toro, T.Gaitanos, H.H.Wolter and V.Baran, Phys.Lett. B562 (2003) 215;
    V.Greco, Ph.D.Thesis 2002.
26. T.Gaitanos et al., Nucl.Phys. A732 (2004) 24.
27. T.Gaitanos, M.Colonna, M.Di Toro, H.H.Wolter, Phys.Lett. B595 (2004) 209
28. G.F.Bertsch, Phys.Lett. B141 (1984) 9.
29. M.Colonna, N.Colonna, A.Bonasera, M.Di Toro, Nucl.Phys. A541 (1992) 295.
30. L.G. Sobotka, Phys.Rev. C 50 (1994) 1272R.
31. M.Colonna, M.Di Toro, A.Guarnera, Nucl.Phys. A589 (1995) 160.
32. M.Di Toro et al., Nucl.Phys. A681 (2001) 426c.
33. V. Baran, M.Colonna, V.Greco, M.Di Toro, M.Zielinska Pfabé, H.H.Wolter, Nucl.Phys. A703 (2002) 603.
34. V.Baran, M.Colonna, M.Di Toro, Nucl.Phys. A730 (2004) 329.

35. M.Colonna, M.Di Toro, G.Fabbri, S.Maccarone, Phys.Rev. C57 (1998) 1410
36. R.Lionti, V.Baran, M.Colonna, M.Di Toro, *Isospin dynamics in fragmentation reactions at Fermi energies*, arXiv:nucl-th/0501012, sub. Phys.Lett.B.
37. G.Casini et al., Phys.Rev.Lett. 71 (1993) 2567.
38. A.A.Stefanini et al., Z.Phys. A351 (1995) 167.
39. J.Toke et. al., Phys.Rev.Lett. 75 (1995) 2920.
40. C.P.Montoya et. al., Phys.Rev.Lett. 73 (1994) 3070.
41. J.Lukasik et al., Phys.Rev. C55 (1997) 1906.
42. T.Lefort et al., Nucl.Phys. A662 (2000) 397.
43. E.Plagnol et al., Phys.Rev. C61 (1999) 0614606.
44. Y.Larochelle et al., Phys.Rev. C62 (2000) 051602(R).
45. L.Gingras et al., Phys.Rev. C65 (2002) 061604.
46. B.Davin et al., Phys.Rev. C65 (2002) 064614.
47. F.Bocage et al., Nucl.Phys. A676 (2000) 391.
48. J.Colin et al., Phys.Rev. C67 (2003) 064603.
49. A.Pagano et al., Nucl.Phys. A734 (2004) 504c.
50. J.F.Dempsey et al., Phys.Rev. C54 (1996) 1710.
51. G.Poggi, Nucl.Phys. A685 (2001) 296c.
52. P.M.Milazzo et al., Phys.Lett. B509 (2001) 204.
53. P.M.Milazzo et al., Nucl.Phys. A703 (2002) 466.
54. D.V.Shetty et al., Phys.Rev. C68 (2003) 021602(R).
55. D.V.Shetty et al., Phys.Rev. C70 (2004) 011601(R).
56. M.B.Tsang et al., Phys.Rev.Lett. 92 (2004) 062701
57. F.Rami et al., Phys.Rev.Lett. 84 (2000) 1120
58. E.De Filippo, A.Pagano, E.Piasecki et al. (Chimera Coll.), *Dynamical Fission in 124Sn + 64Ni Collision at 35 AMeV*, Jan. 2005 submitted
59. J. Wilczynski et al. (Chimera Collab.), *How to calibrate the time scale of emission of intermediate mass fragments*, Int.Jour.Mod.Phys. E (2005) in press
60. Contributions of V.Baran et al. and E.De Filippo et al. at the $IWM2003$ (Int.Work.on Multifragmentation, Ganil 2003)
61. We like to call the Viola-violation-correlation plot as *Wilczynski-2 Plot.* Indeed this correlation, very important to rule-out a statistical fission scenario for fragments produced at mid-rapidity, nicely emerged during hot discussions with Janusz the $LNS - INFN$, Catania. In fact this correlation represents also a *chronometer* of the fragment formation mechanism. In this sense it is the nice Fermi energy complement of the famous *Wilczynski Plot* which gives the time-scales in Deep Inelastic Collisions.
62. V.Baran, M.Colonna, M.Di Toro, Contribution to IWM2003, p.116
V.Baran, M.Colonna, M.Di Toro, M.Zielinska-Pfabe' and H.H.Wolter "Isospin Diffusion at Fermi Energies" in preparation
63. L.-W.Chen, C.M.Ko, B.-A.Li, *Determination of the stiffness of the nuclear symmetry energy from isospin diffusion*, arXiv:nucl-th/0407032, Phys.Rev.Lett. (2005) in press.

THERMODYNAMICAL ASPECTS IN HEAVY ION REACTIONS

M. BRUNO, F. CANNATA, M. D'AGOSTINO, J. DE SANCTIS, S. FABBRI, E. FUSCHINI, E. GERACI, B. GUIOT, G. VANNINI, E. VERONDINI

Dipartimento di Fisica dell'Università and INFN, Sezione di Bologna, Italy

F. GULMINELLI[§]

LPC Caen (IN2P3 - CNRS/EnsiCaen et Université, F-14050 Caen Cédex, France

PH. CHOMAZ

GANIL (DSM - CEA/IN2P3 - CNRS), F-14021 Caen Cédex, France

G. CASINI, M. CHIARI, A. NANNINI

INFN, Sezione di Firenze, Italy

S. BARLINI[†], F. GRAMEGNA, V. KRAVCHUK, A. LANCHAIS[‡], L. VANNUCCI

INFN, Laboratori Nazionali di Legnaro, Italy
[†] *and Dipartimento di Fisica dell'Università, Padova*
[‡] *and Dipartimento di Fisica dell'Università, Bologna*

A. MORONI

INFN, Sezione di Milano, Italy

A. ORDINE

INFN, Sezione di Napoli, Italy

U. ABBONDANNO, G. V. MARGAGLIOTTI

Dipartimento di Fisica dell'Università and INFN, Sezione di Trieste, Italy

---

[§]Member of the Institut Universitaire de France

The excited nuclear systems formed in heavy ion collisions can be studied from a thermodynamical point of view. Charged finite systems have different behaviors with respect to infinite ones. After experimental selection of such equilibrated systems the extraction of thermodynamic coordinates is performed. Different signals compatible with a liquid-gas phase transition have been obtained. In particular a bimodal distribution of the asymmetry between the first two heaviest fragments is presented. Abnormally large fluctuations, which in thermodynamic equilibrium are associated to a negative branch of the heat capacity give indications of a first order phase transition. Perspectives for new generation experiments are indicated.

1. Introduction

Heavy ion collisions are routinely used to explore the phase diagram of nuclear matter. In particular the fact that nuclei in their ground state behave like Fermi liquids, and at high excitation energy can be decomposed in a vapor of light particles and clusters, supports the studies of a possible phase transition in the nuclear bulk (see Fig. 1). Several signals of such a phase

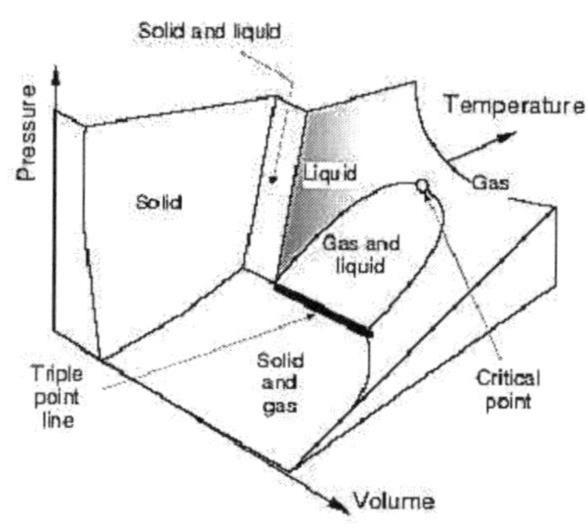

Figure 1. Equation of state for a fluid

change have been found, like a *plateau* in caloric curves[1,2], a sudden onset of multifragmentation and collective expansion, an increased probability for equal sized fragments possibly reminiscent of a spinodal decomposition[3], an abnormal enhancement of partial energy fluctuations tentatively associated to a negative branch of the heat capacity[4], a bimodal distribution of exclusive observables hinting phase coexistence[5], universal scaling laws both for the heaviest fragment and for the droplet distributions[6,7].

In this paper we will present results obtained in peripheral and central collisions of Au projectiles on different targets at 25 and 35 AMeV incident energies[7]. Data have been collected at the NSCL Cyclotron of the Michigan State University with the Multics-Miniball apparatus[8,9].

2. Sorting the events

Fig. 2 shows a schematic representation of a typical collision dynamics as predicted by nuclear transport models, for nearly central collisions. After the projectile hits the target, a fast compression phase ($\approx$ 20 fm/c) starts and pre-equilibrium light particles are emitted. Then the system expands and reaches the so-called freeze-out stage ($\approx$ 100 fm/c) where the interactions among the fragments vanish and fragments are chemically and energetically well defined. Excited fragments then undergo a slow secondary

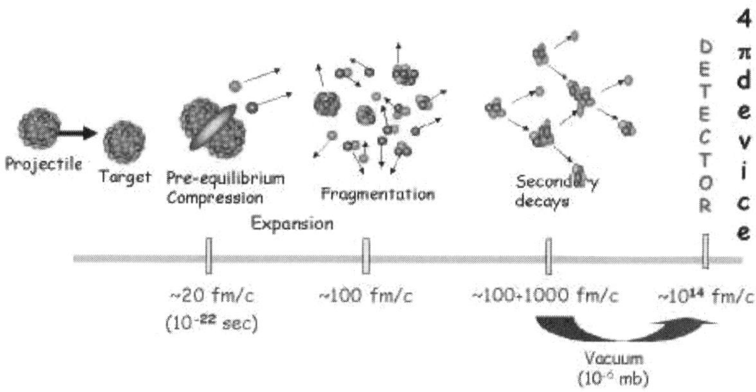

Figure 2. Time evolution of an intermediate incident energy reaction.

decay in vacuum with times of the order of some hundreds of fm/c. Almost all the final products of the reaction are detected by a 4π device after an almost infinite time (nsec) with respect to the previous time scales.

In order to perform a thermodynamic analysis, data have to be selected to obtain systems as close as possible to a thermodynamic equilibrium, and at different excitation energies, which can be experimentally determined by calorimetry:

$E^* + m_0 = \sum_{i=1}^{M}(m_i + k_i) + M_n(m_n + k_n)$,

where m_0, m_i and m_n are the masses of the fragmenting system, of the M fragments and of neutrons, respectively and k_i and k_n are the kinetic energies of fragments and neutrons.

This can be achieved by peripheral reactions, where the excitation function of the Quasi-Projectile (QP) can be studied with only one reaction, or by several measurements of central reactions with different beam energies. Global variables or more sophisticated statistical methods can be used to perform a centrality sorting of the events. One of these is the flow-tensor method[10], where the fragment momenta covariance matrix is diagonalized and the main eigenvector direction provides the average flow direction.

3. Equilibrium partitions

After the events have been sorted, one needs to check the statistical nature of the source emission. This can be done in several ways. From a pure experimental point of view the isotropic emission of all the particles and fragments emitted by the excited sources can be assessed, both in peripheral and in central collisions[11].

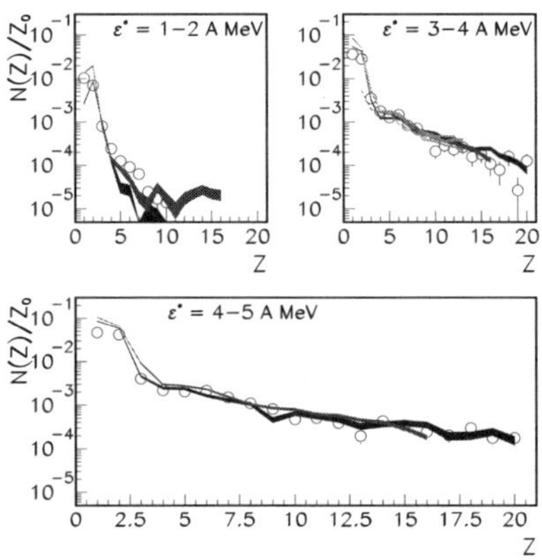

Figure 3. Charge distribution of all detected fragments but the heaviest for Au QP source (symbols) compared to central *Au + C, Au + Cu* collisions (dark grey lines) and the FASA (p,α + Au at 8.1, 4 and 14.6 GeV)[12] and ISIS data (π + Au)[13] (light grey lines) in the same calorimetric energy bins. The distributions are normalized to the charge of the emitting source.

A second check is to compare the measured distributions to predictions of a statistical multifragmentation model, which assumes a freeze-out equilibrium stage and a uniform phase space population with the same constraints as in the data. The comparison of charge and kinetic energy distributions show a very good agreement in the whole range of excitation energies[7].

A third way of assessing equilibrium can be obtained by comparing several observables in the same interval of the excitation energy, independent on the way the excited sources are formed. In Fig. 3 a very good agreement has been obtained between peripheral and central heavy ion reactions, as well as with excited sources obtained with pions or light particles as projectiles[12,13].

4. Self-similarity and scalings

The charge distribution in the 4-5 AMeV excitation energy range exhibits a power law behavior with an exponent close and greater than 2[7] (see Fig. 3), as expected at the critical point of the nuclear liquid gas phase transition[14], or inside the coexistence region of finite systems undergoing a first order phase transition[15].

Two different "critical" analyses have been performed, the first based on the moments of the size distribution[16] and the second on the Fisher droplet model[17]. This last method shows a universal straight line in the scaled yield of fragments, $n_A/(q_0 A^{-\tau})$ *vs.* the scaled temperature $\epsilon A^{\sigma}/T$ in a semi-logarithmic presentation (see Fig. 4). q_0 is a normalization constant depending only on the value of τ[18]. This scaling[7] yields "critical" parame-

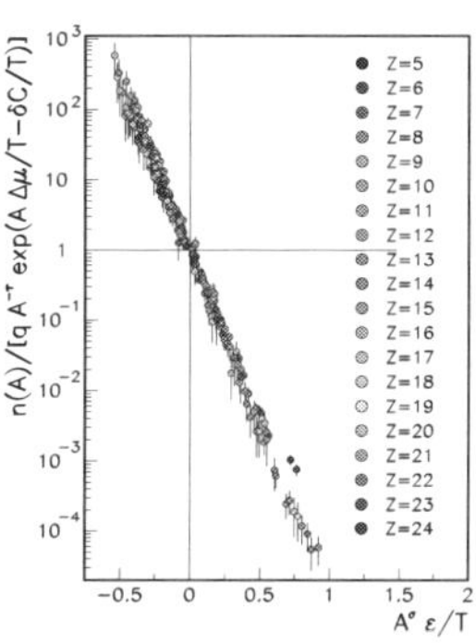

Figure 4. Scaled yield for Au QP events as a function of the scaled temperature.

ters σ and τ very similar to the ones for a liquid-gas phase transition at a "critical" excitation energy $\epsilon_c \approx 4.4\ AMeV$, the very same parameters obtained in a moment analysis[19]. The scaling is consistent with similar results obtained by the EOS[6] and ISIS[20] collaborations.

5. Thermodynamics

To get additional information on the behavior of excited nuclear systems a thermodynamic analysis has to be performed. This can be done in the framework of a "canonical" or "microcanonical" treatment.

5.1. *"Canonical" treatment*

In the thermodynamic limit, the order parameter of a first order phase transition presents an abrupt jump at the transition temperature from the ordered (liquid like) to the disordered (gas like) phase.

For finite systems the transition is smoothed, and in the vicinity of the transition the two phases are populated at the same temperature[21]: the probability distribution of the matter density, and of all other observables correlated to an order parameter like the size of the heaviest fragment, show a bimodal behavior.

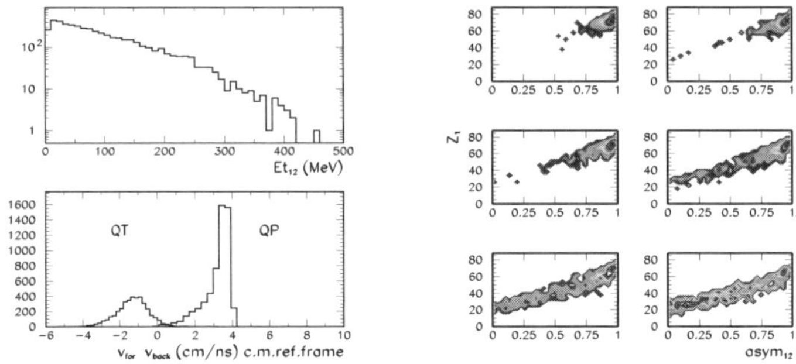

Figure 5. Lower left panel: plot of parallel velocity in peripheral Au + Au at 35 AMeV. QP and quasi-target (QT) emissions are clearly separated. Upper left panel: light particle (Z =1 and Z =2) transverse energy for QT. Right panel: Size of the heaviest fragment as a function of heaviest and second heaviest fragment asymmetry for six different transverse energy bins. The transverse energy increases from the top left to the bottom right panels. Nearly symmetric fission events have been removed[19].

From an experimental point of view, in a sample of binary peripheral events, one cannot fix the temperature of the system, but it is possible to make a canonical-like sorting of the data using, as sorting parameter, the transverse energy of light particles emitted by the quasi-target source. The asymmetry between the two heaviest fragments emitted by the quasi-projectile has been proposed as a possible candidate for the order parameter[5]. In Fig. 5 we show in the left panels the velocity distribution of the fragments indicating a binary collision and the distribution of the transverse energy of $Z = 1$ and 2 particles emitted by the quasi-target. The right panels show the size of the heviest fragment vs. the asymmetry distribution in equally spaced bins of the transverse energy. A bimodal distribution clearly appears at least in the fourth and fifth bins, where the excitation energy is around 4.4 AMeV, as in the "critical" analysis (see sect. 4).

5.2. *"Microcanonical" treatment*

A straightforward way of sorting experimental events is as a function of the total deposited energy. In the limiting case of a perfect detection, this corresponds to a microcanonical sorting.

The excitation energy is composed by a configurational part, due to Coulomb and Q-values of the partitions, and a kinetic part consisting in the translational energy plus the internal energy of fragments at the freeze-out stage:
$E^* = E_{config} + E_{kin} = E_{coul}(V) + Q_v + T_{tr}(T) + E_{int}(T)$.
This decomposition of the measured energy needs an estimation of the average volume V and temperature T of the system at the freeze out stage.

In the absence of collective flows, the average volume at freeze out can be backtraced from the measured mean kinetic energy of the final fragments. The result is a volume about 2.7 times the normal density one, consistent with spherical fragments close to normal density at freeze out[11]. Different thermometers have been used to estimate the temperature [4,11,22] showing a good consistency and a global agreement with other temperature systematics[1,2] and fragment internal excitation energy evaluation[23].

In thermodynamic equilibrium, the fluctuations σ^2_{kin} of the kinetic energy E_{kin} are linked to the heat capacity through the relation
$\frac{C_{kin}}{C} = 1 - \frac{\sigma^2_{kin}}{\sigma^2_{can}}$, where $\sigma^2_{can} = T^2 C_{kin} = T^2 \frac{dE_{kin}}{dT}$.
In a first order phase transition with finite latent heat, σ^2_{kin} exceeds the canonical fluctuations σ^2_{can} and the heat capacity is negative[4,24].

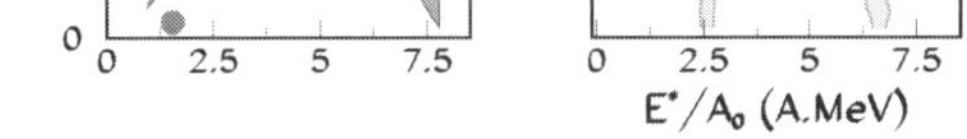

Figure 6. Left panel: normalized partial energy fluctuations for QP events (gray contours) and central Au + C, Au + Cu and Au + Au central events (symbols). Right panel: estimate of the heat capacity per nucleon of the source.

The results are shown in Fig. 6 where two divergences and a negative branch of the heat capacity clearly appear, both for peripheral and central collisions. The separation between the two divergences is related to the latent heat. In the framework of thermodynamic equilibrium these results are a serious evidence of a first order phase transition and have been confirmed by other collaborations[11,25].

6. Perspectives and conclusions

All results of the analyses described in the previous sections, are compatible with a phase coexistence, however the quantitative determination of the nuclear phase diagram is far from being achieved. In particular "contextual" experimental information are needed on the partitioning of the system, fluctuations, and event by event calorimetric excitation energy. To this aim a 4π mass and charge detection system is necessary.

In addition the extra degree of freedom in the nuclear equation of state, the isospin, has to be investigated. New generation devices and exotic beams are needed to fully investigate the phase transition by changing the Coulomb properties and the isospin content (N/Z) of the fragmenting systems. Hot liquid-drop model calculations[26] of the limiting temperature, i.e. the maximal temperature a nucleus can stand while evaporating and a systematic analysis of the saturation temperature (in the caloric curve) as a function of the size of the system[2] lead to the expectation that for $A \approx 100$ proton very rich nuclei the limiting temperature vanishes.

With all this in mind the NUCL-EX collaboration[a] began an experimental campaign aiming to thermodynamically explore systems of

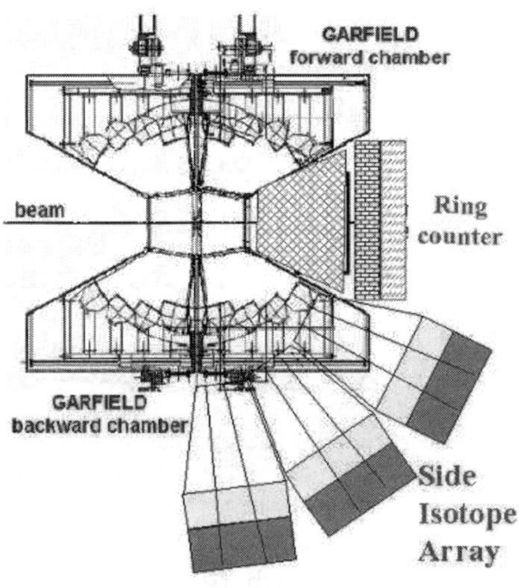

Figure 7. Schematic drawing of the Garfield apparatus

$A \approx 100$ starting from the liquid side. The first experiment with 25 AMeV

[a]INFN Sections and Universities of Bologna, Firenze, Milano, Napoli, Trieste and Laboratori Nazionali di Legnaro

Ni beams on Ca targets has been performed with the Chimera apparatus at the INFN LNS in Catania. The experimental program will continue at the INFN LNL in Legnaro with the GARFIELD apparatus[27] (see Fig. 7). The apparatus consists in two drift chambers covering a large solid angle region where energy loss and residual energies of light particles and fragments are measured with very low thresholds by micro-strip gas chambers and CsI crystals. The two chambers are complemented[28] with a forward annular device (Ring Counter) covering the forward region $6^o \div 18^o$ and another device (Side Isotope Array) which covers from 20° to 90° in the horizontal plane. These two devices consist in three stage (ionization chamber, silicon detector and CsI scintillator) telescopes allowing mass and charge determination up to Oxygen isotopes.

The use of newly designed digital electronics[29] will allows to use a pulse-shape discrimination technique in the CsI crystals in order to have, over the whole solid angle, the isotope determination of light particles and fragments ($Z \leq 4$). First results with a S beam on Ni isotopes at 14.5 AMeV show a very good charge and mass identification[28] and allow also for evaporation residues identification.

In order to have information on fragment multiplicity at the freeze-out

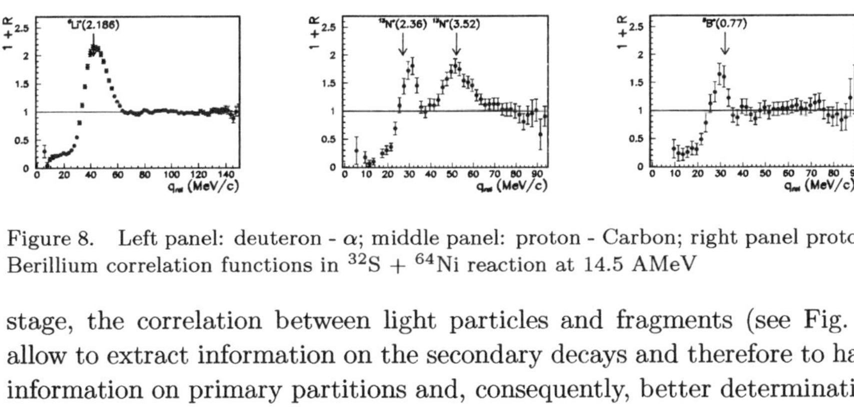

Figure 8. Left panel: deuteron - α; middle panel: proton - Carbon; right panel proton - Berillium correlation functions in $^{32}S + ^{64}Ni$ reaction at 14.5 AMeV

stage, the correlation between light particles and fragments (see Fig. 8) allow to extract information on the secondary decays and therefore to have information on primary partitions and, consequently, better determination of temperatures from isotope abundance.

The analysis is in progress to characterize the source and check the fragment emission time scales.

In conclusion, the physics of hot nuclei is a unique laboratory for the thermodynamics of finite, charged, two component systems and for a quantitative nuclear metrology. Interdisciplinary connections are also important. In Fig. 8 p + ^7Be and p + ^{12}C correlation functions are shown, indicating $^8B^*$ and $^{13}N^*$ excited state formation inside the nuclear medium. This has also astrophysical interests since these processes are important in the stellar evolution.

Acknowledgments

The authors would like to acknowledge the skillful technical assistance of R. Cavaletti, A. Cortesi, L. Costa, P. Del Carmine, M. Giacchini and M. Ottanelli. This work has been supported in part by Grants of Alma Mater Studiorum (Bologna University).

References

1. J. Pochodzalla *et al*, Phys. Rev. Lett. 75 (1995) 1040.
2. J.B.Natowitz *et al.*, Phys. Rev. C65 (2002) 34618.
3. B.Borderie *et al.*, Nucl. Phys. A734 (2004) 495, and ref. therein.
4. M.D'Agostino *et al.*,Phys.Lett. B473(2000) 219; Nucl. Phys. A699(2002) 795.
5. B. Tamain *et al.*, in *Phase transitions in strongly interacting matter* Prague, 2004. Nucl. Phys (to be published).
6. J.B.Elliott *et al.*, Phys. Rev. Lett. 88 (2002) 042701.
7. M.D'Agostino *et al.*, Nucl. Phys. A724 (2003) 455.
8. I. Iori *et al.*, Nucl. Instr. and Meth. A325 (1993) 458.
9. R.T. DeSouza, *et al.*, Nucl. Instr. Meth. A295 (1990) 109.
10. J. Cugnon and D.L'Ĥote, Nucl. Phys. A397 (1983) 519.
11. M.D'Agostino *et al.*, Nucl. Phys A743 (2004) 512 and in *Phase transitions in strongly interacting matter* Prague, 2004. Nucl. Phys (to be published).
12. V. A. Karnaukhov *et al.*, Nucl.Phys. A734 (2004) 520.
13. V. Viola *et al.*, Nucl.Phys. A734 (2004) 487
14. J.E. Finn, *et al.*, Phys. Rev. Lett. 49 (1982) 1321.
15. F. Gulminelli and Ph. Chomaz, Phys. Rev. Lett. 82 (1999) 1402.
16. A. Bonasera *et al.*, La Rivista del Nuovo Cimento vol.23, n.2 (2000).
17. M. E. Fisher, Physics Vol. 3 (1967) 255.
18. H. Nakanishi *et al.*, Phys. Rev. B 22 (1980) 2466.
19. M.D'Agostino *et al.*, Nucl. Phys. A 650 (1999) 329.
20. J. B. Elliot *et al.*, Phys. Rev. C67 (2003) 024609.
21. Ph. Chomaz, F. Gulminelli, V. Duflot, Phys. Rev. E 64 (2001) 046114.
22. P. M. Milazzo *et al.*, Phys. Rev. C60 (1999) 044606.
23. S. Hudan *et al.*, Phys. Rev. C67 (2003) 064613.
24. Ph. Chomaz, F. Gulminelli, Nucl. Phys. A 647 (1999) 153.
25. M. F. Rivet *et al.*, in *Phase transitions in strongly interacting matter* Prague, 2004. Nucl. Phys (to be published).
26. J. Besprovany and S. Levit, Phys. Lett. B217 (1989) 1.
27. F. Gramegna *et al.*, Fizika B12 (2003) 39.
28. M. Bruno *et al.*, in preparation.
29. L. Bardelli *et al.* Nucl. Instr. and Meth. A521 (2004) 480 and contribution to this conference.

THE CHIMERA DETECTOR AT LNS IN CATANIA: RECENT ACHIEVEMENTS AND PERSPECTIVES

A. PAGANO[1], F. AMORINI[2], A. ANZALONE[2], L. AUDITORE[3], R. BASSINI[4], C. BOIANO[4], J. BLICHARSKA[5], V. CAMPAGNA[2], G. CARDELLA[1], C. CALÍ[2], S. CAVALLARO[2], M. D'ANDREA[1], E. DE FILIPPO[1], F. FICHERA[1], E. GERACI[6], N. GIUDICE [1], F. GIUSTOLISI[2], N. GUARDONE[1], A. GRIMALDI[1], A. GRZESZCZUK[5], P. GUAZZONI[4], M. IACONO-MANNO[2], E. LA GUIDARA[2], G. LANZANÓ[1], G. LANZALONE[2], C. MAIOLINO[2], C. MARCHETTA[2], D. NICOTRA[1], M. PAPA[1], S. PIRRONE[1], G. POLITI[1], F. PORTO[2], C. RAPICAVOLI[1], G. RIZZA[1], F. RIZZO[2], S. RUSSO[4], P. RUSSOTTO[2], G. SACCÁ[1], M. SASSI[4], M. L. SPERDUTO[1], A. TRIFIRÓ[3], M.TRIMARCHI[3], S.URSO [1], L. ZETTA[4] AND W. ZIPPER[5]

(CHIMERA COLLABORATION)

[1] *INFN and Dipartimento di Fisica e Astronomia Universitá di Catania , Italy*
[2] *LNS and Dipartimento di Fisica e Astronomia Universitá di Catania, Italy*
[3] *INFN and Dipartimento di Fisica, Universitá di Messina,Italy*
[4] *INFN and Dipartimento di Fisica e Universitá di Milano, Italy*
[5] *Institute of Physics, University of Silesia, Katowice, Poland*
[6] *INFN and Dipartimento di Fisica Universitá di Bologna, Italy*
E-mail: angelo.pagano@ct.infn.it

Since January 2003, the new 4π detector CHIMERA is operational at Laboratori Nazionali del Sud in Catania. The device was designed to study multifragmentation at Fermi energy (10 MeV/nucleon < E/A < 100 MeV/nucleon). Good identification of light charged particles and fragments allow to perform careful investigations of the reaction mechanism. During 2003 and 2004 different experiments were accomplished in the framework of an international collaboration. Triggered by recent increasing interest for isospin physics and future facilities with radioactive beams, in 2006-2007 CHIMERA will be upgraded by a suitable pulse shape analysis in silicon detectors, so that the present identification performance of the apparatus will be extended. CHIMERA fostered the interest of researchers for new developments in the field of particle identification and innovative signal processing methods. In this paper, a report of this activity is given and the experimental capability of the apparatus is illustrated.

1. Introduction

The field of Heavy Ion collisions at Fermi energies (10 MeV/nucleon $<$ E/A $<$ 100 MeV/nucleon) has attracted much interest by nuclear scientists during the last two decades. The reason is that, in this energy domain, the dissipative processes driven by the mean field, typical of the Coulomb-energies regime, are progressively replaced by nucleon-nucleon dynamics. Consequently, a transition in the reaction mechanism is expected[1]. However, due to the complexity of the mechanism involved, it was necessary to develop sophisticated 4π experimental devices, to detect charged particles, neutrons or gamma rays[2]. The Multifragmentation of a transient highly excited nuclear system into several massive clusters of intermediate mass (IMF) and light particles (LP) was the key - phenomenon to be studied. The process was linked to basic properties of the nuclear equation of state (EOS) and, consequently, much importance assumed the concept of Liquid-Gas Phase transition[3]. More recently, isoscaling[4] and isospin degree of freedom[5] have assumed a prominent role, also in view of the production of exotic secondary beams. This last perspective motivates efforts in the direction of improving the identification methods, in order to extend simultaneous mass and charge identification of IMFs in existing 4π detector systems[6] or to plan new designs for future third generation experiments[7]. In this report the recent status of the CHIMERA detector is briefly described and some perspectives in the medium term are discussed. This paper is ideally linked with the early CHIMERA proposal published a few years ago in Ref.[8].

2. The CHIMERA Detector

The CHIMERA(Charged **H**eavy **I**on **M**ass and **E**nergy **R**esolving **A**rray) detector was designed in order to significantly contribute to Multifragmentation reaction studies in Heavy Ion collisions at Fermi energy. Basically, CHIMERA is made of 1192 detection cells, each consisting of a planar 300 μm-silicon detector (200 μm for the most forward ring) followed by a CsI(Tl) scintillator of thickness ranging from 3 cm at backward angles to 12 cm at the most forward angles. In 1997 the first ring of the detector was coupled with the INDRA detector in GANIL for a full test of the experimental procedures involved[9]. In 1999, the forward part of the apparatus, made of 688 detection cells (CHIM688), arranged in nine rings covering the angular range from 1° to 30°, with full 2π azimuthal symmetry around the beam axis, was installed inside the reaction chamber CICLOPE at Laboratori Nazionali del Sud in Catania (LNS). The nine rings were placed, by design, at variable distances from the target, ranging from 3.5 m (1°) to 1 m (30°). In April-June 2000 the experimental campaign REVERSE was abl to study the reactions in reverse kinematics ^{112,124}Sn + ^{58,64}Ni and ^{124}Sn + ^{27}Al at the incident energy E(^{112,124}Sn) = 35 MeV/nucleon[10]. Experimental results and performance of the forward part of the apparatus and the identification methods used for isotopic identification of light fragments, intermediate mass fragments (IMF) and light charged particles (LPC), have been described in detail in Refs. [11]. Figure 1. shows a portion of the forward rings of the detector during its first mounting phase in year 2000.

Figure 1. Forward part of CHIMERA installed inside the CICLOPE reaction chamber at Laboratori Nazionali del Sud in Catania, during its mounting phase

Finally, in January 2003, the apparatus was completed by adding the spherical structure of 40 cm radius housing 504 detection cells and covering the detection angles between 30° and 176°. Figure 2. shows a portion of the spherical configuration with the target framework position. In figure 3. the final configuration of the complete apparatus is shown. During years 2003 and 2004 two separate experimental campaigns were performed. The experiments were performed with the aim to extend the REVERSE campaign by studying the reactions ^{112}Sn + ^{64}Ni and ^{124}Sn + ^{58}Ni at the Sn beam energies of 25 and 35MeV/nucleon, and to further explore the evolution of reaction mechanism. In collaboration with different international institutions several different experiments in the field of energy dissipation, dynamical production of fragments, phase transition, isospin degree of freedom, isoscaling, cluster condensation, and limiting temperature were performed.

Figure 2. Portion of the spherical configuration of the CHIMERA apparatus. The position of the target framework is visible

Figure 3. Recent picture of the CHIMERA multidetector

Three different detection techniques are simultaneously used in CHIMERA. First, the ΔE-E technique is employed for charge identification of heavy ions and for isotopic identification of IMF's with atomic number

$Z < 10$. Second, mass identification is performed with signals from silicon detectors generating the time-of-flight signal (TOF), obtained by comparing the timing of the detector signal and the timing of the high frequency signal (HF) from the cyclotron. Third, energetic light charged particles (LCP), which are stopped in the scintillator crystal, are identified by applying the pulse-shape discrimination method using well shaped amplified signals generated by a 18 mm × 18 mm photodiode, optically coupled to the crystal.

Due to a very compact mounting of the telescopes, the total (nominal) geometrical efficiency of the apparatus is about 95% of the total solid angles (including holes at the forward ($\theta \leq 1°$) and backward direction ($\theta \geq 176°$). However, during the experiments, about 70 telescopes, on average, showed partial or total inefficiency; so, the effective average geometrical efficiency was somewhat reduced to a value around 90% of 4π.

Detectors, preamplifiers and switching system for pulser signals are operated in vacuum and are connected to the main spectroscopic amplifiers located out of the reaction chamber by multiple pin to pin flanges especially developed to reduce electronics cross talk. In figure 4., a portion of the detector array with the electronics connecting detectors and preamplifiers in vacuum and a detail of the cooling system is shown. In figure 5., one of the 600 main motherboards allocating two preamplifiers for silicon detectors (left) and two preamplifiers for photodiodes (right) are shown.

The front-end electronics was integrated in a remote controlled NIM or CAMAC architecture allocating about 300 16-channel compact analogical electronics. The acquisition system was based on a FDL (Fast Data Link) read out VME connection of a 9U VME DAQ system [12].

Chimera benefits of the unique opportunity of the LNS Super-Conducting Cyclotron to deliver beams of suitable intensities for the 4π detector ($\sim 5 \cdot 10^7$ particles/s), high efficiency for physics runs and with good time resolution ($\Delta T \leq 1$ ns) during most of the operation time.

The identification characteristics recently achieved for the forward part of the apparatus (CHIM688) in the recent 2003 and 2004 campaigns were very similar to the ones already described for the 2000 REVERSE experiment [10,11]; so, those results are not shown in this paper. All the basic experimental performances of the spherical backward part of the apparatus such as timing and energy resolution are also very similar to those observed for the forward part. Evidently, as expected, the mass by mass identification properties of the sphere are limited to a value of the atomic mass A ≈ 7 (a factor 4 lower than the one observed in the forward part) according to

Figure 4. Front-end electronics in vacuum. Photodiodes coupled with the rear of the CsI(Tl) crystals are clearly visible

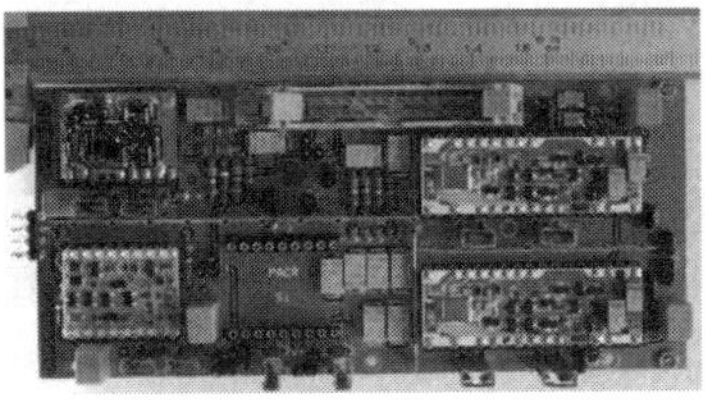

Figure 5. Example of main motherboard of the CHIMERA detector allocating four charge preamplifiers

the reduced flight distance of the detected particles from the target (40cm).

3. Physics with CHIMERA

The physics addressed with CHIMERA is mainly fragment productions in multifragmentation reactions and isospin physics related with phase transition. In these fields, most of the physics results already published are related to the 2000 year campaign where the forward part of the detector was used for the first time in the framework of the REVERSE collaboration. The analysis of the reactions ^{112,124}Sn + ^{58,64}Ni at the beam energy, E(^{112,124}Sn) = 35 MeV/nucleon have revealed very interesting results for both central [13,14] and peripheral collisions [15,16].Other important analyses, such as isoscaling, source reconstruction, angular correlations concerning ^{112,124}Sn induced reactions are in progress and they will be discussed in separate publications. Obviously, the relevant conclusions achieved by the REVERSE and ISOSPIN collaborations, and recently published or communicated in numerous proceedings of conferences [17] are not repeated in the present contribution. However it is useful to discuss some global experimental features underlying these reaction studies in order to illustrate the typical reaction pattern of the Fermy energy domain. In figure 6., a typical global variable event by event analysis is shown. In the left panel, for each of the detected particles, the atomic number is plotted (event by event) as a function of the experimental parallel velocity (i.e., velocity vector projected on the beam direction). The typical classification of fragments in three main groups is easily recognized by distinguishing: a) projectile-like fragments (PLF) having both charge and velocity close the beam ones, b) target-like fragments (TLF), slowly moving in the laboratory rest framework, and having atomic number close to the target one, and c) fragments of intermediate mass (IMF) having also a velocity intermediate between the PLF's and the TLF's velocities. The different plots (from the top to the bottom) correspond to three different selections of the centrality, as roughly evaluated by three different cuts in the charged particle multiplicity. From peripheral (top, $M \leq 6$) to central collisions (bottom, $M \geq 13$) the corresponding Dalitz plots (on the right panel) clearly illustrate the characteristic evolution of the reactions from binary gentle reactions (top) to the most dissipative multifragmentation ones (bottom).

Several different nuclear systems were investigated with the CHIMERA apparatus during 2003 and 2004. For all these systems, an important work of calibration, made in collaboration with all the international groups which have proposed new experiments, is still in progress [17].

Another aspect of the CHIMERA activity is represented by an important work in the field of R&D and detector upgrading. This activity is mainly devoted to pulse shape analysis for both silicon and CsI(Tl) detectors. Important progress was achieved in the field of the pulse shape analysis for the large capacitance planar n-type silicon detectors of CHIMERA mounted in direct - mode configuration [18]. The obtained good results have motivated an upgrading of the apparatus, to be accomplished in 2006-2007. The pulse shape analysis was also performed by a digital sample analysis (DSA). Also in this field, recent important progress of DSA has been successful applied to the signals delivered by a fast charge preamplifier,

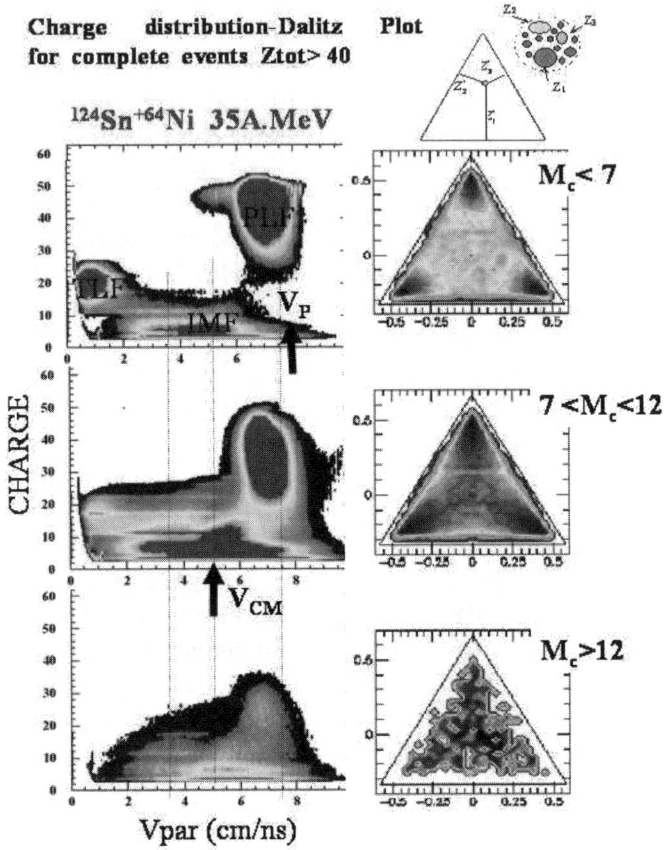

Figure 6. Charge vs parallel velocity scatter plots (left) and Dalitz plots of the three biggest fragments detected event by event (right). The analysis is performed as a function of the particle multiplicity .

especially developed for this purpose. All optimization and filtering procedures of the signals, typical in nuclear spectroscopy, have been performed by software analysis. Very good results have been obtained for energy measurement and charge identification [19]. The digital methods are also very promising for timing applications. Progress of this latter application is crucial in order to design third generation detectors for nuclear reaction studies.

Acknowledgments

The CHIMERA group is grateful to scientists of REVERSE(1998-2202) and ISOSPIN(2003-2005) collaborations: C.Agodi, R.Alba, V.Baran , I. Berceanu, J. Brzychczyk, A. Bonasera, B. Borderie, R. Bougault, M. Bruno, M. B. Chatterjee, A. Chbihi, M. Colonna, R.Coniglione, M. D'Agostino, E. R. Dayras,A.Del Zoppo, M. Di Toro, J. Frankland, E. Galichet, W. Gawlikowicz, D. Guinet, S. Kowalski, N. Le Neindre, E. Migneco, Z. Majka, M. Petrovici, E. Piasecki, P. Piattelli, R. Planeta, M. F. Rivet, E. Rosato, D.Santonocito, P.Sapienza, K. Schmidt, K. Siwek-Wilczynska, I. Skwira, W.U. Schroeder, J. Toeke, G. Vannini, G.Verde, M. Vigilante, J. P. Wieleczko, J. Wilczynski for encouragements, unique guidelines and suggestions during the crucial period of commissioning of the apparatus and physics proposals.

We are grateful to O. Conti, A. Di Stefano, H. Hong, P. Litrico, S. Marino, D. Moisa, G. Peirong, S. Salomone, A. Santonocito, for their invaluable work in assembling CHIMERA. Thanks are due to C. Marketta and E. Costa for the targets of good quality and to D. Rifuggiato, L. Calabretta, and co-workers for delivering beams of very good timing quality. The work was supported in part by the Italian Ministero dell'Istruzione, dell'Università e della Ricerca (MIUR) under contract COFIN2002 and by NATO grants PST.CLG.979417.

One os as (A.P.) is grateful to S.Costa of INFN and Catania university for carefull reading the paper prior publication and for valuable comments.

References

1. A.Bonasera, M. Di Toro and Ch. Gregoire, *Nucl. Phys.* **A 463** (1987) 653; G. Peilert, H. Stoecker and W. Greiner; *Rep. Prog. Phys.* **57** ,(1994) 533
2. *Proceedings of Int. Res. Workshop Poiana Brasov, Romania,* October 7-14, 1996, Ed. M. Petrovici, A. Sandulescu, D.Pelte, H. Stoecker and J.Randrup
3. J.Richter and P.Wagner, *Phys. Rep.* **350** , 1 (2001); A.Bonasera et al., *Rivista del Nuovo Cimento,* **23**,(2002), 1
4. B. Tsang, *Nucl. Phys.* **A 734** (2004), 605-608
5. *Isospin Physics in Heavy Ion Collisions at Intermediate Energies* , Edited by Bao An Li and W. Udo Schroeder, ISBN 1-56072-888-4, Nova Science Publ. Inc., New York (2001)
6. A.Pagano et al., *Proceedings of XLII Int. Winter meeting on Nucl. Physics* , Bormio(Italy), 25-31 January 2004, Ed. by I. Iori, 359-370
7. B.Borderie, *"Summary of the round table on future developments for new calibration 4π detectors", Proceedings of IWM2001,* Catania 28 November , 2001, pag.173; and *Proceedings of IWM2003* GANIL, Caen (France) 5-7 November 2003, ed. by G.Agnello, A. Pagano and S.Pirrone

8. A.Pagano et al., *Proceedings of 2nd Japan-Italy Joint Symposium '95-Perspectives in Heavy Ion Physics,* Ed. M. Ishihara, T. Fukuda. C. sSignorini, World Scientific, (1996) 119-132
9. R.Bougault for CHIMERA and INDRA collaboration, *Nouvelles du Ganil n.60 1997*; J.C.Steckmeyer et al., proc. of the XL Int. Winter Meeting on Nuclear Physics, Bormio, Junuary 21-26 (2002), Ed. I.Iori, 198-208
10. A.Pagano et al.,*Nucl. Phys.* **A 681** (2001), 331c-338c
11. A.Pagano et al.,*Nucl. Phys.* **A 734** (2004), 504-511 and references therein
12. E. De Filippo et al., *Proc. 11th IEEE INPS Real Time Conf.* , Santa Fe, June 1999, p. 78; S. Aiello et al., *IEEE Trans. on Nucl. Sci. 47 no.2* 2000 114-118
13. E.Geraci et al.,*Nucl. Phys.* **A 732** (2004), 173-201
14. E. Geraci et al.,*Nucl. Phys.* **A 734** (2004), 524-527
15. E. De Filippo et al. *Phys. Rev.* **C 71** , 044602 (2005)
16. E. De Filippo et al. *Phys. Rev.* **C**, (2005)in press
17. for a list of recent contribution of the REVERSE/ISOSPIN collaboration see http://www.lns.infn.it/research/chimera
18. M. Alderighi et al., manuscript TNS/00461/2004,R1 *IEEE TNS* to be published
19. M. Alderighi et al., *IEEE TNS* **51** (2004) 1475-1481

DISAPPEARANCE OF COLLECTIVE MOTION IN HOT NUCLEI

D.SANTONOCITO[1)], Y.BLUMENFELD[2)], C.AGODI[1)], R.ALBA[1)], G.BELLIA[1,3)], R.CONIGLIONE[1)], F.DELAUNAY[2)], A. DEL ZOPPO[1)], P.FINOCCHIARO[1)], N.FRASCARIA[2)],F.HONGMEI[1)], V.LIMA[2)], C.MAIOLINO[1)], E.MIGNECO[1,3)], P.PIATTELLI[1)], P.SAPIENZA[1)], J.A. SCARPACI[2)]

1) INFN - Laboratori Nazionali del Sud, Via S.Sofia, 62, Catania, Italy
2) Institut de Physique Nucleaire, IN2P3-CNRS, F-91406, Orsay, France
3) Dipartimento di Fisica e Astronomia dell'Universitá di Catania

The evolution of the GDR γ yield as a function of excitation energy has been investigated in nuclei of mass A $\approx$ 126 - 136 through the reactions ^{116}Sn + ^{12}C at 17 and 23A MeV and the reaction ^{116}Sn + ^{24}Mg at 17A MeV. Hot nuclei produced in incomplete fusion reactions span an excitation energy range between 160 and 290 MeV. Gamma-rays were detected with MEDEA array in coincidence with residues detected in MACISTE. The evolution of the GDR parameters has been investigated as a function of the linear momentum transferred to the fused system. The analysis of the γ spectra and their comparison with CASCADE calculations is presented. A comparison with the gamma spectra measured in the reaction ^{36}Ar + ^{98}Mo at 37A MeV at higher excitation energy is presented. A progressive reduction of γ multiplicity with respect to predictions for 100% of the Energy Weighted Sum Rule is observed above 200 MeV excitation energy.

1. Introduction

Intermediate energy heavy ion collisions are a powerful tool to produce and investigate the behavior of nuclear matter under extreme conditions of density, temperature and excitation energy. Over the last decades a large fraction of the available data was collected measuring evaporation residues, light charged particles and intermediate mass fragments (IMF). Valuable information can also be obtained through the study of gamma-ray emission.

Since photons can be emitted during the collision with different time scales their investigation allows to probe the different stages of the reaction, to follow the evolution of the highly excited system formed towards the equilibration and also to disentangle effects due to nuclear matter properties from those related to collision dynamics.

In recent years extensive studies on the production of high energy gamma-rays has yielded information on the role of 2-body collisions in the dissipation mechanism [1, 2] and the study of thermal photons in coincidence with IMF has shown evidences for the onset of prompt IMF emission [3]. The study of the Giant Dipole Resonance (GDR) built on highly excited states has allowed to collect information on the collective properties of the hot nuclear matter. Its evolution as a function of excitation energy addressed the question of the persistence of collective behavior in extreme conditions. Many years of investigation have clarified the main features of the evolution of the GDR parameters as a function of excitation energy even if the width and the strength dependencies are not yet fully understood.

The most complete systematics has been collected on medium mass nuclei with spherical ground state in the region $A \approx 110 - 120$ [4]. In particular while the centroid energy remains constant with excitation energy to a value of about 15 MeV, the γ multiplicity was observed to increase up to approximately E*=250 MeV in agreement with the 100% of the sum rule strength. At higher excitation energy a pioneering work by Gaardhoje et al. [5] gave the first indication for a saturation of the GDR gamma emission.

The problem of the saturation of the GDR yield with excitation energy was addressed by studying the evolution of the GDR properties in hot nuclei of mass $A \approx 110$ formed in incomplete fusion reactions [6]. Beams of ^{36}Ar at 27 and 37A MeV impinging respectively on ^{90}Zr and ^{98}Mo targets were used to populate hot nuclei in an excitation energy range between 300 and 550 MeV. Gamma-ray spectra, measured in coincidence with heavy residues, were compared with statistical model calculations using CASCADE code. Both reactions showed evidence for a saturation effect. The simplest way to reproduce the data was to introduce a suppression of the gamma emission above a certain excitation energy, the so called sharp cut-off energy. In particular, a cut-off value of about 250 MeV excitation energy allowed to reproduce the data at 27 AMeV. This important result led to the conclusion that an excitation energy per nucleon of about 2.3 MeV, corresponding to a temperature of about 5.5 MeV, represents a limit for the existence of the dipole vibration for $A \approx 110$ nuclei [7]. This can be qualitatively understood assuming that above a certain excitation energy the equilibration time of the collective oscillation becomes longer than the particle emission time leading to a progressive suppression of the GDR gamma emission even if the mechanism that suppresses the collective oscillation at high excitation energies remains still unclear.

In the above mentioned analysis the existence of a limiting excitation energy for the GDR could only be inferred indirectly through statistical model simulations including a sharp cut-off for the gamma emission since the hot nuclei populated in the reaction were in an excitation energy region above the region where the saturation is expected to set in, precluding a detailed study of the onset of the quenching.

The introduction of a sharp cutoff approximation points to a sudden disappearance of the GDR gamma emission while the existence of a progressive quenching of the GDR yield, already below 300 MeV excitation energy, is predicted by different models which show different energy dependencies [8, 9, 10]. In order to investigate the excitation energy dependence of the GDR disappearance we have performed an experiment where hot nuclei with masses $A \approx 126 - 136$ were populated at excitation energies per nucleon between 1.3 and 2.2 MeV. This corresponds to the region where the suppression is expected to set in if one supposes that the parameter governing the GDR quenching is either the excitation energy per nucleon or the temperature.

2. Experimental Method

The experiment was carried out at the Laboratori Nazionali del Sud - Catania using the very asymmetric reactions ^{116}Sn + ^{12}C at 17 and 23AMeV, and ^{116}Sn + ^{24}Mg at 17 AMeV in inverse kinematics. The beams had an average intensity of 1.5 nA, and impinged on 1 mg/cm^2 thick ^{12}C and ^{24}Mg targets.

Light charged particles and gamma rays were detected by the MEDEA multidetector [11] which consists of a ball of 180 BaF2 scintillators 20 cm thick arranged in 8 rings covering polar angles from 30° to 170° and full azimuthal angles. Light charged particles and gamma-rays were identified using the combined information extracted from time of flight and pulse shape analysis technique on the photomultiplier analog signals.

Fusion-like residues populated in the reactions and emitted at polar angle up to 3 degrees were focused by the magnetic field of superconducting solenoid SOLE on the focal plane detector MACISTE [12] placed 16 m from the target. The use of inverse kinematics leads to a very efficient collection of the residues in the forward angle SOLE solenoid. MACISTE consists of four telescopes arranged on a 70×70 cm^2 surface leaving a variable central hole for the beam transit. Each telescope consists, in turn, of a drift chamber used for the energy loss measurements, a multi-wire proportional chamber for the impact position and time of flight determination and, for the energy measurements, a 2 cm thick plastic scintillator coupled to a photomultiplier through a suitable light guide.

Time of flight, measured with respect to the cyclotron RF, was used to identify fusion-like events. The trigger condition of the experiment was given by the coincidence between one of the MACISTE telescopes and at least one MEDEA detector.

The calibrated time of flight spectra for the different reactions are shown in fig. 1. They all exhibit a distribution whose maximum is close to the center of mass velocity indicating the presence of complete or close to complete fusion events. The width of the distribution reflects both the range of linear momentum transfer populated in the reaction and the broadening due to subsequent particle evaporation. This explains why a broader width is observed in the case of ^{24}Mg target compared to ^{12}C one (see Fig. 1).

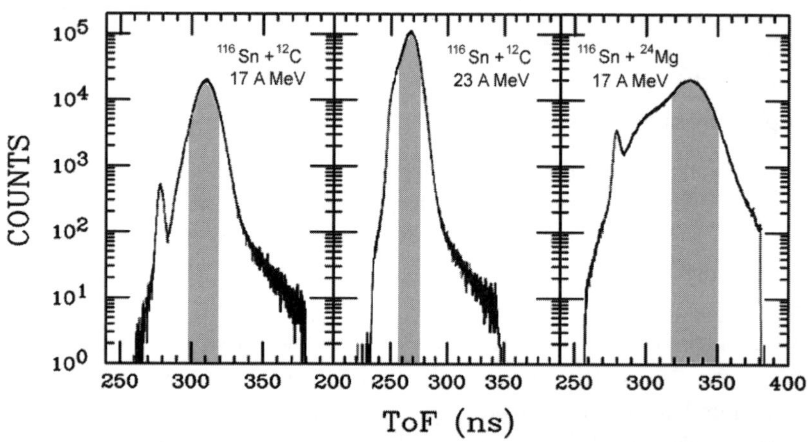

Figure 1. Velocity distributions of the residues detected in MACISTE for the three reaction studied. Filled areas correspond to the velocity regions used in the analysis.

In the analysis events were sorted according to their time of flight and the evolution of the properties of gamma emission was investigated as a function of excitation energy. Events belonging to velocity windows corresponding to a region of $\pm 3\%$ of the center of mass velocity were selected for the three reactions (filled areas in Fig. 1) and retained for the analysis. The massive transfer model was applied to estimate the average excitation energy and masses from residue velocities (see tab.1)[13]. Corrections for energy losses in the target and reaction Q values were applied. This approach has been shown helpful in the determination of the excitation energy of the hot nuclei populated in incomplete fusion reactions which represents a fundamental issue when such a study is undertaken.

Table 1. Values of excitation energy, mass and excitation energy per nucleon for the three reactions estimated using the massive transfer model and taking into account the correction for reaction Q values and beam energy loss in the target.

reactions	E_{beam}	E^* (MeV)	A	E^*/A (MeV)
$^{116}Sn+^{12}C$	17 AMeV	160	127	1.26
$^{116}Sn+^{12}C$	23 AMeV	200	127	1.57
$^{116}Sn+^{24}Mg$	17 AMeV	290	136	2.13

3. Experimental Results

The gamma spectra measured at 90° in coincidence with fusion events selected using the velocity window shown above for all the reactions are presented in Fig.2. They show similar features: at low energies an exponentially decreasing component associated to the statistical emission from the compound nucleus at the end of the decay chain, a pronounced bump around 15 MeV corresponding to the decay of the GDR and, above 35 MeV, an exponentially decreasing component due to the bremsstrahlung radiation arising mainly from first chance np collisions in the first, non equilibrated, stages of the reaction.

The high energy part of the spectra was fitted by an exponential function for energies above 35 MeV [14]. Due to the large statistics available a precise determination of the slope was possible. The values of the slope parameter extracted from the fits are equal to 8.0±0.3 independent of the target and 9.0±0.3 MeV for reactions at 17A and 23A MeV respectively. This beam energy dependence exhibits agreement with the systematics for nucleon-nucleon bremsstrahlung [14]. This contribution, extrapolated down to low energies, was then subtracted from each spectrum in order to obtain, only the statistical gamma component shown in Fig.2 as open symbols. The error bars include the statistical error and the errors on the subtraction of the bremsstrahlung component due to uncertainties on the slope and normalization.

Since the gamma rays from the GDR decay can be emitted at all steps during the decay chain of the hot nucleus the measured gamma spectrum is

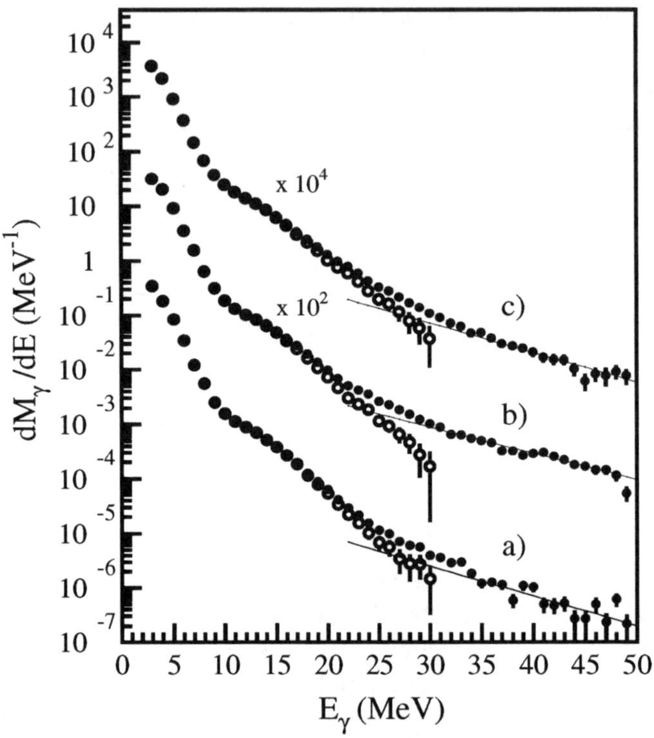

Figure 2. Gamma spectra measured at 90° for the three reactions (full symbols). From bottom to top: a) ^{116}Sn + ^{12}C at 17 AMeV; b) ^{116}Sn + ^{12}C at 23 AMeV; c) ^{116}Sn + ^{24}Mg at 17 AMeV. Open symbols represent the spectra obtained after subtraction of the bremsstrahlung contribution shown as full line.

not a direct reflection of the gamma emission from the compound nucleus at its initial temperature but rather an average over all temperatures from the initial down to zero. Moreover, the various reactions studied here and in previous investigations lead to initial hot nuclei with different masses and spins. Therefore, to study the evolution of the properties of the GDR as a function of the excitation energy we have to rely on a comparison with statistical calculation taking into account the whole decay sequence. The statistical gamma spectra were reproduced using the decay code CAS-

CADE [15] which treats the statistical emission of γ-rays, neutrons, protons and alpha particles from an equilibrated compound nucleus with a given excitation energy and spin.

In the calculation a Lorentzian line shape was assumed for the GDR with a mass dependent centroid energy given by $E_{GDR} = 76.5 \cdot A^{-1/3}$ and a fixed width $\Gamma = 12$ MeV. Besides, the strength was taken equal to 100% of the Thomas Reiche Kuhn (TRK) sum rule and a level density parameter dependent on the temperature of the system was adopted following the parametrization suggested by Ormand et al [16].

As input to the calculation the initial excitation energies and masses inferred from the massive transfer approach were used (see table 1). The results shown as full lines in fig.3 indicate a remarkable reproduction of the data at $E^*/A = 1.26$ and 1.57 MeV over the whole energy range examined while data at $E^*/A = 2.13$ MeV are not reproduced by CASCADE calculation whose prediction markedly overshoots the data in the GDR region indicating the onset of the GDR quenching (fig.3). Note that there is no arbitrary normalization involved in the comparison.

In the same figure, the two upper curves show, as a comparison, the gamma spectra measured in coincidence with heavy residues from the ^{36}Ar + ^{98}Mo reaction at 37 AMeV performed a few years ago at GANIL[17]. The spectra were built selecting incomplete fusion events where hot nuclei were populated at excitation energies per nucleon of 3.24 and 3.87 MeV respectively. The excitation energies and masses were estimated from the residue velocities through the massive transfer model correcting for pre-equilibrium light particle emission as described in ref.[18]. The corresponding Cascade calculations were performed accordingly. The values of the parameters used as input for the calculation are listed in the caption of Fig.3. The comparison shows that at all excitation energies the data are strongly over-predicted by the calculation in the entire GDR region while the low energy region is still reasonably reproduced.

From the comparison of the two set of data a clear picture emerges where a deficit of γ multiplicity with respect to predictions for 100% EWSR is observed. This effect arises in the region around $E^*/A = 2.0$ MeV and increases regularly with increasing excitation energy.

4. Conclusions

The evolution of the GDR properties was investigated through the study of three different reactions. A selection of fusion events in terms of recoil

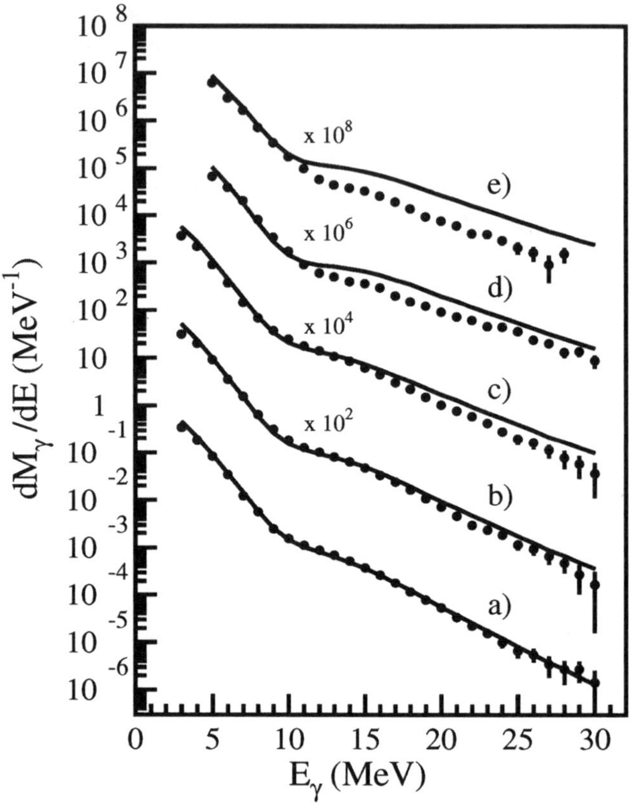

Figure 3. Comparison between statistical γ-ray spectra and CASCADE calculation at different excitation energies. From bottom to top: a) ^{116}Sn + ^{12}C at 17 AMeV, CASCADE input E* = 160 MeV and A = 127; b) ^{116}Sn + ^{12}C at 23 AMeV, CASCADE input E* = 200 MeV and A = 127; c) ^{116}Sn + ^{24}Mg at 17 AMeV, CASCADE input E* = 290 MeV and A = 136; d) ^{36}Ar + ^{98}Mo at 37 AMeV, CASCADE input E* = 350 MeV and A = 108; e) ^{36}Ar + ^{98}Mo at 37 AMeV, CASCADE input E* = 430 MeV and A = 111

velocity was applied and the coincident gamma spectra were analyzed. The gamma multiplicity was found to increase as a function of excitation energy. The comparison of the gamma spectra with CASCADE calculations suggests the onset of a saturation of the gamma yield at an excitation energy per nucleon of about 2 MeV. Some more refined calculations are needed to draw a definitive conclusion on the quenching of the GDR as a function of the excitation energy.

References

1. P.Piattelli et al., *Phys. Lett.* **B442**, 48 (1998)
2. J.H.G.van Pol et al., *Phys. Rev. Lett.* **76**, 1425 (1996)
3. R.Alba et al., *Nucl. Phys.* **A681** 339c (2001)
4. J.J.Gaardhoje et al., *Ann. Rev. Nucl. Part. Sci.* **42**, 483 (1992)
5. J.J.Gaardhoje et al., *Phys. Rev. Lett.* **59**, 1409 (1987)
6. T.Suomijarvi et al., *Phys. Rev.* **C53**, 2258 (1996)
7. J.H. Le Faou et al., *Phys. Rev. Lett.* **72**, 3321 (1994)
8. A.Smerzi et al., *Phys. Lett.* **B320**, 216 (1994)
9. P.F. Bortignon et al., *Phys. Rev. Lett.* **67**, 3360 (1991)
10. P.Chomaz *Nucl. Phys.* **A569**, 203c (1994)
11. E.Migneco et al., *NIM* **A314**, 31 (1992)
12. G.Bellia et al., *IEEE* **43**, 1737 (1996)
13. H.Nifenecker et al., *Nucl. Phys.* **A447**, 553c (1985)
14. H.Nifenecker and J.A.Pinston, *Annu. Rev. Nucl. Part. Sci.* **40**, 113 (1990)
15. F.Puhlhofer, *Nucl. Phys.* **A280**, 267 (1977)
16. W.E.Ormand et al., *Phys. Rev.* **C40**, 1510 (1989)
17. P.Piattelli et al., *Nucl. Phys.* **A649**, 181c (1999)
18. D.Santonocito et al., *Phys. Rev.* **C66**, 044619 (2002)

SECTION III

ULTRARELATIVISTIC HEAVY-ION COLLISIONS

DIMUON AND CHARMONIUM PRODUCTION IN HEAVY-ION COLLISIONS AT THE CERN SPS

E. SCOMPARIN
Istituto Nazionale di Fisica Nucleare (INFN), Sezione di Torino
Via P. Giuria 1,
I-10125 Torino, ITALY
E-mail: scomparin@to.infn.it

Experiments measuring muon pairs have played an important role in the SPS heavy-ion program since its very beginning, in 1986. I start by shortly reviewing the present situation in this field. Then, I describe the NA60 experiment, that has been designed in order to answer specific questions which remained open after the previous SPS experiments. NA60 has studied Indium–Indium collisions in 2003 and proton–nucleus in 2004. The data analysis is presently ongoing. Preliminary results on J/ψ suppression and on the study of low-mass resonances (ϕ, ω) in In-In collisions will be shown. Prospects for the future will also be shortly discussed.

1. Introduction

The studies of high energy nuclear collisions done at the SPS between 1986 and 2000 have given strong indications that, above a certain critical value of the colliding nuclear mass or of the collision centrality, the partons produced in the initial stage of high energy heavy–ion collisions undergo a transition to a system of interacting deconfined partons.

In particular, the J/ψ suppression, as a signature of the formation of a deconfined state, has been studied, via the $J/\psi \to \mu\mu$ decay, by the NA38 and NA50 experiments. In Pb-Pb collisions the J/ψ production pattern, as a function of the collision centrality, shows that above a certain centrality threshold the J/ψ yield is considerably lower than expected from the "nuclear absorption" curve, derived from proton-nucleus and light ions data [1]. One of the current interpretations of this result is that the dense and hot medium formed in the collisions dissolves the χ_c resonance, leading to the disappearance of the fraction ($\sim 30\,\%$) of J/ψ mesons that would otherwise originate from χ_c decays. However, several questions remain open. What is the physical variable driving the J/ψ suppression? Is it the number of participant nucleons? Or the local energy density? Or the average length of nuclear matter, L, traversed by the charmonium state? Are the nuclear dependences of the J/ψ and χ_c yields the same?

Many other physics observables have been studied in the dilepton sector and very interesting results have been obtained; however, as for the J/ψ suppression, their full understanding requires further work, to be still pursued at SPS energies.

In particular, the interpretation of the dimuon mass spectrum between the ϕ and the J/ψ resonances remains unclear. It has been shown by Helios-3 and NA50 that in p-A collisions it is well described by Drell-Yan and simultaneous semi-leptonic decays of D mesons, as expected. On the contrary, in A-A collisions the dimuon mass spectrum shows an excess which grows with the number of nucleons participating in the interaction. Two interpretations of this excess have been considered: it can be due to an unexpected enhancement of charm production or to thermal dimuons emitted from the QGP phase [2].

Furthermore, the CERES experiment measured the dielectron invariant mass spectrum in S-Au and Pb-Au collisions. A comparison with the expected sources (mainly light meson decays, which describe the proton data) shows an excess in the mass range 0.2–0.7 GeV/c^2. This observation [3] has been interpreted as an indication of changes in the mass and decay width of the ρ meson, maybe due to partial restoration of the chiral symmetry. However, this result suffers from lack of statistics and a poor signal-to-background ratio.

Finally, an enhancement of the ϕ meson yield with respect to the $(\omega{+}\rho)$ had been observed by NA50 as a function of centrality in Pb-Pb collisions [4]. This observation is in agreement with the expectation of an increased strangeness yield in a deconfined system, due to the (partial) restoration of chiral symmetry. However, a disagreement had been observed between the inverse slope T of the transverse momentum distribution of the ϕ, measured by NA50 through the decay $\phi \to \mu\mu$, and by NA49 through the channel $\phi \to KK$ [5]. In particular, NA50 observed no centrality dependence for T, while NA49 results showed a significant increase of T with centrality. Furthermore the NA49 T values were higher by about 20%, for central Pb-Pb events.

2. The NA60 experiment

To solve the questions outlined in the previous section, a new dimuon experiment, NA60, has been explicitly designed. With respect to previous experiments, NA60 has strongly improved the dilepton detection techniques [6,7]. In particular, a very good muon pair mass resolution and vertex identification accuracy have been obtained, in order to disentangle the various muon sources and better resolve the signal resonances. Furthermore, by taking data for In–In collisions, a lighter colliding system than Pb–Pb, NA60 investigates the onset of the J/ψ anomalous suppression with the aim of determining the physics variable driving the suppression mechanism.

In terms of detector layout, the NA60 experiment complements the Muon Spectrometer and Zero Degree Calorimeter (ZDC) inherited from the NA50 experiment with a completely new target region, where silicon micro-strip and pixel detectors, integrated in a 2.5 T dipole magnet, are used for the tracking of the charged particles produced in the collisions and for an accurate determination of primary and secondary vertices. By matching the muons measured in the muon spectrometer to tracks in the silicon vertex telescope, simultaneously using information on coordinates and momentum, it is possible to overcome the uncertainties introduced by multiple scattering and energy loss fluctuations, induced by the crossing of the hadron absorber. Besides the improvement in dimuon mass resolution, NA60 can also accurately determine the origin of the muons, and separate prompt dimuons (Drell-Yan, thermal) from the muon pairs due to decays of D mesons.

In the 5 weeks long run of year 2003, NA60 has collected $\sim$230 million dimuon triggers, with a 158 GeV/nucleon Indium beam at intensities around 5×10^7 ions per 5-seconds burst. This statistics is fully adequate for the investigation of the physics topics accessible to the experiment. The preliminary results on the J/ψ suppression and on the study of the yield and of the transverse momentum distributions of the ϕ meson will be summarized in the next sections. Then, in 2004, during a 3 month long data taking, proton–nucleus collisions at 400 and 158 GeV have been studied. The analysis of this data sample is now starting.

3. First results from the study of In-In collisions

3.1. *J/ψ suppression*

The high mass dimuon spectrum (see Fig. 1) shows the J/ψ and ψ' resonances sitting on a continuum composed of Drell–Yan dimuons and of muon pairs from decays of D mesons, besides the combinatorial background from π and K decays, which is determined from the measured like–sign muon pairs. In this analysis the J/ψ yield with respect to the Drell-Yan (DY) events is obtained. The ratio of two dimuon processes has the advantage of being insensitive to most of the experimental inefficiencies and to the integrated luminosity. Furthermore, the Drell-Yan process is known to scale linearly with the product of the projectile and target mass numbers, and is insensitive to the nature of the medium formed in the collisions.

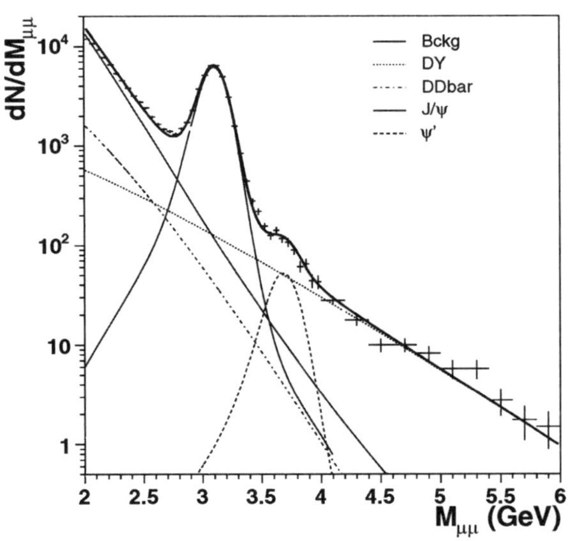

Figure 1. The dimuon mass spectrum ($m_{\mu\mu} > 2$ GeV/c^2) obtained before muon track matching, for the 6500 A event sample. The result of the fit is also shown (see text for details).

The muon track matching is not mandatory for the J/ψ study, since the J/ψ peak is clearly visible on top of the underlying continuum. However, once it is applied, a cleaner dimuon event sample is obtained, with an improved mass resolution and a reduced contribution from the combinatorial background due to π and K decays. Quantitatively, the width of the J/ψ peak decreases from ~ 105 to ~ 70 MeV for the matched sample and the percentage of combinatorial background in $\pm 1\sigma$ around the J/ψ peak drops from $\sim 3\,\%$ to less than $1\,\%$. The only drawback of requiring the track matching is the reduction of the available statistics, since the dimuon matching efficiency in the J/ψ mass region is $\sim$65 %, regardless of centrality. The results given in the following refer to the (J/ψ)/DY ratio integrated over all collision centralities for the sample of events without muon track matching, but the same procedure can be applied to the matched sample.

In order to extract the ratio between the J/ψ and the Drell–Yan production cross–sections, integrated over the range E_{ZDC} <15 TeV, the opposite–sign dimuon mass distribution is fitted to a superposition of the contributions of the various physics processes. The mass distributions are evaluated through a detailed Monte Carlo simulation, using Pythia with MRS–A (low Q^2) parton densities for event generation and GEANT for tracking through the experimental apparatus. The events are then reconstructed like the real data. From the Monte-Carlo calculation it is possible to deduce the J/ψ and Drell–Yan acceptances, A_ψ=12.4% and $A_{DY(2.9<m_{\mu\mu}<4.5)}$=13.4%, in the NA60 phase space window (2.92< y_{lab} <3.92, -0.5< cos$\theta_{\rm CS}$ <0.5). The fit procedure goes through three steps: first one obtains the Drell–Yan yield from the mass region above 4.2 GeV; then the charm normalization is determined from the mass window 2.2<m<2.5 GeV (having fixed the Drell–Yan from the previous step); finally the J/ψ and ψ' yields, and the exact position and width of the J/ψ peak, are extracted from the mass region 2.9<m<4.2 GeV.

In order to compare with previous results, published by the NA38 and NA50 experiments, obtained in p-A, S-U, and Pb-Pb collisions, the In-In result has been referred to the Drell–Yan production cross section integrated in the mass domain between 2.9 and 4.5 GeV. The analysis gives $B_{J/\psi\to\mu\mu}\sigma_{J/\psi}/\sigma_{DY} = 19.2 \pm 1.2$.

Figure 2 shows how the In-In point compares to the previously established J/ψ suppression pattern, as a function of L, the thickness of nuclear matter traversed by the charmonium state. Figure 3 shows the same measurements as a function of the number of nucleons involved in the collision, $N_{\rm part}$. In this figure the data points are divided by the normal nuclear absorption curve, determined from the p–A measurements, which can be seen as a continuous line in Fig. 2. The In–In measurement, when normalized to the absorption curve, gives the value 0.84 ± 0.05. Once the analysis of the centrality dependence of the J/ψ suppression will be finalized, NA60 should be able to probe the J/ψ suppression pattern as a function of centrality, in the ranges 5.5–7.8 fm in L and 50–200 in $N_{\rm part}$. By comparing the In-In and Pb-Pb patterns as a function of different centrality variables, one should understand which is the variable and, therefore, the physics mechanism, driving the J/ψ suppression. Finally, to fully exploit the available J/ψ statistics, NA60 is investigating the possibility of replacing, as a reference process, the low statistics Drell–Yan with an equivalent estimator computed from the large available sample of minimum bias events that has been collected.

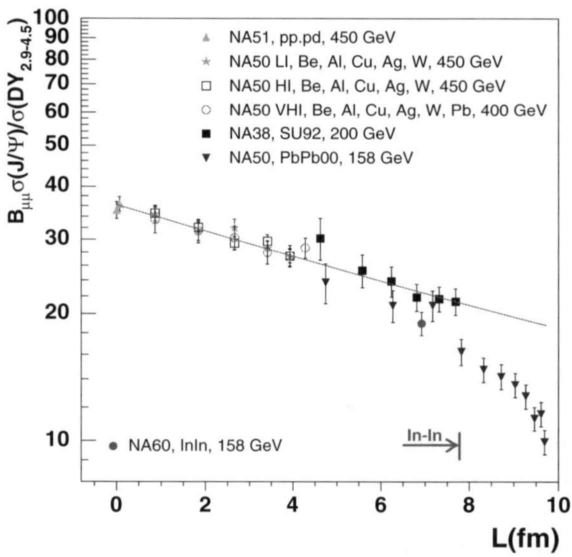

Figure 2. J/ψ suppression pattern as a function of L, including the Indium-Indium measurement. The arrow indicates the maximum centrality that can be reached in Indium–Indium collisions.

3.2. ϕ-meson studies

In the low-mass region of the dimuon invariant mass spectrum the use of the track matching algorithm allows a very significant improvement of the mass resolution. The sample used for this analysis contains 370000 events, with an overall signal to background ratio of 1/4, and corresponds to 35% of the total collected statistics. The ω and ϕ peaks are clearly visible (see Fig. 4). The mass resolution is 23 MeV at the ϕ and does not depend on centrality. The dimuon acceptance at low mass and low p_T has been significantly increased with respect to previous experiments. This major improvement is essentially due to the presence of the dipole field in the target region, which deflects to the angular acceptance of the muon spectrometer the low

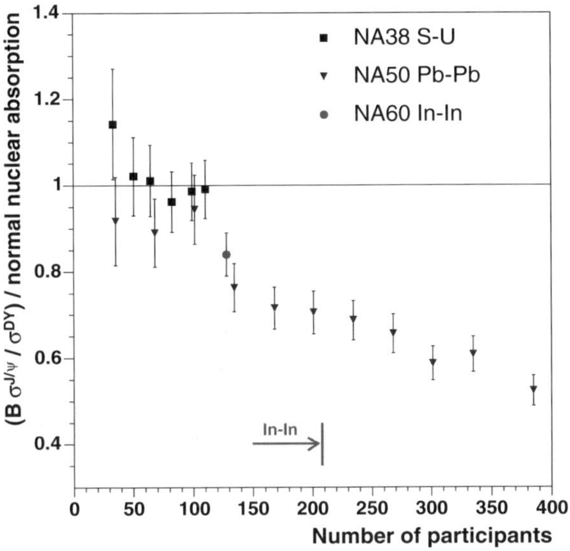

Figure 3. J/ψ suppression pattern as a function of $N_{\rm part}$. The arrow has the same meaning as in Fig. 2.

p_T opposite-sign muon pairs which would otherwise be lost in the dead area surrounding the beam line.

The centrality has been estimated from the measured multiplicity of charged particles. Four centrality bins have been defined, corresponding to an average number of participant nucleons $\langle N_{part} \rangle$ increasing from 20 for the most peripheral to 197 for the most central bin. For each bin, $\langle N_{part} \rangle$ has been evaluated from a Glauber fit to the forward energy spectrum, as measured in our zero-degree calorimeter (ZDC).

Figure 4 shows the signal dimuon mass distributions, after combinatorial background subtraction, for the four centrality classes. In order to more easily evaluate changes in the shape of the dimuon mass distribution between peripheral and central collisions, the spectra have been arbitrarily normalized with respect to each other at the ω peak. It can be seen that the low-mass resonances are very clearly defined in the mass distributions.

In order to study the ϕ/ω cross section ratio, a correction for the detection acceptances and efficiencies, calculated through a detailed Monte Carlo simulation, has been applied. All the pertinent light meson decays, both dimuon and Dalitz, have been generated using the GENESIS code. A con-

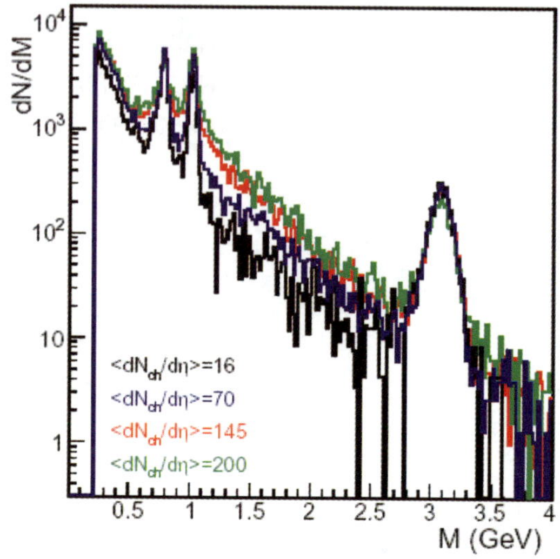

Figure 4. Dimuon mass spectra, after track matching, for different collision centralities.

tinuum source with an exponential fall-off beyond the ϕ was also included. The simulated events were then reconstructed in the same way as the real data. The measured dimuon spectra, for each centrality bin, have been fitted to a superposition of the simulated sources, leaving as free parameters the ratios η/ω and ϕ/ω, besides the continuum normalization. In this analysis the ρ and ω production cross sections are assumed to be identical. The resulting ϕ/ω ratio is shown in Fig. 5 as a function of the number of participants. The same figure includes data points obtained by the NA50 experiment in Pb-Pb collisions at the same energy, 158 GeV per nucleon [4], properly scaled to account for the slightly different phase space coverage of the two experiments. The results obtained in this way for the two different collision systems (Pb-Pb and In-In) are in rather good agreement, suggesting that N_{part} is a good scaling variable to describe the ϕ/ω ratio. However, this conclusion should be seen as very preliminary until all the systematic uncertainties involved in the comparison are well understood.

The ϕ transverse momentum distributions were studied by selecting signal dimuons in a narrow window around the ϕ mass. The background under the ϕ peak was evaluated and subtracted using two side windows symmetrically located around the peak. A two-dimensional acceptance correction (in the y and p_T variables) has been applied to the data. The result-

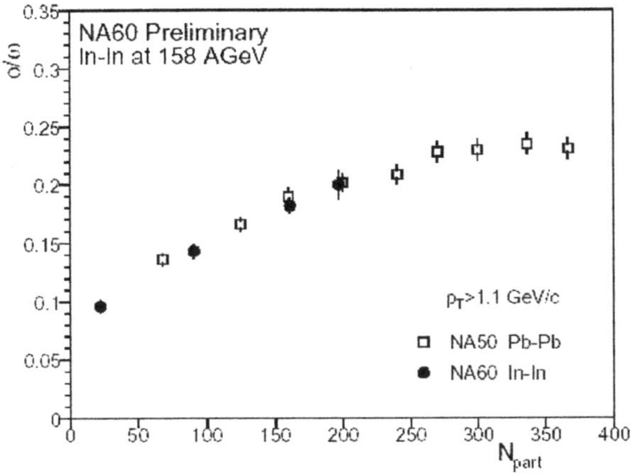

Figure 5. The ϕ/ω production cross section ratio as a function of the number of participants.

ing transverse momentum distributions have been fitted to the expression $1/p_T\mathrm{dN}/\mathrm{dp_T} \sim \exp(-\mathrm{m_T}/\mathrm{T})$, in order to determine the inverse slope parameter, T. Figure 6 shows how the T values extracted from NA60 data vary with the number of participants, in comparison with measurements previously done in Pb-Pb collisions by the NA49 [5] and NA50 [4] experiments, using $\phi \to KK$ and $\phi \to \mu\mu$ decays, respectively. The significant increase of T with N_{part} seen by NA60, and also in the data of NA49 (with bigger uncertainties), seems to be in contradiction with the observations of NA50, which reports T values independent of the collision centrality. To further clarify this issue, a completely independent analysis is currently in progress, to study the ϕ properties through the $\phi \to KK$ decay, thanks to the good quality of the data collected by the silicon vertex telescope.

4. Conclusions

After almost 20 years, dilepton production at the SPS is still a hot and lively physics subject, its study now being carried on by NA60. Preliminary results from the In-In run show a substantial improvement in the quality of the measurements with respect to previous experiments, which should lead to significant impacts in our understanding of the low and intermediate mass dilepton production, as well as of the origin of the anomalous J/ψ suppression.

Furthermore, with the high statistics proton–nucleus data, collected by

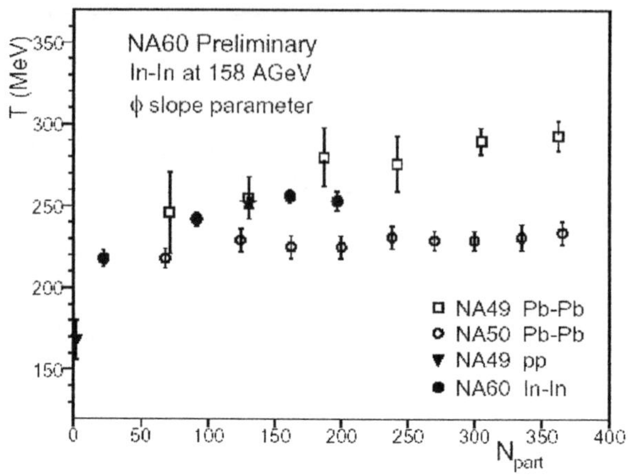

Figure 6. The ϕ inverse slope parameter, T, as a function of the number of participants.

NA60 in 2004, it will be possible to set an accurate reference baseline, which will be compared with the heavy–ion results.

References

1. L. Ramello et al. (NA50 Collaboration), *Nucl. Phys.* A **715**, 243c (2003).
2. M.C. Abreu et al. (NA38 and NA50 Collaborations), *Eur. Phys. J.* C **14**, 443 (2000).
3. J.P. Wessels et al. (CERES Collaboration), *Nucl. Phys.* A **715**, 262c (2003).
4. B. Alessandro et al. (NA50 Collaboration), *Phys. Lett.* B **555**, 147 (2003).
5. V. Friese et al. (NA49 Collaboration), *Nucl. Phys.* A **698**, 487c (2002).
6. Proposal CERN/SPSC 2000-010.
7. A. David et al. (NA60 Collaboration), *J. Phys. G* **30** S1101 (2004).

WHAT IS QUARK-GLUON PLASMA? RECENT PROGRESS *

M. ASAKAWA

Department of Physics
Osaka University
Toyonaka, 560-0043 JAPAN
E-mail: yuki@phys.sci.osaka-u.ac.jp

We analyze correlation functions of charmonia at finite temperature (T) on anisotropic lattices using the maximum entropy method (MEM). It is shown that J/ψ and η_c survive as distinct resonances in the plasma even up to $T \simeq 1.6T_c$ and that they eventually dissociate between $1.6T_c$ and $1.7T_c$ (T_c is the critical temperature of deconfinement). This suggests that the deconfined plasma is non-perturbative enough to hold heavy-quark bound states. The significance of this result in RHIC physics is also discussed.

1. Introduction

Whether hadrons survive even in the deconfined quark-gluon plasma is one of the key questions in quantum chromodynamics (QCD). This problem was first examined in Ref. 1 and Ref. 2 in different contexts. The fate of the heavy mesons such as J/ψ in the deconfined plasma was also investigated in a phenomenological potential picture taking into account the Debye screening[3]. In general, there is no a priori reason to believe that the dissociation of the bound states should take place exactly at the phase transition point[4].

From the theoretical point of view, the spectral function (SPF) at finite temperature T, which has all the information of in-medium hadron properties, is a key quantity to be studied. Recently, it was shown[5] that the first-principle lattice QCD simulation of SPFs is possible by utilizing the

*This work is supported in part by the Grants-in-Aid of the Japanese Ministry of Education, Science and Culture, No. 14540255. Lattice calculations have been performed with the CP-PACS computer under the "Large-scale Numerical Simulation Program" of Center for Computational Physics, University of Tsukuba.

maximum entropy method (MEM). The basic concepts and applications of MEM on the lattice at $T = 0$ and $T \neq 0$ were formulated in Ref. 6. In this article, we first describe the necessity of the maximum entropy method in extracting spectral functions from lattice gauge data and its principles. Then, we present our results of the spectral functions of the J/ψ and η_c channels above the deconfinement phase transition. Finally, we discuss its significance and relation with recent findings at RHIC.

2. Spectral Function

For simplicity, we shall focus on SPFs at vanishing baryon chemical potential. We start by defining the following two-point correlation functions at finite T:

$$\begin{aligned} \tilde{D}^R_{\eta\eta'}(t,\mathbf{x}) &= i\ \mathrm{Tr}\left(\mathrm{R}\left[J_\eta(t,\mathbf{x})\ J^\dagger_{\eta'}(0,\mathbf{0})\right] e^{-H/T}\right)/Z \qquad (1)\\ &\equiv \int \frac{d^4k}{(2\pi)^4}\ D^R_{\eta\eta'}(k^0,\mathbf{k})\ e^{-ikx}\quad ,\\ \tilde{D}_{\eta\eta'}(\tau,\mathbf{x}) &= \mathrm{Tr}\left(\mathrm{T}_\tau\left[J_\eta(\tau,\mathbf{x})\ J^\dagger_{\eta'}(0,\mathbf{0})\right] e^{-H/T}\right)/Z \qquad (2)\\ &\equiv T\sum_n \int \frac{d^3k}{(2\pi)^3}\ D_{\eta\eta'}(i\omega_n,\mathbf{k})\ e^{-i(\omega_n\tau-\mathbf{k}\cdot\mathbf{x})}\quad . \end{aligned}$$

Here J_η and $J^\dagger_{\eta'}$ are composite operators in the Heisenberg representation in real (imaginary) time for $\tilde{D}^R$ ($\tilde{D}$). η and η' denote possible Lorentz and internal indices of the operators. Z is the partition function of the system defined as $Z = \mathrm{Tr}e^{-H/T}$. ω_n is the Matsubara frequency defined by $\omega_n = 2n\pi T$ when J_η and $J^\dagger_{\eta'}$ are the Grassmann even (= bosonic) operators and $\omega_n = (2n+1)\pi T$ when they are Grassmann odd (= fermionic) operators. The retarded product R and the imaginary-time ordered product T_τ are defined, respectively, as

$$\mathrm{R}[A(t)B(0)] \equiv \theta(t)(A(t)B(0) \mp B(0)A(t)), \qquad (3)$$
$$\mathrm{T}_\tau[A(\tau)B(0)] \equiv \theta(\tau)A(\tau)B(0) \pm \theta(-\tau)B(0)A(\tau). \qquad (4)$$

The upper sign is for bosonic operators and the lower sign is for fermionic operators, which is used throughout this article.

The spectral function (SPF) is defined as the imaginary part of the

Fourier transform of the retarded correlation (see, e.g., Ref. 7),

$$A_{\eta\eta'}(k_0, \mathbf{k}) \equiv \frac{1}{\pi}\mathrm{Im}D^R_{\eta\eta'}(k_0, \mathbf{k}) \tag{5}$$

$$= (2\pi)^3 \sum_{n,m} \frac{e^{-E_n/T}}{Z} \langle n|J_\eta(0)|m\rangle\langle m|J^\dagger_{\eta'}(0)|n\rangle$$
$$\times(1 \mp e^{-P^0_{mn}/T})\delta^4(k^\mu - P^\mu_{mn}), \tag{6}$$

where $|n\rangle$ is a state with 4-momentum P^μ_n and the normalization $\langle m|n\rangle = \delta_{nm}$. P^μ_{mn} is defined by $P^\mu_m - P^\mu_n$. $D^R(k_0, \mathbf{k})$ is the Fourier transform of $\tilde{D}^R(t, \mathbf{x})$ with respect to both time and space coordinates.

$A_{\eta\eta'}$ is in some cases directly related to the experimental observables. For example, consider $J_\eta = j^{em}_\mu$ and $J^\dagger_{\eta'} = j^{em}_\nu$ with

$$j^{em}_\mu = \frac{2}{3}\bar{u}\gamma_\mu u - \frac{1}{3}\bar{d}\gamma_\mu d - \frac{1}{3}\bar{s}\gamma_\mu s \cdots, \tag{7}$$

which is the electromagnetic current in QCD. Then, the dilepton production rate from hot matter at finite T, which is an observable in relativistic heavy-ion collisions, is written through Eq. 6 as follows (see, e.g., Ref. 8):

$$\frac{dN_{l^+l^-}}{d^4x d^4k} = -\frac{\alpha^2}{3\pi^2 k^2}\frac{\sum^3_{\mu=0} A^\mu_\mu(k_0, \mathbf{k})}{e^{k_0/T} - 1}, \tag{8}$$

where α is the electromagnetic fine structure constant. The mass of leptons is neglected in this formula for simplicity. Note that this equation is valid irrespective of the state of the system, i.e., whether it is in the hadronic phase or in the quark-gluon plasma.

By calculating the mixed representation of the Matsubara correlator, it can be shown that the following relation holds,

$$D_{\eta\eta'}(\tau, \mathbf{k}) = \int^{+\infty}_{-\infty} \frac{e^{-\tau\omega}}{1 \mp e^{-\beta\omega}} A_{\eta\eta'}(\omega, \mathbf{k})\, d\omega \qquad (0 \le \tau < \beta). \tag{9}$$

When $\eta = \eta'$ and J_η is a quark bilinear operator such as $\bar{q}q$, $\bar{q}\gamma_\mu q$, $\cdots$ etc., one can further reduce the sum rule into a form similar to the Laplace transform:

$$D(\tau, \mathbf{k}) = \int^{+\infty}_0 \frac{e^{-\tau\omega} + e^{-(\beta-\tau)\omega}}{1 - e^{-\beta\omega}} A(\omega, \mathbf{k})\, d\omega$$
$$\equiv \int^{+\infty}_0 K(\tau, \omega)\, A(\omega, \mathbf{k})\, d\omega \quad (0 \le \tau < \beta), \tag{10}$$

where we have suppressed the index η for simplicity. K is the kernel of the integral transform. Eq. 10 is the basic formula, which we shall utilize later.

From now on, we focus on SPFs "at rest" ($\mathbf{k} = 0$) for simplicity and omit the label $\mathbf{k}$.

3. Maximum Entropy Method

Monte Carlo simulation provides $D(\tau)$ in Eq. 10 for a discrete set of points, $\tau = \tau_i$, with

$$1 \leq \tau_i/a \leq N_\tau, \tag{11}$$

where N_τ is the number of the temporal lattice sites and a is the lattice spacing. In the actual analysis, we use data points in a limited domain $\tau \in [\tau_{min}, \tau_{max}]$. Since we can generate only finite number of gauge configurations numerically, the lattice data $D(\tau_i)$ has a statistical error. From such finite number of data with noise, we need to reconstruct the continuous function $A(\omega)$ on the r.h.s. of Eq. 10, or equivalently to perform the inverse Laplace transform.

This is a typical ill-posed problem, where the number of data points is much smaller than the number of degrees of freedom to be reconstructed. The standard likelihood analysis (χ^2-fitting) is obviously inapplicable here, since many degenerate solutions appear in the process of minimizing χ^2. This is the reason why the previous analyses of the spectral functions have been done only under strong assumptions on the spectral shape[9,10]. Drawbacks of the previous approaches are twofold: (i) a priori assumptions on SPF prevent us from studying the fine structures of SPF, and (ii) the result does not have good stability against the change of the number of parameters used to characterize SPF. Both disadvantages become even more serious at finite T, where we have almost no prior knowledge on the spectral shape.

The maximum entropy method is a method to circumvent these difficulties by making a statistical inference of the most probable SPF (or sometimes called the *image* in the following) as well as its reliability on the basis of a limited number of noisy data.

The theoretical basis of the maximum entropy method is Bayes' theorem in probability theory[11]:

$$P[X|Y] = \frac{P[Y|X]P[X]}{P[Y]} \ , \tag{12}$$

where $P[X|Y]$ is the conditional probability of X given Y. The theorem is easily proved by using the product formula, $P[XY] = P[X|Y]P[Y] = P[Y|X]P[X]$. Let D stand for Monte Carlo data with errors for a specific channel on the lattice and H summarize all the definitions and prior knowledge such as $A(\omega \geq 0) \geq 0$. From Bayes' theorem, the conditional probability of having $A(\omega)$ given the data reads

$$P[A|DH] = \frac{P[D|AH]P[A|H]}{P[D|H]}. \tag{13}$$

Here $P[D|AH]$ and $P[A|H]$ are called the *likelihood function* and the *prior probability*, respectively. $P[D|H]$ is simply a normalization constant independent of $A(\omega)$. (Note here that it may be more appropriate to call P "plausibility" instead of "probability", since P does not necessarily have the frequency interpretation[12,13].)

Now, the most probable image is $A(\omega)$ that satisfies the condition,

$$\frac{\delta P[A|DH]}{\delta A} = 0. \tag{14}$$

Furthermore, the reliability of the image satisfying Eq. 14 can be estimated by the second variation, or schematically, $\delta^2 P[A|DH]/\delta A \delta A$.

To proceed further, we need to specify the explicit forms of the likelihood function and the prior probability. This will be discussed below.

3.1. *Likelihood Function*

For large number of Monte Carlo measurements of a correlation function, the data is expected to obey the Gaussian distribution according to the central limit theorem:

$$P[D|AH] = \frac{1}{Z_L}\, e^{-L} \ , \tag{15}$$

$$L = \frac{1}{2}\sum_{i,j}(D(\tau_i) - D_A(\tau_i))C_{ij}^{-1}(D(\tau_j) - D_A(\tau_j)), \tag{16}$$

where i and j run over the actual data points which we utilize in the analysis, $\tau_{min}/a \leq i,j \leq \tau_{max}/a$. For later purposes, we define the number of data points to be used in MEM,

$$N = \tau_{max}/a - \tau_{min}/a + 1. \tag{17}$$

$D(\tau_i)$ is the lattice data averaged over gauge configurations,

$$D(\tau_i) = \frac{1}{N_{conf}} \sum_{m=1}^{N_{conf}} D^m(\tau_i), \tag{18}$$

where N_{conf} is the total number of gauge configurations and $D^m(\tau_i)$ is the data for the m-th gauge configuration. $D_A(\tau_i)$ in Eq. 16 is the correlation function defined by the r.h.s. of Eq. 10.

C in Eq. 16 is an $N \times N$ covariance matrix defined by

$$C_{ij} = \frac{1}{N_{conf}(N_{conf}-1)} \sum_{m=1}^{N_{conf}} (D^m(\tau_i) - D(\tau_i))(D^m(\tau_j) - D(\tau_j)). \quad (19)$$

Lattice data have generally strong correlations among different τ's, and it is essential to take into account the off-diagonal components of C.

In the case where $P[A|H]$ = constant in Eq. 13, maximizing $P[A|DH]$ is equivalent to maximizing Eq. 15 with respect to A, which is nothing but the standard χ^2-fitting. On the lattice the number of lattice data is $O(10)$, which is much smaller than the number of points of the spectral function to be reproduced ($O(10^3)$). Therefore, the χ^2-fitting does not work. This difficulty is overcome in the maximum entropy method, where the existence of a non-constant $P[A|H]$ plays an essential role.

3.2. *Prior Probability*

In MEM, the prior probability is written with auxiliary parameters α and m as

$$P[A|H\alpha m] = \frac{1}{Z_S} e^{\alpha S}, \quad (20)$$

where S is the Shannon-Jaynes entropy,[14,15]

$$S = \int_0^\infty \left[A(\omega) - m(\omega) - A(\omega) \log\left(\frac{A(\omega)}{m(\omega)}\right)\right] d\omega. \quad (21)$$

α is a real and positive parameter, while $m(\omega)$ is a real and positive function called the *default model* or the *prior estimate.* Although α and m are a part of the hypothesis H in Eq. 13, we write them explicitly on the l.h.s. of Eq. 20 to separate them from the other hypotheses. The output image $A_{\rm out}(\omega)$ is given by a weighted average over α:

$$A_{\rm out}(\omega) = \int A_\alpha(\omega) P[\alpha|Dm] d\alpha, \quad (22)$$

where $A_\alpha(\omega)$ is obtained by maximizing

$$Q \equiv \alpha S - L. \quad (23)$$

α is thus eliminated in the final results. α dictates the relative weight of the entropy (which tends to fit A to the default model m) and the likelihood function L (which tends to fit A to the data). m remains in the final results, but one can study the sensitivity of the results against the change of m.

There are some important points regarding the MEM analysis. First, one can prove that the solution of $\delta Q = 0$ is unique if it exists[6]. Therefore, if one solution of $\delta Q = 0$ is found, that is enough. This fact simplifies the numerical analysis quite a lot. Second, since MEM is based on probability theory, it is possible and, at the same time, mandatory to carry out statistical error analyses of the result. This step is crucial in the MEM analysis. No MEM analysis without error analyses carries any information on how reliable the result is.

4. Lattice Simulation and Results

The applications of MEM to $T \neq 0$ system are a big challenge[6]. The difficulty originates from the fact that the temporal lattice size L_τ is restricted as $L_\tau = 1/T = N_\tau a_\tau$, where N_τ is the number of temporal lattice sites. Because of this, it becomes more difficult to keep enough N_{data} to obtain reliable SPFs as T increases. Thus, simulations up to a few times T_c with N_{data} as large as 30^{16} inevitably require an anisotropic lattice.

On the basis of the above observation, we have carried out quenched simulations with $\beta = 7.0$ on $32^3 \times N_\tau$ ($32 \leq N_\tau \leq 96$) lattices. The renormalized anisotropy is $\xi = a_\sigma/a_\tau = 4.0$. We take the naive plaquette gauge action and the standard Wilson quark action. For further details of the lattice simulation, see Ref. 17.

Shown in Fig. 1 (Fig. 2) are $\rho(\omega)$s [a] for J/ψ (η_c) at $T/T_c = 0.78, 1.38,$ and 1.62 (Fig. 1(a) (Fig. 2(a))) and those at $T/T_c = 1.70, 1.78, 1.87,$ and 2.33 (Fig. 1(b) (Fig. 2(b))) (The corresponding numbers of temporal data points used in these figures are 89, 40, 34, 34, 34, 33, and 25 from low T to high T). If the deconfined plasma were composed of almost free quarks and gluons, SPFs would show a smooth structure with no pronounced peaks above the $q\bar{q}$ threshold. To the contrary, we find a sharp peak near the zero temperature mass even up to $T \simeq 1.6T_c$ as shown in Figs. 1(a) and 2(a), while the peak disappears at $T \simeq 1.7T_c$ as shown in Figs. 1(b) and 2(b). The width of the first peak in Figs. 1(a) and 2(a) partly reflects unphysical broadening due to the statistics of the lattice data and partly reflects possible physical broadening at finite T. At the moment, the former width of a few hundred MeV seems to dominate and we are not able to draw definite conclusions on the thermal mass shift and broadening.

---

[a] Following Ref. 6, we define dimensionless SPFs: $A(\omega) = \omega^2\rho(\omega)$ for η_c and $A(\omega) = 3\omega^2\rho(\omega)$ for J/ψ.

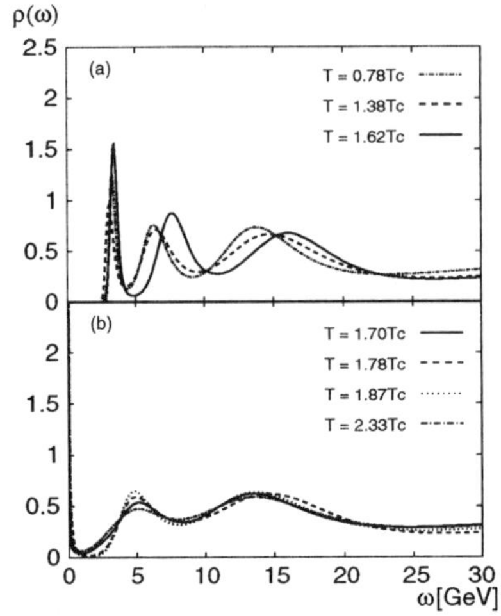

Figure 1. Spectral functions for J/ψ (a) for $T/T_c = 0.78, 1.38$, and 1.62 (b) for $T/T_c = 1.70, 1.78, 1.87$ and 2.33.

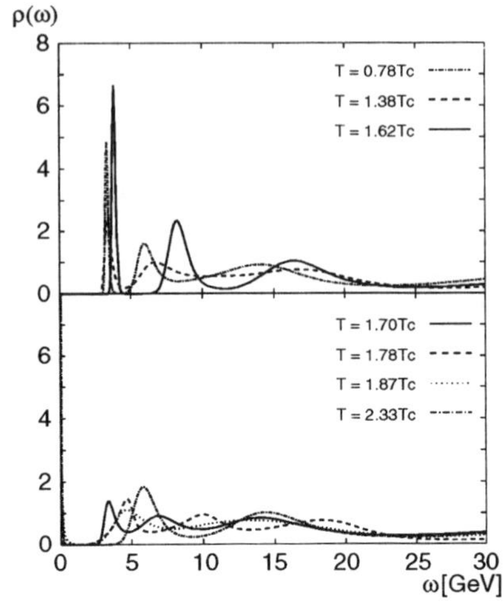

Figure 2. Spectral functions for η_c (a) for $T/T_c = 0.78, 1.38$, and 1.62 (b) for $T/T_c = 1.70, 1.78, 1.87$ and 2.33.

As emphasized earlier, error analyses are absolutely necessary. More specifically, to draw the conclusion on a firm ground, the following two kinds of analyses should be carried out: [I] the MEM error analysis of the resultant SPFs and [II] the sensitivity of the SPFs to $N_{\rm data}$. These tests are crucial to prevent fake generation and/or smearing of the peaks as emphasized in Refs. 6, 16. We have carried out these tests for the above results, and we have confirmed that the above features are indeed statistically significant and are not caused by the artifact of the insufficient number of data points[17].

5. Discussions

In the above, we have found that J/ψ and η_c remain as distinct resonances up to $T \simeq 1.6T_c$ and they disappear between $1.6T_c$ and $1.7T_c$. This suggests that the deconfined plasma is non-perturbative enough to hold heavy-quark bound states unlike the widely believed expectation.

For the yield of J/ψ at RHIC, no conclusive experimental result has been obtained[18]. However, recent other observations strongly suggest that the state of matter above the deconfining phase transition, which has been called quark-gluon plasma, is actually not a weakly interacting gas of quarks and gluons.

One example is the success of the quark recombination model[19] for the hadron yields in the intermediate transverse momentum region[20]. This model assumes that, at the hadronization of the deconfined plasma, hadrons are produced by the recombination of constituent quarks, two quarks for a meson and three quarks for a baryon. There have been criticisms against this assumption, since no gluonic degrees of freedom are taken into account.

There has been a puzzle about the charge fluctuation at RHIC. The charge fluctuation is defined by

$$D = 4\langle(\Delta Q)^2\rangle / N_{\rm ch}, \tag{24}$$

where $\langle(\Delta Q)^2\rangle$ denotes the event-by-event net charge fluctuation within a given rapidity window Δy, and $N_{\rm ch}$ is the total number of charged particles emitted in the rapidity window[21]. For a free plasma of quarks and gluons $D \approx 1$, while for a free pion gas $D \approx 4$. The value of D at RHIC is much larger than the value for the naively expected value for the quark-gluon plasma[22]. The reason for this deviation has been unknown, but recently it was shown that this can be understood if the state just above the phase transition is described as a gas of constituent (anti)quarks with mutual correlations but without gluonic degrees of freedom[23].

The investigation of this strongly interacting plasma has just started. Almost everything needs further refinement, but what is almost for sure now is that there is a new and previously unexpected state of matter just above the deconfinement, strongly interacting quark-gluon plasma (sQGP).

References

1. T. Hatsuda and T. Kunihiro, *Phys. Rev. Lett.* **55**, 158 (1985).
2. C. E. DeTar, *Phys. Rev.* **D32**, 276 (1985).
3. T. Matsui and H. Satz, *Phys. Lett.* **B178**, 416 (1986).
4. An example is the superconductivity in the strong coupling. See,
   P. Noziéres and S. Schmitt-Rink, *J. Low Temp. Phys.* **59**, 195 (1985);
   E. Babaev, *Int. J. Mod. Phys.* **A16**, 1175 (2001).
5. Y. Nakahara, M. Asakawa and T. Hatsuda, *Phys. Rev.* **D60**, 091503 (1999).
6. M. Asakawa, T. Hatsuda and Y. Nakahara, *Prog. Part. Nucl. Phys.* **46**, 459 (2001).
   See also, M. Jarrell and J. E. Gubernatis, *Phys. Rep.* **269**, 133 (1996).
7. J. Negele and H. Orland, *Quantum Many-Particle Systems* (New York: Addison-Wesley) (1988).
8. R. Rapp and J. Wambach, *Adv. Nucl. Phys.* **25**, 1 (2000).
9. M. C. Chu, J. M. Grandy, S. Huang and J. W. Negele, *Phys. Rev.* **D48**, 3340 (1993);
   D. B. Leinweber, *Phys. Rev.* **D51**, 6369 (1995);

D. Makovoz and G. A. Miller, *Nucl. Phys.* **B468**, 293 (1996);
C. Allton and S. Capitani, *Nucl. Phys.* **B526**, 463 (1998).

10. T. Hashimoto, A. Nakamura and I. O. Stamatescu, *Nucl. Phys.* **B400**, 267 (1993); *Nucl. Phys.* **B406**, 325 (1993);
Ph. de Forcrand *et al.* (QCD-TARO Collaboration), *Phys. Rev.* **D63**, 054501 (2001).
11. H. Jeffreys, *Theory of Probability*, (Oxford: Oxford University Press) (1998);
G. E. P. Box and G. C. Tiao, *Bayesian Inference in Statistical Analysis* (New York: John Wiley and Sons) (1992).
12. E. T. Jaynes, *How does brain do plausible reasoning?*, Stanford University Microwave Laboratory report 421 (1957) reprinted in *Maximum-Entropy and Bayesian Methods in Science and Engineering* eds. G. J. Erickson and C. R. Smith, vol. 1 (London: Kluwer Academic Publishers) pp. 1-24 (1988).
13. C. R. Smith and G. Erickson, in *Maximum-Entropy and Bayesian Methods* ed J. Skilling (London: Kluwer Academic Publishers) pp. 29-44 (1989).
14. C. E. Shannon and W. Weaver, *The Mathematical Theory of Communication* (Urbana: University of Illinois Press) (1949).
15. E. T. Jaynes, *Phys. Rev.* **106**, 620 (1957); *Phys. Rev.* **108**, 171 (1957).
16. M. Asakawa, T. Hatsuda and Y. Nakahara, *Nucl. Phys.* **B** (Proc. Suppl.) **119**, 481 (2003).
17. M. Asakawa and T. Hatsuda, *Phys. Rev. Lett.* **92**, 012001 (2004)
18. S. S. Adler *et al.* [PHENIX Collaboration], *Phys. Rev.* **C69**, 014901 (2004).
19. R. J. Fries, B. Müller, C. Nonaka and S. A. Bass, *Phys. Rev. Lett.* **90**, 202303 (2003);
V. Greco, C. M. Ko and P. Levai, *Phys. Rev. Lett.* **90**, 202302 (2003);
R. C. Hwa and C. B. Yang, *Phys. Rev.* **C67**, 034902 (2003).
20. S. S. Adler *et al.* [PHENIX Collaboration], *Phys. Rev. Lett.* **91**, 182301 (2003);
J. Adams *et al.* [STAR Collaboration], *Phys. Rev. Lett.* **92**, 052302 (2004).
21. M. Asakawa, U. W. Heinz and B. Müller, *Phys. Rev. Lett.* **85**, 2072 (2000);
S. Jeon and V. Koch, *Phys. Rev. Lett.* **85**, 2076 (2000).
22. K. Adcox *et al.* [PHENIX Collaboration], *Phys. Rev. Lett.* **89**, 082301 (2002).
23. C. Nonaka, B. Müller, S. Bass and M. Asakawa, in progress.

DESIGN AND PERFORMANCE OF THE ALICE EXPERIMENT AT LHC

E. NAPPI

(on behalf of the ALICE Collaboration)

Istituto Nazionale di Fisica Nucleare,Sezione di Bari,
Via G. Amendola,173 I-70125 Bari, Italy

Head-on collisions between ultrarelativistic heavy nuclei are believed to provide the extreme conditions of energy densities sufficient for the transition of hadronic matter to a short-lived state, called Quark-Gluon Plasma (QGP), where quarks are no longer confined within the nucleon. The forthcoming Large Hadron Collider (LHC) at CERN, scheduled to start commissioning during 2007, will be the ultimate facility for searching QGP. Complementary to the CMS and ATLAS experiments whose physics programmes focus on the search of the Higgs particle and super-symmetric particles in p-p collisions, the design of the ALICE experiment (A Large Ion Collider Experiment), has been optimized to study nucleus-nucleus collisions at LHC energies via a simultaneous measurement of many different observables. The ALICE setup comprises a multipurpose complex of tracking and particle-identification devices installed in a large solenoidal magnet and a muon detection arm in the forward rapidity region. Construction status and anticipated performance of the main ALICE sub-detectors will be discussed.

1. Introduction

According to the lattice-QCD calculations [1], the nuclear matter under extreme conditions of compression and heating evolves into a non-hadronic bulk phase of strongly interacting matter which resembles the conditions of the early Universe, just after the Bing Bang. Inside this plasma of quarks and gluons (QGP), chiral symmetry is approximately restored and colour freely propagates.

The promising results achieved at CERN-SPS and at BNL-RHIC have shown that nucleus-nucleus collisions are a powerful tool to investigate the behaviour of nuclear matter at densities higher than that of the ground state. However, owing to the non-perturbative features of the physics under study, understanding the results of energetic collisions between ultra-relativistic nuclei is a painstaking task requiring many simultaneous observations. The interpretative difficulties met in drawing conclusions from the analysis of SPS and RHIC data will likely be overcome at the LHC in view of the unprecedented energies at which nuclei will be accelerated. In fact this forthcoming facility, scheduled to start operation

in 2007, not only will allow to explore regions of energy and particle density which are significantly beyond those achieved at SPS and RHIC, but will also allow to study new signatures arising from very high momentum transfer processes whose cross sections, at the LHC energies, are large enough to comfortably enable detailed experimental studies. Such hard probes (heavy flavours, direct photons and photon tagged jets, jets and di-jets) will provide valuable information and insights on the early and thermalized stages of the collision phase since their production occurs on so small temporal and spatial scales to entail that any modification of their known properties is caused by the medium they have subsequently traversed. Therefore the main advantage offered by the LHC will be the opportunity to investigate the non-perturbative QGP physics by means of perturbative probes whose predictions are supported by solid theoretical calculations.
The early evolution of Pb-Pb collisions at the LHC energy of 5.5 TeV in the nucleon-pair center-of-mass, about thirty times higher than at RHIC, will mostly be determined by a strong nuclear shadowing. As anticipated by classical QCD, in the regime of low Bjorken-x ($\sim x_T = 2\, p_T/\sqrt{s} = 10^{-4} - 10^{-3}$), made accessible because of the large energy attainable at the LHC, one of the two colliding lead nuclei will resolve the other incoming nucleus as a superposition of a large number of soft gluons which, being so densely packed, will eventually overlap and merge. As a consequence of the large number of energy and momentum exchanges, the thermal equilibration of the medium will be quickly achieved, on a time scale of 0.1 fm/c, leading to a very high initial energy density and temperature. The energy density, of the order of 1000 GeV/fm^3, will be twenty times higher than can be reached at RHIC while the initial temperature [2] will be greater by more than a factor of two. This high initial temperature will extend the lifetime and the volume of the deconfined medium which will subsequently expand while cooling down to the freeze-out temperature. Compared to RHIC, the ratio of the lifetime of the quark-gluon plasma state to the time for thermalization is expected to be larger by an order of magnitude. As a result, the "fireballs" created in heavy ion collisions at the LHC will spend nearly their entire lifetime in a purely partonic state, widening the time window available to experimentally probe the quark-gluon plasma state. In conclusions, the LHC will create a hotter, larger and longer-living QGP state than the present heavy-ion facilities, approaching the most favourable conditions for providing a consistent and uncontroversial experimental evidence of the phase transition. Moreover, since the baryon content of the system after the collision is expected to be concentrated rather near the rapidity of the two colliding nuclei, mid-rapidity nuclear matter at the LHC will be much more baryon-free than at RHIC and closer to the conditions simulated in lattice QCD for a vanishing baryo-chemical potential allowing a more direct comparison with theory predictions.

2. Basic design issues

Currently, more than 1000 physicists from 80 institutions in 28 countries are involved in building ALICE, the only experiment at LHC specifically devoted to investigate heavy ion collisions [3-4]. The uneasy role of being alone for this kind of physics (apart for the CMS and ATLAS experiments which will have some very limited capabilities for nuclear collision studies) conditioned its design. In fact, ALICE is meant as a general-purpose experiment with a single set-up whose design has been optimized to study the full spectrum of observables (hadrons, photons and leptons) together with a global survey of the events over a large phase space.

Designed to cover a wide range of momenta (from 100 MeV/c up to 100 GeV/c) and identified particles (from the electron to the Υ resonances), ALICE must be able to cope with the highest particle density anticipated for Pb-Pb collisions at the LHC. Its apparatus has also been optimized to study interactions between nucleon-nucleon, at lower luminosity than the other LHC experiments, and proton-nucleus, to be used as reference data for the Pb-Pb collisions. Collisions between lighter nuclei than lead are also envisaged to investigate the influence of the energy density and volume on the phase transition properties.

In contrast to p-p physics, which requires high luminosity to search for rare events produced with cross-sections of the order of the pb, the measurement of observables related to the deconfined plasma of quarks and gluons and its time evolution can be performed in nuclear collisions at relatively low luminosity. The LHC design luminosity for Pb beams is $L = 10^{27}$ $cm^{-2}s^{-1}$ corresponding to an interaction rate of about 10^4 Hz of which only a small fraction, approximately 5%, accounts for the most interesting central collisions with maximum particle production. Since fluences and doses roughly scale with luminosity, it follows that requirements on detector electronics and radiation damage issues are much less relevant than those in pp experiments thus allowing to exploit, in the ALICE layout, the outstanding tracking capability of slow devices like the time projection chamber (TPC) and the silicon drift detector. On the other hand, nuclear cross-sections ranging between 10 mb and the barn entail the production of a large amount of particles per Pb-Pb interaction. The high multiplicities anticipated at LHC combined with the large rapidity range to be covered have represented the main technical challenges in designing the ALICE apparatus. The average charged-particle multiplicity per unit rapidity plays an important role not only in assessing the detector performance but also physics-wise, because it largely determines the accuracy with which many observables can be measured, especially in the event-by-event studies of fluctuations. Initial theoretical estimates for the rapidity density of produced charged hadrons for central Pb+Pb collisions ranged from ~2000 to 8000 per unit of rapidity at mid-rapidity. The high multiplicities in central nucleus-nucleus collisions typically arise from the large number of independent and successive nucleon-nucleon collisions,

occurring when many nucleons interact several times on their path through the oncoming nucleus. A scaling function, based on the parton saturation and classical QCD, has recently been derived to predict the charged multiplicity as a function of beam energy. It has been successfully used for evaluating the multiplicity at full RHIC energy (200 GeV) based on the 130 GeV data [5]. Extrapolation of this model to LHC energies yields a multiplicity density at the lower end of the predicted range (2000 or less). However, since the difference in energy between SPS to RHIC is smaller than that between RHIC to LHC, extrapolation to the center-of-mass energy per nucleon pair of $\sqrt{s_{NN}}$ = 5.5 TeV is quite inaccurate and evenly ranges between 1500 and 3000 charged particles per unit of rapidity. In view of this uncertainty, the ALICE detector granularity has been optimized to provide the best performance at a charged particle rapidity density of 4000, assuring a large safety margin, and to maintain the occupancy at the highest predicted multiplicity at such a level to still allow the event analysis.
An irrenounceable feature of heavy ion experiments is the capability to identify the charged hadrons in the full range of momentum because it allows to investigate many new event-by-event observables and plays an important role in the study of the properties of the quark-gluon plasma during its dynamical evolution toward the freeze-out phase when the re-hadronization occurs. At LHC energies, more than 97% of the particles will be produced with a p_t<2GeV/c, the bulk of the remaining 3% will be in the range 2-5 GeV/c although a not negligible fraction (~0.02%) will have a p_t > 5 GeV/c. Therefore ALICE will employ more than one method to perform an efficient and unambiguous particle identification (PID) and envisages two dedicated and complementary systems for di-electrons (a Transition Radiation Detector, TDR) and di-muons (a forward muon arm spectrometer).
Other design considerations are dictated by the need for special physics triggers to select the events of interest for storage. The trigger capabilities become indispensable for the most interesting rare probes in view of the large data volume and slow readout of the central TPC detector (about 25 Hz for central Pb+Pb collisions).

3. The ALICE layout

The ALICE experiment will be housed in the underground cavern, 45 m below ground level, previously used by the LEP L3 experiment, located at Intersection Point (IP) 2 of the LHC machine.
From the artist view of the lay-out, shown in Fig.1, one immediately recognizes that ALICE consists of two main components: a 2π barrel detector system at mid-rapidity (polar angles from 45° to 135°) embedded within the 0.2-0.5 T uniform solenoidal field provided by the large magnet formerly used in LEP's L3 experiment and a forward muon spectrometer (from 2° to 9° polar angles). The following detectors, starting from the beam axis, are sheathed in the L3 magnet:

- an inner tracking system (ITS) [6] with six cylindrical layers of highly accurate position-sensitive silicon detectors designed to track charged particles emerging from the main interaction vertex and to identify decay products of short-lived secondary particles having strange and charm quark content;
- a cylindrical, large volume, time projection chamber (TPC) [7] envisaged to determine the charged particle trajectories curving in the magnetic field, allowing particle momentum and charge to be measured;
- two highly segmented particle identification barrel arrays for hadrons (by means of Time Of Flight, TOF) [8] and electrons (by means of a Transition Radiation Detector, TRD) [9];
- a single arm array of seven Ring Imaging CHerenkov detectors (RICH) optimized to perform an inclusive identification of high momentum charged hadrons (High Momentum Particle Identification Detector, HMPID) [10];
- a single arm high resolution electromagnetic calorimeter (PHOton Spectrometer, PHOS) [11], consisting of 17k lead tungstate crystals ($PbWO_4$) readout by silicon photodiodes designed to search for direct photons and to measure π^0 and η spectra at high momenta. It is located in the bottom part of the central barrel region ($-0.12 < \eta < 0.12$ and $40 < \phi < 140$), at 4.6 m from the vertex and covers 8 m^2. A charged-particle veto detector will be placed in front of the PHOS with the aim to identify charged particles reaching the calorimeter.

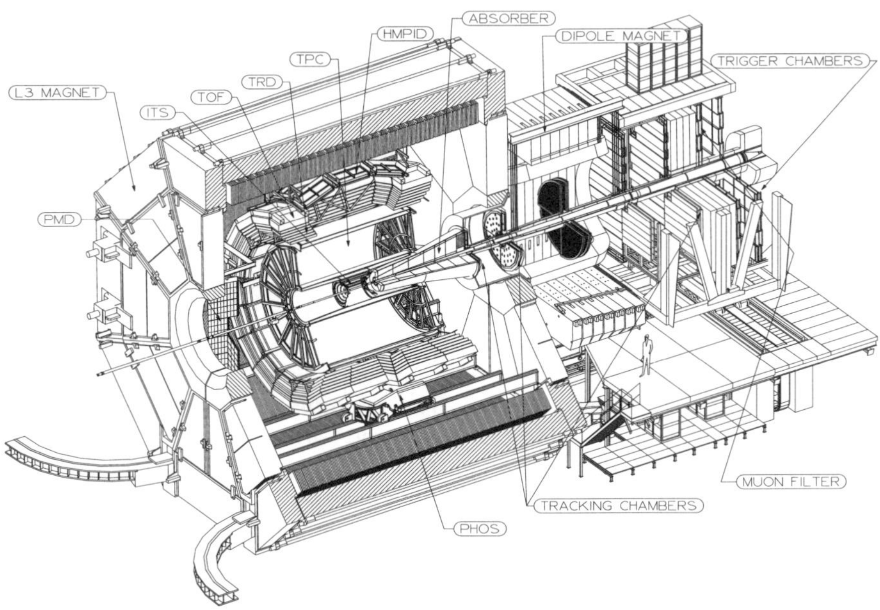

Figure 1. Axonometric view of the ALICE layout.

Several US groups have expressed interest to join ALICE by submitting a financial request to DOE, still pending, for building a large (13,248 channels) acceptance, moderate resolution electromagnetic calorimeter (EMCal). EMCal will extend the measured momentum range for photons and electrons to provide a measurement of jet energy and jet fragmentation functions.
The central detectors are supported inside the L3 solenoid by a cylindrical, metallic structure, 7 m long and with a diameter of 8.5 m, called space-frame.
The forward muon spectrometer [12], featuring measurements of quarkonia states with a Υ mass resolution of about 100 MeV/c^2, consists of a complex arrangement of absorbers (reaching totally about 18 λ_{int}), a large dipole magnet (3 Tm integral field), ten stations of thin multi-wire proportional chambers (MWPC) equipped with highly segmented cathode planes for tracking and four Resistive Plate Chambers (RPCs) for muon identification and triggering.
Four small and very dense zero degree calorimeters [13], made of tungsten and lead with embedded quartz fibres read out by photomultipliers, are located about 100 m downstream the machine tunnels on both sides of the interaction region to measure and trigger upon the impact parameter. A preshower detector called PMD (Photon Multiplicity Detector) [14] has been designed to measure photon multiplicity in the forward region. It consists of a lead converter sandwiched between two planes of high granularity gas proportional counters. Additional detectors [15], designed to provide fast trigger signals and event multiplicity at large rapidity and an array of scintillators (ACCORDE) on top of the L3 magnet, envisaged to trigger on cosmic rays, complete the ALICE set-up.
The experiment was approved in February 1997. The final design of the different detector systems have been comprehensively described in the Technical Design Reports (TDR) [6-15], since mid 1998; the last TDR [15], devoted to the forward trigger detectors (FMD/T0/V0), has been submitted to the LHCC by the end of 2004.
Construction is well advanced for all of the detector systems and the installation of the service structures is in progress. The space-frame was installed in the surface assembly hall at IP-2 in March 2004. This permitted to start the pre-commissioning work related to the installation of the TPC, TOF and TRD detectors. The use of "dummy" detector modules allowed verification of space constraints and installation scenarios. The constituent elements for the large muon absorber were delivered in early 2004 and two out of three absorber units have been completely assembled.
The preassembly of the dipole muon magnet was successfully accomplished in October 2004 and one month later the magnet was ramped up to full power at both polarities. The commissioning tests confirmed the compliance to all design parameters and the long term stability of magnet operation. The dipole is currently being disassembled and transferred to its final location on the muon

arm side. In parallel to the work on the dipole magnet, the so called "back door" of the L3 solenoid magnet was closed in a permanent manner. This allowed to accomplish the construction of the large concrete foundation for the dipole magnet. In addition, platforms and gangways were modified and adapted to the new cavern layout. The refurbishing of the counting rooms has continued during 2004 and the installation of racks is in progress.

4. Features and performance of the main sub-detector systems

4.1. *The inner tracking system*

The ITS has the task to track particles of very low momentum, to identify short-lived particles (mostly hyperons) decaying before they reach the TPC, and to improve the momentum resolution of the tracked particles. It consists of six cylindrical layers: the four innermost ones are made of truly bi-dimensional devices, pixel and drift detectors, owing to the high particle density expected and the necessity to achieve an impact parameter resolution below 100 μm, whereas the two outer planes are made of double-sided silicon microstrip detectors. The main parameters of the constituent subdetectors are shown in Table 1.

The first layer of pixel detectors, located at only 4 cm from the beam axis, covers almost four units of pseudo-rapidity to provide, together with the forward multiplicity detectors, a continuous coverage in rapidity for the measurement of charged particle multiplicity. In order to keep the average layer thickness, all included, below 1% of X_0, a lightweight (28 g/m) but very stiff support structure in carbon fiber has been designed to maintain the sagitta below 100 μm over a meter.

Table 1. Parameters of the various silicon detector types. A module represents a single detector.

	Pixel	Drift	Strip
Spatial precision $r\phi$ (μm)	12	38	20
Spatial precision z (μm)	100	28	830
Two track resolution $r\phi$ (μm)	100	200	300
Two track resolution z (μm)	850	600	2400
Cell size (μm^2)	50x425	150x300	95x40000
Active area per module (mm^2)	12.8x69.6	72.5x75.3	73x40
Total number of modules	240	260	1698
Total number of cells (M)	9.84	23	2.6
Average occupancy (inner layer)	2.1%	2.5%	4%

Drift and strip silicon layers will be equipped with an analog readout for independent particle identification via dE/dx in the $1/\beta^2$ region. The about 8 kW of power dissipated by the over 12 M of electronics channels must be carefully removed to assure the thermal uniformity required by the TPC and the two silicon drift layers. Extensive R&D was required to develop a very efficient cooling system based on a hybrid approach that combines the forced circulation of the liquid C_6F_{14} with a moderate air flow.

4.2. *The Time Projection Chamber*

The TPC is the main tracking detector in the central barrel of the ALICE experiment. Together with the ITS and the TRD, it has to provide charged particle momentum measurement and vertex determination with sufficient momentum resolution (better than 2.5% for electrons with momentum of about 4 GeV/c), tracking efficiency (larger than 90%) and two-track separation (resolution in relative momentum below 5 MeV/c) in the p_t region up to several 10 GeV/c and pseudo-rapidity $|\eta|<0.9$. Furthermore the TPC has to allow the inspection and tracking of electron candidates identified in the TRD, at central collision rates of up to 200 Hz and to provide dE/dx measurements for charged particles with pseudo-rapidities in the range $-1< \eta <+1$.
Although large size TPCs have been successfully employed for the study of heavy ion collisions by the NA49 and STAR experiments at the SPS and RHIC respectively, the extreme particle densities expected at the LHC energies and the large drift volume of the ALICE TPC, the biggest ever built, required a careful design, leading to many individual innovative aspects.
As shown in Fig. 2, it consists of a cylindrical field cage, filled with 88 m^3 gas mixture of 90% Ne – 10% CO_2, divided in two drift regions, back-to-back, by a central membrane. This membrane is a high voltage central electrode which defines in each of these two regions a uniform electrostatic field, to transport the primary electrons over a distance of 250 cm towards the readout end plates. The readout chambers are conventional multi-wire proportional chambers with cathode planes segmented in pads equipped with an electronics chain to amplify, digitize and pre-process the signal before transmission to the DAQ. Each of the 570,000 channels includes a custom digital circuit that, with an innovative approach, performs tail cancellation, digital baseline restoration, data compression and multi-event buffering for event derandomization. The azimuthal segmentation of the readout plane consists of 18 trapezoidal sectors, each covering 20 degrees.

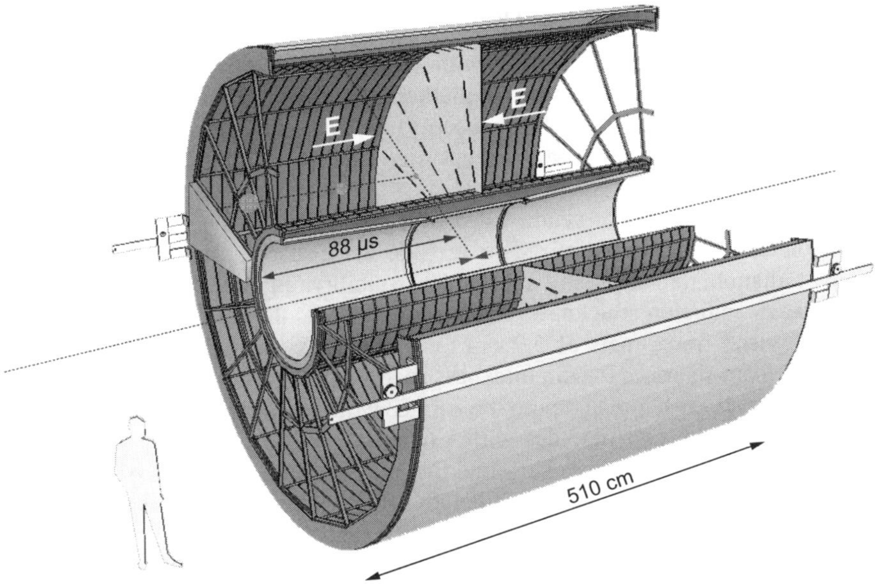

Figure 2. Cut-away view of the ALICE TPC showing the central electrode and the field cage.

Charged particles traversing the field cage volume ionize the gas along their path, liberating electrons that drift towards the end plate of the chamber, where the signal amplification is provided through avalanche effect (gas gain of $2x10^4$) in the vicinity of the anode wires. Moving from the anode wire towards the surrounding electrodes, the positive ions created in the avalanche induce a positive signal on the pad plane.

The field cage and the containment volume are constructed from two concentric cylinders each, sealed by an annular disc (end plate) on either side. The inner radius (r=90 cm) is given by the maximum acceptable density of hits in the inner TPC volume whilst the outer radius (r= 250 cm) is determined by the minimum track length required for achieving a dE/dx resolution better than 10%. Good momentum resolution and the need to keep the distortions small, which are created by the space charge, require a drift gas with low diffusion, low Z and large ion mobility. Extensive investigation of different gas mixtures has been performed, leading to the choice of the adopted mixture (90% Ne-10% CO_2). This drift gas, however, requires a high drift field (400 V/cm) to secure an acceptable drift time of ~88 μs. Therefore, the high voltage on the central electrode has to be as high as 100 kV. An insulating gas envelope (containment) surrounds the actual field cage for operational safety reasons.

4.3. *The Transition Radiation Detector*

The TRD will identify electrons with momenta above 1 GeV/c, near mid-rapidity, to study the suppression of quarkonia decaying in electron pairs and the production of heavy quarks (charm, beauty) in their semileptonic decays. A pion rejection factor of 100% at 90% electron efficiency for momenta larger than 2 GeV/c is required.

It consists of six layers of xenon-filled time expansion chambers with a radiator stack of carbon fibers optimized to provide the best compromise between TR yield, radiation thickness and mechanical stability. In order to keep the total anticipated radiation thickness X/X_0 below 15%, the radiator is made of polypropylene fibres mats of 3.2 cm total thickness, sandwiched between two Rohacell foam sheets of 0.8 cm thickness each. The foam sheets are reinforced by carbon fiber sheets laminated onto the surface with a thickness of 0.1 mm.

540 individual detector modules with a total active area of roughly 750 m^2 and 1.2 million read-out channels surround the TPC starting at a radius of 2.9 m and extending to 3.7 m. With an overall length of 7 m, the TRD barrel is azimuthally segmented in 18 sectors covering a rapidity range between -0.9 to 0.9. Each sector is subdivided in 5 stacks in longitudinal direction. Each section consists of 6 layers in radial direction. The current funding availability will allow to install about half of the modules at LHC start-up.

The electronics has been designed to feature a fast local cluster recognition to produce a Level-1 trigger on high-p_t leptons or hadrons within 6 μs after the collision, thus enriching the statistical sample of the observables under study.

The tracking capability of the TRD, in junction with ITS and TPC, will valuably improve the predicted momentum resolution for charged particles as shown in Figure 3. The results indicate that the ALICE tracking system will reach a good momentum resolution up to ~100 GeV/c, even at the highest multiplicities, to enable detailed jet fragmentation studies.

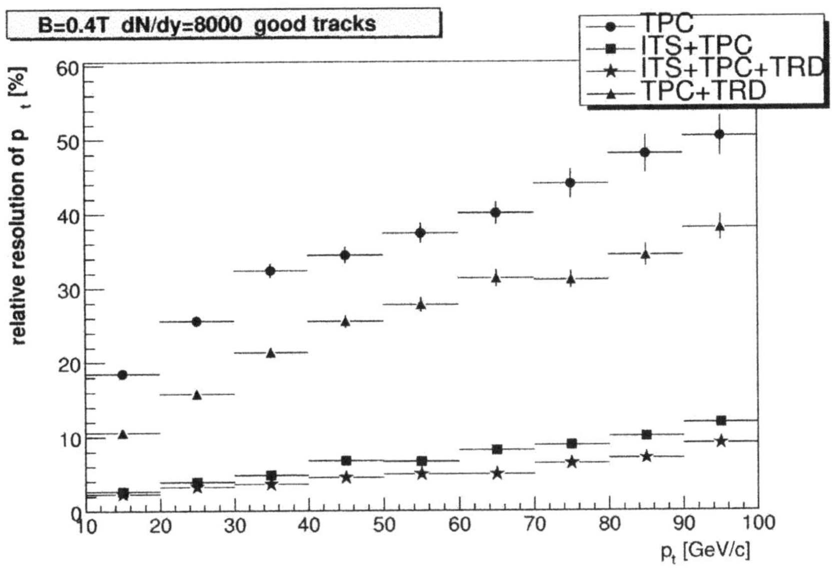

Figure 3. Momentum resolution as a function of the transverse momentum .

4.4. *The Time of Flight Detector*

Low momentum hadron identification will be achieved via measurements of energy loss in the silicon layers and in the TPC gas, while above the $1/\beta^2$ region, identification will be performed by employing the TOF barrel.

Based on 1638 double-stack multigap resistive plate chambers (MRPC) with an intrinsic time resolution of about 60 ps, the TOF barrel covers a surface larger than 160 m^2 at a radius of about 3.7 m. It has been optimized for the identification, on a track-by-track basis, of hadrons with a transverse momentum below 2 GeV/c, with a separation between pions and kaons better than 3 sigmas to keep the contamination below a 10% level in presence of a huge bulk of hadrons at low momentum.

A cross section of the ALICE MRPC is shown in Fig. 4. Each resistive electrode is spaced one from the other with equal sized spacers creating a series of ten gas gaps, each 250 μm thin. Electrodes are connected to the outer surfaces of the stack of resistive plates reading out all gas gaps in parallel while all the internal plates are left electrically floating. The flow of positive ions and electrons generated in the avalanche processes maintain the correct voltage between the intermediate plates. Pickup pads (96 per strip), with an area of 3.5x2.5 cm^2 each, are arranged in two-row arrays. Time resolutions as good as 65 ps have been achieved on large size prototypes. Lifetime and rate capability above 50 Hz/cm^2

have been measured at the CERN Gamma Irradiation Facility (GIF) and proved to be in excess of the ALICE requirements.

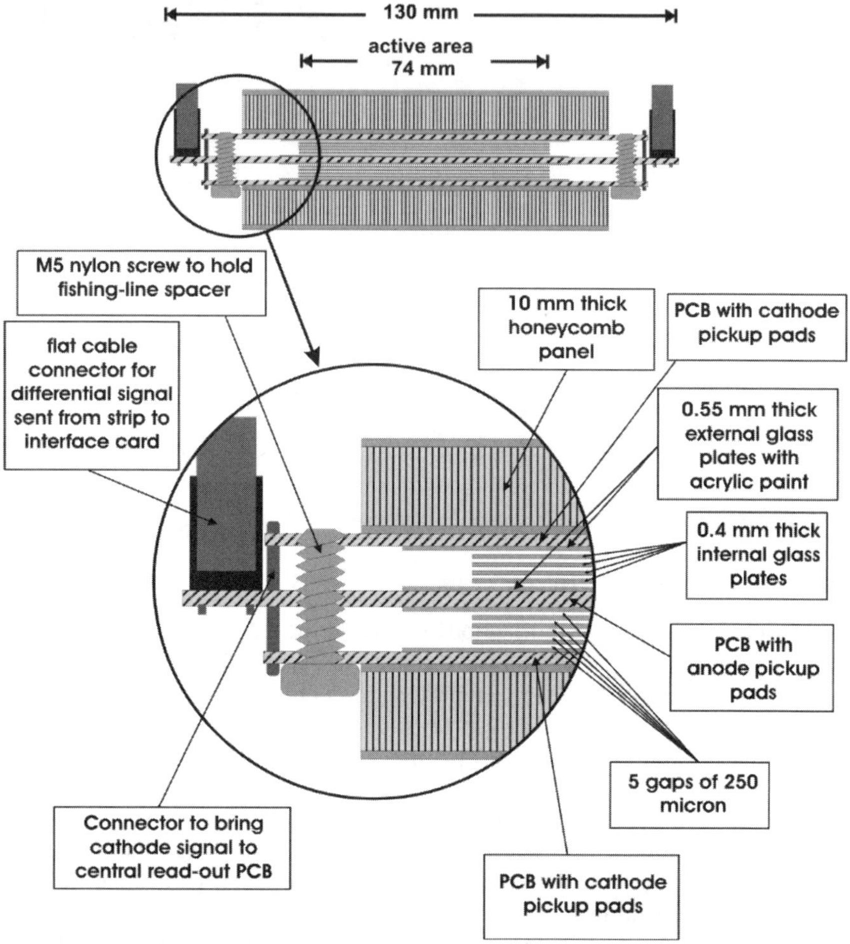

Figure 4. Schematic description of the ALICE TOF.

4.5. *The High Momentum Particle IDentification system*

The HMPID, an array of seven proximity focusing Ring Imaging Cherenkov (RICH) detectors, located at 4.7 m from the beam axis, will further enhance the PID capability of the ALICE layout by exploiting the novel technology of manufacturing large area CsI photocathodes to identify particles beyond the momentum interval covered by the TOF. Namely, the HMPID has been optimised to extend the useful range for the identification of π/K and K/p, on a

track-by-track basis, up to 3 GeV/c and 5 GeV/c respectively. The single-arm geometry with a HMPID acceptance of 5% of the central barrel phase space is motivated by the need to measure the spectra of particles with a momentum above 1 GeV/c in an inclusive manner, with the possibility to build classes of events using the event-by-event characterization performed at lower momenta by the barrel detectors.
The schematic view of the ALICE CsI RICH is shown in Fig.5. The radiator is a 1.5 cm thick layer of low chromaticity C_6F_{14} (perfluorohexane) liquid with an index of refraction of n=1.2834 at λ=175 nm corresponding to β_{min}=0.78 (i.e. a threshold momentum p_{th}(GeV/c)=1.26 m, with m equal to the particle mass in GeV/c^2).

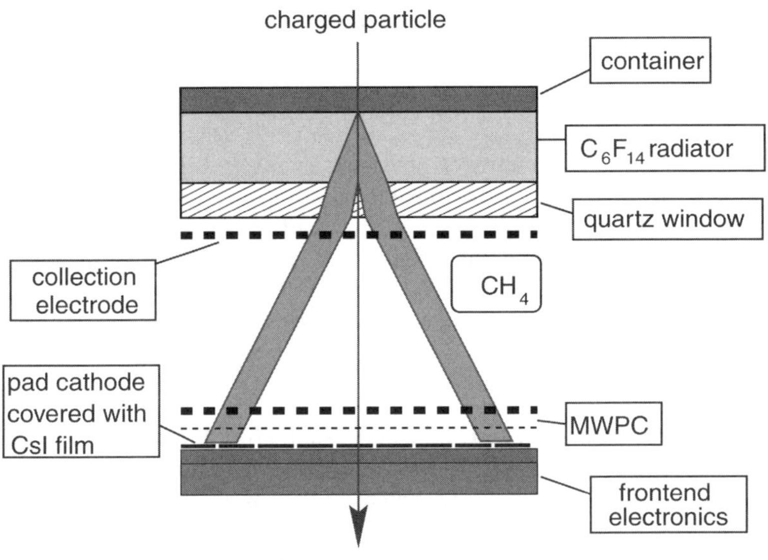

Figure 5. The CsI-based fast RICH detector scheme employed in the HMPID.

4.6. *The Forward Spectrometer*

The Forward Spectrometer is the ALICE subsystem dedicated to study the complete spectrum of heavy quark resonances up to the Υ family. It will measure the decay of these resonances into muons, both in proton-proton and in heavy-ion collisions, with a mass resolution (about 100 MeV/c^2 at quarkonia masses of 10 GeV and better than 70 MeV/c^2 in the J/Ψ region) and momentum precision (about 1%) sufficient to separate all states on a continuum due to B and D meson decays and Drell-Yan processes.

It consists of a complex arrangement of absorbers (reaching totally about 18 λ_{int}) to reduce as much as possible the punchthrough into the spectrometer and the backsplash into the TPC, a large dipole magnet (3 Tm integral field), ten stations of thin MWPC equipped with highly segmented cathodes for tracking and four Resistive Plate Chambers to identify and trigger on the muon tracks escaping from the muon filter through the dipole magnet. A high-density small angle absorber with a central hole will shield the spectrometer from the particles emitted at angles below 2° allowing the beam particles to traverse the spectrometer.
The technology required to build gas-filled chambers with pad read-out is well known, but the large total detector surface over 100 m^2 and the huge number of electronic channels required, almost one million, have demanded to design frameless modules, almost free of dead areas to maximize the acceptance, and low cost frontend chips.

5. Conclusion

A large international community of physicists is engaged in building the ALICE detector at CERN. This new experiment, exploiting the unique possibility offered by the LHC to collide heavy nuclei at unprecedented energies, will look into deconfined strongly interacting matter in conditions significantly more favourable than those achieved at SPS and RHIC, thus allowing to study new signatures and likely draw conclusions on the character of the strong interaction.

References

1. F. Karsch, E. Laermann and A. Peikert, Nucl. Phys. **B 605**, 579, 2001
2. K.J. Eskola and K.Kajantie, Z. Phys. **C75**, 515 (1997).
3. ALICE Collaboration, Technical Proposal, CERN/LHCC/95-71; Addendum 1, CERN/LHCC/96-32 and Addendum 2, CERN/LHCC/99-13.
4. ALICE Collaboration 2004, J. Phys. G **30** 1517-1763.
5. K.K. Eskola, *et al.*, hep-ph/0104010.
6. ALICE Collaboration 1999, CERN/LHCC 99-12, ALICE TDR 4.
7. ALICE Collaboration 2000, CERN/LHCC 2000-001, ALICE TDR 7.
8. ALICE Collaboration 2000, CERN/LHCC 2000-012, ALICE TDR 8 and CERN/LHCC 2002-016, Addendum to ALICE TDR 8.
9. ALICE Collaboration 2001, CERN/LHCC 2001-21, ALICE TDR 9.
10. ALICE Collaboration 1998, CERN/LHCC 98-19, ALICE TDR 1.
11. ALICE Collaboration 1999, CERN/LHCC 99-4, ALICE TDR 2.
12. ALICE Collaboration 1999, CERN/LHCC 99-22, ALICE TDR 5.
13. ALICE Collaboration 1999, CERN/LHCC 99-5, ALICE TDR 3.
14. ALICE Collaboration 1999, CERN/LHCC 99-32, ALICE TDR 6.
15. ALICE Collaboration 2004, CERN/LHCC 04-025, ALICE TDR 11.

IS QGP FOUND AT RHIC?

H. HAMAGAKI
Center for Nuclear Study, Graduate School of Science, University of Tokyo, 7-3-1 Hongo, Bunkyo-ku, Tokyo 113-0033, JAPAN
E-mail: hamagaki@cns.s.u-tokyo.ac.jp

Hadronic matter under extreme conditions have drawn strong attentions, in particular, with many interesting results provided from RHIC. In this article, a selected number of recent results from RHIC are presented and their implications are discussed with main focus on the global aspects of hot and dense matter.

1. Introduction

Since the first collisions between Au nuclei at RHIC (Relativistic Heavy Ion Collider) at Brookhaven National Laboratory (BNL) in June 2000, many interesting results have been obtained. It is intended in this report to pick up some key measurements and discuss on their implications with respect to the quark-gluon plasma (QGP) formation.

This report is organized as follows: First of all, primary goals of studies with high-energy heavy-ion collisions are presented. RHIC and the experiments at RHIC are very briefly explained, and the contributions of Japanese groups are mentioned. Experimental results relevant to the QGP formation are presented, which is a main body of this report. Current status is summarized and near-future prospects are provided in the end.

2. Primary Goal

Primary goal of studies with high-energy heavy-ion collisions is to reveal the existence of a new phase of nuclear matter, often called quark-gluon plasma (QGP), and to study properties of this exotic matter.

Properties of hadronic matter under extreme conditions have drawn strong attention in the last few decades. Hadrons are composite particles consists of (anti-)quarks and bound together via gluon exchange which is described by QCD (Quantum Chromo Dynamics). In the normal conditions, quarks are confined inside hadrons, and cannot be extracted as free particles.

Lattice-QCD simulations predict existence of a new phase of hadronic matter at high temperature region called quark-gluon plasma (QGP) where quarks and gluons are liberated from confinement and moving freely.[1] QGP is a form of existence in the early universe for $\sim 10\mu$sec after the Big Bang. Recent theoretical works have also revealed the existence of very rich phase structure at high baryon density region,[2,3] which may have relevance to the inner core of neutron stars.[4]

High-energy heavy-ion collision is a unique tool to realize hot and dense matter and to study its properties in a laboratory. Studies have started in late 70's,[5] and a new era began when RHIC at BNL started operation in 2000. More detailed descriptions on RHIC and the experiments is provided in the next section.

3. RHIC and the Experiments

RHIC is the first collider-type accelerator of heavy ions ever built, dedicated to the studies of nuclear matter at extreme conditions of high temperature and density. It has two independent rings with circumference of 3.87 km, and is capable of making collisions between different nuclear species. Maximum c.m.s colliding energy is 200 Z/A GeV per nucleon pair. Construction of RHIC started in 1991 and was finally completed in 1999.

A plain view of RHIC accelerator complex is shown in Fig. 1. Several used accelerators, Tandem van de Graaf, proton linac, AGS booster and AGS, are utilized to pre-accelerate and inject ions and protons to the RHIC rings. There are four experimental programs, STAR, PHENIX, PHOBOS and BRAHMS at RHIC. STAR and PHENIX are the large-scale experiments, and the other two are small ones.

June 2000 is a month of epoch making, when the first collisions between Au ions were clearly identified by the experiments at RHIC. Since then, RHIC had four successful runs in the four years, each of which lasted for several months. As for Au+Au collisions, runs at full energy of $\sqrt{s_{NN}} =$ 200 GeV (Year-2 and Year-4), and at a lower energy of $\sqrt{s_{NN}} = 62.4$ GeV (Year-4) were performed. d+Au asymmetric collisions at $\sqrt{s_{NN}} = 200$ GeV were performed in Year-3 run. Polarized p+p runs were also performed in Year-2, Year-3 and Year-4 runs.

A schematic view of the the PHENIX experimental setup is shown in Fig. 2. It consists of two central arms (East and West), two muon arms (North and South) and inner detectors. The PHENIX has a very unique feature of being able to measure photons, electrons and muons as well as hadrons. With this capability, the PHENIX experiment aims to address as many signatures as possible relevant to QGP formation. The Japanese group is participating in the PHENIX experiment, and has been taking leading roles in executing the runs and analyzing data.

Figure 1. Perspective view of the RHIC accelerator complex at Brookhaven National Laboratory, USA.

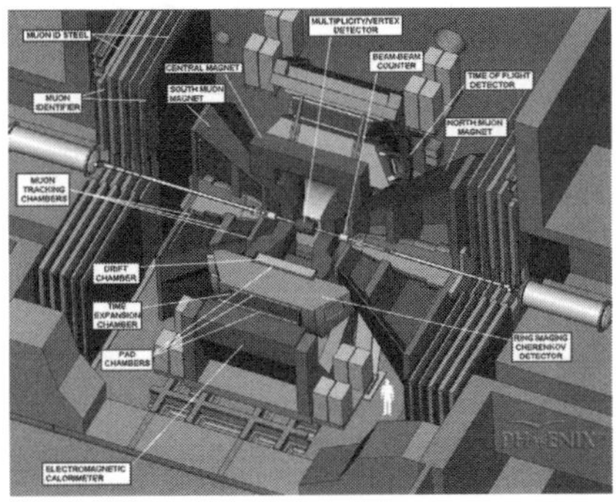

Figure 2. A schematic view of the PHENIX experimental setup.

4. Probing Dense Partonic Matter with Jet

It was surprising that strong suppression was observed in the π^0 yield at high p_T region in central Au+Au collisions.[6] Since most of high-p_T pions are originated from the jet fragmentation process, the yield was expected to be scaled roughly with the yield in p-p collisions multiplied by the number of binary N-N collisions, N_{coll}. Although the observed yield suppression could be most naturally interpreted as jet quenching effect; energy loss of scattered energetic partons while passing through hot and dense matter,[7,8,9] other contributions due to initial-state effects could not be ruled out. Among the models for strong hindrance or saturation of gluon strength, CGC (Color Glass Condensate)[10] is a new one which is being developed to describe a common feature of colliding hadrons at ultra-high energies, that is, strong quenching of gluon strength.

In order to pin down the effects, a controlled experiment using d+Au collisions was performed in the Year-3 RUN. In d+A collisions, initial-state effects should be similar to the case in Au+Au collisions. The final-state effects, however, should be small, since large-scale hot and dense region will not be created in d+Au collisions.

The four experiments at RHIC performed the study for this subject, and published the results.[11] Figure 3 shows the ratio of $N_{\rm coll}$-scaled π^0 yields for Au+Au and d+Au collisions relative to that for p+p collisions as a function of transverse momentum measured by the PHENIX experiment. No noticeable suppression is seen in d+Au collisions, which strongly supports that the yield suppression seen in Au+Au collisions is primarily due to final-state effect, that is, jet energy loss. This conclusion also supports that a state with extremely high density is really formed in the early stage of heavy ion collisions.

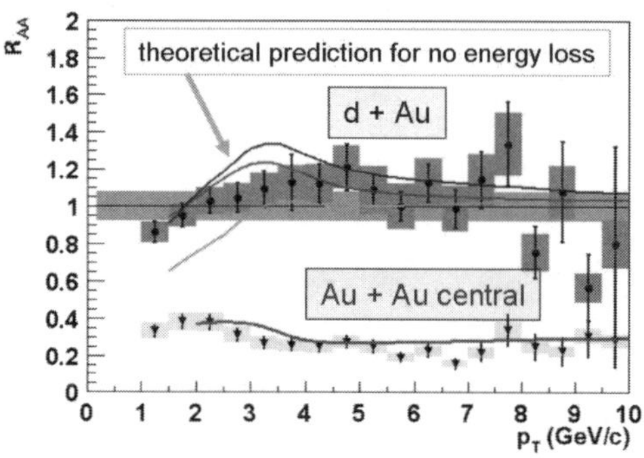

Figure 3. Yield of neutral pions in d+Au and Au+Au central collisions divided by the yield in p+p collisions at $\sqrt{s_{NN}} = 200$ GeV, where the yield in p+p collisions is scaled up with the number of binary nucleon-nucleon collisions calculated using the Glauber model. Apparent difference is seen in the behaviors between d+Au and Au+Au collisions.

5. Thermal Equilibrium

It is not guaranteed a priori that local equilibrium is achieved in the early stage of heavy-ion collisions. With local equilibration, description of space-time evolution of the colliding system is significantly simplified, and comparison of the data with theoretical predictions can be much reliable. In this respect, study of collective behavior of the colliding system has central importance. Idea of radially expanding flow has been utilized to study the collective behavior by measuring momentum spectra of hadrons. Recently, elliptic anisotropy of azimuthal angular distributions draws strong attention.

In A-A collisions with a finite impact parameter, the cross section of the overlapping region of two colliding nuclei normal to the incident direction has an elliptic shape. If thermal equilibration is achieved, larger flux is expected toward the short axis of the ellipsoid, since pressure gradient is larger along the short axis.

Large anisotropy was found.[12,13] An example is shown in Fig. 4,[13] and the result is compatible with a hydro-dynamical calculation assuming a perfect fluid with no viscosity[14].

Recent progress in relativistic hydro-dynamical calculations are found in Ref. 15. Thermalization as early as less than 1 fm is required in order to have such large anisotropy. QGP has been considered to be a weakly coupled system of quasi-freely moving quarks and gluons, but such weakly coupled system may not reach equilibrium in such a short time and will not behave as fluid. QGP, still near T_C in the RHIC energies, could be non-perturbative strongly coupled system. Further studies both in experimental and theoretical, are needed to clarify this important aspects of the QCD matter.

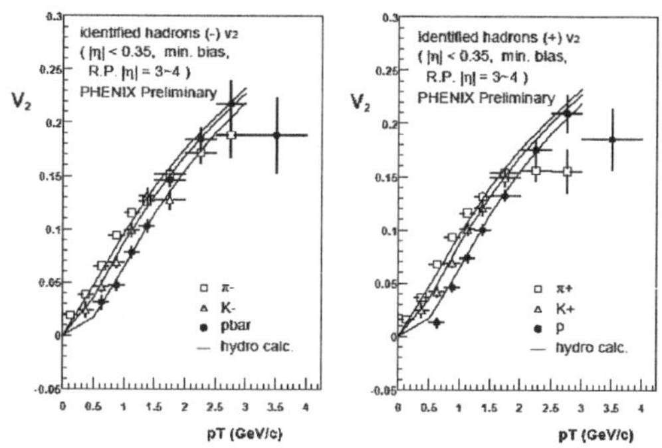

Figure 4. Transverse momentum dependence of v_2, the magnitude of azimuthal anisotropy, for identified particles, π^-, K^-, $\bar{p}$ (left), and π^+, K^+, p (right). The solid lines represent a hydrodynamic calculation for π, K and p from upper to lower curves, respectively, which includes a first-order phase transition with a freeze out temperature of 120 MeV.

6. Chemical Equilibrium

Process toward chemical equilibration advances through inelastic collisions in the early stage of the collisions. The system eventually reaches the chemical freeze out after which elastic collisions become dominant, but it is important to note that there's no guarantee of chemical equilibration achieved at this stage. In order to check the degree of chemical equilibration and to deduce thermo-dynamical variables, such as temperature and chemical potential at the freeze out stage, simultaneous fit to the yield ratios among various particles has been employed with a density function assuming a grand canonical ensemble of hadrons.[16]

As an example, shown in Fig. 5 is a fitting result for Au+Au collisions at $\sqrt{s_{NN}} = 200$ GeV.[17] The simultaneous fit works amazingly well. Obtained freeze out temperature of $\sim$ 157 MeV is close to that predicted for phase transition from hadronic to quark matter. Another point to be noted is that the γ_S factor, strangeness saturation factor, is close to unity, while the γ_S factor is significantly below unity in lower energy collisions at SPS and AGS. The γ_S factor was artificially introduced based on the theoretical results suggesting that the strangeness chemical equilibration may not be achieved during the inelastic collision stage, simply because the reaction rate of strangeness hadrons is too low in hadronic phase. It is also noted here that the production rate is expected to be much higher in deconfined phase, and equilibration time is much shorter.

Natural interpretation for the unity of the γ_S factor and high freeze-out temperature, would be that the deconfined phase is realized in the RHIC collisions, and the chemical equilibration is achieved in the QGP phase.

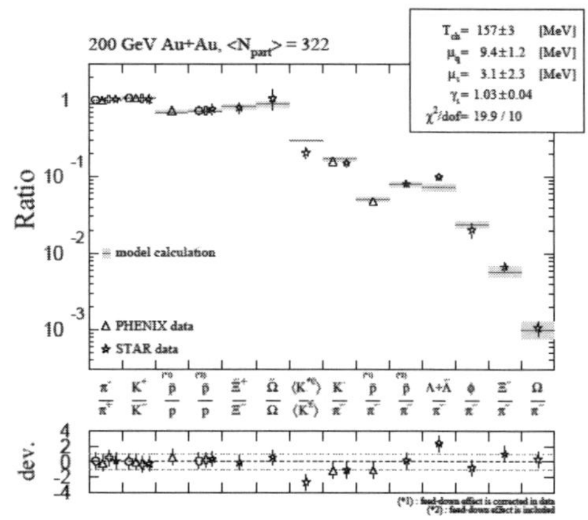

Figure 5. Comparison of fit results to the data in 200 GeV Au+Au central collisions. The points are experimental data, and the vertical bars are experimental errors. The short horizontal lines are fitted results with the model, and the uncertainties in the model is shown with shaded boxes.. The bottom plot shows a deviation of the experimental ratio from the model calculations normalized by the experimental error of the ratio.

7. How to Create Hadrons with Medium pT

As has been already presented, strong suppression of π^0 yield at medium to high momentum region was observed. Behavior of proton yield, however, turned out to be very different from pions. The p/π and $\bar{p}/\pi$ yield ratio in the medium $p_{\rm T}$ region between $2 \sim 5$ GeV/c increases with centrality of collisions for Au+Au collisions at $\sqrt{s_{NN}}$= 200 GeV, as is shown in Fig. 6.[18]

Attribution of this difference to hydro-dynamical expansion of the system is quite natural, since larger mass hadrons suffer more strongly from radial expansion. Centrality dependence of ϕ vector meson yield, however, provided a counter example for this interpretation.[19,20] If the collective flow is a main cause, ϕ meson with mass nearly equal to proton should behave similar to protons, but it was found that the ϕ yield follows that of pions instead of protons.

These results indicate that some of protons at medium $p_{\rm T}$ is produced through a different mechanism other than fragmentation of jets. Quark recombination models, which assumes recombination of valence quarks in the hadronization stage, seem to account for this difference very naturally.[21] It is to be stressed here that quark recombination is a mechanism of hadron formation which becomes important only in the case where a deconfined quark phase with a sea of quarks is realized in the heavy ion collisions.

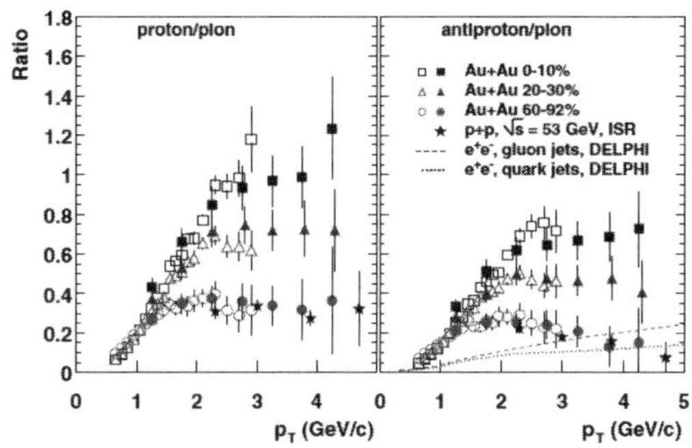

Figure 6. p/π (left) and $\bar{p}/\pi$ (right) ratios for central (0-10%), mid-central (20-30%), and peripheral (60-92%) Au+Au collisions at $\sqrt{s_{NN}} =$ 200 GeV. Open (filled) points are for $\pi^{\pm}$ (π^0), respectively. Data from $\sqrt{s} = 53$ GeV p+p collisions are shown with stars. The dashed and dotted lines are $(\bar{p}+p)/(\pi^+ + \pi^-)$ ratios in gluon and quark jets.

8. Summary and Outlook

The experimental runs at RHIC since the first collisions in 2000 have been extremely successful, and various new and exciting results have been obtained. A selected number of results relevant to QGP formation were introduced and their implications were discussed. It will be safe to summarize the current situation as follows; all the results obtained from RHIC seem to suggest consistently that QGP is realized in the heavy ion collisions at RHIC.

Properties of QGP, however, has not been well understood yet, and the investigation toward this direction will be the next step to proceed. Successful description of hot matter in term of perfect fluid is a surprise, but is surely the major step forward in this direction. Controlled experimental studies, for combinations of various nuclear species at different colliding energies with high statistics, are required for this purpose. In addition, there are several key measurements still to be made or to be extended, which include J/ψ production, light vector mesons, and single photons.

At BNL, discussions of RHIC-II, RHIC upgrade and the experimental programs, have been initiated based upon the current successful operation and experimental results. And in a few years, LHC at CERN will start operation, which will provide Pb-Pb collisions at more than 20 times higher energies than RHIC. Better circumstances for investigating the insight of QGP will be realized soon.

References

1. For recent reviews, see: K. Kanaya, Nucl Phys. **A715**, 233c (2003); F. Karsch, Nucl. Phys. **A698**, 199 (2002); S. Hands, Nucl. Phys. B (Proc. Suppl) **106**, 142 (2002); S. Ejiri, *ibid* **94**, 19 (2001).
2. M. Alford, K. Rajagopal and F. Wilczek, Phys. Lett. B **422**, 247 (1998).
3. R. Rapp *et al.*, Phys Rev. Lett. **81**, 53 (1998).
4. S. Reddy, J. Phys. G **30**, S879 (2004).
5. For a historic review from BEVALAC to RHIC, see: R. Stock, J. Phys. G **30** (2004) S633.
6. K. Adcox *et al.* (PHENIX Collaboration), Phys. Rev. Lett. **88** 022301 (2002).
7. J.D. Bjorken, FERMILAB-PUB-82-59-PHY.
8. M. Gyulassy and X.-N. Wang, Nucl. Phys. B **420**, 583 (1994).
9. R. Baier, Y.L. Dokshitzer, A.H. Mueller, S. Peigne and D. Schiff, Nucl. Phys. B **483**, 291 (1997).
10. For a review of CGC, see: E. Iancu and R. Venugopalan, Quark Gluon Plasma 3, Eds. R.C. Hwa and X.-N. Wang, World Scientific, p249 (106 pages).
11. S.S. Adler *et al.* (PHENIX Collaboration), Phys. Rev. Lett. **91**, 072303 (2003); J. Adams *et al.* (STAR Collaboration), Phys. Rev. Lett. **91**, 072304 (2003); B.B. Bach *et al.* (PHOBOS Collaboration), Phys. Rev. Lett. **91**, 072302 (2003); I. Arsene *et al.* (BRAHMS Collaboration), Phys. Rev. Lett. **91**, 072305 (2003).
12. J. Adams *et al.* (STAR Collaboration), Phys. Rev. Lett. **92**, 052302 (2004).
13. S. Esumi *et al.* (PHENIX Collaboration), Nucl. Phys. **A715** 599-602, (2003).
14. P. Houvinen, P.F. Kolb, U.W. Heinz, P.V. Ruuskanen and S.A. Voloshin, Phys. Lett. **B503**, 58 (2001).
15. For reviews of hydrodynamical description, see: P. Huovinen; Quark Gluon Plasma 3, Eds. R.C. Hwa and X.-N. Wang, World Scientific, p600 (34 pages), and P.F. Kolb and U. Heinz, *ibid*, p634 (81 pages).
16. J. Sollfrank, U. Heinz, H. Sorge, and N. Xu, Phys. Rev. **C59**, 1637 (1999).
17. M. Kaneta and N. Xu, e-Print archive nucl-th/0405068.
18. S.S. Adler *et al.* (PHENIX Collaboration), Phys. Rev. Lett. **91**, 172301 (2003), nucl-ex/0305036.
19. S.S. Adler *et al.* (PHENIX Collaboration), e-Print archive nucl-ex/0410012.
20. K. Schweda (for the STAR Collaboration), J.Phys. G **30** (2004) S693-S700.
21. S.A. Voloshin, Nucl. Phys. **A715**, 379c (2003); D. Molnar and S.A. Voloshin, Phys. Rev. Lett. **91** 092301 (2003); R.J. Fries, B. Mueller, C. Nonaka and S.A. Bass, Phys. Rev. Lett. **90**, 202302 (2003); V. Greco, C.M. Ko and P. Levai, Phys. Rev. Lett. **90**, 202302 (2003); Z.W. Lin and C.M. Ko, Phys. Rev. Lett. **89**, 202302 (2002); Z.W. Lin and D. Molnar, Phys. Rev. C **68**, 044901 (2003).

SECTION IV

HYPERNUCLEI AND EXOTIC ATOMS

MEASUREMENT OF KAONIC HYDROGEN WITH DEAR AT DAΦNE AND FUTURE PERSPECTIVES

C. GUARALDO (on behalf of the DEAR/SIDDHDARTA Collaborations)*
E-mail: guaraldo@lnf.infn.it

G.BEER
University of Victoria, P.O. Box 3055 Victoria B.C. V8W 3P6, Canada

A.M. BRAGADIREANU†, C. CURCEANU (PETRASCU), C. GUARALDO, M. ILIESCU†, V. LUCHERINI, D.L. SIRGHI† AND F. SIRGHI
LNF-INFN, C.P. 13, Via E. Fermi 40 00044 Frascati (Roma), Italy

M. CARGNELLI, H. FUHRMANN, T. ISHIWATARI, P. KIENLE‡, J. MARTON AND J. ZMESKAL
S. Meyer Inst. für subatomare Physik, Boltzmanngasse 3, 1090 Vienna, Austria

J.-P. EGGER
Université de Neuchâtel, Institut de Physique, CH-2000 Neuchâtel Switzerland

K. ITAHASHI, M. IWASAKI AND T. KOIKE
RIKEN, Wako-shi, Saitama, 351-0198, Japan

B. LAUSS
University of California, 366 LeConte Hall, Berkeley, CA 94720. USA

L. LUDHOVA, F. MULHAUSER AND L.A. SCHALLER
Université de Fribourg, Institut de Physique, CH-1700 Fribourg, Switzerland

T. PONTA
IFIN-HH, P.O. Box MG-6 R-76900 Magurele, Bucharest, Romania

R. SEKI§
Department of Physics and Astrophysics, California State University, Northridge, CA 91330, USA

The objective of the DEAR (DAΦNE Exotic Atom Research) and the coming SIDDHARTA (SIlicon Drift Detector for Hadronic Atom Research by Timing Application) experiments is an eV precision measurement of the K_α line shift and width, due to the strong interaction, in kaonic hydrogen and a similar measurement - the first one - in kaonic deuterium. The final aim is a precision determination of the antikaon-nucleon isospin dependent scattering lengths, which allows to better understand the chiral symmetry breaking scenario in the strangeness sector. DEAR has performed the most precise measurement up to now on kaonic hydrogen at the end of 2002. It is for the first time that the K-complex could be clearly identified. The obtained result is presented in this paper. An eV precision measurement of kaonic hydrogen and kaonic deuterium is foreseen in the framework of the newly started SIDDHARTA project, which continues the DEAR scientific line.

1. Introduction

The DAΦNE [1] electron-positron collider at the Frascati National Laboratories has made available a unique negative kaons "beam", providing so unprecedented conditions for the study of the low-energy (anti)kaon-nucleon interaction, a field still largely unexplored.

The DEAR (DAΦNE Exotic Atom Research) experiment [2] at DAΦNE and its successor SIDDHARTA (SIlicon Drift Detector for Hadronic Atom Research by Timing Application) [3] aim at a precision measurement of the strong interaction generated shifts and widths of the fundamental $1s$ level of kaonic hydrogen and kaonic deuterium, via the measurement of the x-ray transitions to this level. The aim is to extract the isospin dependent antikaon-nucleon scattering lengths and, so, to contribute to the understanding of aspects of chiral symmetry breaking in the strangeness sector.

In practice, in studying kaonic hydrogen (deuterium) in order to measure the strong interaction component of the kaon-nucleon force, one measures the shift ϵ of the position of the K_α line ($2p \to 1s$ transition) from the one calculated from a purely electromagnetic interaction:

$$\epsilon = |E_{2p\to 1s}^{measured}| - |E_{2p\to 1s}^{e.m.}| \tag{1}$$

and the width (broadening) Γ of the $1s$ level given by the strong interaction.

*Work partially supported by EU Integrated Infrastructure Initiative HadronPhysics Project under contract number RII3-CT-2004-506078

†also at: Institute of Physics and Nuclear Engineering "Horia Hulubei", IFIN-HH, Particle Physics Department, P.O. Box MG-6 R-76900 Magurele, Bucharest, Romania

‡also at: Technical University Munich, 80290 Germany

§also at: W.E. Kellogg Radiation Laboratory, California Institute of Technology, Pasadena, uppercaseca 91125, usa

The electromagnetic transition energy in kaonic hydrogen is calculated with 1 eV precision by solving the corresponding Klein-Gordon equation and applying the corrections for finite size and vacuum polarization. The resulting value is $E^{e.m.}_{2p\to 1s} = (6480\pm 1)\ eV$ where the 1 eV error is dominated by the uncertainty of the kaon mass.

Until the advent of DAΦNE, the kaonic hydrogen parameters were measured at KEK [4], where the following results were found:

$$\epsilon = -323 \pm 63 \pm 11\ eV;\ \Gamma = 407 \pm 208 \pm 100\ eV \tag{2}$$

This measurement showed clearly that the antikaon-nucleon interaction is of repulsive type, but cannot be considered a precision measurement. The challenging aim of the DEAR/SIDDHARTA experiments is therefore to measure the kaonic hydrogen transition with a precision at the eV level. The kaonic deuterium will be measured for the first time. These results will represent a breakthrough in the study of the low-energy antikaon-nucleon interaction.

In Section 2, the physics of kaonic atoms is briefly discussed. The DEAR experimental setup installed at DAΦNE is presented in Section 3, while experimental results on kaonic atoms are reported in Section 4. The paper ends with the presentation of the coming experiment, SIDDHARTA, in Section 5, followed by Section 6 - Conclusions.

2. Physics of kaonic atoms

A kaonic atom is formed whenever a negative kaon enters an atomic target, for instance hydrogen (deuterium), looses its kinetic energy through ionization and excitations of the medium atoms and molecules and is eventually captured in an excited orbit, replacing an electron. Various collisional cascade processes and radiative transitions deexcite the kaonic atom.

When the kaon reaches low-n states with small angular momentum, it is absorbed through the strong interaction with the nucleus. This strong interaction causes a shift in the energies of the low-lying levels (essentially the $1s$ level) from their purely electromagnetic values, while the finite lifetime of the state is seen in an increase in the observed level width.

The shift ϵ and the width Γ of the $1s$ state of kaonic hydrogen are related to the real and imaginary part of the complex s-wave scattering length, a_{K^-p}. To the lowest order, neglecting isospin-breaking corrections, in the case of kaonic hydrogen these relations are given by the so-called Deser-Trueman formula [5]:

$$\epsilon + i\Gamma/2 = 2\alpha^3\mu^2 a_{K^-p} = (412\ eV\ fm^{-1}) \cdot a_{K^-p} \tag{3}$$

where α is the fine structure constant and μ the reduced mass of the K^-p system. A similar relation applies to the case of kaonic deuterium and to the corresponding scattering length a_{K^-d}.

Recent results by using the non-relativistic effective Lagrangian approach to bound states have shown that the isospin-breaking corrections to the Deser relations might be important [6]. The main source is represented by the unitary cusp in the K^-p elastic amplitude. As far as Coulomb corrections are concerned they are within a few percent.

Further investigations using effective field theories or lattice calculations to predict QCD amplitudes and compare with data from atomic spectra are needed [6,7,8].

The observable scattering lengths a_{K^-p} and a_{K^-d} can be expressed in terms of the $\bar{K}N$ isospin dependent scattering lengths a_0 (I=0) and a_1 (I=1). The kaonic hydrogen scattering length is the average of the two:

$$a_{K^-p} = 1/2(a_0 + a_1) \tag{4}$$

while the kaonic deuterium scattering length a_{K^-d} is related to a_0 and a_1 in the following way:

$$a_{K^-d} = 2(\frac{m_N + m_K}{m_N + m_K/2})a^{(0)} + C \tag{5}$$

where

$$a^{(0)} = \frac{1}{2}(a_{K^-p} + a_{K^-n}) = \frac{1}{4}(3a_1 + a_0) \tag{6}$$

corresponds (in the t-channel) to the isoscalar $\bar{K}N$ scattering length. The first term in eq. (5) represents the lowest-order impulse approximation, i.e. K^- scattering from each (free) nucleon. The second term, C, includes all higher contributions related to the physics associated to the K^-d three-body interaction.

The determination of the $\bar{K}N$ scattering lengths requires the calculation of C. This is a well-known three-body problem, solvable by the use of Faddeev equations. The K^-d three-body problem includes the complication that the K^-p and K^-n interactions involve significant inelastic channels. The K^-p and K^-n scattering lengths are thus complex and so is the K^-d scattering length. Incorporating $\bar{K}N$ scattering data and its sub-threshold behavior, the two-body potentials are determined in a coupled-channel formalism including both elastic and inelastic channels. Three-body Faddeev equations are then solved by the use of the potentials, taking into account the coupling among the multi-channel interactions.

In this framework, the DEAR/SIDDHARTA results will not only contribute to a precise extraction of the scattering amplitude at threshold, but, hopefully, will place strong constraint on the extrapolations to zero energy which are required in any analysis of scattering data. In fact, the s-wave K^-N amplitudes will be determined more accurately, below and above threshold. This will provide a tighter constraint on the p-wave parameters from experimental data and, eventually, reduce the uncertainty in the phenomenological procedure used to calculate the so-called kaon-nucleon sigma terms, fundamental quantities in understanding the chiral symmetry breaking in strangeness sector [9].

3. The DEAR setup on DAΦNE

The principle of the DEAR experiment is the following: low momentum negative kaons produced in the decay of the ϕ-mesons at DAΦNE leave the thin-walled beam pipe, are degraded in energy to a few MeV, enter a gaseous target through a thin window and are finally stopped in the gas. The stopped kaons are captured in an outer orbit of the gaseous atoms, thus forming the exotic kaonic atoms. The kaons cascade down and some of them will reach the ground state emitting X rays. The energy of the X rays emitted in these transitions is measured with a CCD (Charge-Coupled Device) detector system [10].

Fig. 1 shows a schematic of the DEAR experimental setup. The cylindrical cryogenic target cell had a diameter of 12.5 cm and a height of 14 cm. Special care was taken to avoid materials with fluorescence X rays in the region of the kaonic atoms transitions. Therefore, a light target was chosen, made only of aluminium (top-plate and entrance ring), kapton (side wall and entrance window) and a support structure in fiberglass. The target was operated with hydrogen at 2 bar and 25 K ($\rho = 2.1$ g/l).

16 CCD detector chips (Marconi Applied Technologies, CCD55-30) with a total area of 116 cm^2 were placed around the cryogenic target cell. Each chip has 1242 x 1152 pixels with a pixel size of 22.5 μm x 22.5 μm and a depletion depth of about 30 μm. The working temperature was stabilized at 165 K to achieve the energy resolution of 150 eV at 6 keV with a readout every 90 seconds. The CCD front-end electronics and controls, and the data acquisition system, were specially made for this experiment [11].

4. Kaonic hydrogen measurement

The DEAR experiment was installed at DAΦNE at the beginning of 1999. After a period of machine and setup optimization, with a continuous in-

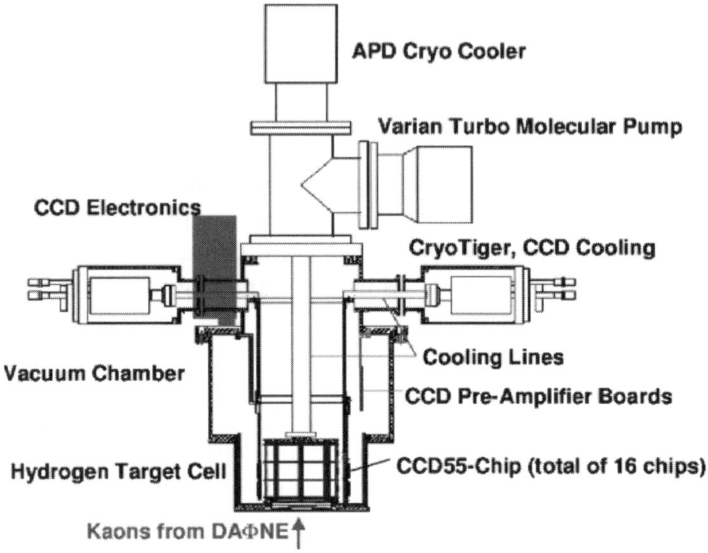

Figure 1. *Schematic representation of the DEAR setup*

crease of luminosity and decrease of background, in 2002 DEAR performed two kind of kaonic atoms measurements: the kaonic nitrogen and the kaonic hydrogen one. The measurement of kaonic nitrogen [12] had multiple tasks and deliverables: a feasibility study of the DEAR technique to produce and detect kaonic atoms at DAΦNE; study of the machine background and of the setup performance and the optimization of the signal to background ratio; the first measurement of kaonic nitrogen transition yields.

In the period November-December 2002 the kaonic hydrogen measurement was performed. Data for 58.4 pb^{-1} were collected in this period. At the end of the period, a background measurement with separated electron and positron beams and intentionally high X-ray background was performed (no-collision spectrum).

Two independent analyses were performed in order to obtain the kaonic hydrogen lines from which to extract the strong interaction shift and width. In both analyses Voigt functions with Gaussians for the detector resolution were used for the kaonic hydrogen lines. Fit parameters were the intensities of K_α, K_β and K_γ, the energy of K_α and the Lorentzian width for K_α, equal for all the K-transitions.

In one of the analyses a simultaneous fit of the kaonic hydrogen and no-collision spectra was performed. The same function fitted the continuous background and the electronic peaks, apart a normalization factor. The energy region corresponding to K_{high} (higher than K_γ) was excluded from the fit, since the lines in this region could not be distinguished by the fit and no precise information exists for the relative yields. It was estimated by Monte Carlo that the systematic error introduced by this cut is at the level of the eV and was included in the final result.

In the second analysis the kaonic hydrogen spectrum was analyzed together with a spectrum built as a sum of the kaonic nitrogen [12] one and a subset (low CCD occupancy) of the no-collision one. A constrained fit based on the ratios of the electronic transitions in the two spectra, was then performed. The K_{high} region was dealt with analogously to the first analysis. Some tests of the stability of the results and of the systematic errors were done by considering various cascade models of kaonic hydrogen transitions giving the transition yields [13].

The "pure" kaonic hydrogen spectrum with the background (continuous and structured) subtracted is shown in Fig. 2.

The results coming from the two independent analyses are in good agreement and the weighted averages of the two analyses for the shift and width of the $1s$ ground state of kaonic hydrogen are:

$$\epsilon = -193 \pm 37(stat.) \pm 6(syst.)\, eV;\ \Gamma = 249 \pm 111(stat.) \pm 39(syst.)\, eV \quad (7)$$

These results confirm the KEK values and the repulsive character of the K^-p interaction at threshold. They differ however significantly from the KEK results (eq. (2)) in two aspects: the errors are a factor 2-3 smaller; moreover, DEAR was able, for the first time, to obtain a full pattern of the K-series lines of kaonic hydrogen: K_α, K_β and K_γ were clearly identified with an overall statistical significance of 6.2 σ.

5. Future perspectives: the SIDDHARTA experiment

DEAR has performed the most precise measurement of kaonic hydrogen; the precision which was achieved, however, is at the level of tens of eV, whilst the goal of a precision measurement should push this precision in the eV range. The DEAR precision was limited by a signal/background ratio of about 1/70. In order to go beyond this ratio a jump of quality is necessary. Consequently, an accurate study of the background sources present at DAΦNE was re-done. The background includes two main sources: synchronous background: coming together with the K^- - related to the interaction of K^- on the nuclei of the setup and also to the ϕ-decay processes;

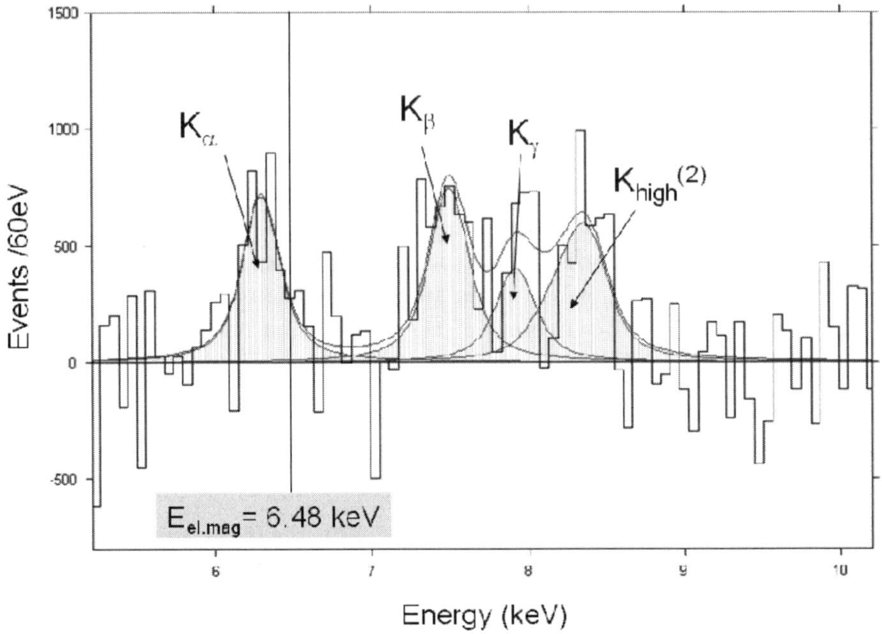

Figure 2. *The kaonic hydrogen background (continuous and structured) subtracted spectrum*

it can be defined hadronic background; asynchronous background: final products of electromagnetic showers in the machine pipe and in the setup materials originated by particles lost from primary circulating beams either due to the interaction of particles in the same bunch (Touschek effect) or due to the interaction with the residual gas.

Accurate studies performed by DEAR showed that the main background source in DAΦNE is of the second type, which shows the way to reduce it. A fast trigger correlated to the negative kaon entrance in the target would cut the main part of the asynchronous background. This was not possible in DEAR, which used as X-ray detector the CCDs - which are slow devices.

A new detector was then identified, which preserves all the good features of the CCDs (resolution, linearity and stability) but is fast enough to supply a trigger at the level of one μs. It was estimated to cut in such way the background present in DEAR by 2-3 orders of magnitude. The detector is the newly developed large area Silicon Drift Detector (SDD) [14], based on which a new experiment, which continues the DEAR scientific line, was born. The new experiment is SIDDHARTA (SIlicon Drift Detector for Hadronic Atom Research by Timing Application) [3].

Presently, construction and testing of the SDD detectors and of the electronics and mechanical structures are in progress. A preliminary version of the SIDDHARTA setup is shown in Fig. 3. With this setup, it was estimated that a measurement at eV level for kaonic hydrogen becomes feasible with an integrated luminosity of about 100 pb^{-1}. The setup will be installed at DAΦNE in the end of 2006 and start taking data for kaonic hydrogen, followed by kaonic deuterium.

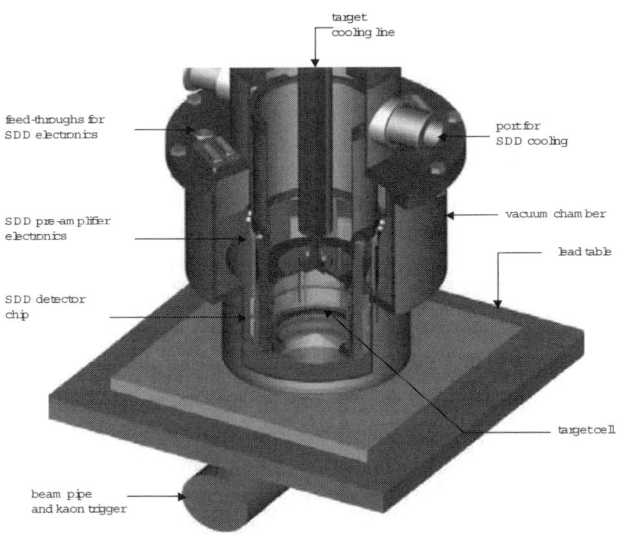

Figure 3. *SIDDHARTA preliminary setup.*

6. Conclusions

DAΦNE collider has unique features as kaons source: intrinsically clean and low momentum, a situation unattainable with fixed target machines, especially suitable for kaonic atom research.

The DEAR/SIDDHARTA experiments combine the newly available techniques with the good kaon beam quality to initiate a renaissance in the investigation of low-energy kaon-nucleon interaction.

DEAR has performed the most precise measurement of kaonic hydrogen; the eV precision measurement of the strong interaction parameters of the fundamental level in kaonic hydrogen will be performed by SIDDHARTA. The first measurement of kaonic deuterium is as well planned. These results will open new windows in the study of the kaon-nucleon interaction, in particular the chiral symmetry breaking in the strangeness sector.

Acknowledgments

The DAΦNE and BTF (Beam Test Facility) groups are warmly acknowledged for the very good cooperation and team-work.

References

1. G. Vignola, Proc. of the "5th European Particle Accelerator Conference", Sitges (Barcelona), Eds. S. Myres *et al.*, Institute of Physics Publishing, Bristol and Philadelphia (1996) 22 .
2. S. Bianco *et al.*, Rivista del Nuovo Cimento **22**, No. 11 (1999) 1.
3. J. Zmeskal, SIDDHARTA Technical Note IR-2 (2003); C. Curceanu (Petrascu), SIDDHARTA Technical Note IR-3 (2003).
4. M. Iwasaki *et al.*, Phys. Rev. Lett **78** (1997) 3067;T.M. Ito *et al.*, Phys. Rev. **A58** (1998) 2366.
5. S. Deser *et al.*, Phys. Rev. **96** (1954) 774; T.L. Truemann, Nucl. Phys. **26** (1961) 57; A. Deloff, Phys. Rev. **C13** (1976) 730.
6. U.-G. Meissner, U. Raha and A. Rusetsky, E. Phys. J. **C35** (2004) 349.
7. J. Gasser, Mini-proceedings of the 4th International Workshop on *Chiral Dynamics 2003: Theory and Experiment*, Bonn 2003, edited by U.-G. Meissner, H.-W. Hammer and A. Wirzba, p.126.
8. J. Gasser, in Frascati Physics Series, Proceedings of the *DAFNE 2004: Physics at meson factories*, Frascati 2004 - to appear.
9. E. Reya, Rev. Mod. Phys. **46** (1974) 545; H. Pagels, Phys. Rep. **16** (1975) 219.
10. J.-P. Egger, D. Chatellard, E. Jeannet, Part. World. **3** (1993) 139.
11. M. Iliescu *et al.*, Nucl. Instr. Meth., in preparation.
12. T. Ishiwatari *et al.*, Phys. Lett. **593** (2004) 48.
13. T. Jensen, DEAR Technical Note IR-45 (2003).
14. E. Gatti, P.Rehak, Nucl. Instr. and Meth. **225** (1984) 608; E. Gatti, P.Rehak, Nucl. Instr. and Meth. **A235** (1985) 224.

ATOM-AT-A-TIME CHEMISTRY OF THE TRANSACTINIDE ELEMENT, RUTHERFORDIUM (ELEMENT 104) TOWARDS EXPERIMENTAL VERIFICATION OF RELATIVISTIC EFFECTS IN CHEMICAL PROPERTIES

Y. NAGAME,[1] K. TSUKADA,[1] M ASAI,[1] A. TOYOSHIMA,[1] K. AKIYAMA,[1] Y. ISHII,[1] I. NISHINAKA,[1] T. K. SATO,[1] M. HIRATA,[1] T. ICHIKAWA,[1] H. HABA[2], S. GOTO[3] AND M. SAKAMA[4]

[1] *Japan Atomic Energy Research Institute, Tokai, Ibaraki 319-1195, Japan*
[2] *Cyclotron Center, RIKEN, Wako, Saitama 351-0198, Japan*
[3] *Department of Chemistry, Niigata University, Niigata 950-2181, Japan*
[4] *Department of Radiologic Science and Engineering, The University of Tokushima, Tokushima 770-8509, Japan*
E-mail: nagame@popsvr.tokai.jaeri.go.jp

Chemical properties of the transactinide element, rutherfordium (Rf), produced in the reaction ^{248}Cm(^{18}O,5n) have been studied at an atom-at-a-time scale. Ion-exchange experiments of Rf together with the lighter homologues in the periodic table of the elements, group-4 elements Zr and Hf, in hydrofluoric acid solutions have been conducted with a rapid ion-exchange separation apparatus. From the systematic study of the anion-exchange behavior of Rf, we have observed an unexpected chemical behavior of Rf; the fluoride complex formation of Rf is significantly different from those of the homologues. Characteristics of the complexing strength of the Rf fluoride are briefly discussed by comparing with those of Zr and Hf and also with theoretical predictions by relativistic molecular density-functional calculations.

1. Introduction

Presently, we know of 25 man-made transuranium elements. According to the actinide concept,[1] the $5f$ electron series ends with element 103, lawrencium (Lr), and a new $6d$ transition series is predicted to begin with element 104, rutherfordium (Rf), and is called transactinide elements or superheavy elements.[2] The currently discovered 14 transactinide elements, elements 104 through 118 except for element 117, are placed in the periodic table under their lighter homologues in the $5d$ electron series, Hf to Hg, and in the successive groups 13 to 18, Tl - Rn (see Fig.1).

Studies of chemical properties of the transactinide elements offer unique opportunities to obtain information about trends in the periodic table of the elements at the limits of nuclear stability, and to assess the magnitude of the influence of relativistic effects on chemical properties. According to the calculations of electron configurations of heavier elements, it is predicted that sudden changes in the structure of electron shells may appear due to relativistic effects which originate from the increasingly strong Coulomb field of the highly charged atomic nucleus.[4] Therefore, it is expected that heavier elements show a drastic rearrangement of electrons in their atomic ground states, and as electron configurations are responsible for chemical behavior of elements, such relativistic effects can lead to surprising chemical properties. Increasing deviations from the periodicity of chemical properties based on extrapolation from lighter homologues in the periodic table are consequently predicted.[5–7] It would be no longer possible to deduce detailed chemical properties of transactinide elements simply from the position of the periodic table.

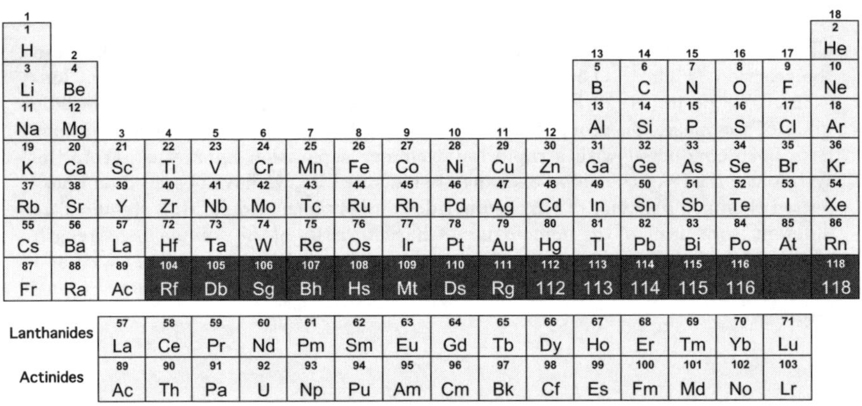

Figure 1. Periodic table of the elements. Element 111 has been named roentgenium (Rg).[3]

The transactinide elements must be produced at accelerators using reactions of heavy-ion beams with heavy target materials and must be identified by measurement of their decay or that of their known daughter nuclei with unambiguous detection techniques. Because of the short half-lives and the low production rates of the transactinide nuclides, however, each atom produced decays before a new atom is synthesized. This means that any chemistry to be performed must be done on an "atom-at-a-time" basis.[2,8] Thus, rapid, very efficient and selective chemical procedures are indispensable to isolate desired transactinide nuclides.

In order to carry out chemical experiments of the transactinide elements with single atoms, we have developed some experimental apparatuses: a beam-line safety system for the usage of a gas-jet coupled radioactive target and recoil chamber, a rotating wheel catcher apparatus for the measurement of α and spontaneous fission (SF) decay of transactinide nuclei and an automated rapid ion-exchange separation apparatus based on high performance liquid chromatography coupled with an on-line α-particle detection system.[9]

The transactinide nuclide, 78-s ^{261}Rf, that is commonly used for recent chemical studies of element 104, has been produced in the reaction ^{248}Cm(^{18}O,5n) at the JAERI tandem accelerator. An excitation function of the reaction has been measured with the rotating wheel catcher apparatus MANON (**M**easurement system for **A**lpha particle and spontaneous fissio**N** events **ON**-line), and the maximum production cross section has been determined to be 13 nb at around 94-MeV ^{18}O energy.[10] Atom-at-a-time chemistry of Rf together with the group-4 elements Zr and Hf in acidic solutions have been conducted with the rapid ion-exchange separation apparatus, AIDA (**A**utomated **I**on-exchange separation apparatus coupled with the **D**etection system for **A**lpha-spectroscopy). AIDA enables us to perform cyclic discontinuous column chromatographic separations of short-lived nuclides in aqueous phases and automated detection of α-particles within a typical cycle of 1 - 2 min. The adsorption behavior of Rf as a function of the acid concentration has been systematically studied with AIDA.[11,12]

In this paper, we report on the production of ^{261}Rf and an atom-at-a-time chemistry of Rf in hydrofluoric acid (HF) solutions at JAERI.

2. Production of the transactinide nuclide ^{261}Rf

Longer-lived transactinide nuclides required in chemical studies are existing in the neutron-rich region of the chart of the nuclides and they are produced in the so-called hot fusion reactions of such as ^{18}O, ^{19}F, ^{22}Ne, and ^{26}Mg projectiles with actinides targets of ^{244}Pu, ^{248}Cm, ^{249}Bk, and so on.

Figure 2 shows the schematic of the experimental set-up for the production of ^{261}Rf. A ^{248}Cm target of 590 μg/cm^2 in thickness prepared by electrodeposition onto a 2.2-mg/cm^2-thick beryllium backing foil was bombarded by ^{18}O^{6+} beams delivered from the JAERI tandem accelerator. The reaction products recoiling out of the target were stopped in He gas ($\approx$1 bar), attached to KCl aerosols, and were transported through a gas-jet transport system to the apparatus MANON, where the reaction products were deposited on polyethylene terephthalate foils of 120 μg/cm^2

in thickness and 20 mm in diameter at the periphery of an 80-position stainless steel wheel of 80-cm diameter. The wheel was periodically rotated to position the foils between six pairs of Si PIN photodiodes for α-particle detection.

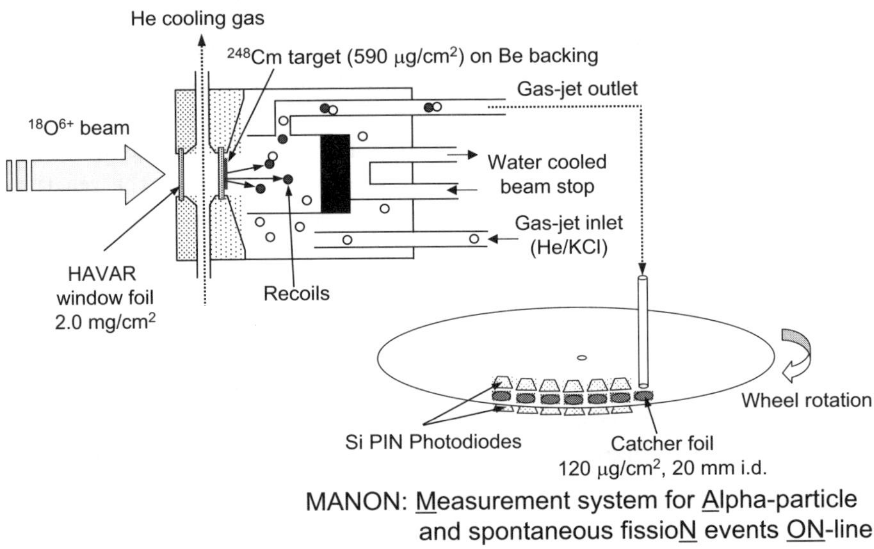

Figure 2. Schematic representation of a target and recoil chamber arrangement with a He/KCl gas-jet system and the apparatus MANON (**M**easurement system for **A**lpha-particle and spontaneous fissio**N** events **ON**-line).

The sum of α-particle spectra measured in the six top detectors in the bombardment of ^{248}Cm with 94-MeV ^{18}O ions is shown in Fig. 3(a). The total beam dose was 2.35×10^{16}. In the α-energy range of 8.10 - 8.40 MeV, α lines from 78-s ^{261}Rf (8.28 MeV) and its daughter 25-s ^{257}No (8.22, 8.27, and 8.32 MeV) are clearly seen. A total of 166 events were registered both in the top and bottom detectors in the singles measurement and 57 $\alpha(^{261}\text{Rf})$-$\alpha(^{257}\text{No})$ correlation events were detected. Assuming a 100% α-decay branch for both ^{261}Rf and ^{257}No, the production cross sections of ^{261}Rf were evaluated. In Fig. 3(b), the measured cross sections are plotted as a function of the ^{18}O bombarding energy together with the literature data,[13,14] where the relative cross section values by Silva *et al.*[14] are normalized to the present results. The data are smoothly connected with the maximum cross section of about 13 nb at around 94 MeV.[10] This results in a production rate of about 2 atoms per min under the given conditions.

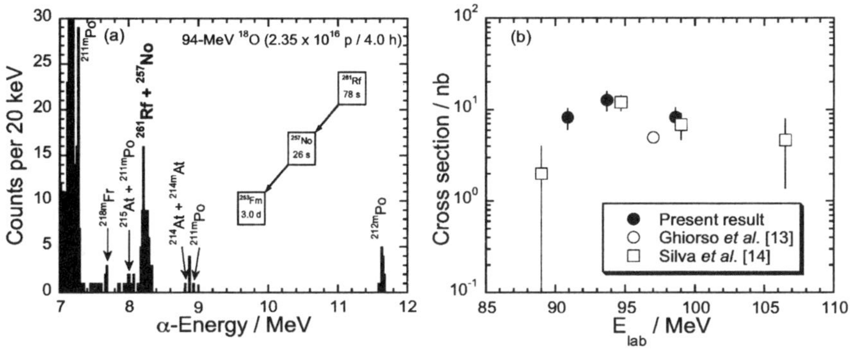

Figure 3. (a) Sum of α-particle spectra measured in the bombardment of the ^{248}Cm target with 94-MeV ^{18}O ions, and (b) cross sections of the ^{248}Cm(^{18}O,5n)^{261}Rf reaction as a function of the bombarding energy of ^{18}O. The data taken from the literature[13,14] are also shown.

3. Chemical studies of Rf

3.1. *Atom-at-a-time chemistry*

The chemical study needs to be carried out on the phenomena that gives the same results for only a few atoms and for macro amounts of atoms; the question arises as to whether a meaningful chemistry with single atoms is possible. For singe atoms, the classical law of mass action is no longer valid, because the atom cannot exist in the different chemical forms taking part in the chemical equilibrium at the same time. Guillaumont *et al.* suggested for single-atom chemistry to introduce a specific thermodynamic function, the single-particle free enthalpy.[15] An expression equivalent to the law of mass action is derived, in which activities (concentrations) are replaced by probabilities of finding the species in the appropriate phases. According to this law, an equilibrium constant (distribution coefficient) of the atom between two phases is correctly defined in terms of the probabilities of finding the atom in one phase or the other. If a static partition method is used, this coefficient must be measured in repetitive experiments. Since dynamical partition methods can be considered as repetitive static partitions, the displacement of the atom along the chromatographic column gives a statistical result.[15]

For short-lived atoms, the partition equilibrium must be reached during the life-time of the isotopes, which requires fast reaction kinetics. Based on the kinetics of a single-step exchange reaction, Borg and Dienes suggested that at certain conditions a measurement of the partition of the atom between the phases with very few atoms will already yield an equilibrium constant close to the "true" value provided that both states are rapidly sampled.[16] The above interpretation indicates that chromatographic systems with fast reaction kinetics are ideally suited for single-atom separation as there is rapid and multiple sampling of the adsorbed or mobile chemical species.[8]

To perform rapid and cyclic ion-exchange chromatographic experiments with Rf, we developed the apparatus AIDA that consists of a modified ARCA (**A**utomated **R**apid **C**hemistry **A**pparatus) which is the miniaturized computer-controlled liquid chromatography system[17] and an automated on-line α-particle detection system. Figure 4 shows the schematic of AIDA.

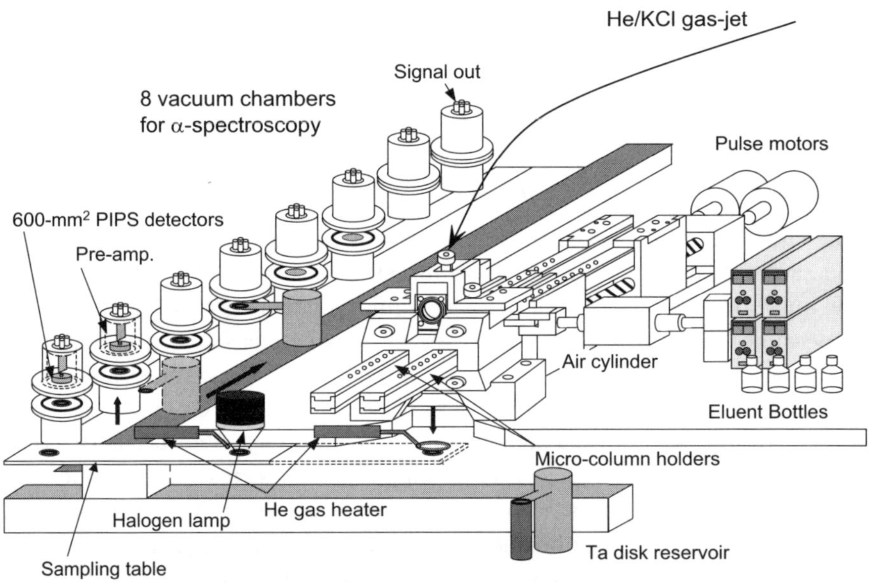

Figure 4. Schematic of AIDA (**A**utomated **I**on-exchange separation apparatus coupled with the **D**etection system for **A**lpha spectroscopy).

3.2. *On-line production of Rf, Zr and Hf*

Due to the influence of relativistic effects, deviations from the periodicity of the chemical properties based on extrapolations from the lighter homologues are expected. Thus, the experimental approach should involve a detailed comparison of the chemical properties of the transactinide elements with those of their lighter homologues under strictly identical conditions. For the on-line production of Rf and the homologues, we used two kinds of targets; one is the mixed target of ^{248}Cm (610 μg/cm^2 in thickness) and Gd (39.3%-enriched ^{152}Gd of 36 μg/cm^2 in thickness) prepared by electrodeposition onto a 2.4-mg/cm^2 beryllium foil for the simultaneous production of ^{261}Rf and 3.24-min ^{169}Hf via the ^{248}Cm(^{18}O,5n) and Gd(^{18}O,xn) reactions, respectively. Another is the ^{nat}Gd (370-μg/cm^2 thickness) and ^{nat}Ge (660-μg/cm^2 thickness) mixed target to produce ^{169}Hf and 7.86-min ^{85}Zr through the ^{nat}Gd(^{18}O,xn) and ^{nat}Ge(^{18}O,xn) reactions, respectively.[11,12]

3.3. *Atom-at-a-time chemistry of Rf - Fluoride complex formation of Rf*

In the following, we outline the experimental procedures and a part of the results on the anion-exchange behavior of Rf in HF solutions. There are some advantages for the study of fluoride complexes. The fluoride anion F^- strongly combines with transition metal cations, i.e., strong ionic bonds are formed between metal cations and the F^- ions. Thus, information about such as a charge density and an ionic radius of the metal cation is obtained through the fluoride complex formation process. Also the fast reaction kinetics of the fluoride complex formation is the advantage of studying the chemical behavior of short-lived nuclides.

^{261}Rf and ^{169}Hf were produced in the 94-MeV ^{18}O-induced reactions of the ^{248}Cm/Gd target. The reaction products recoiling out of the target were transported by the He/KCl gas-jet system to the collection site of AIDA (see Fig. 5). After collection for 125 s, the site was mechanically moved on the top of one of the micro columns, where the products were dissolved with 240 μL HF solution of various concentrations (1.9 - 13.9 M) and were fed onto the chromatographic column filled with the anion-exchange resin MCI GEL CA08Y (particle size of about 20 μm) at a flow rate of 0.74 mL/min. (In the present experiment, we used two different micro columns, 1.6 mm i.d. $\times$ 7.0 mm and 1.0 mm i.d. $\times$ 3.5 mm, to measure a wide range of distribution coefficients (K_d values). The values given here refer to the condition when the 1.6 mm i.d. $\times$ 7.0 mm column was used.[12]) The effluent was collected on a tantalum (Ta) disk as fraction 1 and evaporated to dryness using hot helium gas and a halogen heat lamp. Then, the products remaining in the column were eluted with 250 μL of 4.0 M HCl at a flow rate of 1.0 mL/min. The effluent was collected on another Ta disk and was evaporated to dryness as fraction 2. The pair of disks was automatically transported to the α-spectroscopy station equipped with eight 600-mm^2 passivated ion-implanted planar silicon (PIPS) detectors. After the α-particle measurement, the 493 keV γ-radiation of ^{169}Hf was monitored with Ge detectors to determine the elution behavior of Hf and its chemical yield. The anion-exchange experiments with ^{85}Zr and ^{169}Hf produced from the Ge/Gd mixed target were conducted under the same conditions as those with ^{261}Rf and ^{169}Hf. The effluents were collected in polyethylene tubes and were assayed by γ-ray spectroscopy.[12] Each separation was accomplished within 20 s and the α-particle measurement was started within 80 s after the collection of the products at the AIDA collection site. The chemical yield of Hf including deposition and dissolution efficiencies of the aerosols was approximately 60%.[12]

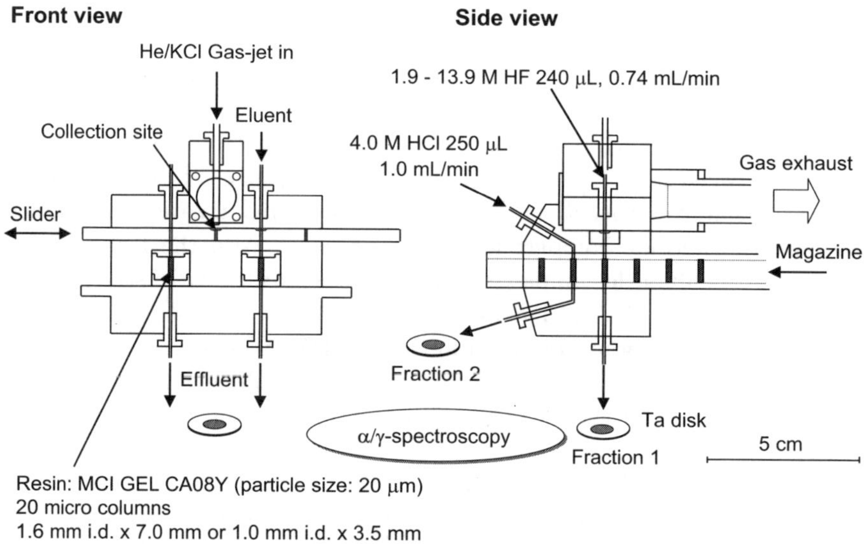

Figure 5. Schematic of the ion-exchange part of AIDA.

From the 4226 cycles of the anion-exchange experiments, a total of 266 α events from ^{261}Rf and its daughter ^{257}No were registered including 25 time-correlated α pairs of ^{261}Rf and ^{257}No. Figure 6 shows the variation of the adsorption probability of Rf, Zr, and Hf as a function of the concentration of HF, [HF].[12] The ordinate shows the adsorption probabilities of these elements, $\%\text{ads} = 100A_2(A_1 + A_2)$, where A_1 and A_2 are the eluted radioactivities observed in fractions 1 and 2, respectively. The adsorption values of Zr and Hf were also measured in the separate experiments using ^{89m}Zr and ^{167}Hf produced in the ^{89}Y(p,n) and ^{nat}Eu(^{19}F,xn) reactions, respectively.[12] As shown in Fig. 6, the adsorption values of Zr and Hf obtained in the different experiments agree well, and the adsorption of Zr and Hf is absolutely equal in a wide range of HF concentrations and steeply decreases with [HF], while that of Rf decreases at much lower [HF]. The lower adsorption of Rf indicates that the fluoride complex formation of Rf is weaker than those of Zr and Hf.

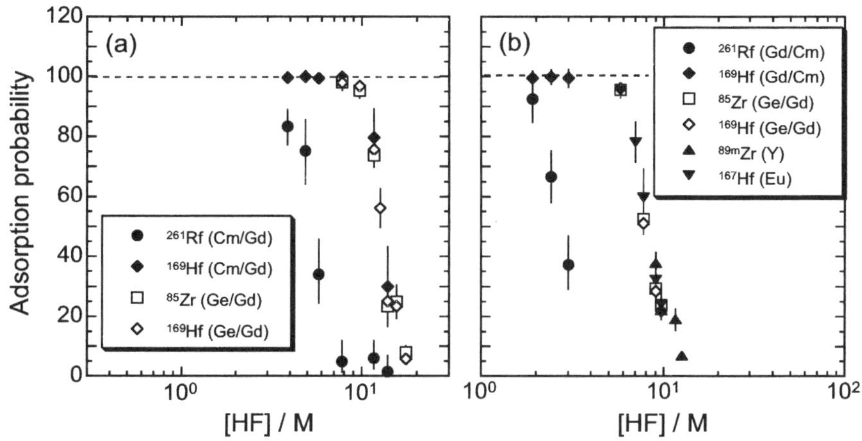

Figure 6. Adsorption behavior of Rf, Zr and Hf on the anion-exchange resin CA08Y as a function of the concentration of HF, [HF], obtained with (a) the 1.6 mm i.d. × 7.0 mm column and with (b) the 1.0 mm i.d. × 3.5 mm one. The data for ^{261}Rf and ^{169}Hf obtained from the Cm/Gd target are shown by closed symbols, while those for ^{85}Zr and ^{169}Hf from the Ge/Gd target are by open ones. The adsorption values of ^{89m}Zr and ^{167}Hf produced from the ^{89}Y and ^{nat}Eu targets, respectively, are also included in (b). Adapted from Ref.[12]

Figure 7 shows the distribution coefficients (K_d) for Rf, Zr, and Hf as a function of the concentration of HF. (The distribution coefficient K_d denotes the ratio of the ion concentration in the resin to that in the solution.) The K_d values of these elements decrease linearly with [HF] in the log K_d vs. log [HF] plot. It should be noted that the slopes for Zr and Hf are clearly -3 (dashed line), while that for Rf is significantly different, i.e. -2 (solid line). According to the equilibration of ion-exchange reactions, the slope in this plot, logarithm K_d against logarithm of the ligand concentration indicates the charge state of the metal compounds. Thus, the results demonstrate that Rf is likely to be present as the hexafluoro complex, $[RfF_6]^{2-}$ similar to the well known $[ZrF_6]^{2-}$ and $[Hf_6]^{2-}$ at lower [HF], while Zr and Hf are likely to be present in the form of the heptafluoro complexes, $[ZrF_7]^{3-}$ and

$[HfF_7]^{3-}$ as suggested by Kim *et al.*[18] and Korkisch.[19] The activity coefficients of the involved chemical species, however, are not available in the case of the higher concentration of HF, thus, the definite identification of the anionic fluoride complexes is presently an open question.

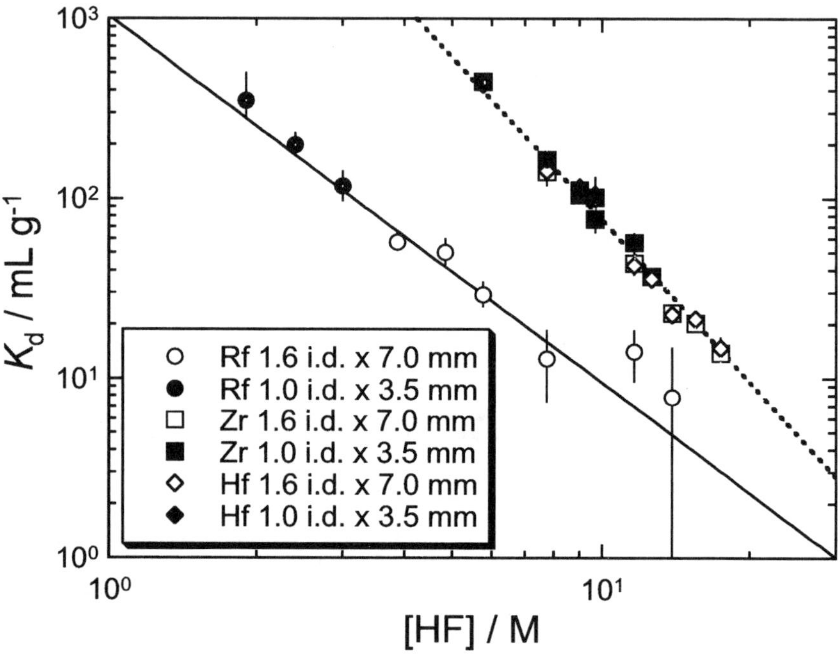

Figure 7. Distribution coefficients (K_d) for Rf, Zr and Hf on the anion-exchange resin CA08Y as a function of the concentration of HF. The data from the 1.6 mm i.d. × 7.0 mm and 1.0 mm i.d. × 3.5 mm columns are depicted by open and closed symbols, respectively. Adapted from Ref.[12]

Recently, we have studied the anion-exchange behavior of Rf in the mixed HF/HNO_3 solutions to further understanding of the fluoride complexation of Rf. The K_d values for Rf and the homologues Zr and Hf have been measured as a function of the fluoride ion concentration $[F^-]$. The results indicate that at constant concentration of the nitrate ion $[NO_3^-] = 0.01$ M, the formation of the hexafluoro anionic complexes of Zr and Hf occurs at a fluoride ion concentration of about 10^{-5} M, while that of Rf started at around 10^{-3} M. There was about two-orders of magnitude difference in the fluoride ion concentration between Rf and the homologues for the formation of the hexafluoro complexes. This clearly demonstrates that the formation of the fluoride complexes of Rf is much weaker than those of the homologues Zr and Hf.[20]

Relativistic density-functional calculations of the electronic structures of hexafluoro complexes ($[MF_6]^{2-}$, M=Rf, Zr, and Hf) have been performed to evaluate the stability of the complexes. The results indicate that the sequence in the overlap populations between the valence d orbitals of M^{4+} and the valence orbitals of F^- is found to be Zr $\approx$ Hf $>$ Rf, suggesting that the Rf complex is less stable than those of Zr and Hf in the $[MF_6]^{2-}$ structure.[21] Although the calculations qualitatively agree with the experimental results, further theoretical approaches are needed to quantitatively understand the present interesting results.

According to the HSAB (Hard and Soft Acids and Bases) concept,[22] the fluoride anion is a hard anion and interacts stronger with (hard) small cations. Thus, a weaker fluoride complexation of Rf as compared to Zr and Hf would be reasonable if the size of the Rf^{4+} ion is larger than those of Zr^{4+} and Hf^{4+} as expected by Johnson and Fricke.[23] (It is well known that the radii of Zr^{4+} and Hf^{4+} are very similar). To further understanding of the chemical properties of Rf, however, the complexation study with the softer halide anions (Br^- and I^-) will give some valuable information.

4. Conclusions

From the chemical behavior of Rf on an atom-at-a-time basis, we observed large difference in the fluoride complex formation between Rf and the lighter homologues Zr and Hf. The result indicates the fluoride complexing strength of Rf is much weaker than that of Zr and Hf. The ionic radius of Rf is expected to be larger than those of Zr and Hf, which causes the weak ionic bonding of the fluoride complex of Rf. The present result is qualitatively consistent with the prediction of relativistic effects. However, to understand quantitatively the unexpected behavior of Rf, further theoretical calculations which include relativistic effects are also required.

Acknowledgments

The present study at JAERI has been carried out in collaboration with Niigata University, Osaka University, Tokyo Metropolitan University, University of Tokushima, Kanazawa University, Tsukuba University, Gesellschaft für Schwerionenforschung (GSI), and Mainz University. This work was partly supported by the JAERI tandem accelerator collaboration program and the program on the scientific cooperation between JAERI and GSI in research and development in the field of ion beam application. We would like to acknowledge the crew of the JAERI tandem accelerator for providing the stable and intense ^{18}O beam.

References

1. G. T. Seaborg, *Chem. Eng. News* **23**, 2190 (1945).
2. M. Schädel (ed.), *The Chemistry of Superheavy Elements*, Kluwer Academic Publishers, Dordrecht, 2003.
3. J. Corish and G. M. Rosenblatt, *Pure Appl. Chem.* **76**, 2101 (2004).
4. B. Fricke and W. Greiner, Phys. Lett. **30B**, 317 (1969).
5. P. Pyykkö, Chem. Rev. **88**, 563 (1988).
6. V. Pershina, *Theoretical Chemistry of the Heaviest Elements.* In: *The Chemistry of Superheavy Elements* (M. Schädel ed.), Kluwer Academic Publishers, Dordrecht, 2003, pp.31-94.
7. P. Schwerdtfeger and M. Seth, Encyclopedia of Computational Chemistry, Vol. 4, P. von R. Schleyer et al. (eds.), Wiley, New York, 1998, pp. 2480-2499.
8. J. V. Kratz, *Chemistry of Transactinides.* In: *Handbook of Nuclear Chemistry, Vol. 2, Elements and Isotopes* (A. Vértes, S. Nagy, and Z. Klencsár eds.), Kluwer Academic Publishers, Dordrecht (2003), pp. 323-395.
9. H. Haba *et al.*, Radiochim. Acta **89**, 733 (2001).
10. Y. Nagame *et al.*, J. Nucl. Radiochem. Sci. **3**, 85 (2002).
11. H. Haba *et al.*, J. Nucl. Radiochem. Sci. **3**, 143 (2002).
12. H. Haba *et al.*, J. Am. Chem. Soc. **126**, 5219 (2004).
13. A. Ghiorso *et al.*, Phys. Lett. **32B**, 95 (1970).
14. R. J. Silva *et al.*, Nucl. Phys. **A216**, 97 (1973).
15. R. Guillaumont *et al.*, Radiochim. Acta **46**, 169 (1989); R. Guillaumont *et al.*, Radiochim. Acta **54**, 1 (1991).
16. R. J. Borg and G. J. Dienes, J. Inorg. Nucl. Chem. **43**, 1129 (1981).
17. M. Schädel *et al.*, Radiochim. Acta **48**, 171 (1989).
18. J. I. Kim *et al.*, Anal. Chim. Acta **64**, 29 (1973).
19. J. Korkisch, *Handbook of Ion Exchange Resins: Their Application to Inorganic and Analytical Chemistry*, CRC Press, Boca Raton, FL, 1989.
20. A. Toyoshima, Ph. D thesis, Osaka University, 2004.
21. M. Hirata, private communication.
22. R. G. Pearson, J. Am. Chem. Soc. **85**, 3533 (1963).
23. E. Johnson and B. Fricke, J. Phys. Chem. **95**, 7082 (1991).

THE KAONIC HELIUM CASE

C. CURCEANU (PETRASCU) (on behalf of the SIDDHDARTA Collaboration)*
E-mail: petrascu@lnf.infn.it

A.M. BRAGADIREANU[†], C. CURCEANU (PETRASCU), F. GHIO[‡],
B. GIROLAMI[‡], C. GUARALDO, M. ILIESCU[†], P. LEVI SANDRI,
V. LUCHERINI, D.L. SIRGHI[†] AND F. SIRGHI
LNF-INFN, C.P. 13, Via E. Fermi 40 00044 Frascati (Roma), Italy

M. CARGNELLI, H. FUHRMANN, T. ISHIWATARI, P. KIENLE[§],
J. MARTON AND J. ZMESKAL
Stefan Meyer Institut für subatomare Physik, Boltzmanngasse 3, 1090 Vienna, Austria

C. FIORINI, A. LONGONI AND T. FRIZZI
Politecnico di Milano, Sez. di Elettronica, Via Golgi 40, 2133, Milano, Italy

K. ITAHASHI, M. IWASAKI AND T. KOIKE
RIKEN, Wako-shi, Saitama, 351-0198, Japan

T. PONTA
IFIN-HH, P.O. Box MG-6 R-76900 Magurele, Bucharest, Romania

H. SOLTAU AND P. LECHNER
PNSensors GmbH, Rönnerstr. 28, D-80803 München, Germany

L. STRUDER
MPI for Extraterrestrial Physics, Giessenbachstr. D-85740 Garching, Germany

The only three existent kaonic helium X-ray transition measurements at present are referring to the transitions to 2p level. These measurements are more than 30 years old and the obtained results, affected by big errors, are much larger than those predicted by optical models. It is thought that the optical model is inadequate, due to the presence of the $\Lambda(1405)$ resonance, not properly taken into account. Because the nucleons in the helium nucleus are tightly bound, the effective energy of the K^-p interaction (1432 MeV at threshold) is in helium much closer to the energy of the resonance than in other nuclei. It is then planned to measure the kaonic helium X-ray transitions to the 2p level in the framework of the SIDDHARTA (SIlicon Drift Detector for Hadronic Atom Research by Timing Application) experiment, at the DAΦNE collider of Frascati National Laboratories, and to confirm or not the discrepancy reported by the previous experiments with a much smaller error.

1. Introduction

A kaonic atom is formed whenever a negatively charged kaon enters a target, loses its kinetic energy and is eventually captured into a high atomic Bohr orbit, replacing one of the outer electrons. The kaon then cascades down through its own series, initially by Auger transitions in which orbital electrons are ejected (not valid for kaonic hydrogen) and in later stages by emission of X rays. Finally the particle in a state of low angular momentum will be absorbed by the nucleus via the strong interaction. This strong interaction is the reason for a shift of the energy of the lowest atomic levels from their purely electromagnetic values, whilst the absorption reduces the lifetime of the states, broadening the X-ray transition to this final atomic levels.

The study of kaonic hydrogen atom can give information about the strong interaction parameters (shift and width) of the $1s$ level. Such kind of measurement was performed lately at the DAΦNE accelerator by the DEAR experiment - and will continue, together with the measurement of kaonic deuterium, in the framework of SIDDHARTA Collaboration [1].

For what concerns the kaonic helium, the situation today is rather ambigous: the only 3 existent experiments, performed more than 20 years ago, give, within few σs, results, for the $2p$ strong interaction parameters, which are about 2 order of magnitude different with respect to the theoretical

*Work partially supported by EU Integrated Infrastructure Initiative HadronPhysics Project under contract number RII3-CT-2004-506078

†also at: Institute of Physics and Nuclear Engineering "Horia Hulubei", IFIN-HH, Particle Physics Department, P.O. Box MG-6 R-76900 Magurele, Bucharest, Romania

‡also at: INFN Sszione di Roma I and Istituto Superiore di Sanita, Viale Regina Elena, 299 - 00161 Roma, Italy

§also at: Technical University Munich, 80290 Germany

predictions, performed in the framework of an optical model calculation. A new precise measurement is then badly needed; this measurement might be performed as well in the framework of the SIDDHARTA experiment.

In Section 2 the present experimental situation is reviewed, while in Section 3 the results are discussed together with the theoretical predictions. In Section 4 the case of kaonic helium measurement in SIDDHARTA is presented while in Section 5 some conclusions are drawn.

2. Review of the kaonic helium experimental results

There are three experiments which performed the kaonic helium measurement up to now [2,3,4]. All these experiments used liquid helium as target and Si(Li) detectors and measured the strong interaction parameters (shift and width) for the $2p$ level. The obtained results are reported in Table 1.

Table 1. Strong interaction parameters for the $K^{-4}He$ atoms

ΔE_{2p} (eV)	Γ_{2p} (eV)	Ref.
-41 ± 33	-	2
-35 ± 12	30 ± 30	3
-50 ± 12	100 ± 40	4

The average values extracted from these results are:

$$\Delta E_{2p} = -43 \pm 8 \ eV \tag{1}$$

$$\Gamma_{2p} = 55 \pm 34 \ eV \tag{2}$$

These results are more than 20 years old and by far of being precise measurements. However, as discussed in the next Section, there are indications from these results of a strong disagreement with the theoretical predictions based on simple Optical Models.

3. Theoretical predictions for kaonic helium

The optical model calculations, using a 3 parameter Fermi distribution for the nuclear density $\rho(\mathrm{r})$ and a mean value for the scattering length $\bar{a} = 0.34 \pm 0.03 + i(0.84 \pm 0.03)\mathrm{fm}$, obtained from a fit [5] to all available measurements for nuclei with $Z > 2$, give for the strong interaction parameters fr kaonic helium the following results [6]:

$$\Delta E_{2p} = -0.13 \pm 0.02 \; eV \tag{3}$$

$$\Gamma_{2p} = 1.8 \pm 0.05 \; eV \tag{4}$$

With more sofisticated forms for ρ(r) the obtained results are very similar.

Comparing the experimental with theoretical results, obtained from the optical model calculations, there are two orders of magnitude of difference. Many speculations were and are being done concerning the reason of this disagreement. One of the reasons might be related to the fact that the nucleons inside the ^{4}He are tightly bound, resulting in an effective energy for the kaon-nucleon interaction (1411 MeV) very close to the $\Lambda(1405)$ resonance. The consequence of this might be the fact that the parameters which fit the data for heavier nuclei (further away from the $\Lambda(1405)$ resonance) might not fit the kaonic helum values.

As pointed out by Batty [6] no simple modification of the value for $\bar{a}$ however gives a good fit to the helium data, unless a kaon-nucleus bound state is invoked, since these bound states can give rise to large energy shifts and widths in the atomic states. Better (but not perfect) agreement with experimental data can be obtained for $\bar{a} = 1.203 \pm 0.006 + i(0.010 \pm 0.001)$fm or $\bar{a} = 3.718 \pm 0.018 + i(0.032 \pm 0.002)$fm, so when the imaginary part is small.

Recent coupled channel calculations [7] are showing, however, the way for a possible agreement of the theory with experimental results.

At this stage, it is of utmost importance for the understanding of the underlying physics to perform new experiments, with an improved accuracy. This will become possible in the framework of the SIDDHARTA experiment at the DAΦNE accelerator at Frascati National Laboratories.

4. Kaonic helium measurement in SIDDHARTA

SIDDHARTA (SIlicon Drift Detector for Hadronic Atom Research by Timing Application) [8] experiment is a continuation of the DEAR (DAΦNE Exotic Atom Research) [9] experiment - which performed the most precise measurement of kaonic hydrogen to day at the DAΦNE Collider [10]. The primary goal of SIDDHARTA is an eV precision measurement of kaonic hydrogen and kaonic deuterium transitions, measurement which will allow to extract isospin dependent antikaon-nucleon scattering lengths at a precision unattained before.

The new feature to be used in SIDDHARTA is the detector: the newly developed large area Silicon Drift Detector (SDD) [11], which should allow the reduction of the background (by applying a μs level trigger) present in DEAR by some orders of magnitude.

In the next Sections, a brief description of the SDD detector and of the tests performed with a preliminary setup, are given.

4.1. *Large area silicon drift detectors*

The Silicon Drift Detectors (SDD) were developed as position sensitive detectors which operate in a manner analogous to gas drift detectors [11]. Recently, SDD started to be used as X-ray detectors in X-ray fluorescence spectroscopy, electron miniprobe analysis systems and synchrotron light application. SDDs with sensitive areas of up to 10 mm^2 are commercially available since several years. The typical energy resolution is better than 140 eV (FWHM) at 5.9 keV and -20 degrees centigrade.

The outstanding property of a SDD is its extremely small anode capacitance, which is independent of the active area. Thus, the electronics noise is very low, and much shorter shaping times than for PIN diodes or Si(Li) detectors can be used. The two main noise contributions of a detector are the leakage current and the capacitance. The equivalent noise charge increases linearly with the capacitance but only with the square root of the leakage current, which is correlated to the temperature. The leakage current depends on the detector area, but due to the drift principle the anode size and thus the capacitance is constant. Then, it can be expected that SDDs with much more than 10 mm^2 area still have a good energy resolution. Taking then into account the small shaping time, a trigger application of large area SDD as an X-ray detector, with a time window limited by its active area (drift time) can be envisaged. One of the ideal applications of such a triggered large area SDD detector is the measurement of exotic atoms X-ray transitions . The X-ray energies, ranging from few to tens of keV, are well in the range of maximum efficiency of SDDs, while a trigger based on a time-window of the order of μs was demonstrated by Monte Carlo simulation to dramatically reduce the background and to allow a precision measurement of exotic atom transitions. The trigger in this case is given by the specific process which generated the K^- at DAΦNE, namely a back-to-back reaction of the type:

$$\phi \to K^+K^-. \tag{5}$$

A program to develop such kind of large area (1 cm^2) triggerable SDD detectors, with integrated electronics (JFET) on it, started as a collaboration of MPI (Max-Planck Institut), PNSensors, Politecnico di Milano and LNF.

4.2. *Experimental requests*

The experimental requests put forward for the new large area triggerable SDD detectors with integrated JFET and equipped with a newly under development electronics, to be used for X-ray measurements, are:

- energy range of interest: 0.5 - 20/40 keV, with a selectable gain;
- capability to operate under mainly high energy events (background), with an event rate of the order of KHz/channel;
- energy resolution: better than 140 eV (FWHM) at 6 keV of energy;
- stability and linearity better than 10^{-4} for a precision measurement;
- trigger at the level of $1\mu s$.

Total number of channels to be processed: 200 (for a total area of 200 cm^2); eventual use of multiplexing - to be optimized. This is the first application in which SDDs are used with a trigger system.

4.3. *Preliminary measurements*

4.3.1. *BTF tests with a 7 chips array prototype*

Preliminary tests were performed at the Beam Test Facility (BTF) at Frascati with a prototype SDDs array: 7 chips of $5mm^2$ each. A trigger was implemented and tested with a time window of $1\mu s$. A synchronous (with BTF beam) as well as an asynchronous background (Fe and Sr sources) were implemented and it was checked that the rejection factor is in agreement with what expected. In Fig. 1 the results of the tests with the trigger are shown [8].

4.3.2. *Preliminary laboratory tests with a 30 mm^2 chip*

Preliminary measurements in laboratory with a SDD chip of 30 mm^2, shown in Fig. 2, were performed in June 2004. The chip was cooled at -40^0 C (by the use of a cryotiger) and a shaping time of 0.75 μs was used. The energy resolution measured was of 139 eV (FWHM) at 5.9 keV, as seen in Fig. 3.

The tests will be continued with the trigger measurements on the BTF.

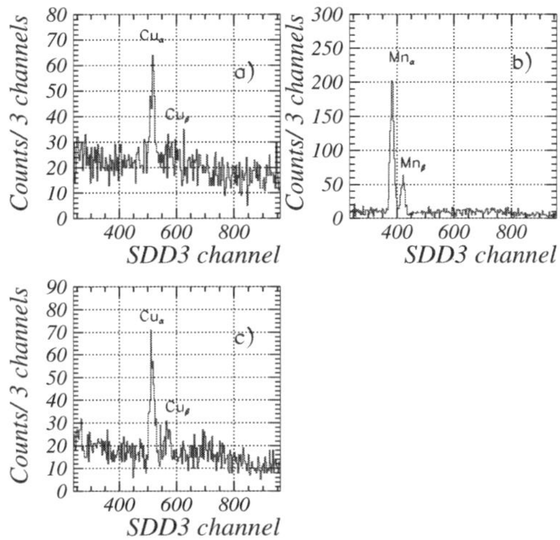

Figure 1. *a) No trigger, only BTF signal which excites the Cu-line, 5 Hz rate - 16 hours of DAQ; b) No trigger, 60 Hz, BTF signal (Cu) covered by Sr plus Fe radioactive sources as asynchronous background - 20 minutes DAQ; c) same as b) but trigger on, 5Hz as in a) - 16 hours of DAQ.*

Figure 2. *The 30 mm^2 SDD chip used for testing in the laboratory.*

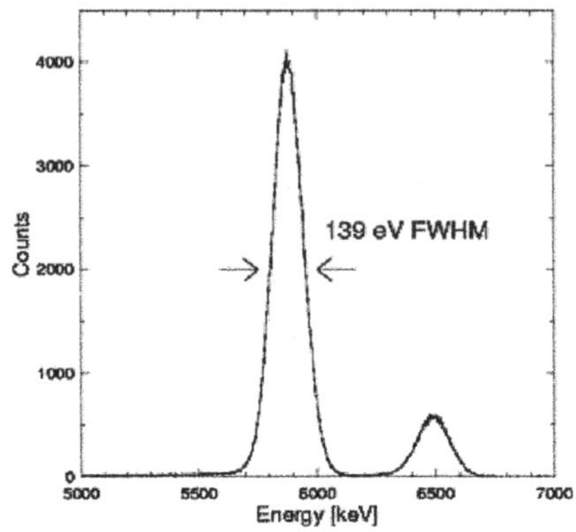

Figure 3. *The X-ray spectrum from an Iron source as measured in the laboratory with the 30 mm*^{2} *SDD chip.*

4.4. *The 1 cm*^{2} *SDD chip*

The 1 cm^2 SDD chip to be used in the SIDDHARTA experiment is under production, at PNSensors (Germany). The electronics is under development. The SDD layout on the readout side is shown in Fig. 4.

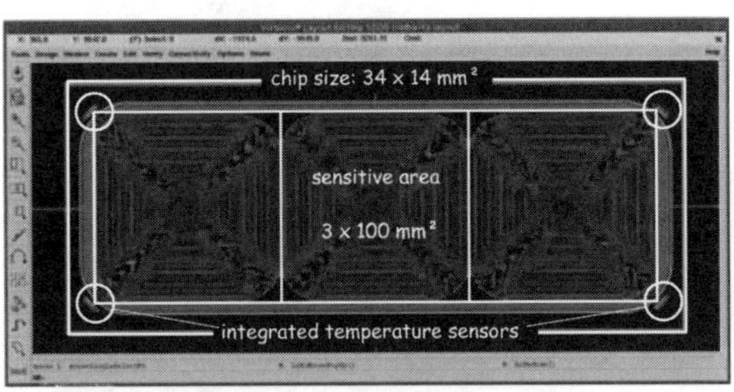

Figure 4. *SDD layout on the readout side*

The sensitive area per chip (containing 3 SDDs) is 3 cm^2; the SDDs are squared with round corners. The maximum diagonal drift length is 5.15 mm (centerlines) and 6.4 mm (diagonal).

Presently, construction and testing of the SDD detectors and of the electronics and mechanical structures is in progress. The setup was presented in [1].

With SIDDHARTA, it was estimated that a measurement at eV level for kaonic helium 2*p* strong parameters becomes feasible with an integrated luminosity of about 100 pb^{-1}. The setup will be installed at DAΦNE in the end of 2006 and start taking data.

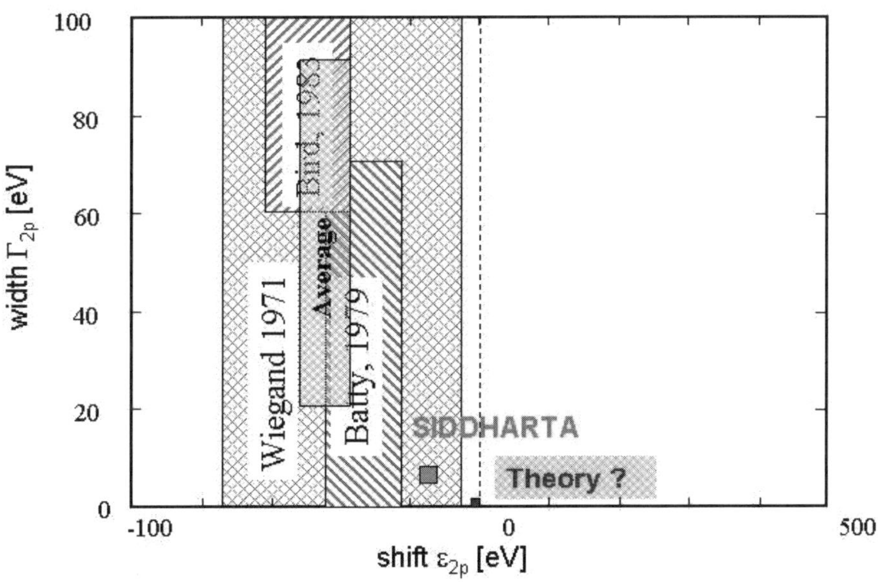

Figure 5. *Kaonic helium case: experiment versus theory and the SIDDHARTA projected result.*

5. Conclusions

There is, apparently, a strong discrepancy between the experimental results, based on 3 experiments and theoretical predictions for the kaonic helium $2p$ strong interaction parameters. On experimental side, the three experiments are more than 20 years old and the overall precision, on the average values, is within few σs . New experimental results, with a better accuracy, are badly needed. Such an experiment is planned in the future in the framework of the SIDDHARTA Collaboration, at the DAΦNE accelerator in Frascati. A measurement at eV precision level can be reached. A summary of the actual situation concerning kaonic helium experiment/theory - with a projection of the SIDDHARTA precision, is shown in Fig. 5.

Acknowledgments

The BTF (Beam Test Facility) group is warmly acknowledged for the very good cooperation and team-work.

References

1. C. Guaraldo, contribution to the present Proceedings.
2. C.E. Wiegand and R.H. Pehl, Phys. Rev. Lett. **27** (1971) 1410.
3. C.J. Batty *et al.*, Nucl. Phys. **A326** (1979) 455.
4. S. Baird *et al.*, Nucl. Phys. **A392** (1983) 297.
5. C.J. Batty, Nucl. Phys. **A372** (1981) 418.
6. C.J. Batty, Nucl. Phys. **A508** (1990) 89c.
7. M. Iwasaki, contribution to the present Proceedings.
8. J. Zmeskal, SIDDHARTA Technical Note IR-2 (2003); C. Curceanu (Petrascu), SIDDHARTA Technical Note IR-3 (2003).
9. S. Bianco *et al.*, Rivista del Nuovo Cimento **22**, No. 11 (1999) 1.
10. G. Vignola, Proc. of the "5th European Particle Accelerator Conference", Sitges (Barcelona), Eds. S. Myres *et al.*, Institute of Physics Publishing, Bristol and Philadelphia (1996) 22
11. E. Gatti, P.Rehak, Nucl. Instr. and Meth. **225** (1984) 608;
    E. Gatti, P.Rehak, Nucl. Instr. and Meth. **A235** (1985) 224.

HYPERNUCLEAR PHYSICS WITH FINUDA NOW AND IN THE FUTURE *

T. BRESSANI

Dipartimento di Fisica Sperimentale, Università di Torino, via P. Giuria 1 Torino, Italy, and INFN Sez. di Torino, via P. Giuria 1 Torino, Italy

Selected results following the first analyses of the data collected during the first run of FINUDA at DAΦNE will be presented and discussed. They concern the spectroscopy of ${}^{12}_{\Lambda}C$, for which a considerable production of the core excitation states in (K^-_{stop}, π^-) reaction was observed for the first time and the evidence for a K^-pp deeply bound state, observed for the first time with the invariant mass method. Finally, the perspectives of Hypernuclear Physics at a machine with a luminosity increased by at least one order of magnitude will be outlined.

1. Introduction

The FINUDA (acronym for "FIsica NUcleare a DAΦNE") experiment can be considered an experiment of third generation in Hypernuclear Physics. The original design of the FINUDA apparatus and, in particular, the large angle covered for the detection of charged and neutral particles emitted in the formation and decay of hypernuclei, allows for the simultaneous measurement of observables like excitation energy spectra, lifetimes and partial decay widths for mesonic and non-mesonic decay, with high statistics and good energy resolution (better than 1 MeV). Furthermore, these observables can be measured for different targets at the same time, thus reducing systematic errors in comparing properties of different hypernuclei. In addition, search for new exotic systems, like $\bar{K}$–nucleus bound states can be addressed.

The first FINUDA data taking at DAΦNE started in December 2003 and was successfully concluded in March 2004. In the following the initial results

*THIS RESEARCH IS PART OF THE EU INTEGRATED INFRASTRUCTURE INITIATIVE HADRON PHYSICS PROJECT UNDER CONTRACT NUMBER RII3-CT-2004-506078

from the experiment will be reported, which fully confirm the expected capability of FINUDA to perform high quality Hypernuclear Physics at the DAΦNE collider.

DAΦNE (Double Annular Φ-factory for Nice Experiments) consists of two rings, one for electrons and the other for positrons, that overlap in two straight sections where the beams collide. The energy of each beam is 510 MeV in order to produce the $\phi(1020)$ meson in the collisions.

At the luminosity $\mathcal{L}$=10^{32}cm^{-2}s^{-1}, the ϕ meson is produced at a rate $\sim 4.4 \cdot 10^2$ s^{-1}. The $\phi \to K^+K^-$ branching ratio is $\sim$ 49% and therefore, since the ϕ is produced almost at rest, DAΦNE is a source of $\sim 2.2 \cdot 10^2$ (K^+K^-) pairs/s, collinear, background free and of very low energy ($\sim$ 16 MeV). The low energy of the kaons is the key-feature for performing hypernuclear physics experiments at the DAΦNE ϕ-factory.

The main idea of FINUDA [1] is to slow down to rest the negative kaons from the $\phi \to K^+K^-$ decay in thin solid targets, so as to study the successive formation and decay of hypernuclei produced by the strangeness exchange reaction:

$$K^-_{stop} + {}^AZ \to {}^A_\Lambda Z + \pi^- \tag{1}$$

where ${}^AZ$ indicates a target nucleus and ${}^A_\Lambda Z$ the produced hypernucleus.

I will not repeat here, even in a short version, the main features and performances of the detector and of the machine, reported recently elsewhere [2,3], but I will report the more interesting results so far obtained for hypernuclear spectroscopy [4] and for the search of deeply bound $\bar{K}$–states [5].

2. ${}^{12}_{\Lambda}$C spectroscopy

Eight nuclear targets (3 of ^{12}C, 2 of ^6Li, 1 of ^7Li, ^{27}Al and ^{51}V) were installed for the first data taking and the events resulting from the interaction of stopped K^- with them were collected simultaneously.

In order to evaluate the capabilities of FINUDA to yield relevant spectroscopic parameters, the analysis started on ^{12}C targets. For ${}^{12}_{\Lambda}$C an excitation spectrum with a 1.45 MeV FWHM resolution was recently obtained at KEK using the (π^+, K^+) reaction at 1.05 GeV/c by the E369 Collaboration [6].

The requirement of high quality tracks (long tracks crossing the whole spectrometer, with a hit on each tracking detector) reduced the analysed data to about the 40% of the whole available ^{12}C sample.

The raw momentum spectrum of the π^- coming from the analysed ^{12}C targets is shown in Fig. 1. Different processes produce π^- after K^-

absorption and reproduce well the experimental spectra [7]:

- *a)* quasi-free Σ^+, Σ^0 and Λ production: $K^-p \to \Sigma^+\pi^-$, $K^-n \to \Sigma^0\pi^-$, $K^-n \to \Lambda\pi^-$;
- *b)* quasi-free Λ decay: $\Lambda \to p\pi^-$;
- *c)* quasi-free Σ^- production: $K^-p \to \Sigma^-\pi^+$, followed by $\Sigma^- \to n\pi^-$;
- *d)* two nucleon K^- absorption: $K^-(NN) \to \Sigma^-N$, followed by $\Sigma^- \to n\pi^-$.

All the mentioned reactions were simulated in detail in the FINUDA Monte Carlo program. However, in the momentum region where the bound states of ^{12}C are expected (beyond $\sim$ 260 MeV/c), only process *d)* is contributing. The dashed line in Fig. 1 represents this contribution, normalized in the $(275 \div 320)$ MeV/c momentum region, beyond the physical region for the production of Λ-hypernuclei via reaction (1).

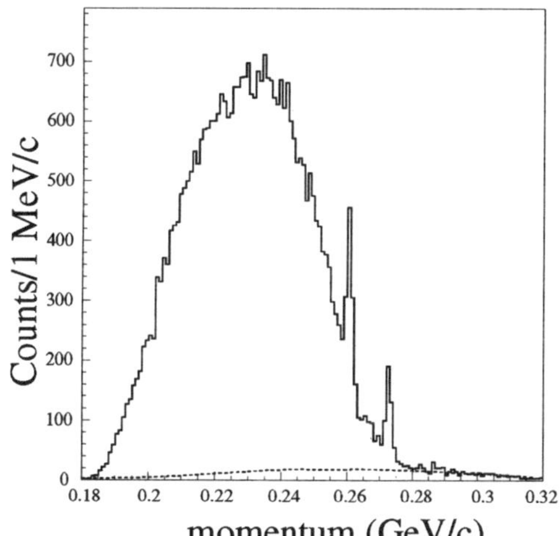

Figure 1. Spectrum of the momentum of the π^- emitted from the interaction vertex of a K^- onto a carbon target. The dashed line represents the contribution from K^- absorption by two nucleons (see text).

In order to obtain the Λ binding energy distribution the contribution of the *d)* process is subtracted from the π^- momentum distribution, and the momentum units converted into binding energies ($-B_\Lambda$). The two prominent peaks, at B_Λ around 11 MeV (ground state) and 0 MeV, were already observed in previous experiments [7] and interpreted as $(\nu p_{\frac{1}{2}}^{-1}, \Lambda s)$ and $(\nu p_{\frac{3}{2}}^{-1}, \Lambda p)$ (ν= nucleon). The experimental energy resolution was determined by fitting the $B_\Lambda \simeq 11$ MeV peak with a gaussian curve, and amounts to 1.29 MeV FWHM. The experimental spectrum closely resembles the one from E369 experiment [6]. This is expected, as the production of hypernuclear states is, in first approximation, determined by the momentum trasferred to Λ's, which is grossly comparable for both experiments ($\sim$ 250 MeV/c for FINUDA, $\sim$ 350 MeV/c for E369). The $\sim$ 100 MeV/c difference may account for the different yield of the two main peaks.

The value obtained for the ${}^{12}_{\Lambda}$C ground state formation is $(1.01 \pm 0.11_{stat} \pm 0.10_{sys}) \times 10^{-3}/(K^-$ stopped). It agrees very well with the value $(0.98 \pm 0.12) \times 10^{-3}$ previously measured at KEK [7,8].

In between the two main peaks, there are also indications of other states produced with weaker strengths. In order to reproduce this spectrum six gaussian functions were used, centered at the B_Λ values reported in Ref. [6]; the widths were fixed, for all of them, to $\sigma = 0.5$ MeV, corresponding to the experimental resolution. The result of this fit is shown in Fig. 2a).

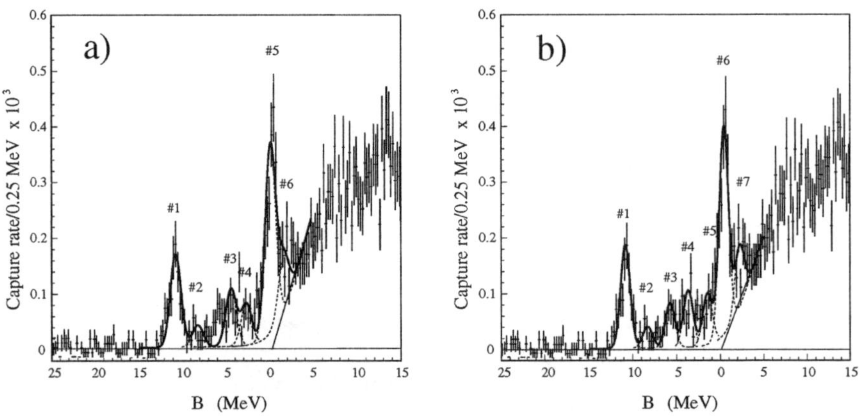

Figure 2. Λ binding energy spectrum of ${}^{12}_{\Lambda}C$ measured by the FINUDA experiment. a) The solid line represents the result of a fit with 6 gaussian functions (#1 ÷ #6), as explained in the text; b) the solid line represents the result of a fit with 7 gaussian functions as explained in the text. The solid line starting at $B_\Lambda = 0$ MeV represents the contribution from the quasi-free Λ production.

The spectrum is not well reproduced, the resulting reduced χ^2 is 3.8, and in particular the region -10 MeV $< -B_\Lambda < -5$ MeV is poorly fitted.

The capture rates for these different contributions normalized to the ground state capture rate are reported in the second column of the upper part of Table 1.

Table 1. Results from $-B_\Lambda$ spectrum fits: the upper part of the table corresponds to a fit performed with the same peaks layout of E369 experiment[6], with 6 gaussian functions. The lower part corresponds to a fit with 7 hypernuclear levels. The last column reports the capture rates corresponding to each peak. The errors reported for peak (#2 ÷ #6) in the upper part and (#2 ÷ #7) in the lower part of the table do not include the error on the ${}^{12}_{\Lambda}C$ ground state capture rate.

Peak number	$-B_\Lambda$ (MeV) *(Fixed at E369 values)*	Capture rate/(stopped K^-)[$\times10^{-3}$]
1	−10.76	$1.01 \pm 0.11_{stat} \pm 0.10_{syst}$
2	−8.25	0.23 ± 0.05
3	−4.46	0.62 ± 0.08
4	−2.70	0.45 ± 0.07
5	−0.10	2.03 ± 0.14
6	1.61	0.58 ± 0.11
Peak number	**$-B_\Lambda$ (MeV)**	**Capture rate/(stopped K^-)[$\times10^{-3}$]**
1	-10.94 ± 0.06	$1.01 \pm 0.11_{stat} \pm 0.10_{syst}$
2	-8.4 ± 0.2	0.21 ± 0.05
3	-5.9 ± 0.1	0.44 ± 0.07
4	-3.8 ± 0.1	0.56 ± 0.08
5	-1.6 ± 0.2	0.51 ± 0.08
6	0.27 ± 0.06	2.04 ± 0.15
7	2.1 ± 0.2	0.59 ± 0.11

A better χ^2= 2.3 is obtained by adding a further contribution, and leaving the positions of the seven levels free. Their values are reported in the second column of the lower part of Tab. 1. The capture rates for these different contributions are again normalized to the capture rate for the ${}^{12}_{\Lambda}C$ ground state formation. The result of the fit is shown in Fig. 2b). A contribution from the quasi-free Λ-production, starting from $B_\Lambda = 0$, was included in both fits.

The peaks #2 and #3 can be naively attributed to the ${}^{11}C$ core excited states at 2.00 and 4.80 MeV. The excitation of these peaks was expected in several theoretical calculations; the peak energies should give information on the spin-dependent Λ-N interaction. However, the peak #4 and a newly observed peak #5 are not explained in such a simple way. Motoba [9] suggested that such highly excited states could be explained in an extended shell model with intershell couplings. There exist several positive-parity excited states of the ${}^{11}C$ core in this energy region which could contribute to these structures. It can be noticed that the integrated strength for the excitation of all these weakly excited states compared to that of the two main peaks is more than twice as large than the one reported by E369.

The sum of the capture rates for the B_Λ = 0.27 MeV and B_Λ = 2.1 MeV states is $(2.63 \pm 0.14_{stat}) \times 10^{-3}$, and agrees with the KEK result $(2.3 \pm 0.3) \times 10^{-3}$, in which the contributions for the two states were not resolved. It is remarkable that the relative intensities for the contribution at B_Λ = 0.27 MeV and B_Λ = 2.1 MeV are close to the values found by Dalitz *et al.* [10] in an emulsion experiment.

In conclusion, a disagreement between the experimental capture rates and the theoretical predictions [11,12,13] is confirmed; also the relative intensities for the production of the two main peaks as well as those from the core excitation [14] are not well reproduced by theoretical calculations.

3. Evidence for a kaon–bound state K^-pp

The information on K^- nucleus interaction has been obtained with the kaonic atom data [15]. Unfortunately, there are qualitatively two different predictions for the depth of the K^- nucleus potential at normal nuclear matter density: very deep attractive potentials ($-Re\ V_{opt}(\rho_0) \sim 150 - 200$ MeV) and much shallower potentials ($-Re\ V_{opt}(\rho_0) \sim 50 - 75$ MeV). Both types of potentials reasonably describe the shifts and widths of the X–ray data.

Recently, an interesting idea has been proposed suggesting the possible existence of deeply bound nuclear $\bar{K}$ states in light nuclei [16], for which the potential is very deep. Since the potential is strongly attractive in the $I = 0$ $\bar{K}N$ interaction, proton–rich $\bar{K}$–nuclei such as K^-pp and K^-ppn are predicted to have large binding energies of 50–100 MeV. A further interesting suggestion is that the nucleus might be shrunk because of the large binding energy and form a high–density state.

The FINUDA spectrometer enables the clear identification of the formation of $\bar{K}$–nuclear bound states through their decays to $\Lambda + X$.

Λ particles can be identified by reconstructing the invariant mass of a proton and a negative pion as shown in Fig. 3 a). The peak position agrees well with the known Λ mass, and the width of the peak is as narrow as 6 MeV/c^2 FWHM.

When a K^- interacts with two protons, one expects that a hyperon–nucleon pair ($\Lambda + p$, $\Sigma^0 + p$ or $\Sigma^+ + n$) is emitted in the opposite direction, ignoring a final state interaction inside the nucleus. The angular correlation between a Λ and a proton from the same point in the target [Fig. 3 b)] clearly indicates the existence of this kind of reaction. Even for heavy

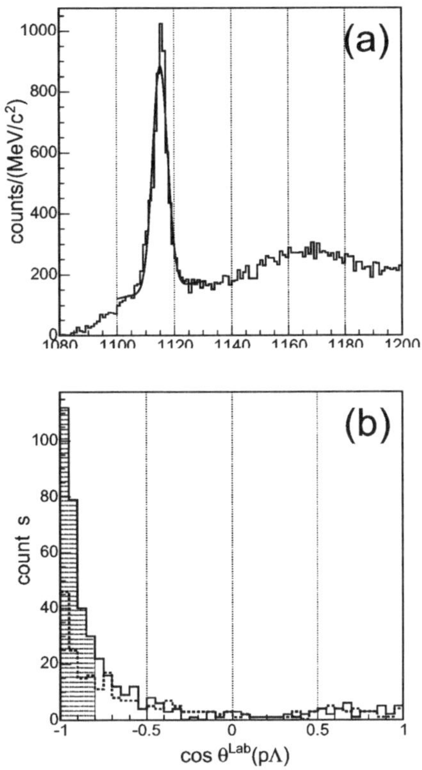

Figure 3. a) Invariant mass distribution of a proton and a π^- for all the events in which these two particles are observed, fitted by a single gaussian together with a linear background in the invariant mass range 1100–1130 MeV/c^2. b) Opening angle distribution between a Λ and a proton (solid line: ^6Li, ^7Li and ^{12}C, dashed line: ^{27}Al and ^{51}V). The shaded area ($cos\theta^{Lab} < -0.8$) is selected as the back–to–back events.

nuclei such as ^{27}Al and ^{51}V similar correlations were observed although less pronounced as compared to those for light nuclei.

Only the $\Lambda - p$ pairs emitted in opposite directions ($cos\theta^{Lab} < -0.8$) coming only from the light nuclear targets (^6Li, ^7Li and ^{12}C) are considered. In this way, the possible contamination from other processes can be minimized, and the signal–to–noise ratio as well as the available statistics are kept to a good level.

Since the back–to–back angular correlation between a Λ and a proton is so clear, it is naturally expected that the two particles are emitted from a "K^-pp" intermediate system. If the reaction process was simply a two–nucleon absorption process, the mass of the system should be close to the

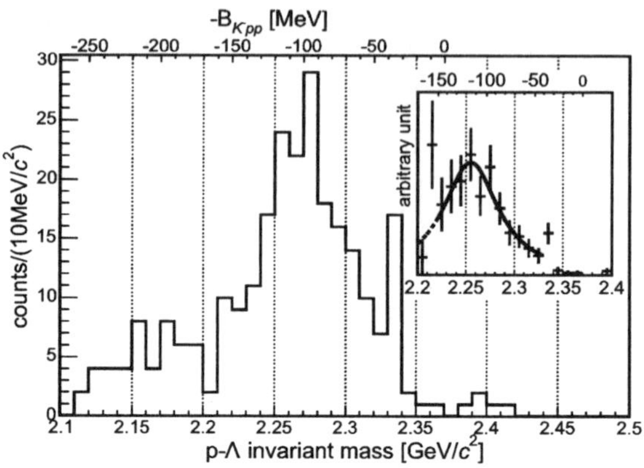

Figure 4. Invariant mass of a Λ and a proton in back–to–back correlation ($cos\theta^{Lab} < -0.8$) from light targets before the acceptance correction. The inset shows the result after the acceptance correction for the events which have two protons with well–defined good tracks. Only the bins between 2.22 and 2.33 GeV/c^2 are used for the fitting.

sum of a kaon and two proton mass, namely 2.370 MeV/c^2. The invariant mass distribution of the $\Lambda - p$ pairs is shown in Fig. 4. A significant mass decrease of the K^-pp system with respect to its expected mass is observed. It can be interpreted as a bound state composed of a kaon and two protons, hereafter abbreviated as K^-pp.

In the inset of Fig. 4 the acceptance corrected invariant mass distribution for events with two well–defined long–track protons is shown. Since the trigger and detection acceptance are monotonically increasing functions of the invariant mass in this mass region, the peak further shifts to a lower mass side. The binding energy $B_{K^-pp} = 115^{+6}_{-5}$(stat.)$^{+3}_{-4}$(sys.) MeV and the width $\Gamma = 67^{+14}_{-11}$(stat.)$^{+2}_{-3}$(sys.) MeV are obtained from the fitting with a Lorentzian function (folded with a gaussian with $\sigma = 4$ MeV/c^2, corresponding to the detector resolution, estimated with a Monte Carlo simulation) in the region of 2.22–2.33 GeV/c^2. Although we still have ambiguities on absolute normalization, a rough estimate on the yield of $K^-pp \to \Lambda + p$ is of the order of 0.1% per stopped K^-.

I notice that this is the first observation of a K–bound state by the invariant mass method. Previous observations, in particular for a K^-nnp state, were reported by using the missing mass method in the reaction (K^-_{stop}, pX) [17].

4. Future plans for the FINUDA experiment

The luminosity of DAΦNE is slowly but constantly improving with time. At present it is a factor about 2 compared with that of the first run. FINUDA is scheduled to run for a long period (7–8 months) in 2006, and likely also in 2007/2008. The 2006 run will be dedicated to a high statistics study of the spectra of charged particles and neutrons emitted following the capture at rest of K^- in targets of ^6Li, ^7Li, ^9Be and ^{16}O (water). A very rich and interesting Physics program will be pursued, including Spectroscopy of Hypernuclei, Mesonic and Non–Mesonic Decay, Rare two–Body Decays, Neutron–Rich Hypernuclei and Deeply–Bound $\bar{K}$ states. The expected statistics per target will increase by a factor 5–10 depending on the target/observable (^6Li and ^7Li) and new information will be gained for the light targets ^9Be and ^{16}O. In a subsequent run (2007/2008), targets of larger A could be examined. The major improvement needed to cope with the increased luminosity is the DAQ speed. An increase of a factor 4 is under way and shortly will be operational. The internal barrel of scintillators will be also replaced by a new one, equipped with PhotoMultipliers instead of Hybrid PhotoDiodes, under construction at KEK and ready by Spring 2005.

Following the good Physics results obtained at DAΦNE, ideas for a high luminosity machine, DAΦNE2 ($\mathcal{L}$ from 10^{33} $cm^{-2}s^{-1}$ to 10^{34} $cm^{-2}s^{-1}$) were put forward and are under active discussion. New experiments on Hypernuclear Physics could be performed at DAΦNE2.

The use of intrinsic Ge–detectors for the detection of de–excitation γ–rays from excited hypernuclear states allowed a spectacular step forward in Hypernuclear Spectroscopy [18]. The improvement of the energy resolution by three orders of magnitude (from MeV to keV) allowed the starting of very interesting investigations covering several topics of Hypernuclear Physics.

Thanks to the modularity of the FINUDA apparatus, it has been possible to envisage a detector upgrade in order to extend its activity beyond the completion of its original physics program. The idea is that of removing a couple of drift chambers both from the inner and the outer arrays. With a reduction of the present solid angle covered by the spectrometer ($\sim 2\pi\ sr$) of only $\sim 28\%$, such a modification provides room to install a couple of segmented γ-ray detectors inside the FINUDA magnet coil, directly looking at the nuclear stopping targets.

The renewed FINUDA2 apparatus will then resume its activity with a physics program centered on γ-ray hypernuclear spectroscopy, but still maintaining the unique capability to detect with the same apparatus and in coincidence the prompt π^- from the hypernucleus formation reaction as well as its decay products.

The main interest of the new physics program will be focussed on the γ-ray spectroscopy of a wide range of both light and medium-heavy hypernuclei, provided that the corresponding chemical element can be machined in solid slices[a].

R&D work on the use of segmented clover Ge–detectors operating in strong magnetic fields and in presence of a huge background of m.i.p. has already been started with funds provided by the VI FP EU framework.

References

1. Bressani T., in *Proc. Workshop on Physics and Detectors for DAΦNE*, April 9-12, 1991, Frascati, Italy, edited by Pancheri G. (Laboratori Nazionali di Frascati), p. 475.
2. Bressani T., *Acta Phys. Pol.* **B35** (2004), 959.
3. Zenoni A., in *Proceedings Int. School of Physics "E. Fermi", Course CLVIII*, to be published.
4. Agnello M. *et al.*, submitted for publication to *Phys. Lett.* **B**
5. Agnello M. *et al.*, submitted for publication to *Phys. Rev. Lett.*
6. Hotchi H. *et al.*, *Phys. Rev.* **C64** (2001), 044302.
7. Tamura H., *PhD Thesis*, University of Tokyo, UT PN-220 (1987)
8. Tamura H. *et al.*, *Prog. Th. Phys. Suppl.* **117** (1994), 1
9. Motoba T., *Nucl. Phys.* **A639** (1998), 135c
10. Dalitz R.H. *et al.*, *Nucl. Phys.* **A450** (1986), 311c
11. Gal A. and Klieb L., *Phys. Rev.* **C34** (1986), 956
12. Matsuyama A. and Yazaki K., *Nucl. Phys.* **A477** (1988), 6
13. Cieply A. *et al.*, *Nucl. Phys.* **A696** (2001), 173
14. Itonaga K. *et al.*, *Prog. Theor. Phys.* **84** (1990), 291
15. Batty C.J., Friedman E. and Gal A., *Phys. Rep.* **287** (1997), 385
16. Akaishi Y. and Yamazaki T., *Phys. Rev.* **C65** (2002), 044005
17. Suzuki T. *et al.*, *Phys. Lett.* **B597** (2004), 263; Iwasaki M., this Conference
18. Tamura H. *Nucl. Phys.* **A639** (1998), 283c; Tamura H. *et al.*, *Phys. Rev. Lett.* **84** (2000), 5963

[a]The FINUDA/FINUDA2 arrangement is such that only solid targets are allowed to be installed around the DAΦNE collider beam pipe.

STRANGE MULTI-BARYON AND KAONIC ATOMS

M. IWASAKI[A,D], H. BHANG[C], G. FRANKLIN[B], K. GOMIKAWA[E], R.S. HAYANO[E], T. HAYASHI[D], K. ISHIKAWA[D], S. ISHIMOTO[F], K. ITAHASHI[A], M. IWAI[F], T. KATAYAMA[D], Y. KONDO[D], Y. MATSUDA[A], T. NAKAMURA[D], S. OKADA[D,A], H. OUTA[F,A], B. QUINN[B], M. SATO[D], M. SHINDO[E], H. SO[C], T. SUGIMOTO[D], P. STRASSER[A,F], K. SUZUKI[E], S. SUZUKI[F], T. SUZUKI[E,A], D. TOMONO[D,A], A.M. VINODKUMAR[D], E. WIDMANN[E], T. YAMAZAKI[A] AND T. YONEYAMA[D]

[A]*DRI, RIKEN, Wako-shi, Saitama, 351-0198, Japan*

[B]*Department of Physics, Carnegie Mellon University, Pittsburgh, PA 15213, USA*

[C]*Department of Physics, Seoul National University, Shikkim-dong, Kwanak-gu, Seoul 151-742, Korea*

[D]*Department of Physics, Tokyo Institute of Technology, Ookayama, Meguro-ku, Tokyo 152-8551, Japan*

[E]*Department of Physics, University of Tokyo, Hongo, Bunkyo-ku, Tokyo 113-0033, Japan*

[F]*IPNS/IMSS, KEK, Oho, Tsukuba-shi, Ibaraki 305-0801, Japan*

Recently, we have performed an experimental search for deeply bound kaonic states by the kaon absorption reaction at rest in a liquid helium target. We observed very distinctive mono-energetic peak formation in a proton missing-mass spectrum. We denote it as a strange tribaryon, $S^0(3115)$, with baryon number 3, charge 0, isospin 1 and strangeness -1. If we attribute the mono-energetic peak to the formation of a deeply bound kaonic state, the separation energy of the kaon should be as deep as about 200 MeV. In the present paper, we will overview the present experimental data, and discuss the experimental program for the near future.

1. Introduction

Our knowledge of the interaction between the K-meson (especially $\overline{K}$) and the nucleus is far from satisfactory. Experimentally, the interaction has been studied either by the cross section of the in-flight $K^{\pm}N$ reaction, or the energy shift and broadening (due to the strong interaction) of the cascade x rays from kaonic atoms. The experimental approach is the same as πN, although the lower yield, shorter lifetime and the variety of reaction products make the kaon experiment much more difficult. Furthermore, the existence of the subthreshold resonance, $\Lambda(1405)$, 27 MeV below K^-p energy, makes the situation more complicated to understand.

Actually, x-ray data of kaonic hydrogen were puzzling for a long time because the x-ray observation and the subthreshold $\Lambda(1405)$ pole conflict. The strongly attractive nature between K^- and proton was confirmed recently[1]. There are two possible ways to understand the present experimental data; i) conventional optical-model calculations give a consistent result with the experimental data (excluding kaonic helium x-ray data) using an attractive but strongly absorptive interaction. On the other hand, ii) a coupled channel calculation by Akaishi and Yamazaki [2], in which the $\Lambda(1405)$ is treated as a bound state between K^- and proton, suggests that the $\overline{K}N$ interaction is extremely attractive and less absorptive, thus stable K-mesonic state formation is allowed in light nuclei.

Therefore, if we embed a kaon into the nucleus, a kaon is promptly absorbed by a nucleon (converted to a hyperon), in the former framework. On the other hand, in the latter framework, a kaon can survive longer than the typical time scale of the strong interaction, without loosing its identity as a meson in the nucleus. Thus, if one embeds a kaon into ^{3}He, a high density baryon system will be formed in which the kaon separation energy is as large as about 100 MeV with an absorption width of about 20 MeV.

The discovery of such a system will open a new research area to study hyper-dense matter in objects such as "compact stars", with densities greater than neutron stars, in the laboratory framework. In such a system, one may study the precursor effect toward the totally new $\langle\overline{q}q\rangle$-condensation phase, color super-conductivity, which may realize at the high density and low temperature limit. Because the constituent-quark or hadron mass is expected to be a function of $\langle\overline{q}q\rangle$-condensation strength, one may obtain a hint as to how hadronic mass is realized after the big bang. The quantum state of the bound kaon is well defined, so it will give us less ambiguous data compared to the high-density and high-temperature objects which can be realized in heavy ion collisions.

2. KEK PS-E471 Experiment

Very recently, we performed an experimental search for the predicted deeply-bound kaonic state by kaon absorption at rest in a liquid helium target. In the experiment, we observed nucleon emission from the reaction, and discovered distinct mono-energetic proton formation[3]. This can be produced by the two-body reaction of a proton together with the unknown object, S^0, whose mass is about 3115 MeV/c^2, namely,

$$(K^- \ {}^4\mathrm{He})_{\mathrm{atomic}} \rightarrow \mathrm{S}^0(3115) + p. \tag{1}$$

The observed system should have baryon number B=3, charge Z=0, isospin T=1 and strangeness S=-1.

The discovery of peak structure in the proton spectrum was quite astonishing to us, because we were searching for mono-energetic neutron formation. The theoretical prediction of a deeply-bound kaonic state by Akaishi and Yamazaki is the formation of the state at $M = 3194$ MeV/c^2 with $T = 0$ and $Z = 1$[2], which can only be studied via the

$$(K^- \ {}^4\mathrm{He})_{\mathrm{atomic}} \rightarrow \mathrm{S}^+ + n \tag{2}$$

reaction. Since the present experimental setup is not optimized for proton spectroscopy, we need to be very careful to check whether the proton peak in the spectrum has been formed for trivial reasons, such as experimental bias, other reaction processes, *etc.*

2.1. *E471 Experimental Setup*

Let us briefly describe our experimental setup, shown in Fig. 1. The negative kaon of 660 MeV/c is momentum degraded by a graphite block (not shown), its trajectory recorded by a fine segmented drift chamber (BDC), and stopped in the liquid ^{4}He target. We required at least one hit in the double-layered trigger counters (TC) to track one of the reaction/decay particles (mostly pions and protons) in a large area drift chamber (VDC). The pulse height information of the double-layered TC, together with injection angle obtained by VDC, enables us to perform a qualitative analysis of the p/π separation as shown in Fig. 2. A rough evaluation of the momentum can also be done if β is well below one, because the pulse heights are given by the partial integration of the Bethe-Bloch equation.

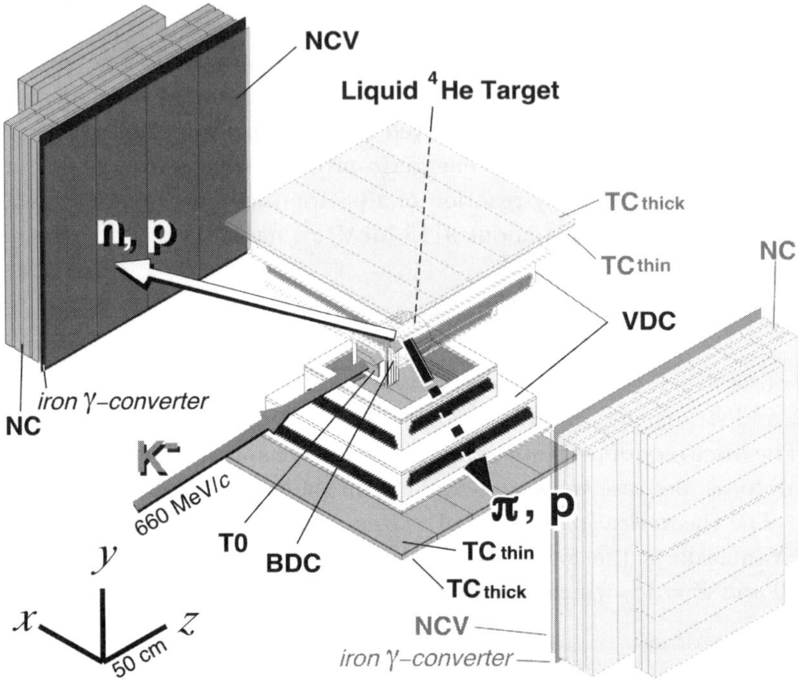

Figure 1. Schematic figure of the experimental setup. The notations are: T0, beam timing counter; BDC, beam-line drift chamber (16 layers); VDC, vertex drift chamber (12 layers); TC_{thin} and TC_{thick}, thin (0.6 cm) and thick (3 cm) trigger counter; NCV, neutron-counter charged-particle veto; NC, neutron-counter array (5 cm thick). Both tracking devices, BDC and VDC, are highly redundant to remove possible ghost tracks, important in the detailed vertex analysis. To simplify, several counters (mostly for veto) are omitted. All the scintillation counters shown here are viewed by PMTs at both ends.

The kaon reaction point is obtained by a closest-approach of the two trajectories between incoming kaon and outgoing reaction/decay particle. Kaon at-rest events in the target are selected by the correlation between the pulse height of the final beam line counter (T0 : placed between the degrader and the BDC) and the stopping range in the helium target.

The protons and neutrons are detected by a double-arm segmented plastic scintillator array (NC), which is placed 2 m away from the target center. A thin layer of scintillation counters (NCV) covering the NC array is for neutron separation for NC-detected particles, and p/π separation is done by using the energy sum of the NC array. The thin iron plate is to produce a γ-ray shower, which is used for the time-zero calibration of the NC array system. The momentum of the protons and neutrons is calculated

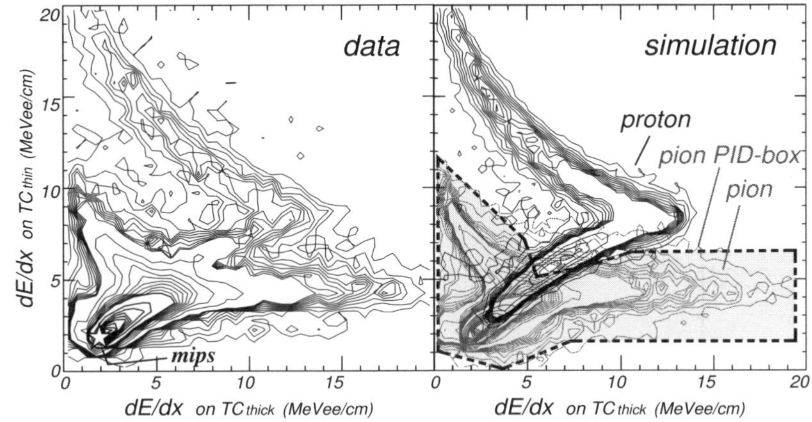

Figure 2. Contour plot of dE/dx on TC_{thick} (horizontal-) and TC_{thin} (vertical-axis) counters obtained by data (left). Minimum-ionizing-particles (*mips*) are expected to distribute around the star mark. In the simulation (right), proton- and pion- components are separately plotted, and the pion PID-box applied to the data is also shown as dashed line. The step of the contour is in logarithmic scale.

by means of time-of-flight (TOF) from the reaction vertex to the detection point in the NC array.

2.2. *Proton Spectrum*

Fig. 3 shows the missing-mass spectrum of the protons obtained after the energy loss correction in the target. To demonstrate the stability of the particle identification (PID) boxes on the TC contour plot (shown in Fig. 2 (right)), we used slightly different PID criteria from those given in Ref. 3, in which the box shape is simplified with the resultant acceptance of high momentum proton contamination into the pions.

A clear signal is observed over the expected background protons, which mainly come from non-mesonic kaon absorption; $KNN \rightarrow \Lambda N$ (Br. $\sim$ 20 %). We have studied all other possible experimental biases which may produce a fake peak in the spectrum, such as data comparison between left and right NC, and between the former and latter half of the data taking period, reaction-vertex fiducial cut, *etc.*

We requested a VDC-hit at the hardware level, so the missing mass spectrum is not an unbiased inclusive spectrum. One may wonder whether the trigger condition might bias the spectrum to form a spurious peak in the missing mass spectrum. However, it is impossible to form a peak structure

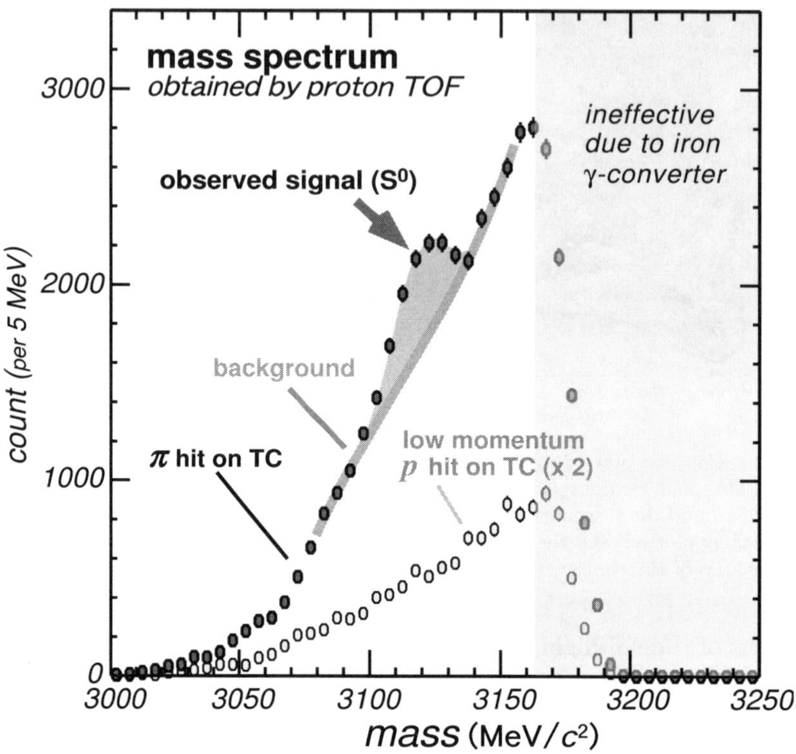

Figure 3. Missing mass spectrum of the semi-inclusive proton events. Close and open circle are for pion- and proton- triggered event on TC counter, respectively. The event-selection is obtained by the PID-box defined on the scatter plot shown in Fig. 2.

because the angular acceptance in the x-y plane (the coordinate is shown in Fig. 1) is as large as $\pm\ 45^0$, so that any kinematic bias should be smeared out.

As described in Ref. 3, the only possible background process is the chain reaction of hypernuclear formation and one of its non-mesonic decay branches

$$K^- + {}^4\text{He} \;\rightarrow\; {}^4_\Lambda\text{He} + \pi^- \quad (255\ \text{MeV}/c) \quad \text{and} \tag{3}$$

$${}^4_\Lambda\text{He} \;\rightarrow\; {}^3t + p \quad (508\ \text{MeV}/c). \tag{4}$$

How about the yield of the peak formation? The hypernuclear formation probability is known to be about 2 %. On the other hand, the total non-mesonic decay branch is known to be about 16 %, but the partial branching ratio to the specific non-mesonic decay channel (4) is not known. The upper limit of the peak formation probability from this chain reaction is about 0.3 % per stopped kaon, and the actual branch could be one order of magnitude smaller. The yield is expected to be very small, however the momentum of a proton produced in reaction (4) is very close to that of the signal, so that one should be extremely careful.

In Ref. 3, we have already excluded this possibility by using the TC pulse height information. Actually, if the chain reaction (3) and (4) is the reason for peak formation, then the trigger particle on TC should be a pion from reaction (3), so that it is easy to check whether all the peak events are associated with the *minimum-ionizing-particles* (*mips*; the energy loss should be about 2 MeV/cm in both the $TC_{\rm thin}$ and $TC_{\rm thick}$ counter, marked as a star in Fig. 2). It is shown that the peak events are dominantly much slower than *mips* in the reference.

In the present paper, let's study the peak by a different approach. The tribaryon $S^0(3115)$ is expected to be triggered mostly by the pion from the following decay chain,

$$S^0(3115) \rightarrow YNN \quad \text{and} \tag{5}$$

$$Y \rightarrow \pi^{\pm}N. \tag{6}$$

In this decay chain, the pion is produced by the weak decay of the hyperon (Y), then the motion of Y before the decay can be studied by the vertex inconsistency between incoming kaon and outgoing pion trajectories. Namely, in the formation process, the distance-of-closest-approach of the two trajectories may lie beyond the experimental vertex resolution because of hyperon motion. Moreover, the hyperon should be boosted opposite to the proton motion due to the formation reaction (1), so that the pion trajectory should reflect the boost.

To study this hyperon motion, let us introduce a scalar product $\mathbf{v}_{\rm CA} \cdot \hat{\mathbf{v}}_p$, where $\mathbf{v}_{\rm CA}$ is the vector from the kaon trajectory to that of the pion at the closest approach between the two trajectories, and $\hat{\mathbf{v}}_p$ is the unit vector of the proton motion. If vector $\mathbf{v}_Y$, which represents the hyperon track, is parallel to $\hat{\mathbf{v}}_p$, then $\mathbf{v}_Y$ is equal to $\mathbf{v}_{\rm CA}$ so that the scalar product $\mathbf{v}_{\rm CA} \cdot \hat{\mathbf{v}}_p$ gives the distance of the hyperon motion relative to the proton motion. Therefore, one can use the scaler product for the qualitative analysis of the motion of the hyperon relative to that of proton, although it looses sensitivity if the hyperon motion is orthogonal to $\hat{\mathbf{v}}_p$.

Fig. 4 shows the missing mass spectrum classified by $\mathbf{v}_{\rm CA} \cdot \hat{\mathbf{v}}_p$. The peak yield evaluation in the three spectra is difficult because the background shape changes with $\mathbf{v}_{\rm CA} \cdot \hat{\mathbf{v}}_p$, and the lines in the spectra are simple spline

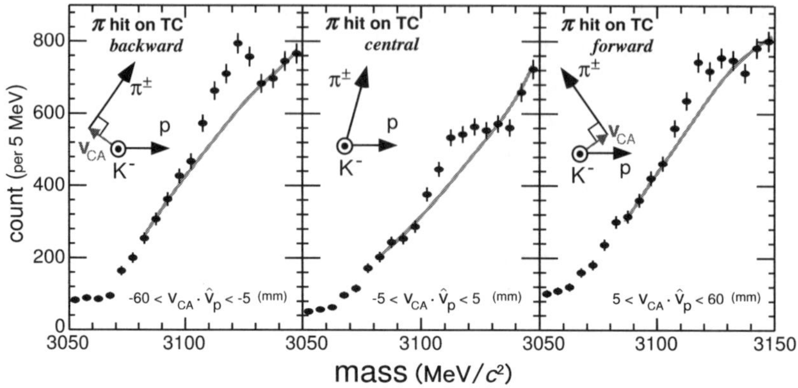

Figure 4. Missing mass spectrum of the pion triggered proton events, classified by three $\mathbf{v}_{\rm CA} \cdot \hat{\mathbf{v}}_p$ regions. $\mathbf{v}_{\rm CA} \cdot \hat{\mathbf{v}}_p$ regions and corresponding event topologies are shown as insets. The vertex in central $\mathbf{v}_{\rm CA} \cdot \hat{\mathbf{v}}_p$ region is consistent with the experimental vertex resolution.

functions with a node at 3200 MeV/c^2. It is clear that the function is not sufficient for the quantitative yield evaluation. However, the tendency is clear that the peak is mostly in the backward region, which is consistent with $S^0(3115)$ formation. On the other hand, if the peak is formed by the non-mesonic weak decay of the hypernucleus, reaction (3) ensures that the yield should be located dominantly in the central region.

3. Forthcoming Experiments

It is clear that further experimental study of $S^0(3115)$ is needed, both for more information on the state itself, and the $\overline{K}N$ interaction. There are two experimental programs scheduled in 2005 at KEK-K5 before the KEK PS shutdown. These experimental groups, with most members in common, are planning to use the E471 experimental setup as a common platform for the experiments introduced below.

3.1. *KEK-PS E549 Experiment*

As described, the E471 experimental setup is optimized for neutron spectroscopy. In the hardware trigger, we required at least one charged track detected in VDC (VDC-hit), which is not needed for proton spectroscopy if we have a tracking device for protons. It should also be noted that the iron γ-ray converter prevents detecting low momentum protons so that the detector is inefficient for protons below ~300 MeV/c.

The most straightforward way to improve the experimental setup for better proton measurement is the installation of tracking devices together with start- and stop- counters for proton TOF in the direction of the NC array, which allows triggering the unbiased inclusive event from the K^-p reaction. The statistics of the proton spectrum will be drastically improved, because a VDC hit is not required. The measurement of the proton trajectory gives improved accuracy for the path length, and the dedicated TOF counters also improve the time resolution of the proton by a factor of about three, which may allow us to deduce the width of the state. Removal of the iron γ-ray converter enlarges the acceptance of the proton spectrum, which enables us to search for the excited state of $S^0(3115)$.

These improvements will present two additional and confirmatory studies for the assertion of the existence of $S^0(3115)$. One is the vertex consistency study between kaon and proton. If the peak is formed by the primary reaction 1, two trajectories should coincide within the experimental resolution. Another is the reaction timing study. The improved timing resolution enables us to discriminate between whether the peak is formed promptly (strong interaction) or delayed due to hypernuclear decay (weak interaction).

In the present paper, we haven't described the neutron missing-mass spectrum, but it also gives promising structure at around 3140 MeV/c^2. However, the statistics of the neutron data are not high enough to give a definitive argument. Therefore, we enlarged the volume of the NC array to improve the statistical accuracy. For the better study of the tracks in VDC, we also added two more layers for TC. In E549, we will take data before summer 2005, using these improvements of our experimental apparatus.

3.2. *KEK-PS E570 Experiment*

Whether $S^0(3115)$ can be interpreted as a kaon bound state is another important question. To answer this question, we wish to determine the $\overline{K}N$ interaction more precisely by an x-ray measurement of the kaonic helium atom. Previously observed x-ray energy shift and width of the $3d \to 2p$ transition of kaonic helium[4,5] is quite anomalous in view of conventional optical potential calculations in which almost no shift can happen because of the large imaginary part in the potential. In other words, the observed repulsive shift of about 40 eV requires an unreasonably small imaginary part for the optical potential. Therefore, it is extremely important that the experimental results of the anomalous energy shift of kaonic-helium x rays be re-examined with much improved accuracy. The confirmation of the anomalous energy shift of the kaonic atom $2p$ level requires a weakly absorptive interaction, and strongly indicates the existence of an s-wave deeply-bound kaonic state in the nucleus.

In contrast to the optical potential calculation, Akaishi's coupled channel calculation gives a finite shift of about 10 eV, which is rather close to the previous x-ray experimental results. In his calculation, the pole of the p-wave kaon-nuclear bound state locates slightly below the kaonic atom $2p$ level[6], and the nuclear bound state pushes up the level energy of the atomic state. Akaishi pointed out that the biggest difference between the coupled-channel and optical-potential approaches is the treatment of the final state of the kaon absorption reaction from the atomic level. Optical potentials (obtained by the global fit to kaonic atom data) have a strongly absorptive imaginary part, while the less absorptive interaction is realized due mainly to the final state pion's centrifugal effect, which pushes the pion wave function away from the nucleus.

Presently, we are preparing a kaonic atom x-ray measurement on a ^{4}He target with an expected accuracy $\sim$ 1 eV (experiment E570) before the end of 2005. To achieve this absolute energy calibration, we plan to use a recently developed high resolution x-ray detector (silicon drift detector; SDD) with several foils for the in-beam calibration to produce well-known fluorescence x rays. Because the SDD is an order of magnitude thinner than a conventional Si(Li) detector, we are expecting a better signal-to-noise ratio in this experiment. It should be noted that the x-ray measurement of kaonic ^{3}He is also very important. In the coupled-channel calculation, the shift is expected to be attractive, in contrast to kaonic ^{4}He, so that it gives a cross check of the calculation and, at the same time, one can deduce the $\overline{K}N$ interaction more precisely, if both data sets are available. Therefore, the kaonic ^{3}He x-ray measurement should also be done in the near future.

References

1. M. Iwasaki et al., *Phys. Rev. Lett.* **78**, 3067 (1997).
2. Y. Akaishi and T. Yamazaki, *Phys. Rev.* **C65**, 044005 (2002).
3. T. Suzuki et al., *Phys. Lett.* **B597** 263 (2004).
4. C.J. Batty et al., *Nucl. Phys.* **A326** 455 (1979).
5. S. Baird et al., *Nucl. Phys.* **A392** 29 (1983).
6. Y. Akaishi, *private communication.*

RECENT RESULTS ON HIGH RESOLUTION HYPERNUCLEAR SPECTROSCOPY BY ELECTROPRODUCTION AT JEFFERSON LAB, HALL A

F. GARIBALDI, E. CISBANI, S. COLILLI, F. CUSANNO, R. FRATONI, S. FRULLANI, F. GIULIANI, M. GRICIA, M. LUCENTINI, F. SANTAVENERE, P. VENERONI

INFN Roma1, gr. collegato Sanità and Istituto Superiore di Sanità, Viale Regina Elena 299, I-00161 - Rome, Italy

H. BREUER, C. C. CHANG

University of Maryland, College Park, Maryland 20742, USA

P. BYDZOVSKI, M. SOTONA

Nuclear Physics Institute, Rez near Prague, Czech Republic

P. BRINDZA, C. W. DE JAGER, R. FEUERBACH, E. FOLTS, D. W. HIGINBOTHAM, B. KROSS, J. J. LE ROSE, R. MICHAELS, B. REITZ, J. SEGAL, B. WOJTSEKHOWSKI, C. ZORN

Jefferson Lab, Newport News, Virginia 23606, USA

R. DE LEO, G. DE CATALDO, L. LAGAMBA, S. MARRONE, E. NAPPI

INFN Sezione di Bari, Via Amendola 173, I-70126 Bari, Italy

P. MARKOWITZ

Florida International University, Miami, Florida 33199, USA

M. IODICE

INFN Sezione di Roma Tre,

Via della Vasca Navale 84, I-00146 Rome, Italy

G. M. URCIUOLI
INFN Sezione di Roma,
Piazzale A. Moro 2, I-00185 Rome, Italy

Y. QIANG
Massachusetts Institute of Technology,
Cambridge, Massachusetts 02139, USA

The first "systematic" study of 1 p shell hypernuclei with electromagnetic probes has started in Hall A at Jefferson Lab [5]. The aim is to perform hypernuclear high resolution spectroscopy by the electroproduction of strangeness on four 1p-shell targets: ^{12}C, ^{9}Be, ^{16}O, ^{7}Li. The first part of the experiment on ^{12}C and ^{9}Be has been performed in 2004, the second part (^{16}O and ^{7}Li) is scheduled for June 2005. To overcome the major experimental difficulties, namely the low counting rate and the challenging Particle IDentification (PID), two septum magnets and a Ring Imaging CHerenkov (RICH) detector had to be added to the existing apparatus. After underlining the particular role the electroproduction reaction plays in hypernuclear physics we describe the challenging modifications of the Hall A apparatus. Preliminary results on ^{12}C and ^{9}Be are presented.

1. Introduction

The Lambda hypernucleus, a nuclear system with strangeness S=-1 in which the Lambda hyperon replaces one of the nucleons, is a long-living baryon system and provides a variety of phenomena. The strange quark is an impurity in the system and the study of its propagation can reveal configurations or states not seen in other ways. The study also gives interesting important insight into the structure of ordinary nuclear matter. The hyperon is not excluded from the filled nucleon orbitals by the Pauli principle and can penetrate deep inside the nucleus. Moreover the high resolution spectroscopic study of light hypernuclei allows access to important information on the nature of the force between nucleons and strange baryons, i.e. the $\Lambda - N$ interaction. The nucleus provides a unique laboratory for this study. From low-energy $\Lambda - p$ scattering studies we obtained information about the s-state $\Lambda - N$ interaction, but the non-central part of the $\Lambda - N$ force is not well established. Especially useful information comes from the spin dependent part of the interaction[1]. Hypernuclear experimental studies up to now have been carried out essentially by hadron-induced reactions with limited energy resolution (1.5 MeV at the best). Moreover the spin flip transition excitation was weak so doublets have not been detected. In the electromagnetic case, the spin flip transitions are strong, so doublet splitting detection is possible in principle because both members are populated. The disadvantage of smaller electromagnetic cross sections is partially compensated by the high current, continuous beam available at JLab[15]. The feasibility of these experiments has been recently demonstrated by the Hall C E-89-009 experiment [3]. The aim of the Hall A experimental program is the first "complete" study in 1p shell nuclei: ^7Li, ^9Be, ^{12}C, ^{16}O[5]. A missing mass resolution as good as 350 keV - 400 keV (FWHM) can be attained, with beam energy stability of 2.5×10^{-5} (σ_E/E) and the High Resolution Spectrometers (HRS) momentum resolution, $\Delta p/p \simeq 10^{-4}$ (FWHM)[15]. In the following, after presenting the physics information that can be extracted from the experiment, we describe the experimental difficulties and the challenging modifications made to the Hall A apparatus. Then preliminary results on ^{12}C and ^9Be will be presented.

2. Hypernuclei

In the Lambda hypernucleus the Λ is weakly coupled to the nuclear core, and the shell model can be used to describe this system. In this model the Λ resides in the 1s shell in the hypernuclear ground state. Excited states are obtained by promoting a Λ to higher orbits or coupling a Λ in the ground state to an excited nuclear core. The shell model is a precise and predictive tool used to describe nuclear spectra. Based on the experimental knowledge of some light hypernuclei, scattering data of hyperons from nucleons, and flavor SU(3) symmetry, an effective ΛN interaction for the shell model can be constructed [12]. Consider a hypernucleus in which a nucleon in a p shell is replaced by a Λ hyperon in the s shell (the Λ particle not being subjected to Pauli blocking). This system can be described, by analogy with the spectroscopy of light nuclei, in terms of the $|s^4\, p^{(A-5)} * s^{\Lambda}; JT\rangle$ configurations, in which the Λ particle couples to a nuclear state of the parent nucleus with spin equal to J^{A-1}, creating a doublet of states $J = J^{A-1} \pm 1/2$ (weak coupling model). The Λ-nucleon interaction can be described by the potential $V_{\Lambda N} = V + \Delta + s_\Lambda + s_N + T$, where $V = V(r)$ is the central part, $\Delta = V_\sigma(r) s_\Lambda \cdot s_N$ is the spin-spin term, $s_\Lambda = V_\Lambda(r) s_\Lambda \cdot l_{N\Lambda}$ is the Λ spin orbit term, $s_N = V_N(r)\, s_N \cdot l_{N\Lambda}$ is the N spin orbit-term. $T = V_T(r) S_{12}$ is the tensor part, where $S_{12} = 3(\vec{s}_N \cdot \hat{r})(\vec{s}_\Lambda \cdot \hat{r}) - \vec{s}_N \cdot \vec{s}_\Lambda$ The potential is particularly sensitive to the spin dependent part of the interaction [2].

2.1. *Electromagnetic versus hadronic probes*

The spectroscopic information that can be obtained from (K^-,π^-), (π^+,K^+) and $(e,e'K^+)$ reactions can be summarized as follows. In the strangeness exchange reaction: $K^- + n \rightarrow \pi^- + \Lambda$, the momentum transferred to the target nucleus is small ($q <$ 100 MeV/c). The spin flip transitions for forward pion angles ($\theta_\pi < 10°$) are negligible. Due to the strong absorption of the kaon and pion, the reaction takes place on peripheral nucleons. As consequence, in this reaction, $\Delta l = \Delta s = 0$ transitions dominate and substitutional states are predominantly populated. In the stopped kaon (K_{stop}, π) reaction, the momentum transferred to the nucleus is 250 MeV/c, hence the Δl = 1, 2 transitions dominate, but spin flip is very weak. In the strangeness associated production $\pi^+ + n \rightarrow K^+ + \Lambda$, the momentum transferred is as high as q = 350 MeV/c, but the spin flip amplitude is still weak at $\theta_k < 10°$. As a consequence, Δl = 1, 2 and $\Delta s = 0$ transitions are favoured. The energy resolution of the processes described above is typically of the order of 2 MeV.

In the electromagnetic production of strangeness: $e + p \rightarrow e' + K^+ + \Lambda$, the momentum transfer to the hypernucleus is rather large (q about 350 MeV/c for light nuclei). The spin-flip production is strong so both $\Delta l \neq 0$ and $\Delta s \neq 0$ transitions are available. Because the production of the $K^+ - \Lambda$ pair goes on the proton, the hypernuclei generated are different from those produced by hadronic probes. The reactions are complementary: different hypernuclei and different states are generated,and different energy resolutions are obtained[4]. The experimental advantage of the electromagnetic probe is evident: much better energy resolution. Improving the missing mass resolution has obvious advantages. Two, three or four peaks have been found for ^{12}C, as the resolution improves from 4 to 2 MeV, and to 1.5 MeV in the hadronic case[7]. For the electromagnetic probe it has been shown that doublet splitting can be detected in principle, if the energy resolution is about 500 keV or better, in the case of $^9Li_\Lambda$ (see Fig.1[4]). For the electromagnetic probe it has been shown that doublet splitting can be detected in principle, if the energy resolution is about 500 keV or better, in the case of $^9Li_\Lambda$ (see Fig.1[4]).

3. The experiment

The experiment took place in Jefferson Lab Hall A. The HRS spectrometers were used to detect the electrons and kaons. To get reasonable counting rates we have to consider the following. The electron scattering angle has

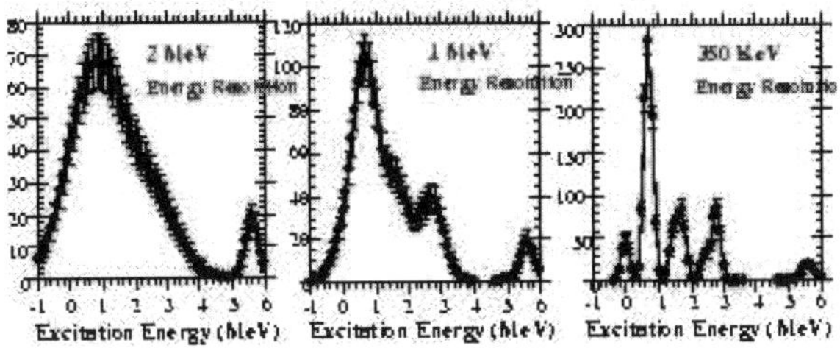

Figure 1. Be missing mass spectrum with three different missing mass resolution.

to be small to get a high virtual photon flux and the kaon angle has to be close to the virtual photon direction to minimize the momentum transfer. The momentum transfer to the hypernucleus in electroproduction is rather large and decreases steadily with increasing energy of the virtual photon so high energies are preferable. Since cross sections depend strongly on Q^2, the measurements have to be made at low Q^2. To keep a reasonable kaon survival fraction the kaon momenta have to be fairly high. Scattering angles of 6° for both for electron and kaons, an incident energy of 3.77 GeV, a photon energy of 2.2 GeV, a scattered electron momentum of 1.86 GeV/c, a kaon momentum of 1.96 GeV/c, and $Q^2 = 0.079$ (GeV/c)^2 have been used.

The expected counting rates for the two nuclei are reported in Table 1 . The C4 model [8] of the elementary process is used in the calculation. The electron-nucleus luminosity, obtained with a constant beam current

Background	(Hz)	$^{12}C(e,e'K)^{12}B_\Lambda$			$^9Be(e,e'K)^9Li_\Lambda$		
		E (MeV)	J	counts (h^{-1})	E (MeV)	J	counts (h^{-1})
(e,e')	$1.5 \cdot 10^5$	0	1^-	5.9	0	$3/2^+$	1.78
(e,π)	$2.0 \cdot 10^5$	0.03	2^-	34.6	0.69	$5/2^+$	9.7
(e,K)	$3.6 \cdot 10^3$	2.54	1^-	14.9	1.42	$1/2^+$	1.95
(e,p)	$1.8 \cdot 10^5$	5.46	2^-	4.5	1.71	$3/2^+$	2.8
(e,e'π)	$\sim 1.0 \cdot 10^2$	6.04	1^-	1.0	2.43	$5/2^+$	1.07
(e,e'K) (QF)	$\sim 0.1 \cdot 10^1$	10.03	3^+	5.8	2.78	$7/2^+$	3.04
(e,e'p)	$\sim 5.0 \cdot 10^1$	10.63	3^+	27.1			

of 100 μA and target thickness of 100 mg/cm^2, ranges from 2.4×10^{36} to 5.4×10^{36} cm^{-2}s^{-1}. The single and accidental coincidence counting rates are about constant for all the investigated nuclei and the numbers reported in Table 1 represent upper limits for our conditions.

3.1. *Experimental challenges*

3.1.1. *Forward angle. Septum magnets*

In order to allow experiments at very forward angles, smaller than the actual HRS minimum angle (12.5^o), two septum magnets were added to the HRS's. The two magnets bend particles scattered at angles as small as 6° in order to make their trajectories overlap with trajectories detectable by the HRS's. This overlapping as well as the room for the septum magnets has been achieved by moving the target upstream. The chosen target shift makes the new HRS angular acceptance $\sim$ 4.5msr and the maximum field $\sim$ 3.3 Tesla. The septum magnetic length is 84 cm. This new spectrometer configuration provides a general purpose device that extends the HRS features at small scattering angles[13] while meeting the HRS optical performances.

3.1.2. *The challenge of Particle IDentification*

As previously mentioned a very powerful PID system is mandatory for this experiment. In the electron arm Cherenkov counters were expected to give pion rejection ratios up to 10^3. The dominant background (knock-on electrons) can be reduced another 2 orders of magnitude by the lead glass shower counters, giving a total pion rejection ratio $\geq 10^5$. In the hadron arm two aerogel Cherenkov counters with $n = 1.015$ and $n = 1.055$ have been used. Fig.2 shows the velocity β vs the momentum of pions, kaons and protons. With the choice of $n = 1.05$ it is possible to separate kaons (above threshold) from protons (below threshold) in the range 1.6 - 3.0 GeV/c. Using n = 1.01 allows the separation between pions and kaons so that with the combination of the two, kaons can be identified. Nevertheless, it has been shown that TOF and threshold Cherenkov counters are not sufficient for the unambiguous kaon identification needed to obtain "background free" missing mass spectra.

4. Improving the PID: RICH detector

Simulations performed to understand how to improve the kaon identification capability showed that a Ring Imaging CHerenkov (RICH) detector would provide the added PID capability. Fig.3 shows the advantage of using the RICH. It can be clearly seen, in the case of $^9Li_\Lambda$ how the low signal

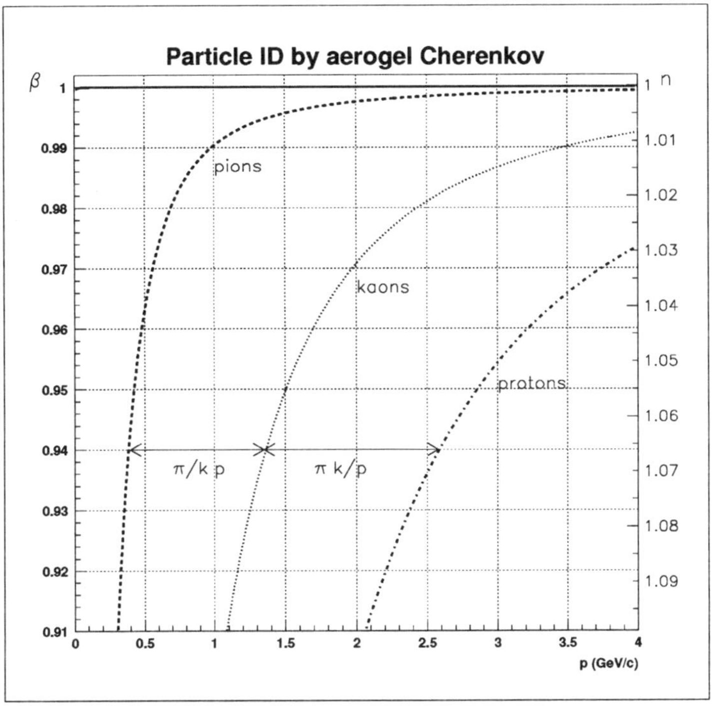

Figure 2. PID with Threshold aerogel Cherenkov counters.

to noise ratio makes the spectroscopic measurements particularly hard. A clear assignment can be only given to the second component of the first doublet, loosing any (important) information on the position of the first component. It should be realized that a standard PID system capable of rejecting pions and protons only at a 95% level is probably not enough to extract outstanding physics information. The choice was a proximity focusing CsI/freon gaseous detector. The RICH discriminates particles by differentiating between different values of Cherenkov emission angle. The Cherenkov radiation, emitted in a trasparent medium (the radiator) whose refractive index is appropriate for the range of particle momentum being specifcally studied, is transmitted through an optical elements, which could be either focusing with a spherical or parabolic mirror or not focusing (proximity focusing), onto a photon detector that converts photons into photoelectrons with high spatial and time resolution (Fig.4-left). In our case the proximity focused geometry was chosen. The Cherenkov photons, emitted along a conic surface in the radiator (C_6F_{14} with n=1.29), chosen because

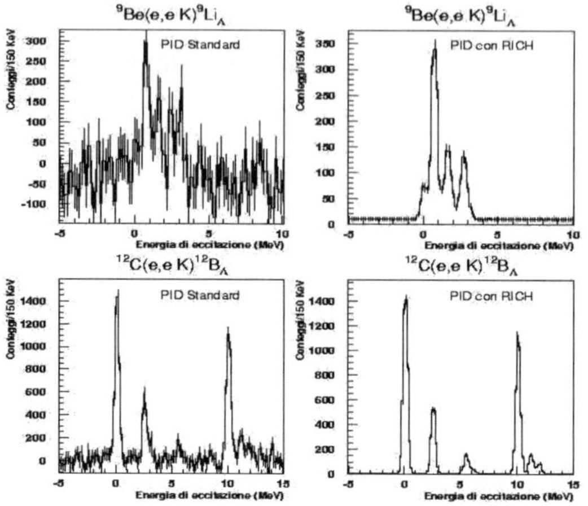

Figure 3. ^{12}C and ^9Be with and without RICH

of the momentum of the particles to be identified (2 GeV/c) are refracted by the freon-quarz methane interfaces and strike a pad plane after traveling through a proximity gap of 10 cm filled with methane. A dedicated evaporation facility, with on-line Quantum Efficiency (QE) measurement capability was built for this [19].

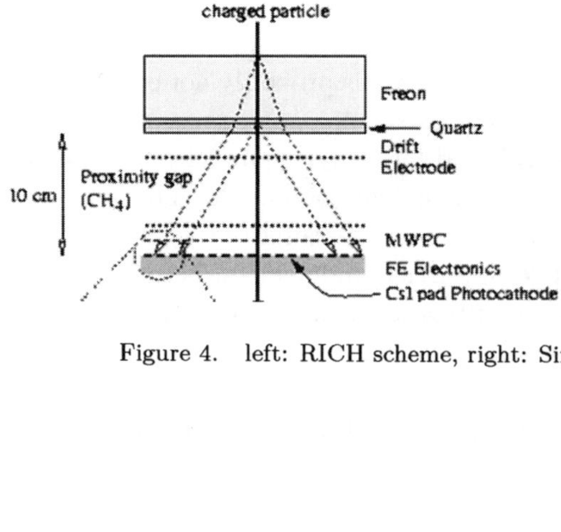

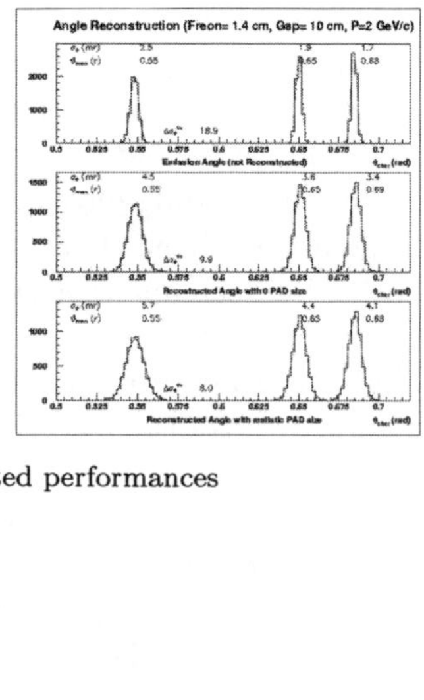

Figure 4. left: RICH scheme, right: Simulated performances

4.1. *Detector performances*

In Fig.5 the Time of Coincidence (TOC) spectra between the electron and the hadron arm are shown. The top left spectrum is obtained without any PID selection. Kaons are not visible. In the top right plot the aerogel kaon selection is applied showing that kaons are barely visible. When the RICH selection is applied (bottom right plot) only the kaon contribution survives. In Fig.6 the good RICH performance is shown: a separation of pions from kaons at level of 6 sigma has been attained to be compared with MonteCarlo caluclations (Fig.4-right). The role of the detector in cleaning up the pion and protons is evident. A rejection factor of 1000 for pions has been measured [4]. It should be considered that the RICH analysis has not

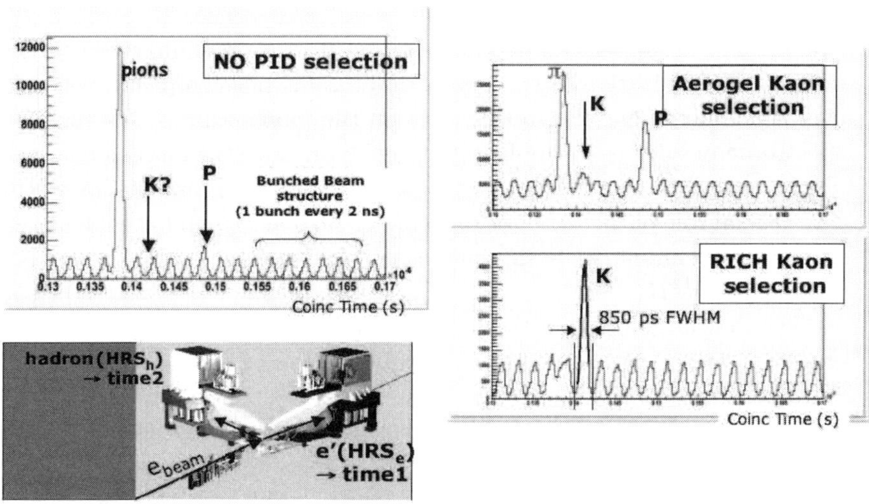

Figure 5. The role of the RICH for Kaon identification: many pions and protons are still present if RICH cuts are not applied

5. Preliminary results

The physics information that can be extracted from the spectra strongly depends on the energy resolution, determined by the beam energy spread (σ_E/E); by the spectrometer momentum resolution, and by the capability of reducing background with PID. Data were taken for both ^{12}C and

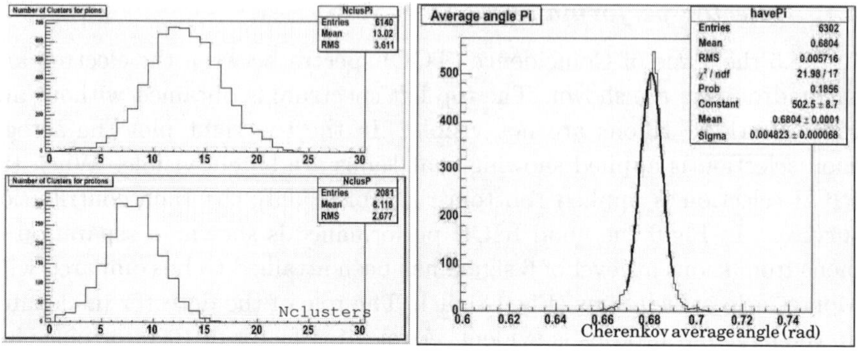

Figure 6. RICH performances. On the left side the number of photoelectrons for pions and protons is shown. On the right side the distribution of Cherenkov angle reconstruction is reported, showing a resolution of 5 mr

$^9Be(e,e'K^+)$ reactions. The analysis is still in a preliminary stage. First of all the optimization of the optics database is not completed yet. Moreover the analysis has shown that cuts on the momentum acceptance of the spectrometer have to be applied in order to improve the missing energy resolution. Fig.7 shows the missing mass spectrum with and without RICH. The crucial role of the detector in cleaning the background is evident. Fig.8 shows the preliminary physics analysis of the missing energy spectra of the $^{12}C(e,e'K^+)^{12}B_\Lambda$ (left plot) and $^9Be(e,e'K^+)^9Li_\Lambda$ reactions (right plot).

The first large peak, in the left plot, and another peak at $\sim$ 10 MeV are clearly visible, corresponding to the substitution of a p shell proton with a Λ in a s state and p state, respectively. In between two levels at $\sim$ 2.4 MeV and $\sim$6 MeV are also evident. The background is evaluated by fitting the 'not-physical' region (negative region of the missing energy spectrum). The resulting values of the Signal to Noise Ratio (SNR) for the two identified core-excited-state peaks are 7.5 at 2.6 MeV and 6.5 at 5.4 MeV.

Two theoretical curves have been superimposed on both spectra, differing in the model used for the elementary $K^+ - \Lambda$ production on protons. The hypernuclear wave function is the same for the two curves as computed by M. Sotona. The solid line uses the model of Bennhold-Mart(K-MAID)[12], the dashed line the one by Saghai Saclay-Lyon (SLA)[11]. Both curves have been normalized to the first (ground state) experimental peak. In the $^{12}C(e,e'K^+)^{12}B_\Lambda$ reaction, the relative intensities (with respect to the ground state) of the first excited peak at 2.6 MeV and of the strongly

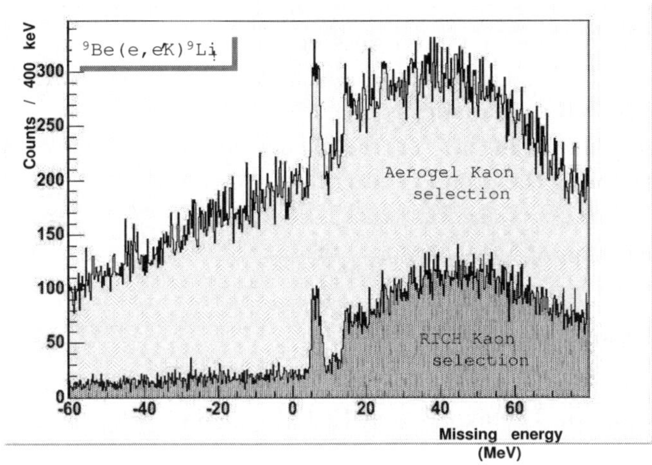

Figure 7. $^{9}Li_{\Lambda}$ missing energy spectrum with and without the RICH selection.

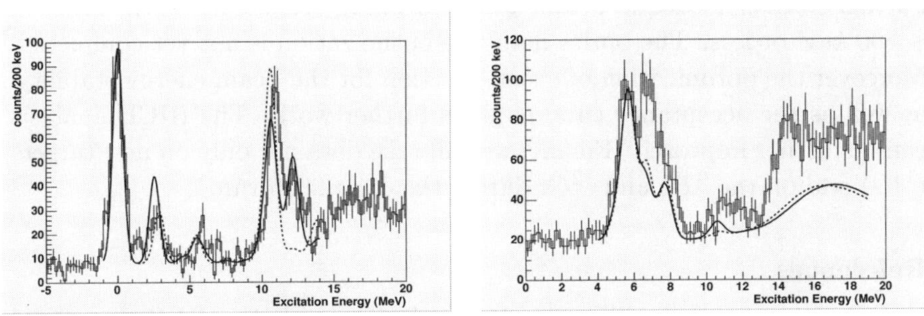

Figure 8. Left: $^{12}B_{\Lambda}$ missing energy spectrum, the solid and dashed lines represent the theoretical data (see text). Right: $^{9}Li_{\Lambda}$ missing energy spectrum, the solid and dashed lines represent the theoretical data (see text).

populated $p - \Lambda$ state at 11 MeV seem to be better reproduced by the MAID model than the SLA one. Another peak at $6 - 7$ MeV is underestimated by both models. Of course, for any conclusion one has to wait for the final results of the analysis.

In the reaction $^{9}Be(e, e'K^{+})^{9}Li_{\Lambda}$ the relative strength of the doublets is different from theoretical prediction. At the present stage of the analysis is not possible to select the "best" model. In fact the analysis and the interpretation of the measured missing mass spectra is still at a preliminary stage and not ready for final conclusions. Improvements in the energy resolution as well as statistics is crucial.

A first evaluation of the cross section has been extracted from the carbon data. Considering the ground state of the $^{12}C(e,e'K^+)^{12}B_\Lambda$, the cross section is $\sigma_{g.s.}(^{12}B_\Lambda) = 5.0 \pm 0.2\ (stat) \pm 1.0\ (syst)\ \frac{nb}{(msr)^2 GeV}$, to be compared with a prediction of [5],$\sigma_{g.s.} \sim 5.4 \frac{nb}{(msr)^2 GeV}$.

6. Conclusion

The first 'systematic' study of $1p$ shell hypernuclei with an electromagnetic probe has begun. The experiment E94-107 in Hall A at Jefferson Lab, designed to perform hypernuclear spectroscopy of light hypernuclei, took data on ^{12}C and ^{9}Be targets. The new experimental devices (septum magnets and RICH detector) have proven to be very effective. The RICH detector provided excellent kaon identification and a clean kaon signal over large pion and proton backgrounds. The analysis is still at a preliminary stage. Further work is needed to attain a missing energy resolution of the order of $\sim 500\ keV$ or less. The optics database optimization is not yet completed. Moreover the optimization of event selection for the beam energy stability, as well as the acceptance cuts, requires further work. The RICH analysis can be further improved. Finally, we will take data not only on new targets (^{16}O) and on the ^{9}Be and ^{12}C, during the June 2005 run.

References

1. C.B. Dover and D. J. Millener, in Modern Topics in Electron Scattering (Word Scientific Singapore, 1990, B. Frois and I. Sick eds.).
2. H. Bando, T. Motoba, J. Zofka, Internatinal Journla of Modern Physics A5, 1990, 4021
3. T. Myioshi *et al.* Phys. Rev. Lett. 90, 232502-1
4. F. Garibaldi, High Resolution Hypernuclear Spectroscopy, Proeceedings of EPJ
5. F. Garibaldi, S. Frullani, P. Markowitz, J. LeRose, T. Saito. E-94-107 JLab proposal
6. F. Garibaldi *et al.*, "Status Report for Experiment E94-107, High Resolution 1p shell" Hypernuclear Spectroscopy in Hall A at TJNAF
7. H. Hotchi et al., Phys. Rev C 64 , 044302
8. R.A.Williams. Chueng-Ryong Ji and C. R. Cotanch, Phys. Rev. C46 (1992), 1617
9. G.M. Urciuoli, E. Cisbani, S. Frullani, F. Garibaldi, M. Iodice, L.Pierangeli et

al., Proceedings of the "Workshop on the spectroscopy of hypernuclei", World Scientific proceedings, 121-132.

10. P. Brindza et al., "Superconducting Septum Magnet Design for Jefferson Lab Hall A", IEEE Trans. Appli. Supercond., Vol II, N. 1.
11. J.C. David, C. Fayard, G.-H. Lamot, and B. Saghai, Phys. Rev. C **53** (1996) 2613; Mizutani, C. Fayard, G.-H. Lamot, and B. Saghai, Phys. Rev. C **58** (1998) 75.
12. T. Mart and C. Bennhold, Phys. Rev. C **61** (2000) 012201.
13. A. Deur, JP Chen, F. Garibaldi, "The GDH Sum Rule and the Spin Structure of ^3He and Neutron using Nearly Real Photons", JLab E-97-107 experiment; K. Kumar, D. Lhuillier, "Constraning the Nucleon Strangeness Radius in Parity Violating Electron Scattering", JLab E-99-115 experiment.
14. R. Perrino *et al.*, Nucl. Instr. Meth. A 457 (2001), 571.
15. J. Alcorn *et al.*, Nucl. Instr. Meth A 522 (2004), 294.
16. L. Lagamba *et al.*, Nucl. Instr. Meth. A 471 (2001), 325.
17. F. Garibaldi *et al.* "Hadron Identification at Jefferson Lab, Hall A", Proceedings of the International Conference on New Detectors, Erice, November 1997, 452
18. F. Garibaldi *et al.*, Nucl. Instr. Meth. A 502 (2003) 117.
19. F. Cusanno *et al.*, Nucl. Instr. Meth. A 502 (2003) 251.
20. F. Cusanno *et al.*, Nucl. Instr. Meth. A 525 (2004), 163.
21. CERN/LHCC 98-19, ALICE TDR 1, 14 August 1998.
22. http://www.jlab.org/exp_prog/experiment_schedule/index.html

SECTION V

NUCLEAR ASTROPHYSICS

RECENT DEVELOPMENTS IN NUCLEAR ASTROPHYSICS IN ITALY

F. TERRASI
Dipertimento di Scienze Ambientali, Seconda Universitá di Napoli and INFN section of Naples
E-mail: filippo.terrasi@unina2.it

L. GIALANELLA, G. IMBRIANI
Dipartimento di Scientze Fisiche, Universitá Federico II di Napoli and INFN section of Naples
E-mail: lucio.gialanella@na.infn.it, gianluca.imbriani@na.infn.it

FOR THE ERNA AND LUNA COLLABORATIONS

1. Introduction

The detailed description of most astrophysical and cosmological scenarios requires a very accurate knowledge of many relevant nuclear processes([1]). This knowledge can be gained in laboratories on Earth, whereas the very small cross sections typical at the astrophysical relevant energies challenge the experimentalists.

Essentially the experiments performed in this field can be divided into two classes. The first one, referred to as "direct methods", aims to a direct measurement of the relevant physical quantities, essentially the reaction cross section, and a good understanding of the reaction mechanisms, on the basis of which one can determine the cross section at the astrophysical relevant energies extrapolating the measured data. One has to mention that in few selected cases it was possible to measure the cross section directly at the astrophysical relevant energy([2]).

The second kind of measurements relies on the determination of the astrophysical relevant quantities from some different measured observables, based on some model, that determines the name of "indirect methods". In the community of nuclear astrophysicists in Italy there are groups performing excellent experiments using both direct and indirect methods. We will focus our contribution on the direct methods, since the experiments exploiting indirect methods are the subject of another contribution(3). The main problem of direct measurements is determined by the background signals, which, together with the low cross sections, set a limit to the energy range that can be investigated with a simple setup on the Earth surface.

Essentially there are three sources of background, i.e. cosmic rays, enviromental radioactivity and beam-target induced nuclear reactions. Each of these sources produces background of different nature and energy, so that each reaction to be studied deserves a special care in suppressing the relevant background component. In the last years two different approaches, in which Italian groups gave a relevant contribution, have provided very good results. In the first case, the suppression of the cosmic background provided by a huge passive shielding, like the one that can be obtained at the INFN underground laboratory at Gran Sasso (LNGS), is exploited; the LUNA (Laboratory Underground for Nuclear Astrophysics) collaboration set up in the last decade a worldwide still unique facility for measuring low energy cross sections of astrophysical interest, installing two accelerators, respectively 50 and 400 kV(4), at LNGS.

The second approach is based on quite complex apparatuses in order to optimize the detection efficiency, selectivity and the background suppression. An example of that is provided by Recoil Mass Separators, that allow to measure the cross section of radiative capture reactions by means of the detection of the residual nucleus, without the need of gamma measurements, that can be eventually performed in order to gain information on the transitions involved in the reaction. As a follow up of the pioneering work done of the NaBoNA (Naples Bochum Nuclear Astrophysics)(5) collaboration to study the ^{7}Be(p, γ)^{8}B a new separator has been designed and installed at the Dymanitron Tandem Laboratorium of the Ruhr-Universitaet Bochum to study the ^{12}C(α, γ)^{16}O reaction.

2. ERNA

The reaction ^{12}C(α, γ)^{16}O (Q = 7.16 MeV) takes place during helium burning in Red Giants(1). The cross section at the relevant Gamow-energy, $E_0 \simeq 0.3$ MeV, determines, together with the convection mechanism in the Helium stellar core, the abundances of Carbon and Oxygen at the end of Helium burning. This, in turn, influences the nucleosynthesis of elements up to the iron region for massive stars([6]) and the composition of C/O White Dwarfs in the case of intermediate mass stars([7]). For these reasons, the cross section $\sigma(E_0)$ should be known with a precision of at least 10%.

In spite of experimental efforts over nearly 30 years, one is still far from this goal. All previous efforts have focused on the observation of the capture γ-rays, including one experiment that combined γ-detection with coincident detection of the ^{16}O recoils. Due to the low cross section and various backgrounds depending on the exact nature of the experiments, γ-ray data with useful, but still inadequate, precision were limited to center-of-mass energies 1.2 MeV $< E_{cm} <$ 3.2 MeV. To improve the situation, a new experimental approach has been undertaken at the 4 MV Dynamitron tandem accelerator in Bochum, called ERNA. In this approach, the reaction is initiated in inverted kinematics, ^{12}C$(\alpha,\gamma)^{16}$O, i.e. a ^{12}C ion beam is guided into a windowless ^{4}He gas target and the ^{16}O recoils are counted in a ΔE-E telescope placed in the beam line at the end of the separator, where the separator filters the intense ^{12}C projectiles from the ^{16}O recoils. ERNA is designed to study the reaction over the energy range E = 0.7 to 5.0 MeV.

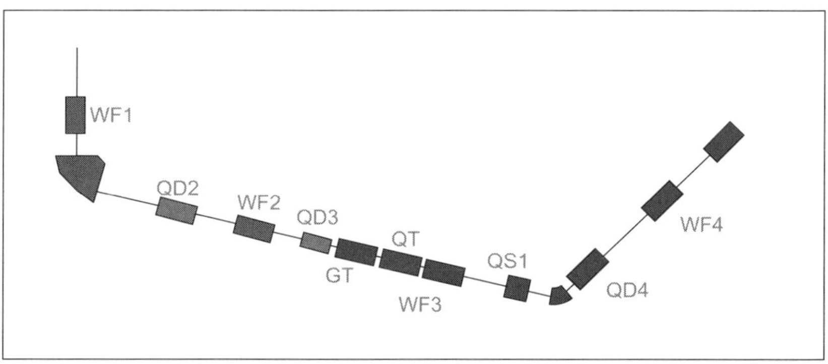

Figure 1. The ERNA set-up.

The ERNA setup (Fig. 1) is described elsewhere([8,9,10,11,12]). Briefly, the ion beam emerging from the tandem is focused by a quadrupole doublet (QD1), filtered by a 52° analysing magnet, and guided into the 75° beam line of ERNA by a switching magnet (these elements are not shown in

Fig. 1). A quadrupole doublet (QD2) after the switching magnet is used to focus the beam on the gas target. For the purpose of beam purification, there is one Wien filter (WF1) before the analysing magnet and one (WF2) between QD2 and the gas target. For the simulation of the angular spread of the ^{16}O recoils, there is a quadrupole doublet (QD3) shortly before the gas target (GT). After the gas target, the separator consists sequentially of the following elements: a quadrupole triplet (QT), a Wien filter (WF3), a quadrupole singlet (QS1), a 60° dipole magnet, a quadrupole doublet (QD4), a Wien filter (WF4), and the ΔE-E telescope. Finally, several steerers, Faraday cups, slit systems, and apertures are installed along the beam line for setting-up and monitoring purposes.

During 2004, the total cross section for E_{cm}=1.9 to 5.0 MeV has been measured. Fig. 2 shows the matrix collected at E_{cm}=2.4 MeV. These data, whose analysis is in progress([13]), represent a new information, since measurements at energies above 3.2 MeV were in the past hindered by the high beam induced neutron background, which plays an important role in the determination of the astrophysical $S(E_0)$. Measurements at lower energies and measurements of γ-ray angular distributions are planned for 2005.

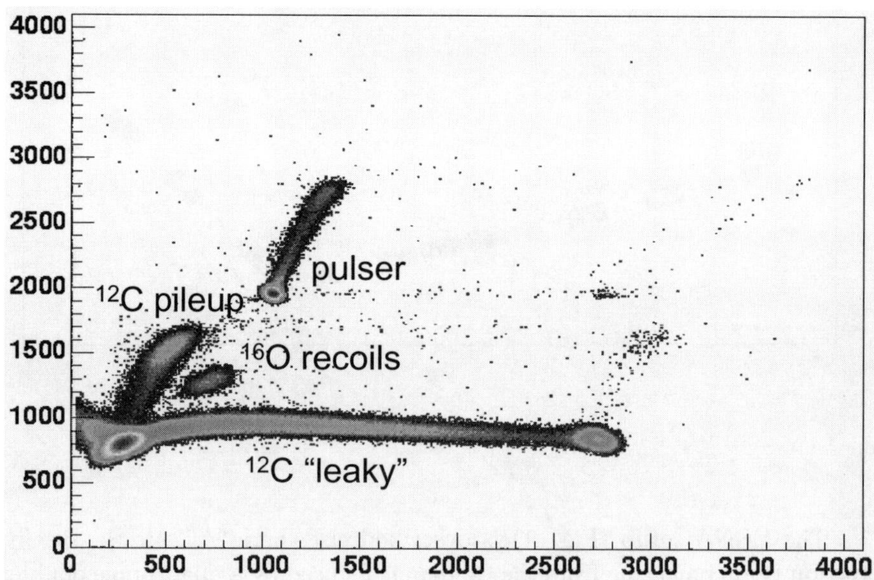

Figure 2. ΔE-E identification matrix for $^{12}C(\alpha,\gamma)^{16}O$ at E_{cm}=2.4MeV.

3. LUNA

During the last decade the LUNA collaboration has studied several reactions of astrophysical interest. Here we present some results on the more recent experiment performed at LNGS, i.e. the study of $^{14}N(p, \gamma)^{15}O$.

The capture reaction $^{14}N(p, \gamma)^{15}O$ ($Q = 7297\ keV$) is the slowest process in the hydrogen burning CNO cycle. It has a high astrophysical interest since its reaction rate influences the age determination of globular clusters as well as the energy production and neutrino spectrum of the sun.

Globular Clusters (GCs) represent the oldest resolved stellar populations. Their age practically coincides with the time elapsed since the epoch of the formation of the first stars in the Universe and provides an independent check of the reliability of standard (and non-standard) cosmological models. The ages may be derived by comparing the observed Color-Magnitude diagrams with theoretical isochrones([14]). The location in the HR diagram of both the Main Sequence and the Red Giant Branch do not change with age. On the contrary the Turn Off and the Subgiant Branch substantially depend on the age. During most of its life, a low mass star burns H in the center via the pp chain. However, when the central H mass fraction reduces down to 0.1, the nuclear energy produced by the H-burning becomes not sufficient and the stellar core must contract to extract some energy from its gravitational field. Then, the central temperature (and the density) increases and the H-burning switches from the pp-chain to the more effcient CNO-burning. Thus the escape from the Main Sequence is powered by the onset of the CNO burning, whose bottleneck is the $^{14}N(p, \gamma)^{15}O$ reaction. A modification of the rate of this reaction alters the turn off luminosity, but leaves almost unchanged the stellar lifetime, which is mainly determined by the rate of the pp reaction([14]).

In the center of the sun the CNO cycle is also partially active. Then a fraction of the total neutrino flux comes from the β-decay belonging to the cycle. Obviousuly the total amount of CNO solar neutrino directly depends on the value of the $^{14}N(p, \gamma)^{15}O$ reaction rate. In spite of the fact that this component represents only about 1% of the total flux, it plays an important role in some solar neutrino experiments where the detected neutrinos have energy of about 1 MeV, e.g. BOREXINO, where the CNO neutrino flux accounts for about 10 % of the expected detected neutrinos.

The typical temperature of a stellar core which is burning hydrogen via the CNO cycle is about $20 * 10^6 K$. In the case of ^{14}N+ p reaction, the corresponding Gamow energy, that includes the effect of the penetrability of the Coulomb barrier of the two interacting nuclei, is about 50 keV. At solar energies the cross section of $^{14}N(p,\gamma)^{15}O$ is dominated by a subthreshold resonance at -504 keV. At energies higher than 100 keV the cross section is dominated by the resonance at $E_R = 278$ keV with transitions to the excited states at energies of 5.18 MeV, 6.18 MeV and 6.79 MeV and the ground state in ^{15}O. According to Schröder et al. 1987 that extended the measurements over the energy range $E_p = 0.2$ to 3.6 MeV, the main contribution to the total S-factor at zero energy comes from transitions to the ground state in ^{15}O and to the subthreshold state at $E_x = 6.79$ MeV.

They found([15]) the total S-factor at zero energy to be $S(0) = 3.20 \pm 0.54$ keV-b. Angulo et al.([16]) re-analyzed Schröder's experimental data using a R-matrix model; they found for the total S factor at zero energy $S(0) = 1.77 \pm 0.20$ keV-b, which is a factor 1.7 lower than the values used in the recent compilations. The main difference between the two analyses concerns the S(0) factor for capture to the ^{15}O ground state, where Angulo et al. found a factor of 19 lower than the value of Schröder.

The LUNA collaboration started then a new experiment based on two different approaches. First, measurements were performed using a high resolution HPGe detector in combination with a solid, high purity ^{14}N target to identify the contributions of the different transitions. Then a different setup was used, consisting in a large volume high efficiency BGO summing detector in combination with a ^{14}N gas target, in order to extend the measurements of the total cross section of the reaction to an energy as close as possible to the Gamow energy.

A recent letter reported on the results of the first experiment, that covers an energy range down to 120 keV. The analysis of the transitions to the 6.7 MeV and to ground states confirms the prediction of Angulo et al.. A new value $S_{tot}(0) = 1.7 \pm 0.1 \pm 0.2$ $keV\ b$ for the extrapulated S-factor is obtained by means of an R-matrix fit([17]). To perform the fit it was necessary to use the high energy data from([15]), corrected for the summing effect in the detectorfor cascade transitions, which was not considered in the analysis of([15]). Since the high energy data have a relevant influence on the extrapolation, a new measurement has been started at the 500 kV and 4 MV accelerators at the Ruhr-Universitaet Bochum in order to obtain new data([18]). A final analysis of these data will be given in a forthcoming paper([19]).

After the completion of the first experiment, the gas target experiment was started. The measurements have been extended down to E_{cm}=80 keV, that is the upper edge of the range of interest for stellar CNO burning, and the results will be published in a forthcoming paper. The two experiments cover two energy overlapping regions.

Figure 3 shows the preliminary results of both measurements in form of S-factor([20,19]). As a consequence of the new determination so far obtained by the LUNA collaboration, a new estimate of the age of Globular Cluster was established, that has also consequences for the age of the Universe. Indeed the age of the oldest GCs provides an estimate of the age of the Universe, if one take into account the time elapsed between the Big Bang and the formation of the first stars. Gratton et al.([21]), by means of the TOL-A relation conclude that the age of the oldest Galactic Clusters is 13.4 Gyr (± 0.8 random, ± 0.6 systematic). The revision of the $^{14}N(p,\gamma)^{15}O$ rate increases the estimate of the age of the oldest GCs, and therefore of the age of the Universe, by about 0.6 Gyr([14]).

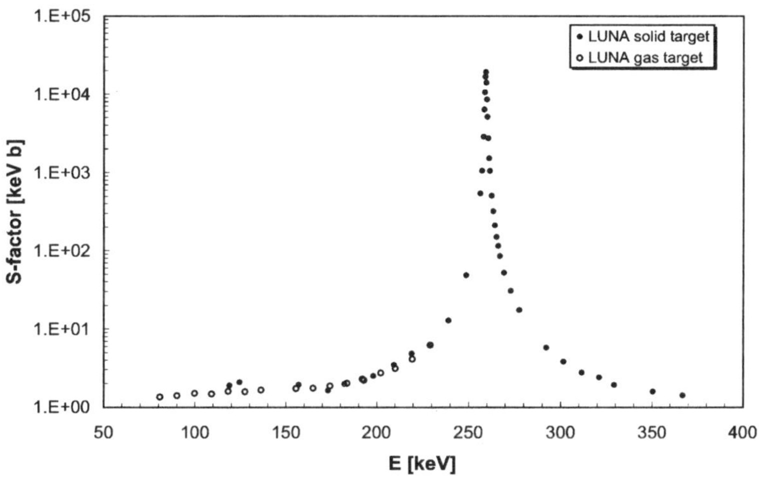

Figure 3. Excitation function for $^{14}N(p,\gamma)^{15}O$ cross section measured at LUNA by means of solid target and gas target set-up.

References

1. Rolfs, C.E. and Rodney, W.S. 1988, Cauldrons in the Cosmos, University of Chicago Press
2. R. Bonetti, C. Broggini, L. Campajola, P. Corvisiero, A. DAlessandro, M. Dessalvi, A. DOnofrio, A. Fubini, G. Gervino, L. Gialanella, U. Greife, A.

Guglielmetti, C. Gustavino, G. Imbriani, M. Junker, P. Prati, V. Roca, C. Rolfs, M. Romano, F. Schuemann, F. Strieder, F. Terrasi, H. P. Trautvetter, and S. Zavatarelli, 1999 Phys. Rev. Lett., 82, 5205

3. Cherubini, S. et al, 2005, this volume
4. Formicola, A., Imbriani, G., Junker, M., et al, 2003, Nucl.Instr.Meth. A 507, 609
5. Terrasi, F., Gialanella, L., Imbriani, G., Strieder, F., Campajola, L., D'Onofrio, A., Greife, U., Gyurky, G., Lubritto, C., Ordine, A., Roca, V., Rolfs, C., Romano, M., Rogalla, D., Sabbarese, C., Somorjai, E., Trautvetter, H. P., Nuclear Physics A, 688(2001)539-542
6. Imbriani, G., Limongi, M., Gialanella, L., Terrasi, F., Straniero, O., and Chieffi, A., ApJ, 558(2001)903
7. Straniero, O., Dominguez, I., Imbriani, G. and Piersanti, L., ApJ 583(2003)878
8. D.Rogalla, S.Theis, L.Campajola, A.D'Onofrio, L.Gialanella, U.Greife, G.Imbriani, A.Ordine, V.Roca, C.Rolfs, M.Romano, C.Sabbarese, F.Schümann, F.Strieder, F.Terrasi, H.P.Trautvetter, Nucl.Instr.Meth. A437(1999)266
9. D.Rogalla, M.Aliotta, C.A.Barnes, L.Campajola, A.D'Onofrio, E.Fritz, L.Gialanella, U.Greife, G.Imbriani, A.Ordine, J.Ossmann, V.Roca, C.Rolfs, M.Romano, C.Sabbarese, D.Schürmann, F.Schümann, F.Strieder, S.Theis, F.Terrasi, H.P.Trautvetter, Eur.Phys.J., A6(1999)471
10. D.Rogalla, D.Schürmann, F.Strieder, M.Aliotta, N.DeCesare, A.DiLeva, C.Lubritto, A.D'Onofrio, L.Gialanella, G.Imbriani, J.Kluge, A.Ordine, V.Roca, H.Röcken, C.Rolfs, M.Romano, F.Schümann, F.Terrasi, H.P.Trautvetter: Nucl.Instr.Meth., A513(2003)573
11. L.Gialanella, D.Schürmann, F.Strieder, A.Di Leva, N.De Cesare, A.D'Onofrio, G.Imbriani, J.Klug, C.Lubritto, A.Ordine, V.Roca, H.Röcken, C.Rolfs, D.Rogalla, M.Romano, F.Schümann, F.Terrasi, H.P.Trautvetter, Nucl.Instr.Meth., A522(2004)432
12. D.Schürmann, F.Strieder, A.Di Leva, L.Gialanella, N.De Cesare, A.D'Onofrio, G.Imbriani, J.Klug, C.Lubritto, A.Ordine, V.Roca, H.Röcken, C.Rolfs, D.Rogalla, M.Romano, F.Schümann, F.Terrasi, H.P.Trautvetter, Nucl.Instr.Meth. A531(2004)428S
13. D.Schürmann, et al., in preparation
14. Imbriani, G., Costantini, H., Formicola, A., et al, A&A, 420(2004)625-629
15. Schröder et al, 1987, Nucl. Phys. A, 467, 240
16. Angulo, C., & Descouvemont, P., 2001, Nucl. Phys. A, 690, 755
17. Formicola, A., Imbriani, G., Costantini, H., et al, Phys.Lett.B, 591(2004)61-68
18. Klug, J. et al, in preparation
19. Imbriani, G. et al, in preparation
20. Lemut, A. et al, in preparation
21. Gratton et al, 2003, AA, 408, 529

DIRECT APPROACH TO EXPLOSIVE HYDROGEN BURNING PROCESS WITH CRIB

S. Kubono[a], T. Teranishi[b], M. Notani[c], H. Yamaguchi[a], A. Saito[a], J.J. He[a], M. Wakabayashi[a], H. Fujikawa[a], G. Amadio[a], H. Baba[a], T. Fukuchi[a], S. Shimoura[a], S. Michimasa[d], S. Nishimura[d], M. Nishimura[d], Y. Gono[d], A. Odahara[e], S. Kato[f], J.Y. Moon[g], J.H. Lee[g], C.S. Lee[g], J.C. Kim[h], K.I. Hahn[i], T. Ishikawa[j], T. Hashimoto[j], H. Ishiyama[j], Y.X. Watanabe[j], M.H. Tanaka[j], H. Miyatake[j], Zs. Fülöp[k], V. Guimarães[l], and R. Lichtenthaler[l]

[a]Center for Nuclear Study (CNS), University of Tokyo, Wako Branch at RIKEN, Hirosawa 2-1, Wako, Saitama, 351-0198 Japan
E-mail: kubono@cns.s.u-tokyo.ac.jp
[b]Department of Physics, Kyushu University, Fukuoka, 812-8581 Japan
[c]Physics Division, Argonne National Laboratory, Illinois 60439, USA
[d]RIKEN, Saitama, 351-0198 Japan
[e]Nishi-Nippon Institute of Technology, Fukuoka, 800-0394 Japan
[f]Department of Physics, Yamagata University, 999-8560 Japan
[g]Department of Physics, Chung-Ang University, Seoul, 156-756 Korea
[h]Department of Physics, Seoul National University, 151-742 Korea
[i]Ewha Womens' University, 120-750 Korea
[j]IPNS, KEK, Tsukuba, 305-0801 Japan
[k]Institute of Nuclear Research (ATOMKI), Debrecen, H-4001 Hungary
[l]Departmento de Fysica Nuclear, Universidade de São Paulo, Brazil

Experimental efforts to nuclear astrophysical problems by the direct method are discussed including our new data for the explosive hydrogen burning process. We first present our new extensive, low-energy in-flight RI beam separator CRIB, which has been installed primarily for nuclear astrophysics. This is delivering RI beams of 10^6 pps for some ions and will be increased by one or two orders of magnitudes in near future. Two experiments performed with CRIB are discussed. One of the most crucial stellar reactions $^{14}O(\alpha,p)^{17}F$ for the onset of the high-temperature rp-process was directly investigated and the transitions through the states at around 6.2 MeV in ^{18}Ne were first observed, confirming the importance of the process predicted before. The reaction leading to the first excited state in ^{17}F was also found to have a considerable contribution. A proton resonance search experiment of ^{23}Mg + p was also discussed, which is a part of our series of resonance search studies relevant to the early stage of the rp-process.

1. INTRODUCTION

The hydrogen burning process is one of the most interesting subjects among nuclear astrophysical problems because hydrogen is the most abundant element and has a variety of occasions where this process plays a major role. Specifically, this process attracts a lot of interest in relation to observations of elements and isotopic abundances. The hydrogen burning under explosive conditions, such as novae and X-ray burst, involve many unstable nuclei on the proton-rich nuclear region where nuclear physics is still poorly known. ^{22}Na and ^{26}Al are the interesting targets for astronomical gamma-ray observation. One possibility of ^{26}Al production considered is a production in the occasion of novae. To investigate such possibility, one needs to know the nuclear reactions that involve proton-rich unstable nuclei in that mass region. Although there are several simulation methods available for nuclear astrophysics, the most reliable approach is the direct method with the low-energy RI beams at the energy of the site of interest.

An extensive low-energy in-flight RI beam separator, named CRIB, has been installed[1,2], and now is running successfully for nuclear astrophysics experiments at the Center for Nuclear Study, University of Tokyo (CNS).

The design and the performance of CRIB will be described in section 2, and some experimental results in the following sections. Summary and the scope are discussed in sec. 5.

2. DIRECT METHOD APPROACH IN EXPERIMENTS

For the direct approach to nucleosynthesis under explosive burning conditions, one need high-intensity and high-quality RI beams at very low energies. We have adopted the in-flight method for the simplicity and the cost effectiveness, although RI beam productions at low energy was not pursued so widely because available production target thickness is small, giving low intensity RI beams. However, because of the rapid development of the heavy ion source technology, we can get high intense beams, with which one can investigate stellar reactions directly. Figure 1 displays the plane view of the separator CRIB, which consists of /F0/QMDQ/F1/DMQ/F2/QQWQQ/F3/, where Q is a quadrupole magnet, M a multipole magnet, D a dipole magnet, W Wien filter section, and Fi the i-th focal plane of the separator. The first half (F0→F2) is a double achromatic separator, basically same as an ordinary fragment separator, and the second half (F2→F3) is a Wien filter system that cleans up the beam of interest.

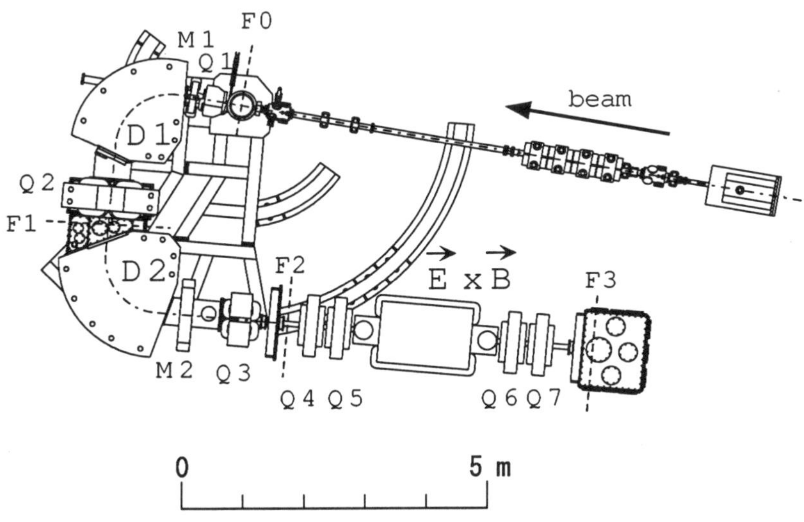

Figure 1. The plane view of the low-energy in-flight RI beam separator, CRIB.

The CRIB also has been equipped with high-power Faraday cup plates which were set along the walls of the first dipole magnet and cooled by water. Another investment was made for an installation of a window-less gas target system for the RI beam production. This has, however, never been used for experiment, because we are still working on development of the primary beam quality.

Table 1 displays some test results of RI beam productions with CRIB. When (p,n) reactions were applied in the inverse kinematics, one can obtain

Table 1. RI beam intensities obtained at CRIB

RI beam	Primary beam	Reaction	Intensity (pps)	Purity (%)
^{7}Be	^{7}Li	^{1}H(^{7}Li, ^{7}Be)	1x10^{6}	90
^{8}Li	^{8}Li	^{2}H(^{7}Li, ^{8}Li)	1x10^{6}	100
^{10}C	^{10}B	^{1}H(^{10}B, ^{10}C)	1.6x10^{5}	90
^{14}O	^{14}N	^{1}H(^{14}N, ^{14}O)	1.6x10^{6}	90

the RI beam intensity of a little over 10^6 pps with primary beam intensities of 200 – 500 pnA. These numbers are currently limited by the count rate of the beam-monitoring PPAC counters being used for the position and particle identification, and also by the production target. By optimizing the system at this stage, we may increase the RI beam intensities roubhly by a factor of ten. We have an on-going project, called the AVF Upgrade Project, for maximizing the system to attain the RI beam intensity of the order of 10^8 pps. We may need at least these intensities for studying directly radiative capture reactions of proton with proton-rich nuclei.

The optimum degrader thickness was checked experimentally for heavy ions at 4 – 10 MeV/u, and was found to be around dR/R = 0.1, which should be compared with dR/R = 0.3 often used at 50 – 100 MeV/u. Here, R is the stopping range of the RI ions.

Under the AVF Upgrade Project, this low-energy RI beam facility has been developed including the ion source, the AVF cyclotron, the beam line, and CRIB. Several experiments were already performed at the double achromatic focal plane F2. Some of the experimental results are discussed in this paper.

The Wien filter section gives a capability of better particle separation and also provides some interesting features for other studies. The velocity separation section has 1.5 m-long electric parallel plates that have the maximum voltages of ± 200 kV for a gap of 8 cm, giving 50 kV/cm. The maximum velocity dispersion designed was about 0.8 cm/%. The separation capability was verified with an ^{14}O beam produced from the ^{1}H(^{14}N,^{14}O) reaction at F0. It gave almost 100 % purity at F3. Another favorable feature of the RI beam from the filter is the beam quality. Under some condition, one may need only the Wien filter without the degrader, and obtain a small beam spot since the major factor for the RI beam sizes is the straggling at the degrader.

The present low-energy in-flight method is very useful for nuclear and nuclear astrophysics. We obtained quite reasonable beam intensities like a little over 10^6 pps for ^{14}O with a primary beam of 200 pnA. This intensity is currently limited by the production target because the window foils, 2.2-μ Havar, of the gas target do not stand with high beam currents, which is one of the major subjects under development.

3. THE DIRECT MEASUREMENT OF THE ^{14}O(α,p)^{17}F REACTION

The high-temperature (high-T) rp-process may typically take place in a X-ray burst, which is considered to be an event on the surface of a neutron star with accretion of hydrogen gas from the companion star in a main sequence phase. Here, one of the most critical stellar reactions is ^{14}O(α,p)^{17}F for the ignition of the high-T rp-process. There are many experiments performed by indirect methods[3,4] for the problem, but not by the direct method with an ^{14}O beam. Two experiments were made previously using the time-reverse reaction ^{17}F(p,α)^{14}O. Only some transitions through resonances above E_{cm} = 3 MeV in ^{18}Ne were reported[5,6]. Note that the most critical energy region is around E_{cm} = 1 - 2 MeV for the present problem. This reaction has been successfully investigated recently for the first time using a high intensity ^{14}O beam from CRIB.

A low-energy ^{14}O beam was produced by the ^{1}H(^{14}N,^{14}O) reaction at 8.4 MeV/u, and separated. The intensity and the purity of the beam was 1.6 x 10^6 pps and 85 %, respectively, at F2. The momentum spread of the beam was defined to 1 % by setting an aperture at F1, the momentum dispersive focal plane. The secondary target of He was cooled down to about 30 K, so that the target length was shortened roughly by a factor of 10, which made the present experiment possible. This is described in detail in ref. 7.

For the measurement of the ^{14}O(α,p)^{17}F cross section, we applied the thick-target method[8], which has been developed in the last decade for low-energy RIB experiments and applied for proton elastic scattering experiments. This method was successfully applied in the present case. Figure 1 displays a proton spectrum measured at 0° with a Silicon counter telescope. Several peaks are clearly seen that correspond to the (α,p) reaction mostly leading to the ground state in ^{17}F. The transitions through the 6.15 and 6.29 MeV states in ^{18}Ne were seen for the first time. These transitions were considered to be the main contributions to the stellar reactions under the X-ray burst condition[3]. The cross sections are roughly the same as predicted in ref. 3, thus confirming the primary importance of the two contributions. The transitions through the states at 7 - 8 MeV are also clearly observed. The peak at around 6.5 MeV is considered to be the transition through the state at 7.1 MeV in ^{18}Ne decaying to the first excited state at 0.495 MeV in ^{17}F. Since there is no state of a large proton width in the ^{17}F + p scattering[6] and the states in this energy region in ^{18}Ne cannot have a large α width as it is so close to the α threshold, the peak around 6.5 MeV cannot be explained by a state in ^{18}Ne. This implies that the transition through the 7.1 MeV state in ^{18}Ne increases the reaction rate roughly by 50 %. Note that the reaction study with the time-reverse reaction cannot access this process.

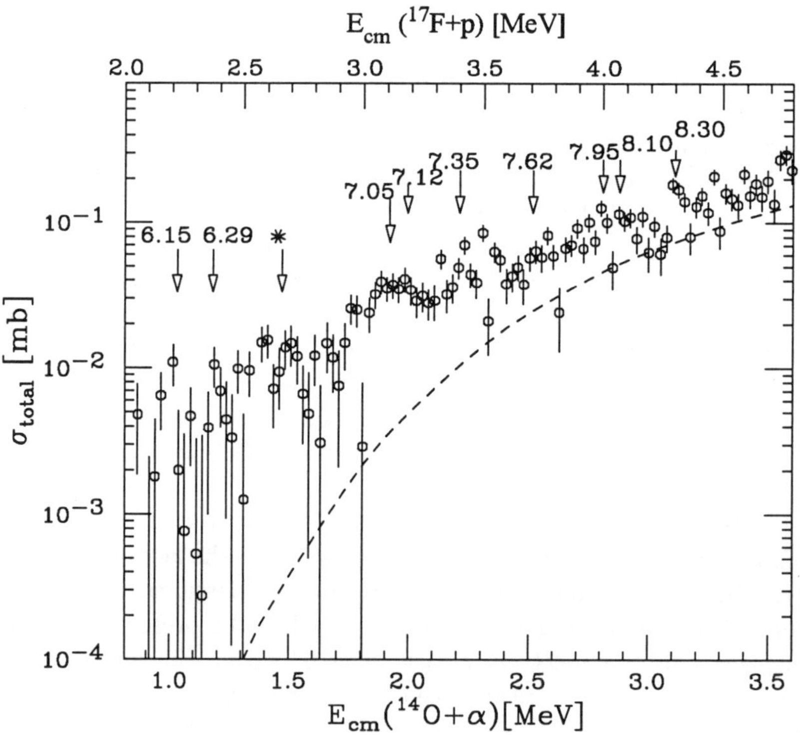

Figure 2. Proton spectrum from the ^{14}O(α,p)^{17}F reaction measured at $\Theta_{Lab} = 0°$. The asterisk indicates a transition to the 0.495-MeV first excited state in ^{17}F.

4. SEARCH FOR PROTON RESONANCES RELEVANT TO THE EARLY STAGE OF THE rp-PROCESS

The mechanism of the early stage of the rp-process is of great interest. Previously, we studied by indirect methods the excited states near and above the proton threshold in the proton-rich nuclei, relevant to the early stage of the rp-process[9], where many new states were identified. However, the reaction rates are not determined yet because the resonance properties are not known. Thus, we have started to investigate the properties of these proton resonances by the direct method. So far, we studied the proton resonant scattering of ^{21}Na+p, ^{22}Mg+p, ^{23}Mg+p, ^{25}Al+p and ^{26}Si+p as well as ^{24}Mg+p for testing the thick target method[8] with the present experimental setup. The present data of ^{24}Mg+p are very well reproduced by the R-matrix calculation[11] with known resonance parameters, confirming the validity of the method. Preliminary results on ^{25}Al+p and ^{26}Si+p are presented elsewhere[10].

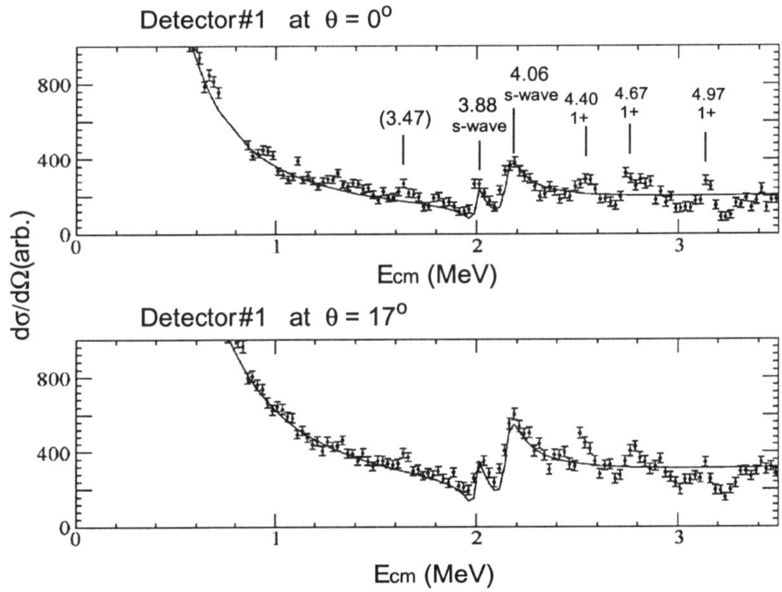

Figure 3. Elastic scattering of ^{23}Mg+p measured at $\Theta_{Lab} = 0°$ and 17°. The solid line is a fit with R-matrix analysis. Possible states are indicated by the excitation energies in ^{24}Al.

Figure 3 displays the proton excitation functions of ^{23}Mg+p. This is the first experiment to investigate ^{24}Al by a proton resonance scattering. No resonance parameters were known before for the states in ^{24}Al[12]. We can see clearly two resonances at 3.88 and 4.06 MeV, and they are fitted well by s-wave resonances and are in agreement with previously known states at 3.885 and 4.059 MeV[12]. We can also see three resonances that probably correspond to the 1^{+} states known by the beta decay study of ^{24}Si[13]. Detailed analysis is in progress.

5. SUMMARY

The present result demonstrates that one can investigate stellar reactions of the rp-process very efficiently at CRIB. We have shown that an in-flight RI beam production method at low energies is very useful, and has a possibility to obtain RI beam intensity of the order of 10^{8} pps for light nuclides near the line of stability. The Wien filter gives a better condition for the property of RI beams as one does not always have to use the degrader that deteriorates the RI beam quality considerably. The present work demonstrates that even small accelerator laboratories can devote to extensive RI beam programs including nuclear astrophysics.

References

1. S. Kubono, et al., Eur. Phys. J. **A 13** (2002) 217.
2. Y. Yanagisawa, S. Kubono, et al., Nucl. Instr. Meth. **A 539** (2005) 74.
3. K.I. Hahn, et al., Phys. Rev. **C54** (1996) 1999.
4. I.S. Park, et al., Phys. Rev. **C59** (1999) 1182.
5. B. Harss, et al., Phys. Rev. Lett. **82** (1999) 3964.
6. J.C. Blackmon, et al., Nucl. Phys. **A688** (2001) 142c.
7. M. Notani, S. Kubono, T. Teranishi, et al., Nucl. Phys. **A738** (2004) 411.
8. S. Kubono, Nucl. Phys. **A 693** (2001) 221, and references therein.
9. S. Kubono, Prog. Theor. Phys. **96** (1996) 275.
10. J.Y. Moon, et al., to be published in Nucl. Phys. A.
11. J.M. Blatt and L.C. Biedenharn, Rev. Mod. Phys. **24** (1952) 258.
12. S. Kubono, T. Kajino, and S. Kato, Nucl. Phys. **A588** (1995) 521.
13. V. Banerjee, et al., Phys. Rev. **C67** (2002) 024307.

APPLICATIONS OF THE TROJAN HORSE METHOD IN NUCLEAR ASTROPHYSICS

S. CHERUBINI, C. SPITALERI, A. MUSUMARRA, S. ROMANO AND A. TUMINO

Dipartimento di Metodologie Fisiche e Chimiche per l'Ingegneria - Università di Catania and Laboratori Nazionali del Sud-INFN, Catania, Italy

R.G. PIZZONE

Laboratori Nazionali del Sud-INFN, Catania, Italy

The Trojan Horse Method allows for the measurements of cross sections in nuclear reactions between charged particles at astrophysical energies. The basic features of the method are discussed and recent applications are presented. Information on the electron screening potential for various reactions is also obtained by comparison with direct measurements.

1. INTRODUCTION

Reactions between nuclei in stellar environments take generally place at energies much lower than the Coulomb barrier,E_{CB}, existing between the colliding particles. The relevant center of mass energy range is called "Gamow window", E_G, and is typically of the order of a few tens of keV (non-explosive burning) or hundreds of keV (explosive stellar burning) at most, while E_{CB}, is in the MeV region.

Nuclear fusion reactions proceed then via tunnel effect with an exponential decrease of the cross section: $\sigma(E) \sim exp(-2\pi\eta)$, where η is the Sommerfeld parameter([1,2]). This implies that the cross sections values are generally very small, i.e. of the order of micro- and even nanobarns, resulting in serious difficulties if one wants to make a measurement in the Gamow window region.

Given these difficulties, measurements were performed at higher energies and then extrapolated to the Gamow ones. In performing this extrapolation, owing to the strong exponential suppression of the cross section mentioned above, the so called astrophysical factor S(E) is introduced:

$$S(E) = E\sigma(E)exp(2\pi\eta) \quad (1)$$

where the inverse of the Gamow factor $exp(2\pi\eta)$ removes the dominant energy dependence of $\sigma(E)$. The extrapolation is then easier thanks to the much smoother behaviour of S(E) with respect to the original excitation function $\sigma(E)$. Nonetheless, this procedure is not completely safe and often errors occurred in the evaluation of S(E) at E_G energies, e.g. owing to the presence of unknown resonances in the energy range of interest.

A number of experimental solutions were suggested in oder to avoid extrapolations and to measure the cross sections of interest directly in the Gamow energy region. The state of the art in this field is represented by the LUNA experiment performed at the underground LNGS INFN laboratory([3]).These enormous experimental efforts eventually led to pointing out a new problem.

Actually, nuclei in a nuclear reaction experiment are generally surrounded by electron clouds. On the other hand, in astrophysical enviroments the colliding nuclei face completely different plasma conditions that are simulated in stellar models. These models need the so called bare nucleus cross-section (i.e. cross section between completely ionized atomic nuclei) as an input. Thanks to the direct measurements performed in the Gamow region, the presence of electrons around nuclei in laboratory experiment turned out to give non negligeable effects at astrophysical energies: the screened cross section $\sigma_s(E)$ is larger than the bare nucleus one, $\sigma_b(E)$ ([3,4,5,6,7,8,9]). This effect is usually described by introducing an enhancement factor $f_{lab}(E)$ ([4])

$$f_{lab}(E) = \sigma_s(E)/\sigma_b(E) \approx exp(\pi\eta U_e/E) \quad (2)$$

U_e is the electron screening potential energy in the laboratory. As mentioned above this is different from the one that could describe the stellar environment. Assuming one knows U_e, then $\sigma_b(E)$ can be obtained and the effective cross section in stellar plasma conditions $\sigma_{pl}(E)$ is given by $\sigma_b(E) \times f_{pl}$, where f_{pl} is the electron screening enhancement factor.

Hence, though measurements were performed at Gamow energies, the values of the bare nucleus cross section are again obtained by extrapolating direct data measured at higher energies, where electron screening effects are negligible. Moreover, electron screening potentials obtained from the analysis of these measurements were higher than those expected on the base of atomic physics models ([8]). This fact deserves a special attention, as if one wants to have reliable models for astrophysical environements, laboratory results must be understood on first instance.

2. INDIRECT METHODS

In order to overcome the above mentioned difficulties (the experimental ones and those connected with electron screening problem) additional information is required. Methods that allow for the indipendent measurement of the electron screening potentials for reactions where direct data in the Gamow window are available could be then particularly usefull.

To this end, a number of indirect methods for measuring cross sections of interst for astrophysics were developed. Among them, the Coulomb dissociation method ([10,11]) is used to study radiative capture reactions, the Asymptotic Normalization Coefficient one ([12,13,14,15,16]) is useful in extracting direct capture cross sections using peripheral transfer reactions while the Trojan Horse Method, THM, ([19 - 35]) allows for the extraction of cross section for processes induced by charged particles.

The THM is a powerful tool where a specific reaction mechanism, namely the quasi-free one, is selected in a three body reaction that takes place at energies above the Coulomb barrier between the colliding nuclei. From this, information on a two body process of astrophysical interest (at very low and even vanishing energies) can be deduced as explained in the following. Moreover the cross section obtained in this way will be unefected by the electron screening problem. The THM has already been used in a number of experiments and the results of some of them are reported here (Table I).

3. QUASI-FREE MECHANISMS AND THE TROJAN HORSE METHOD

The idea of extracting two body cross sections at astrophysical energies from the study of three body reactions where a quasi-free reaction mechanism is selected ([17,18]) came out from previous researches on the quasi-free mechanism at low energies ([36 - 51]) In particular, excitation functions for the two body cross section of the ^{7}Li(p,α)α and ^{6}Li(p,α)^{3}He reactions were obtained indirectly at energies of a few MeV ([36,37]).

If a three body cross section $A+B \rightarrow c+d+S$ proceeds via a quasi-free mechanism, it can be described by a Feynman diagram of the type shown in Fig. 1. A is assumed to have a strong $x \oplus S$ cluster structure. The upper pole in the figure describes the virtual break-up of the nucleus A into clusters x and S; S is then considered to act as a spectator to the $B+x \to c+d$ reaction that takes place in the lower pole. This approximation is called Impulse Approximation (IA)[52].

In Plane Wave Impulse Approximation (PWIA) the cross section of the three body reaction can be factorized as follows:

$$\frac{d^3\sigma}{dE_c d\Omega_c d\Omega_C} \propto KF \left(\frac{d\sigma}{d\Omega}_{cm}\right)^{off} \cdot |\Phi(\vec{p}_s)|^2 \tag{3}$$

This factorization keeps memory of the two poles of Fig.1 ([53,54]), namely:
- $[(d\sigma/d\Omega)_{cm}]^{off}$ is the off-energy-shell differential cross section for the two body B(x,c)C reaction at the center of mass energy $\mathrm{E_{cm}}$ given in post collision prescription by:

$$\mathrm{E_{cm}} = \mathrm{E_{c-d}} - \mathrm{Q_{2b}} \tag{4}$$

where $\mathrm{Q_{2b}}$ is the two body Q-value of the $B+x \to c+d$ reaction and $\mathrm{E_{c-d}}$ is the relative energy between the outgoing particles c and d;
- KF is a kinematical factor containing the final state phase-space factor and it is a function of the masses, momenta and angles of the outgoing particles:

$$KF = \frac{\mu_{AB} m_c}{(2\pi)^5 \hbar^7} \frac{p_d p_c^3}{p_{AB}} \left[\left(\frac{\vec{p}_{Bx}}{\mu_{Bx}} - \frac{\vec{p}_{dc}}{m_c}\right) \cdot \frac{\vec{p}_c}{p_c}\right]^{-1} \tag{5}$$

- $\Phi(\vec{p}_s)$ is the Fourier transform of the radial wave function $\chi(\vec{r})$ for the $x - S$ inter-cluster motion, usually described in terms of Hänkel, Eckart and Hultheń functions depending on the $x - S$ system properties.

This phenomenological approach resembles that of the theory of the THM proposed by Baur ([19]), whose basic idea is to extract a $B+x \to c+d$ two-body reaction cross section, at astrophysical energies from a suitable $A+B \to c+d+S$ three-body reaction.

Under appropriate kinematical conditions, the the "Trojan Horse" A brings cluster x inside the nuclear field of B at energies higher than the Coulomb barrier while the reaction $B+x \to c+d$ takes place at the desired astrophysical energies. This is possible because x is a virtual particle.

Owing to the approximations involved in the THM, one cannot extract the absolute value of the two-body cross section. However, the absolute value can be extracted through normalization to the direct data available at energies above the Coulomb barrier. Moreover, since we select the region of low momentum p_s for the spectator (in general $p_s < 40$ MeV/c), where PWIA and DWIA momentum distributions have very similar shapes ([55]), the PWIA approach was often used for the analysis of the experimental results.

It has to be stressed that the two body cross section obtained in this way is insensitive to both the Coulomb barrier and to the electron screening effects. Hence, if one wants to compare it with the cross section of the $B + x \to c + d$ process obtained from a direct measurement, the THM results have to be multiplied by the appropriate penetration function P_l. This takes into account the penetrability effects present in the direct data ([20,22]). The direct and the THM cross sections formula are then related by:

$$\left(\frac{d\sigma}{d\Omega}\right)_{direct} \propto \left[\frac{d\sigma}{d\Omega}\right]_{THM} \cdot P_l \tag{6}$$

As the THM data are not affected by electron screening effects, the cross section obtained using this method is the bare nucleus one, $\sigma_b(E)$. From this, the $S_b(E)$ factor can be obtained easily and a model-independent estimate of the screening potential U_e can be derived by comparison with the direct measured, screened, $S_{direct}(E)$ factor.

4. DATA ANALYSIS AND RESULTS

Data analysis consists primarily in identifying the phase space region where the QF mechanism is dominant and in selecting coincidence events whose spectator momenta are close to zero. The presence of this mechanism can be tested in various experimental ways ([56,57,58]). It has to be stressed, however, that these are just necessary conditions to be fulfilled before extracting the astrophysical S(E) factor by means of the THM.

Monte Carlo calculations were used to get the $KF \cdot |\Phi(\vec{p}_s)|^2$ product that appears in the PWIA formula. The momentum distribution entering the calculation was that given in [25,17]. The geometrical efficiency of the experimental setup as well as the detection thresholds of the detectors (in most of the experiments PSDs were used) were also taken into account in these calulations.

Table 1. Two-body reactions studied via Trojan Horse Method.

	Direct reaction	Indirect reaction	E_{inc} (MeV)
[1]	^{7}Li + p $\rightarrow \alpha + \alpha$	^{7}Li + d $\rightarrow \alpha + \alpha$+n	19-22
[2]	^{6}Li + d $\rightarrow \alpha + \alpha$	^{6}Li + ^{6}Li $\rightarrow \alpha + \alpha + \alpha$	5.9
[3]	^{6}Li + p $\rightarrow ^3He + \alpha$	^{6}Li + d $\rightarrow ^3He + \alpha$+n	14, 25
[4]	^{11}B + p $\rightarrow^8$Be + α	^{11}B + d $\rightarrow^8$Be + α+n	27
[5]	d + ^{3}He $\rightarrow$ p + α	^{6}Li + ^{3}He $\rightarrow$ p + $\alpha + \alpha$	5-6
[6]	p + ^{9}Be $\rightarrow \alpha$ + ^{6}Li	d + ^{9}Be $\rightarrow \alpha$ + ^{6}Li+ n	22
[7]	d + d $\rightarrow t + p$	^{6}Li + d $\rightarrow t + p + \alpha$	14
[8]	α + ^{12}C $\rightarrow \alpha$ + ^{12}C	^{6}Li +^{12}C $\rightarrow \alpha$ + ^{12}C+ d	20

Table 2. Astrophysical factor S(0) and electron screening potential U_e.

	Reaction	$U_e^{Adiab.}$ (eV)	U_e^{Dir} (eV)	U_e^{THM} (eV)	$S(0)^{Dir}$ (MeVb)	$S(0)^{THM}$ (MeVb)
[1]	^{7}Li + p $\rightarrow \alpha + \alpha$	186	300±160	330± 40	0.059	0.055±0.003
[2]	^{6}Li + d $\rightarrow \alpha + \alpha$	186	330±120	340±50	17.4	16.9±0.5
[3]	^{6}Li + p $\rightarrow ^3He + \alpha$	186	440±150	450±100	2.86	3.0±0.3
[4]	^{11}B + p $\rightarrow^8$Be + α	304	430±80	—	2.1	0.31±0.05
[5]	d + ^{3}He $\rightarrow$ p + α	115	219±7	180±40	6.51	6.08±1.42

Following the PWIA prescription, the two-body cross-section $d\sigma/d\Omega_{cm}$ was derived dividing the selected three-body coincidence yield by the result of the Monte Carlo calculation, inserting the proper penetration coefficient and normalizing the results to the direct cross section.

Table 1 presents a partial list of the reactions studied via the THM and the corresponding three body reactions investigated.

In all of these cases, the behavior of the THM excitation function, after normalization, is similar to the direct one in the whole investigated range except for energies below 100 keV roughly where electron screening effect is no longer negligible in the direct data.

Figures 2 to 4 show the astrophysical S(E) factors extracted via THM for the $^6Li+p \rightarrow \alpha+^3He$ (Fig.2), $^7Li+p \rightarrow \alpha+\alpha$ (Fig.3), $^3He+d \rightarrow \alpha+p$ (Fig.4).

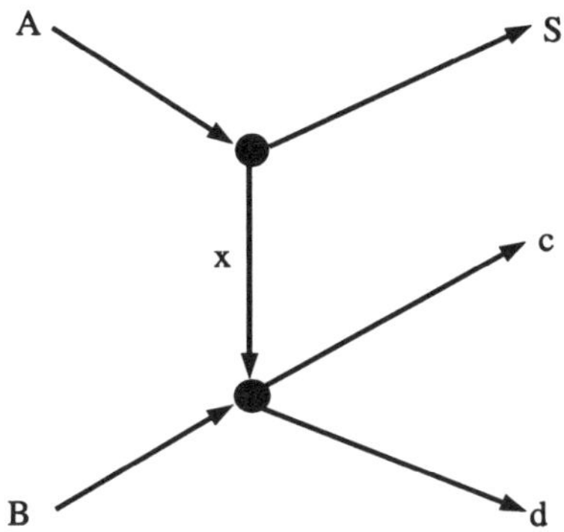

Figure 1. The pole diagram for a quasi-free reaction A(B,cd)S. The upper pole represents the virtual decay of nucleus A: $A \to x + S$. The lower pole shows the virtual reaction $B + x \to C + c$.

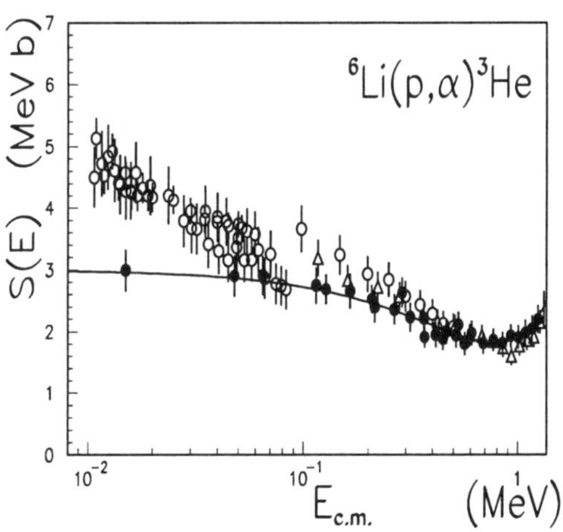

Figure 2. The S(E) factor for the ^{6}Li+ p $\to \alpha +^3 He$ reaction obtained using the THM is compared with those obtained by direct measurements ([29] and ref. therein).

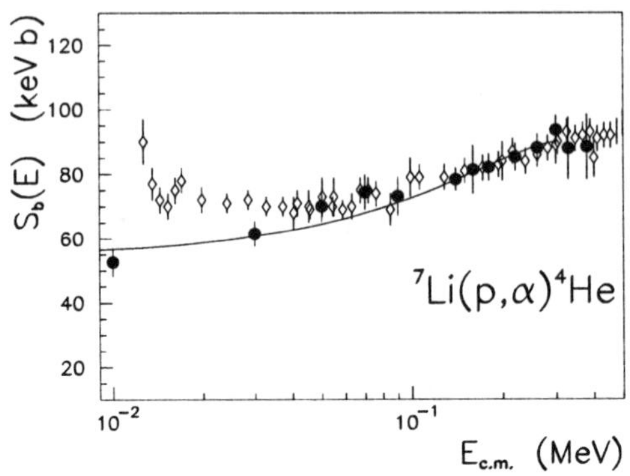

Figure 3. ^{7}**Li + p**$\rightarrow \alpha + \alpha$. The S(E)− extracted with the Trojan Horse method (full dots)([27]) is compared with the direct data from Ref([6])(open dots); a fit to the indirect data with a second-order polynomial is also shown as a solid line.

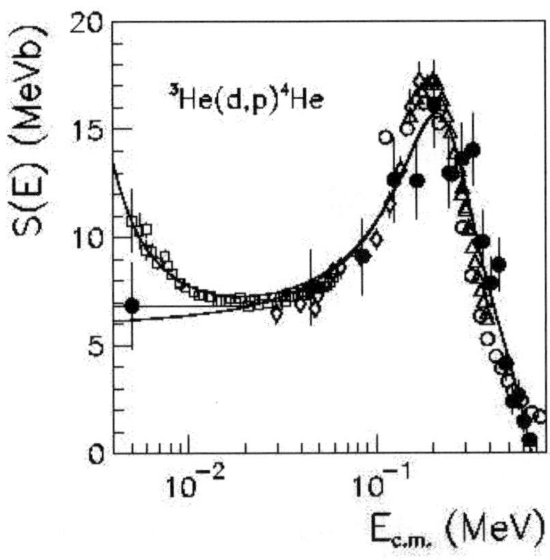

Figure 4. ^{3}**He +d** $\rightarrow \alpha + p$. Astrophysical S(E)-factor (right) for the reaction ^{3}He + d $\rightarrow \alpha + p$ as obtained by THM (full dots) ([31]) and direct measurements (open dots). The dashed line represents a polynomial fit to the indirect S(E).

The values of the electron screening potential U_e obtained from direct and THM measurements are shown in Table 2. It has to be noted that some of the THM results have lower uncertainties than the direct measurements. Moreover, the puzzle of the U_e values higher than those estimated theoretically is still there and it need an explanation.

5. CONCLUSIONS

The present paper reports on the basic features of the THM and a review of recent applications to several reactions that are of interest in astrophysics.

In particular the possibility of extracting the bare nucleus two-body cross section via THM has to be remarked.

However, a lot remains to do in the future to achieve reliable information for many key reactions and processes, including those where radioactive species are involved.

New theoretical developments are also strongly needed for the study of electron screening effects in fusion reactions that could affect significantly also the field of applications (e.g. thermonuclear fusion reactors).

References

1. C.Rolfs, W.S. Rodney, Cauldrons in the cosmos, University of Chicago Press, Chicago (1988)
2. C.Rolfs, Prog. Part. Nucl. Phys., **153** , 23 (2001)
3. F. Streider, C.Rolfs, C.Spitaleri,P.Corvisiero Naturwissenschaften **88**, 461, (2001)
4. H.J Assenbaum, K.Langanke, C.Rolfs, Z.Phys., **A327**, 461 (1987)
5. S.Engstler et al.Phys. Lett.**B202**(1988)179
6. S.Engstler et al., Phys. Lett.**B 279**(1992)20
7. Luna Collaboration, Nucl. Phys., **A706**,203 (2002)
8. R.Bonetti et al., Phys. Rev.Lett., **82**, 5205 (1999)
9. M.Junker et al., Phys. Rev., **C57** ,2700 (1998)
10. G. Baur and H. Rebel, J. Phys. G **20**, 1 (1994) and references therein
11. G. Baur and H. Rebel, Annu. Rev. Nucl. Part. Sci. **46**, 321 (1996)
12. X.D.Tang et al. Phys. Rev. C**67**, art.n.015804 (2003)
13. A.M.Mukhamedzhanov et al. Phys. Rev. C **63**, art.n.024612 (2001)
14. A.Azhari et al. Phys.Rev., **C63** , 055803 (2001)
15. A.M.Mukhamedzhanov, R.E.Tribble Phys. Rev. C**59**, 3418(1999)
16. A.M.Mukhamedzhanov et al. Phys. Rev. C **56**, 1302 (1997)
17. C.Spitaleri et al. Nucl. Phys. A **719**,99 (2003)
18. C. Spitaleri, Problems of Fundamental Modern Physics II, World Scientific, p. 21-35, 1990
19. G. Baur, Phys. Lett. B **178**, 135 (1986)

20. S. Cherubini et al. Ap. J. **457**, 855 (1996).
21. G. Calvi et al. Nucl. Phys. **A621**, 139c (1997)
22. C. Spitaleri et al. Phys. Rev. C **60**, 055802 (1999)
23. C. Spitaleri et al. Eur. Phys. J **A7**, 181 (2000)
24. A. Aliotta et al. Eur.Phys. J., A9, 435(2000)
25. M. Lattuada et al. Ap.J. **562**, 1076 (2001)
26. C. Spitaleri et al. Phys. Rev. C **63**, 005801 (2001)
27. A. Musumarra et al. Phys. Rev., **C64** , 068801 (2001)
28. A. Tumino et al. Nucl.Phys.A (2003)
29. A. Tumino et al. Phys. Rev. **C 67), 065803 (2003)**.
30. C.Spitaleri et al Phys. Rev. C **69**, 055806 (2004)
31. M.La Cognata et al., proc. to Cluster 2003 conference, published on Nucl. Phys. A (2004)
32. R.G. Pizzone et al, proc. to Cluster 2003 conference, published on Nucl. Phys. A (2004)
33. A. Rinollo et al. Report INFN-LNS, 2002
34. S. Typel and. H. Wolter, Few Body Syst.**29**, 7 (2000)
35. S. Typel and G. Baur, Ann. Phys. **305**, 228 (2003)
36. M. Zadro et al. Phys. Rev. C **40**, 181 (1989)
37. G. Calvi et al. Phys. Rev. C **49**, 1848 (1990)
38. M.Furic et al. Phys. Lett. **39 B**, 629 (1972)
39. Dj. Miljanić et al. Nucl. Phys. **A 215**, 221 (1973)
40. Dj. Miljanić et al. Phys. Lett. **50 B**, 330 (1974)
41. J.Kasagi et al. Nucl. Phys. **A 239**, 233 (1975)
42. I.Slaus et al. Nucl. Phys. **A 286**, 67 (1977)
43. N.Arena et al. Nuovo Cimento **45**, 405 (1978)
44. M.Lattuada et al. Nuovo Cimento **62**, 165 (1981)
45. M.Lattuada et al. Nuovo Cimento **69**, 1 (1982)
46. M.Zadro et al. Z.Phys.A-Atomic Nuclei **325**, 119 (1986)
47. M.Lattuada et al. Nucl. Phys. **A 458**, 493 (1986)
48. M.Zadro et al. Nucl. Phys. **A 474**, 373 (1987)
49. M.Lattuada et al. Z.Phys.A-Atomic Nuclei **330**,183(1986)
50. S.Blagus et al. Z.Phys.A-Atomic Nuclei **337**, 297(1990)
51. V.Soic, et al. Eur. Phys. J. **A 3**, 303 (1998)
52. G.F. Chew, Phys.Rev. **80**, 196(1950)
53. U.G.Neudatchin, Y.F.Smirnov, At. Energy Rev., **3** , 157 (1965)
54. G.Jacob et al., Rev. Mod. Phys. **38** , 121 (1966)
55. N.S.Chant , P. G. Roos, Phys. Rev. C **15**, 57 (1977)and referencences therein
56. A. Guichard et al., Phys. Rev., C4, 493 (1986)
57. M.Jain et al. Nucl.Phys. A, **153**,49,(1970)
58. M. Chevalier et al., Nucl. Phys., **A216**, 519 (1973)

DIRECT MEASUREMENT OF THE ASTROPHYSICAL $^{8}Li(\alpha, n)^{11}B$ REACTION

H. ISHIYAMA, T. HASHIMOTO, Y.X. WATANABE, H. MIYATAKE, M-H. TANAKA, N. YOSHIKAWA, S. C. JEONG, Y. FUCHI, I. KATAYAMA AND T. NOMURA

Institute of Particle and Nuclear Studies,
High Energy Accelerator Research Organization(KEK),
Tsukuba, Ibaraki, 305-0801 Japan

T. ISHIKAWA AND K. NAKAI

Faculty of Science and Technology, Tokyo University of Science
Noda, Chiba, 278-8510 Japan

S.K. DAS, Y. MIZOI AND T. FUKUDA

Osaka Electro-Communication University,
Neyagawa, Osaka, 572-8530 Japan

S. MITSUOKA, P.K. SAHA, K. NISHIO, M. MATSUDA, H. IKEZOE AND S. ICHIKAWA

Japan Atomic Energy Research Institute(JAERI)
Tokai, Ibaraki, 319-1195 Japan

T. FURUKAWA, H. IZUMI AND T. SHIMODA

Department of Physics, Osaka University
Toyonaka, Osaka, 560-0043 Japan

M. TERASAWA

Tsuchiura Nihon University Junior High School
Tsuchiura, Ibaraki, 300-0826 Japan

T. SASAQUI

National Astronomical Observatory
Mitaka, Tokyo, 181-8588 Japan

The excitation function of the ^{8}Li$(\alpha, n)^{11}$B reaction was measured in the energy region of E_{cm} = 0.7 MeV − 2.6 MeV by using a highly efficient detector system and a highly pure low-energy ^{8}Li−beam. The obtained reaction cross section is roughly 1.6 times smaller than the one by a previous inclusive measurement. The excitation function of the ^{8}Li$(\alpha, n)^{11}$B to the ^{11}B ground state and the excitation energy spectrum in the residual nucleus ^{11}B are also shown.

1. Introduction

The (α, n) reactions of light neutron-rich radioactive nuclei play important roles at the precedent stage of the r-process in the neutron rich environment, for example, at a "hot bubble" formed in a supernova explosion [1]. A systematic study for these astrophysical reaction rates with using low-energy (E = 1 − 2 MeV/u) radioactive nuclear beams (RNBs) has been started at the tandem facility in Japan Atomic Energy Research Institute (JAERI) [2-5]. Especially, the ^{8}Li$(\alpha, n)^{11}$B reaction is one of key reactions for heavy element synthesis in going across the stability gap at A = 8.

Some measurements concerning the cross section of the ^{8}Li$(\alpha, n)^{11}$B reaction have been so far performed[6-9]. Paradellis et al. measured for the first time the excitation function of the ^{8}Li$(\alpha, n)^{11}$B reaction using the inverse reaction, ^{11}B$(n, \alpha)^8$Li. However, this technique measures only the cross section of ^{8}Li$(\alpha, n)^{11}$B to the ^{11}B ground state, while the reaction can proceed also to excited bound states of ^{11}B. The direct measurements of the reaction cross section using ^{8}Li-beam have been performed in two different methods. One is the inclusive measurement [7, 8] without neutron detection, and the other is exclusive one [9]. There is a significant difference, however, in the magnitude of the reported cross sections; the cross sections measured in ref. [9] are, on the average, roughly a half of those reported in refs. [7, 8]. An exclusive measurement is generally regarded as more reliable because it is possible to determine the contribution to the reaction cross section at each final excited state. But we should note the poor statistics reported in ref. [9].

Therefore, we have performed again the exclusive measurement with higher statistics and also have aimed at coming closer to the energy region around 0.5 MeV, which corresponds to the Gamow window at T_9 = 1. For this purpose, we prepared a high-purity and low-energy ^{8}Li radioactive nuclear beam and constructed a new detector system with high efficiency. In this report, the excitation function of the ^{8}Li$(\alpha, n)^{11}$B reaction in the energy region from 0.7 MeV to 2.6 MeV in the center-of-mass system is presented together with some information of the contribution to the reaction cross section at each final excited state of ^{11}B.

2. Experiment

2.1. ^{8}*Li-beam production*

Low-energy neutron-rich beams have been produced by using transfer reactions on light targets. In this case, it is important to avoid impurities originating from primary beam particles. Therefore, we have utilized a recoil mass separator (RMS) [10, 11] existing at the tandem facility in JAERI, which consists of two electric dipoles and a magnetic dipole as shown Fig.1. The RNB can be separated from the primary beam using the difference of the magnetic rigidity and electric rigidity.

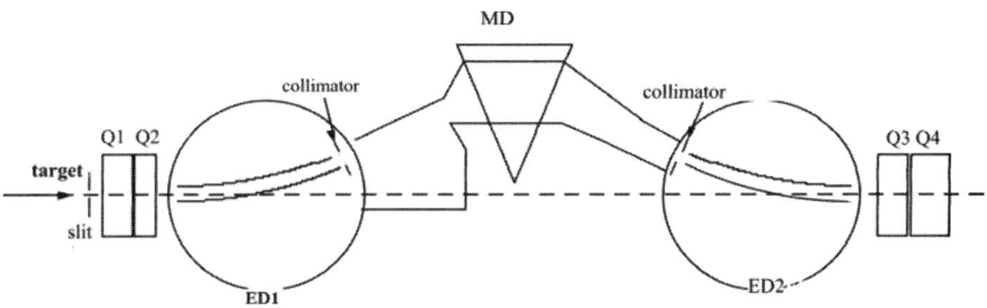

Figure 1. An ion optical configuration of the JAERI-RMS. The Q, ED and MD stand for magnetic quadrupole, electric dipole, and magnetic dipole, respectively.

The ^{8}Li-beam has been produced via ^{9}Be(^{7}Li, ^{8}Li) reaction. A 42μm ^{9}Be foil was set at the target position. In order to make the energy resolution of the resultant ^{8}Li-beam better, the target foil was tilted at 40° with respect to the beam axis. The maximum intensity of the primary ^{7}Li^{3+}-beam was about 200 enA, its initial energy being 24 MeV. The intensity of ^{8}Li-beam is 4.8×10^{3} pps / 10 enA ^{7}Li^{3+} beam at the focal plane of the RMS. Its energy and energy resolution were 14.6 MeV and 5%, respectively. Although a little amount of ^{6}He particles are mixed in the secondary ^{8}Li-beam obtained, no ^{7}Li impurities have been observed. The purity of the ^{8}Li-beam is 99%.

2.2. *Detector system*

The present detector system is schematically shown in Fig. 2. It consists of a time-of-flight (TOF) counter, a "multi-sampling and tracking proportional chamber" (MSTPC) [12] placed at the focal plane of the RMS, and a neutron detector array, in which the first one is composed of a multichannel plate (MCP) and a parallel plate avalanche counter (PPAC). The absolute energy of the ^{8}Li-beam is determined by TOF between the MCP and the PPAC. Then, the ^{8}Li-beam is injected into the MSTPC filled with the gas of He + CO_2 (10%) which works as counter gas and gas target.

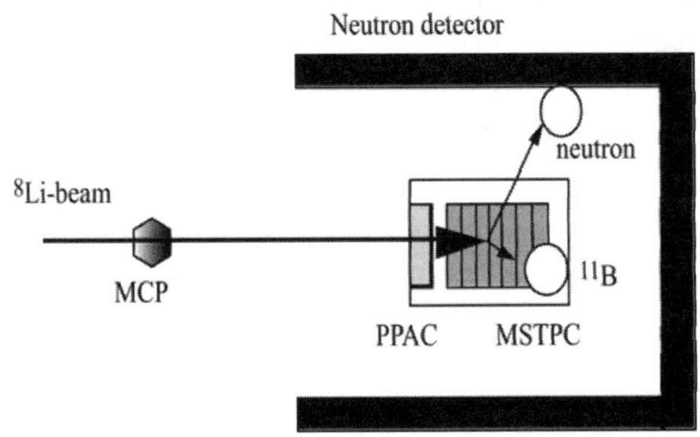

Figure 2. A schematic view of the experimental setup. The ^{8}Li-beam provided from the RMS is directly injected into the MSTPC filled with He + CO_2 gas, which acts as counter gas and gas target. A neutron detector array consisting of 28 plastic scintillators is placed around the MSTPC.

Fig. 3 shows the cross-sectional view of the MSTPC composed of a drift space and a proportional region. The latter region consists of 24 cathode pad cells with anode wires. The MSTPC has high detection efficiency almost 100%, and can measure three-dimensional tracks of charged particles and energy losses along their trajectories. When a nuclear reaction takes place inside the MSTPC, the energy loss (dE/dx) changes largely due to the change of the relevant atomic numbers. Therefore, the reaction point can be determined by detecting dE/dx change, and the reaction energy at this point can be evaluated by summing up the energy losses at upstream pad cells. The excitation function in a wide energy region can be measured in this set up. The gas pressure was set at the 29.3 kPa so as to measure the energy region from E_{cm} = 0.7 MeV to 2.6 MeV.

A neutron detector array to detect neutrons emitted from nuclear reactions is placed to surround the MSTPC as much as possible. It consists of 28 pieces of BC408 plastic scintillators, covering 31.4% solid angle of 4π. The absolute energy of a neutron is obtained by TOF measurement

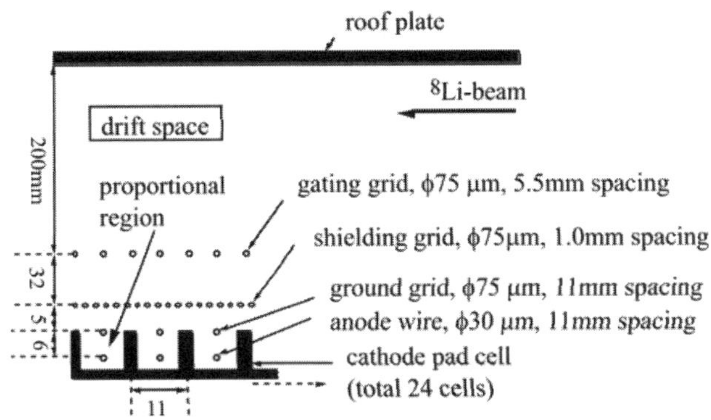

Figure 3. A schematic cross-sectional view of the MSTPC (not scaled).

between the PPAC and a plastic scintillator. The vertical position of the neutron is determined by the position of a plastic scintillator itself, and the horizontal position is evaluated by the time difference of both side signals of the scintillator. The absolute efficiency was calibrated with neutrons from a ^{252}Cf fission source. It is about 40% at 5.0 MeV.

An overall efficiency of the present system turned out to be about 10%, so that we can measure the cross section down to roughly 10 mb in a few days under 10^4 pps beam rate.

3. Analysis

The ^{8}Li beam-injection rate into the MSTPC was set at 5×10^3 pps. This rate was restricted by the upper limit of the present data acquisition system. The trigger for data acquisition was generated by the coincident signal between the PPAC and one of plastic scintillators. The maximum trigger rate was 20 cps, which was mainly due to accidental coincidences between the injected ^{8}Li-beam and background signals of plastic scintillators. In order to select only true events, the obtained data are processed through following 4 steps.

The first step is the particle identification of the beam. ^{6}He particles mixed in the secondary beam are rejected using a TOF- ΔE plot, in which the latter is measured by the first cathode pad.

In the second step, the multiple track events are rejected, because the true event gives a single track event, whereas the multiple tracks are generated mainly by coincident events between background neutrons or γ-rays and elastically scattered ^{8}Li particles.

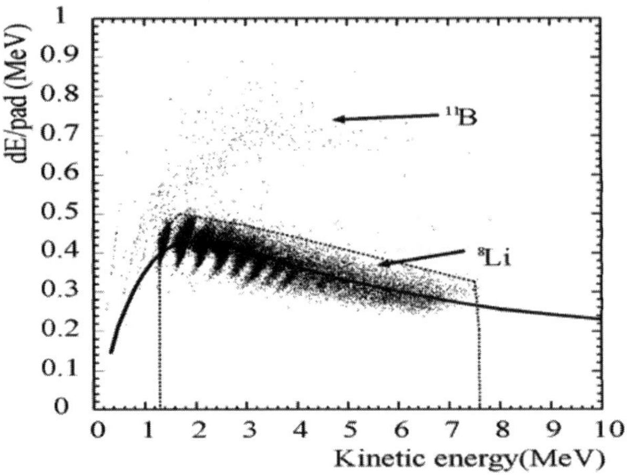

Figure 4. An E-ΔE plot of emitted particles from reaction points. The solid line is the measured energy loss of the ^{8}Li particle. Accidental events in the region surrounded with dashed lines are rejected.

The third step is to detect any sudden change of the energy loss between a certain pad and the next pad, for the determination of the reaction point. The threshold of the energy loss difference between two pads is set at 30 keV by considering the energy loss of low-energy ^{11}B emitted from the reaction. After this step, it is possible to obtain the E-ΔE plot of emitted particles from reaction points as shown in Fig. 4. Accidental events of ^{8}Li-beam particles still exist, because the threshold of the energy loss difference is comparable to the energy straggling of the energy loss signal. But it is easily to select only true events by rejecting ^{8}Li-beam particles in this E-ΔE plot, based on the calibrated energy loss curve of ^{8}Li, as shown in Fig. 4.

The selected events are finally checked if they fulfill the kinematical conditions for true events; the accidental and background reaction events such as ^{12}C(^{8}Li, n) do not satisfy the condition.

4. Results

The obtained excitation function is shown in Fig. 5 by solid circles together with early works, which are the cross sections obtained by the inclusive measurements [7, 8] (triangles) and by the previous exclusive measurement

[9] (squares). The present data indicates ten times better statistics than the previous exclusive measurement and we have achieved the excusive measurement at the energy region down to 0.7 MeV, which is just above the Gamow window ($T_9 = 1$) as shown in Fig. 5.

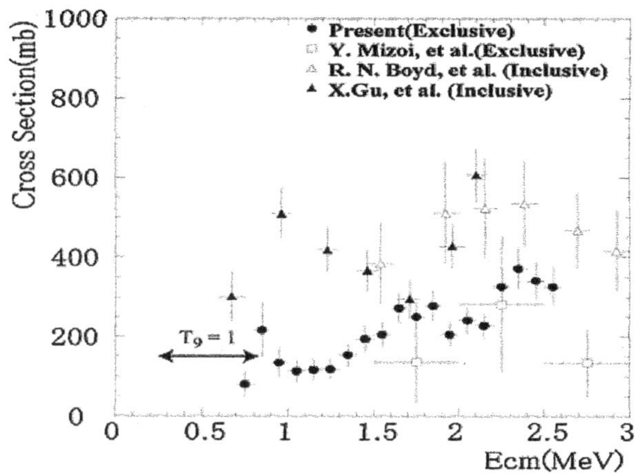

Figure 5. The obtained excitation function of $^{8}\text{Li}(\alpha, \text{n})^{11}\text{B}$ reaction (black circles) together with previous measurements.

The present data are consistent with ones of the previous exclusive measurement within their error bars. In the energy region below 1.5 MeV, the present result is roughly 1.6 times smaller than one of the previous inclusive measurement. This result strongly suggests that the previous data included background events, for example, elastic scattering events. The exclusive measurement has an advantage to distinguish the reaction events from background events with taking account of the neutron information simultaneously obtained.

The excitation energy spectrum of ^{11}B has been obtained simultaneously in the present experiment as shown in Fig. 6. Arrows indicate excited states of ^{11}B with their energies in unit of MeV. The averaged neutron branching ratios in the energy region from 0.7 MeV to 2.6 MeV are tabulated as shown in Table 1 and are almost consistent with ones of the previous exclusive measurement within error bars, although the energy region in the previous case is different as from 1.7 MeV to 6.7 MeV.

The excitation function to each excited state of ^{11}B has been also deduced. Fig. 7 shows the excitation function of $^{8}\text{Li}(\alpha, \text{n})^{11}\text{B}$ to the ^{11}B

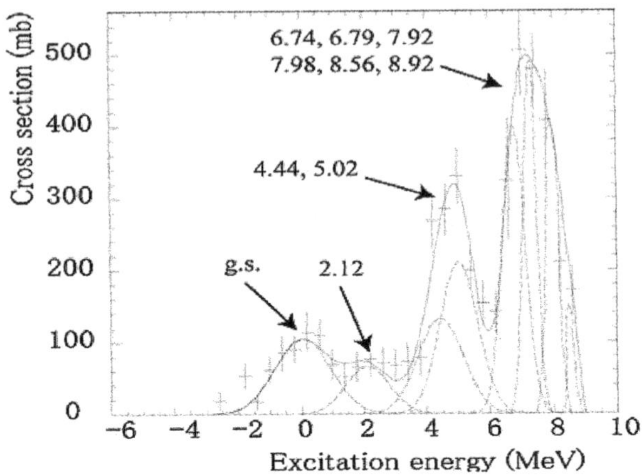

Figure 6. The excitation energy spectrum in ^{11}B. The solid line is the result of a fit and the dashed lines show individual components for levels of ^{11}B.

Table 1. Averaged branching ratios of neutron transition to excited states in ^{11}B. The first column shows excitation energy and the second shows the spin and parity of each excited state. Each branching ratio together with its error is shown in the third column as well as the previous result [9] in the forth column.

E_x in ^{11}B [MeV]	$2J^\pi$	present	previous [9]
0.0	3^-	20.1 (21)	24.3 (126)
2.12	1^-	6.6 (12)	15.2 (104)
4.44	5^-	10.1 (22)	4.2 (72)
5.02	3^-	17.8 (27)	6.1 (57)
6.74 + 6.79	$7^-, 2^+$	18.9 (27)	13.0 (80)
7.29	5^+	13.2 (15)	12.5 (104)
7.98	3^+	11.7 (12)	10.6 (71)
8.56	3^-	2.4 (05)	0.0 (21)
8.92	5^-	–	14.2 (67)

ground state, indicated by solid circles. White circles indicate the data of the inverse reaction [6]. The present data are consistent with the previous ones in the energy region below 1.0 MeV. But in the energy region above 1.0 MeV, they do not agree with the previous ones. Especially, in our data the resonant structure around E_{cm} = 1.3 MeV does not exist. The previous data consist of only the transition to the ^{8}Li ground state in the energy region below 0.94 MeV, but in the above energy region, they may include the other transition to the first excited state ($E_x = 0.94$ MeV, 1^+) in ^{8}Li. The J^π of the corresponding ^{12}B excited state ($E_x = 11.3$ MeV) at $E_{cm} = 1.3$ MeV is not identified. But if the J^π of the 11.3 MeV state is, for example, 0^- or 1^+, the transition probability to the ^{8}Li ground state (2^+) would be suppressed.

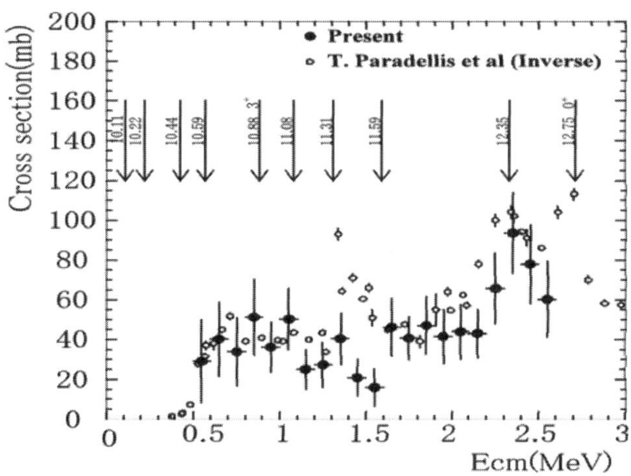

Figure 7. The excitation function of ^{8}Li$(\alpha, n)^{11}$B to the ^{11}B ground state together with the previous data. Arrows indicate energy positions of the possible excited states in the compound nucleus ^{12}B, with their known excited energies, spines and parities [13].

5. Summary

We have succeeded in producing the pure low-energy ^{8}Li-beam and in construction of the new detector system with high detection efficiency, which enable the measurement of the ^{8}Li$(\alpha, n)^{11}$B reaction cross section down to the level of 10 mb.

The excitation function of the ^{8}Li$(\alpha, n)^{11}$B reaction has been measured in the energy region from 0.7 MeV to 2.6 MeV, corresponding to the Gamow window at $T_9 = 2-3$. The present result is consistent with one of the previous exclusive measurement and has at least ten times better statistics than the previous one. In the energy region below 1.5 MeV, the present result is roughly 1.6 times smaller than one of the previous inclusive measurement. The excitation function of the ^{8}Li$(\alpha, n)^{11}$B to the ^{11}B ground state and the excitation energy spectrum in the residual nucleus ^{11}B are also deduced. The excitation function to the ^{11}B ground state is consistent with one of the previous measurement in the energy region below $E_{cm} = 1.0$ MeV.

A subsequent experiment in the energy region below $E_{cm} = 0.7$ MeV has been already performed and its analysis is in progress.

References

1. M. Terasawa, *et al., Astrophys. J.* **562**, 470 (2001).
2. H. Ishiyama, *et al.,, Ameri. Insti. Phys.* **704**, 453 (2003).
3. T. Ishikawa, *et al., Nucl. Phys.* **A718**, 484 (2003).
4. H. Miyatake, *et al., Nucl. Phys.* **738**, 401 (2004).
5. T. Hashimoto, *et al.,, Nucl. Phys.* **A746**, 330 (2004).
6. T. Paradellis, *et al., Z. Phys.* **A337**, 211 (1990).
7. R. N. Boyd, *et al., Phys. Rev. Lett.* **68**, 1283 (1992).
8. X. Gu, *et al., Phys. Lett.* **B343**, 31 (1995).
9. Y. Mizoi, *et al., Phys. Rev.* **C62**, 065801 (2000).
10. H. Ikezoe, *et al., Nucl. Instrum. Meth. Phys. Res.* **A376**, 470 (1996).
11. T. Kuzumaki, *et al., Nucl. Instrum. Meth. Phys. Res.* **A437**, 107 (1999).
12. Y. Mizoi, *et al., Nucl. Instrum. Meth. Phys. Res.* **A431**, 112 (1999).
13. F. Ajezenberg-Selove, *et al., Nucl. Phys.* **A506**, 1 (1990).

PERSPECTIVES IN NUCLEAR ASTROPHYSICS

P.PRATI, ON BEHALF OF LUNA COLLABORATION
Istituto Nazionale di Fisica Nucleare and Dipartimento di Fisica dell'Universita' di Genova,
Via Dodecaneso 33, 16146 Genova, Italy

It is known that the chemical elements and their isotopes were created by nuclear fusion reactions in the hot interiors of remote and long-vanished stars over many billions of years [1]. The present picture is that all elements from carbon to uranium have been produced entirely within stars during their fiery lifetimes and explosive deaths. The detailed understanding of the origin of the chemical elements and their isotopes combines astrophysics and nuclear physics, and forms what is called nuclear astrophysics. In turn, nuclear reactions are at the heart of nuclear astrophysics: they influence sensitively the nucleosynthesis of the elements in the earliest stages of the universe and in all the objects formed thereafter, and control the associated energy generation, neutrino luminosity, and evolution of stars. A good knowledge of the rates of these fusion reactions is essential to understanding this broad picture. Some of the most important experimental techniques to measure the corresponding cross sections, based both on direct and indirect methods, will be described in this paper.

1. INTRODUCTION

Stars generate the energy which allows them to shine from millions to billions of years, by nuclear reactions occurring in their interior. Simultaneously, the network of nuclear reactions operating in the hot, dense stellar medium is believed to be the source of all the chemical elements existing in the universe.

Stars spend most of their lifetime in the quiescent burning stage: here energy is produced through the nuclear reactions of the pp chain and CNO cycle, where four protons are transformed into a Helium nucleus according to the many-step reaction: $4p \rightarrow {}^4He + 2e^+ + 2\nu$ with an energy release of 26.7 MeV. When hydrogen fuel has been used, star moves towards the Red Giant region, where Helium burning becomes effective through the two step "triple alpha process": ${}^4He + {}^4He \rightarrow {}^8Be^*$; ${}^8Be + {}^4He \rightarrow {}^{12}C + \gamma$, followed by radiative α captures: ${}^{12}C + \alpha \rightarrow {}^{16}O + \gamma$ and ${}^{16}O + \alpha \rightarrow {}^{20}Ne + \gamma$. From now on, the fate and the evolution of the stars strongly depend on their mass at birth. Stars with masses less than $\approx 8M_\odot$ do not reach temperature and densities in the center sufficient to ignite further burning stages involving heavier elements. Mainly because of their enormously shorter lifetimes, more massive stars become the most efficient breeders of the heaviest elements. After helium burning, these stars go through periods of carbon, oxygen, neon and silicon burning in their central core, forming higher Z elements, till eventually iron is reached. At this stage, no further energy can be produced via fusion reactions: nuclear core burning stages ceases, resulting in a collapse of the stellar core and the final explosion of the star as a Type II supernova. This paper is focussed on the experimental study of the quiescent burning stage reactions, where energies, and consequently reaction rates, are extremely low. Understanding energy production in stars and nucleosynthesis processes requires knowledge of a large number of charged particle reaction cross sections at interaction energies usually far below the Coulomb barrier. For this reason accurate measurement of the relevant reaction cross sections (ranging from nb down to fb) are severely hampered. Therefore only in very few favourable cases they have been measured within, or close to, the relevant Gamow energy window. Alternatively the energy dependence of $\sigma(E)$ at stellar energies has been extrapolated from higher energies using the definition of the $S(E)$-factor: $\sigma(E) = S(E)\frac{1}{E}exp(-2\pi\eta)$, where $\sigma(E)$ and $S(E)$ are the cross section and the astrophysical S-factor respectively, and η is the Sommerfeld parameter. However, such extrapolation, even if guided by theory, can lead to a considerable uncertainty. Many techniques were proposed to overcome this severe experimental problem. Two complementary approaches are possible, based on both direct and indirect methods "to know" the S-factor value at the Gamow energy. Some of them will be discussed in this paper and their most relevant results will be shown. Concerning direct methods, improvements aimed to increase the detection statistics or to reduce the background were among the best proposed solutions: in this sense it is worth noticing the wide experimental work done by LUNA Collaboration in Gran Sasso underground laboratory. Another direct technique is very promising: the use of a sophisticated recoil separator to allow the detection of the compound nucleus in coincidence with the other reaction products (typically radiative capture photons). ERNA recent results on the measurement of the cross section of the fundamental reaction: $^{12}C(\alpha,\gamma)^{16}O$ will be presented. Various indirect approaches, as the Asymptotic Normaliazation Coefficients (ANC), the Coulomb dissociation and the Trojan Horse Method (THM)[2] have been applied over the years to reactions of astrophysical relevance whenever direct approaches alone could not provide definitive information. In this paper the THM method will be briefly reported as a clever complementary tool for studying reactions of astrophysical interest.

2. TROJAN HORSE METHOD

The basic idea of the THM relies on the assumption that a three-body reaction $A(B, cd)S$ can proceed via a quasi free reaction mechanism that, under particular kinematical conditions, can be dominant. In these conditions, the reaction $A(B, cd)S$ can be described by the diagram reported in Fig. 1. The target nucleus A is assumed to break-up into clusters x and S; S is then considered to be a spectator of the $B + x \rightarrow d + c$ two-body reaction, where c and d are the outgoing particles. In this picture,

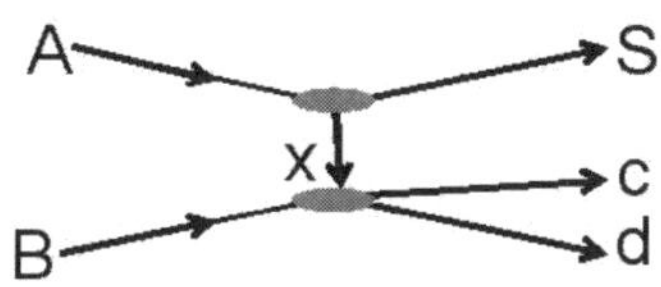

Figure 1. Diagram for the polar approximation of the THM method

the cross section of the three-body reaction can be factorized into the two terms corresponding to the two vertices of the diagram of Fig. 1. In order to apply the THM method a suitable three-body reaction and proper kinematical conditions (angular region and energy in c.m. frame) should be found. In order to describe the Quasi Free process, a simple model based on the PWIA can be used, where the triple-differential cross section is given by: $\frac{d^3\sigma}{dEd\Omega_1 d\Omega_2} \propto KF(\frac{d\sigma}{d\Omega})|\Phi(p_s)|^2$ Here, KF is a kinematical factor, $\Phi(p_s)$ is the momentum distribution of the cluster x inside the nucleus A, $(\frac{d\sigma}{d\Omega})$ is the two-body off-shell differential cross section of the virtual $B(x, c)d$ reaction. In this approach, if $\Phi(p_s)$ is known and KF is calculated, it is possible to derive $(\frac{d\sigma}{d\Omega})$ from a measurement of $\frac{d^3\sigma}{dEd\Omega_1 d\Omega_2}$. This method of course cannot provide absolute values of the differential two-body cross section of astrophysical interest. But, properly normalizing data to previously existing absolute measurements taken at higher energies via direct methods, it is possible to obtain information on the cross section at typical Gamow energies. Moreover, the obtained data represent the "bare" nuclei cross section and are not affected by the electron screening enhancement. This technique, which seems very promising represents therefore a powerful tool to obtain information both of astrophysical interest (S-factor of fusion reactions at very low energies) and of atomic effects (electron screening potential). The most relevant results obtained using the THM technique are reported in the following. The S-factor of the reaction $^6Li(d,\alpha)^4He$[3,4] obtained through $^6Li(^6Li,\alpha\alpha)^4He$ (where $^6Li = d \oplus \alpha$ and α is the spectator) is shown in Fig. 2a. S-factor values are normalized to previous data[5]. It is clearly seen the electron screening effect: the obtained value

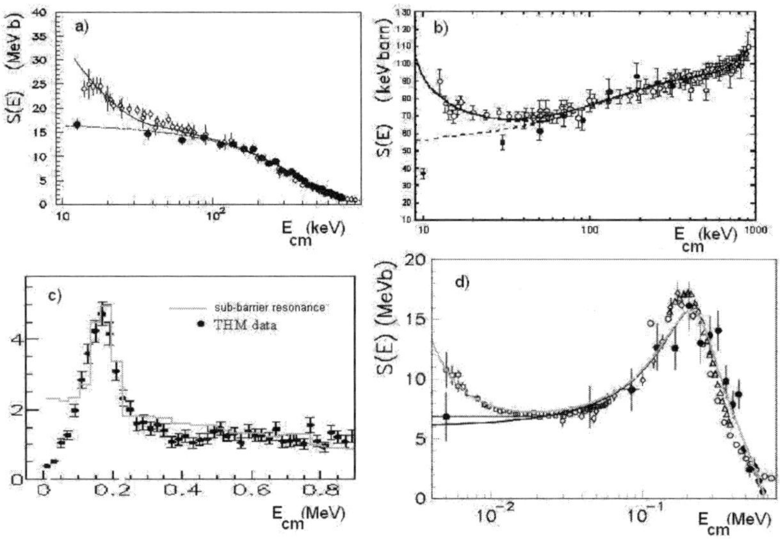

Figure 2. S-factor of the reactions $^6Li(d,\alpha)^4He$ (4a), $^7Li(p,\alpha)^4He$ (4b), $^{11}B(p,\alpha_0\alpha)^8Be$ (4c) and $^3He(d,p)^4He$ (4d) obtained through the THM method

for the screening potential is $U_e = 340 \pm 51 eV$, a factor two larger than the theoretical adiabatic limit ($U_e = 186eV$). Results obtained for the reaction $^7Li(p,\alpha)^4He$[6] are shown in Fig. 2b: the three-body reaction used in this case is $^7Li(d,\alpha\alpha)n$, where deuteron is considered a cluster of proton (interacting particle) and neutron (spectator). Also for this reaction the obtained potential screening $U_e = 350eV$ is twice the adiabatic limit ($186eV$). Important consistency tests of the THM method have been recently done studying reactions which present a resonant behaviour rather than a smooth trend. Fig. 2c and 2d show respectively the encouraging results obtained on the reactions $^{11}B(p,\alpha_0\alpha)^8Be$ and $^3He(d,p)^4He$ where the THM method is able to reproduce the resonant behaviour of the reactions.

3. RECOIL MASS SPECTROMETER

This technique can be fruitfully applied to many radiative (p,γ) and (α,γ) reactions of astrophysical interest. The basic idea is to detect, together with the de-excitation photon, the compound (fused) nucleus as well. There are many advantages of such a coincidence measurement: the use of a gas, and therefore very pure and stable target; high detection efficiency; background free coincidence γ-ray spectra; a measurement of the total cross section. Presently the ERNA[7,8] recoil mass spectrometer is mounted at the Bochum 4 MV dynamitron tandem accelerator, where the measurement of the fundamental reaction $^{12}C(\alpha,\gamma)^{16}O$ is carried on. This reaction ($Q =$ 7.16 MeV) takes place during helium burning in Red Giants. The cross section at the relevant Gamow energy, $E_0 = 0.3$ MeV, determines not only the nucleosynthesis of elements up to the iron region, but also the subsequent evolution of massive stars, the dynamics of supernovae and the kind of remnants after supernova explosion. For this reason, the cross section $\sigma(E_0)$ should be known with a precision at least 10%. In spite of experimental efforts over nearly 30 years, we are still far from this goal. ERNA recoil separator has been designed to measure this reaction down to $E_{cm} = 0.8$ MeV with a precision better than 10%. The huge background reduction in in γ-ray spectra is clearly evident in Fig. 3a.

Presently the lowest achieved energy is $E_{cm} = 2.1$ MeV, as seen in Fig. 3b: clearly visible are the contribution of the narrow $E2$ and the broad $E1$ resonances to the S-factor.

It is also evident the difficulty to extrapolate the cross section down to the Gamow energy, due to the presence, below threshold, of two other resonances, respectively $E1$ and $E2$, and their interference (see insert in Fig. 3b).

4. UNDERGROUND EXPERIMENTS

We have seen that low-energy studies of thermonuclear reactions in a laboratory at the earth surface are hampered predominantly by background effects of cosmic rays in the detectors, leading typically to more than 10 background-events per hour in common detectors. Conventional passive or active shielding around the detectors can only partially reduce the problem of cosmic ray background. The best solution was to install an accelerator facility in a laboratory deep underground. As a pilot project in 1993, a 50 kV accelerator facility has been installed in the underground laboratory Laboratori Nazionali del Gran Sasso (LNGS), where the flux of cosmic-ray

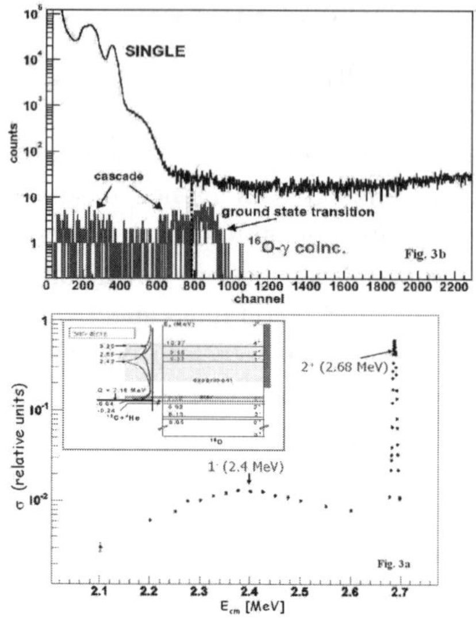

Figure 3. 3a - Background reduction in γ-ray spectra obtained in coincidence with the recoil; 3b Preliminary results of the measurement of the S-factor (in a.u.) of the reaction $^{12}C(\alpha,\gamma)^{16}O$ obtained with ERNA recoil separator

muons is reduced by a factor of 10^6 compared with the flux at the surface. This unique project, called LUNA (Laboratory for Underground Nuclear Astrophysics), was designed[9] primarily for a renewed study of the reaction $^3He(^3He,2p)^4He$, at low energies, aiming to reach the solar Gamow peak at $E_0 \pm \Delta/2 = 21 \pm 5$ keV. This goal has been reached[10,11] with a detected reaction yield of about 1 event per month at the lowest energy, $E_{cm} = 16$ keV, with $\sigma = 20fb$ or $2 \cdot 10^{-38} cm^2$. Thus, the cross section of an important reaction of the hydrogen-burning proton-proton-chain has been directly measured for the first time at solar thermonuclear energies (Fig. 4a); in principle, extrapolation is no longer needed in this reaction. A second reaction of the pp-chain has been fully measured below the solar Gamow peak: $d(p,\gamma)^3He$[12]. Since the rate of this reaction is high, with respect to that of the deuterium producer $p(p,e^+\nu)d$, it only affects the equilibrium abundance of deuterium in an H-burning low mass star. However, the knowledge of its cross section at the Gamow energy is very important because well before the onset of the H-burning (during the so called pre-main sequence phase), an important d-burning takes place in proto-stars. Fig. 4b shows the successful results obtained for the $d(p,\gamma)^3He$ reaction. The work done put in evidence in a clear way the research potential of LUNA and demonstrated that all of the experimental requirements in such low-rate, time-consuming experiments can be fulfilled. In 2001 a

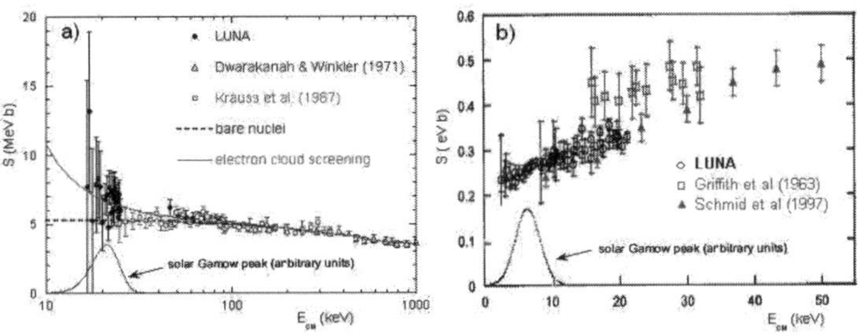

Figure 4. S-factor of the reaction $^3He(^3He,2p)^4He$ (7a) and $d(p,\gamma)^3He$ (7b). In both cases the Gamow energy window has been fully explored

400 kV high-current accelerator[13] has been installed at LNGS, which opens the possibility of improving our knowledge of other key reactions by shifting their measured low-energy limit significantly closer or even to within the Gamow peak. The first reaction measured with this new accelerator is $^{14}N(p,\gamma)^{15}O$[14]. This is the slowest reaction of the hydrogen-burning

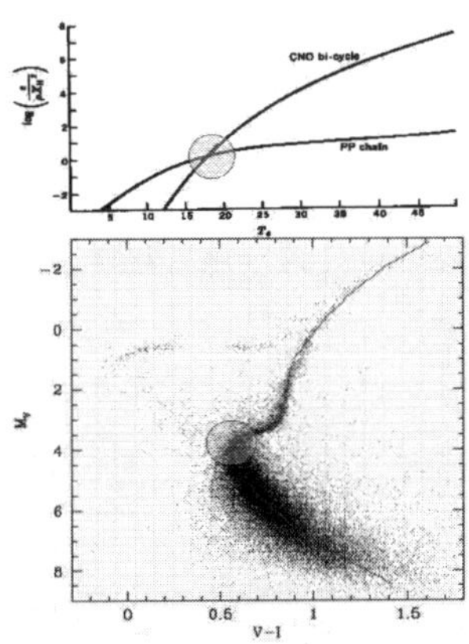

Figure 5. The onset of CNO cycle and the turnoff point of stars from the Main Sequence

CNO cycle, and therefore its cross section, $\sigma_{1,14}$, is directly responsible for the yield of the cycle. In turn, it fixes oxygen and nitrogen solar neutrino fluxes since $\Phi_\nu(^{15}O) \propto \sigma^{1}_{1,14}$ and $\Phi_\nu(^{13}N) \propto \sigma^{0.85}_{1,14}$, whose exact knowledge is extremely important for solar neutrino experiments. This reaction also has a cosmological importance[15]: its cross section tunes the age of globular clusters. The Main Sequence stars presently observed in Globular Clusters have masses smaller than of the Sun. As in the Sun, these low mass stars burn H in the center, mainly through the pp chain. However, towards the end of their life, the nuclear energy relased by the H burning becomes insufficient and the stellar core must contract to extract some energy from its own gravitational field. Then, the central temperature increases and the H burning switches from the pp chain to the more efficient CNO cycle. Thus, the departure from the Main Sequence is powered by the CNO cycle, whose bottleneck is the $^{14}N(p,\gamma)^{15}O$ reaction (see Fig. 5). The luminosity of the turnoff depends on the rate of this key reaction: the larger the rate, the fainter the turnoff and viceversa. On the contrary, the total lifetime is only marginally affected by a change in CNO, because it is mainly determined by the rate of $p(p,e^{+}\nu)d$. As a consequence, a decrease of the CNO rate would imply brighter turnoff points, for a given age, or older stars for a given turnoff luminosity. In spite of the importance of this reaction, only a few measurements existed[16] at a significantly higher energy than the Gamow window. Moreover, extrapolation fit of existing data[17] were

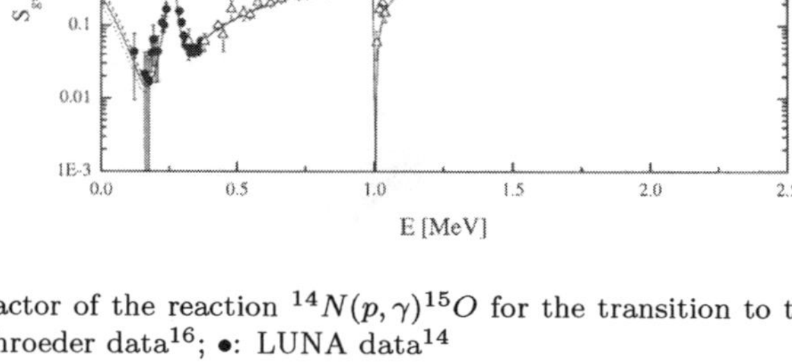

Figure 6. S-factor of the reaction $^{14}N(p,\gamma)^{15}O$ for the transition to the ground state of ^{15}O. $\triangle$: Schroeder data[16]; $\bullet$: LUNA data[14]

controversial, differing S_0 by nearly a factor three. LUNA collaboration repeated the measurement applying two different experimental techniques, depending on the explored [keV] energy range:140 keV $\leq E_{cm} \leq$ 380 keV: pointlike solid target and high resolution HPGe detector were used, to measure the single gamma transitions, including their angular distributions; 70 keV $\leq E_{cm} \leq$ 220 keV: gas target and $\approx 4\pi$ BGO summing crystal for the measurement of the total S-factor at the lowest achievable energy. The astrophysical S-factor for the transition to the ground state in ^{15}O is shown in Fig. 6. This was found to be nearly an order of magnitude lower than measured in previous experiment[16]: the discrepancy was understood in terms of summing effects in γ-spectra not properly taken into account in[16]. As a consequence, the total S(0)-factor results: $S_{tot}(0) = 1.7 \pm 0.2$ keV $\cdot b$. Two major consequences of this work are: the age of Globular Clusters (and therefore of our Galaxy) should increase by $0.8 \div 1$ Gyear; CNO neutrino fluxes would reduced by a factor ~ 2.

In the next future the activity at the underground LUNA facility will be devoted, as well as at ERNA, to study the $^{3}He + \alpha \rightarrow {}^{7}Be + \gamma$ with the goal to measure the reaction cross section with a total uncertainty lower than 5

Future developments of the LUNA facility could be based on a new accelerator able to bring on target proton and helium beams with energy up to 2.5 MeV and intensity in the milliampere range. With such a machine, experiments on reactions involved in the hotter stellar environments will be possibile, in particular on $^{12}C + \alpha \rightarrow {}^{16}O + \gamma$. The possibility to install at LNGS an accelerator in the MeV range depends on the solution of several and challenging problems. In particular technologically advanced solution will be necessary to fit the facility in the quite limited space available in the underground halls while very efficient shieldings will be required to "protect" other experiments running at LNGS from possibile beam-induced neutron background.

References

1. C.Rolfs, W.S.Rodney, Cauldrons in the cosmos, University of Chicago Press, Chicago (1988).
2. C.Spitaleri et al. Nucl Phys A719(2003)99c.
3. S.Cherubini et al. Ap. J., 457(1996)855.
4. C.Spitaleri at al. Intern. Conf. Nucleus Nucleus collision - Strasburg 2000.
5. S.Engstler et al., Z. Phys., A342 (1992) 471.
6. C.Spitaleri et al. Phys. Rev. C60 (1999) 55802.
7. D.Rogalla et al. Nucl. Intr. Meth A513 (2003) 573.
8. L.Gialanella et al. Nucl. Intr. Meth A522 (2004) 432.
9. U.Greife et al. Nucl. Instr. and Meth. A350(1994)327.

10. M.Junker et al. Phy. Rev. C , vol. 57, N. 5, 1998, 2700 -2710.
11. R.Bonetti et al. Phys. Rev. Lett. 82, 26, (1999), 5205-5208.
12. C.Casella et al. Nucl. Instr. and Meth. A706(2002)203.
13. A.Formicola et al. Nucl. Instr. and Meth. A507(2003)609.
14. A.Formicola et al. Phys Lett. B591 (2004)61.
15. G.Imbriani et al. A & A 420 (2004) 625.
16. U.Schrőder et al. Nucl Phys A467(1987)240.
17. C.Angulo and P. Descouvement Nucl. Phys. A690 (2001) 755.

SECTION VI

APPLICATIONS OF NUCLEAR PHYSICS

MEDICAL APPLICATIONS OF HEAVY ION ACCELERATORS

M. TORIKOSHI
National Institute of Radiological Sciences,
4-9-1 Anagawa,
Inage-ku, Chiba-shi 263-8555, JAPAN
E-mail: torikosi@nirs.go.jp

A heavy ion medical accelerator HIMAC was completed in 1993 at National Institute of Radiology Sciences, and clinical trials for cancer therapy were started with carbon-ions in the next year. The rationale for use of the carbon-ion for cancer therapy lies in the great advantages over photon or low-LET radiations, those are dose localization, higher ability to kill cells even if they are hypoxic. About 2,000 patients were treated until August 2004. HIMAC has been working with excellent stability and reliability since after the beam commissioning.

1. Introduction

X-ray was used first in medicine in 1896 that was one year after Roentgen's discovery of x-ray. Since then photon therapy has become one of the most popular modalities for cancer therapy. Particle radiotherapy also started about seventy years ago; Fast neutron therapy was started in 1938, and LBL started charged particle therapy in 1954. National Institute of Radiological Sciences (NIRS) started the fast neutron therapy using Van de Graff in 1970 and also started proton radiotherapy for treatment of eye-melanoma with a cyclotron in 1978. In the same year, the fast neutron therapy was shifted to the cyclotron facility to use higher energy neutrons. In 1993, a heavy ion accelerator facility was completed, which was the first heavy ion accelerator dedicated to medical applications in the world. Following beam commissioning and biology experiments for about seven months, cancer therapy was started with carbon-ion as a clinical trial in June 1994. Until August 2004, about 2,000 patients were treated. In this paper, some topics were reported: features of the carbon-ion for radiotherapy, the outline of the medical accelerator HIMAC and the present status of the clinical trials.

2. Features of Heavy-Ion Ratiotherapy

It is well known that the distribution of energy deposited by heavy-ions in materials shows Bragg peak at the end of the range. If adjusting the incident energy of the ions, the range of the ions can be controlled precisely to deliver a large amount of radiation dose to cancerous tissues, while surrounding tissues in which the ions penetrate before stopping receive fewer doses. It is also well known that the heavy-ion scatters much less than electrons or x-rays. These features of the heavy-ion result in excellent dose localization that is one of the most different points from photon. Linear energy transfer (LET) of carbon-ion reaches about 150 keV/μm at the Bragg peak. Radiations of higher LET are known to have greater biological effectiveness per unit absorbed dose than the radiations of lower LET such as x-ray or proton.[1] It comes from the fact that higher-LET radiations cause denser ionization, so cells receives more easily serious lesion. This effect appears in change in the number of survived cells in a colony after being exposed to a radiation. Fig. 1 of cell survival rates against the radiation dose of various LET radiations shows that at a certain dose the survival rates decrease as the LET becomes higher.[2] Biological effectiveness of an interesting radiation was defined comparing with that of a reference radiation, which was referred to be relative biological effectiveness (RBE).[3] RBE strongly depends on LET. It maximizes at about from 100 to 150 keV/μm, but it decreases as LET becomes higher than those.

The high-LET radiation has also another advantage that it effectively kills hypoxic cells that are generally resistant for a low-LET radiation. Since a core of cancerous tissue tends to be hypoxic, x-ray is not effect for such a cancer. This effectiveness is referred to be oxygen enhancement ratio (OER) that was defined in ICRU Report 30.[3]

3. Outline of HIMAC

Heavy Ion Medical Accelerator in Chiba, HIMAC consists of three ion sources, a cascade of linear accelerators (linacs), two identical synchrotron rings installed on a lower and upper floors of underground level (lower ring and upper ring, respectively) and vertical and horizontal beam lines. The bird's eye view the HIMAC facility is shown in Fig.2. There are three treatment rooms; one is equipped with vertical and horizontal beam ports, and the others have a vertical port and a horizontal port, respectively. There are also four experiment rooms. HIMAC provides carbon beam for the cancer therapy in the daytime while it provides various ion-beams for basic

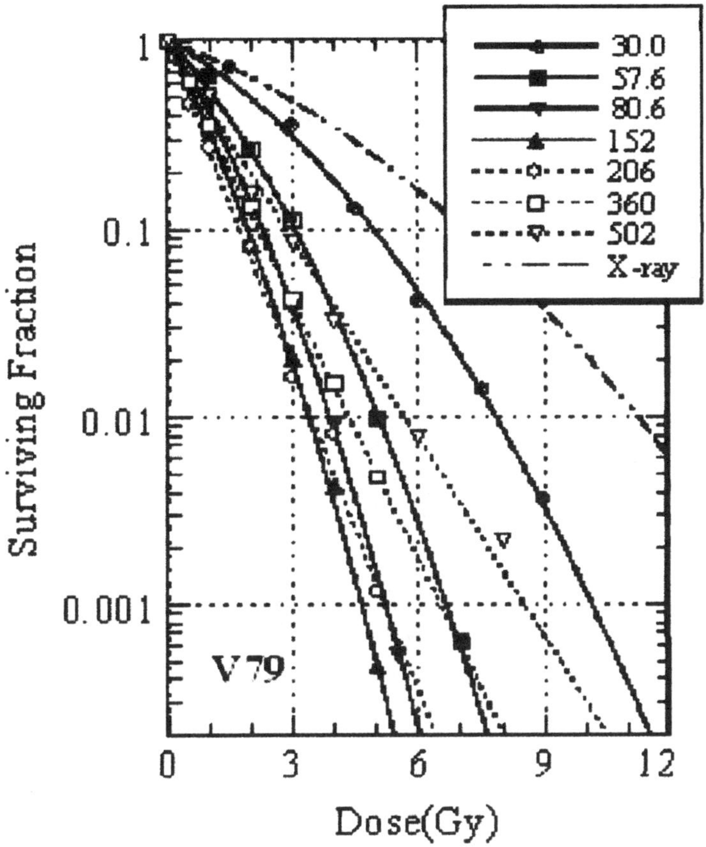

Figure 1. Cell survival rates versus dose delivered by various LET radiations.

experiments in the nighttime and weekend. Each synchrotron ring provides beams for about 5,000 hours a year. Only less than 0.1 % of all the scheduled machine-time was lost by machine troubles. HIMAC works very stably and reliably.

3.1. *Ion-sources and Linear Accelerators*

There are two ECR and one PIG ion-sources. One of ECR ion-sources are operated by 10 GHz radio frequency (RF) and another is operated by 18 GHz RF. The 10GHz-ECR is mainly used for the radiotherapy. It generates about 250 μA C^{2+} ions. The other ion-sources are usually used

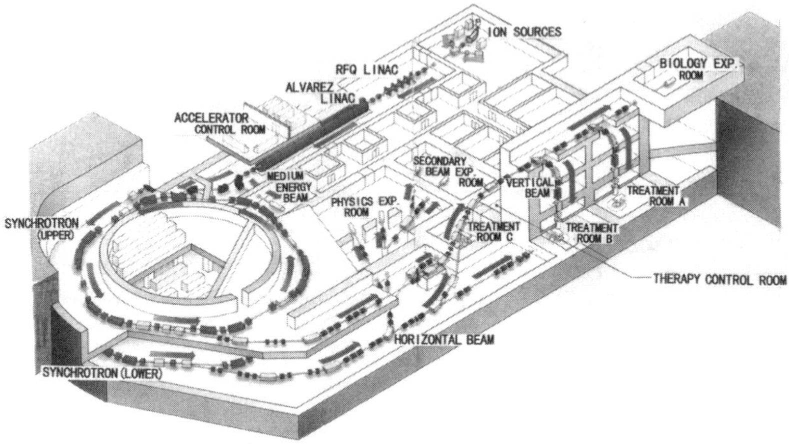

Figure 2. Bird's eye view of HIMAC.

for generating of various ion-species for basic experiments. Ion species that the ion-sources have ever generated range from proton to Xe. The ions are accelerated up to 8 keV/u after being extracted from the ion-sources. The ions with the charge-mass-ratio of larger than equal to 1/7 are accelerated up to 800 keV/u by RFQ and up to 6 MeV/u by DTL in sequence. Both of the linacs are operated by 100 MHz RF. The ions of 6 MeV/u are stripped of the residual electrons by a carbon foil furnished at the exit of the DTL. Light ions are fully stripped efficiently. The de-bunching system installed downstream from the DTL compresses the momentum spread into less than 0.1 %. The combination of a deflection magnet and a slit system analyzes momentum of ions and separates the ions with a specific charge state before the synchrotron rings receive them.

Since this system is operated in a pulse mode, it provides individually three different kinds of ion-species to two synchrotron rings and one experimental room where 6 MeV/u beams are available. It is referred as a time-sharing system.

3.2. *Synchrotron and Beam Transport Lines*

Both synchrotron rings have an ability to accelerate particles with a charge-mass-ratio of 1/2 up to 800 MeV/u at the maximum. The maximum energy of the upper ring is, however, limited to 600 MeV/u due to magnetic-rigidity of magnets used for the vertical beam line. Both rings are of a strong-focus separation-function type and their beam optical structure is FODO. They are operated 180° out of phase at a repetition rate of 1/2 or 1/3 Hz. The beams are slowly extracted in a third-resonance method or an RF Knock-Out method.[4] The latter method enables us to switch ON/OFF of beam extraction in less than one millisecond. Since it is easy to extract the beam synchronized with patient's breath in this method, it is applied to the beam irradiation of organs that moves with patient's breath, those are lung and liver as typical cases.

The upper ring provides beams to the vertical beam-lines for providing beams to the therapy rooms A and B and the biology experiment room. The lower ring provides beams mainly to the rooms B and C and the experiment rooms through the horizontal beam-lines; moreover, the lower ring also provides the vertical beam-lines via a beam line connecting of the horizontal beam line and the vertical beam line. The lower ring can therefore play the role for auxiliary upper ring when the upper ring does not work well due to troubles. When switching the connecting beam line with a pulse magnet, a DC-like heavy-ion beam with the duty factor of more than 80 % can be obtained.

3.3. *Beam Delivery System*

In Fig. 3, the irradiation system of the room B is shown as a typical example. Both of the horizontal port and the vertical port consist of identical devices and have almost same structure. The incident beam focuses on an isocenter (a center of rotation) without momentum dispersion when not modulating the beam. When irradiating the carbon beam for treatment, the beam must deliver a dose biologically homogeneously all over the target volume. A biological dose is defined by a physical dose multiplied by RBE. Its unit is called Grey equivalent and referred as GyE. In orer to realize the biologically homogeneous radiation field at the target volume in the broad beam method,[5] the beam is modified as follows: (1)a pair of wobbler magnets and a metallic scatterer laterally widen a pencil beam, (2)a ridge filter spreads the energy dispersion of the monochromatic beam, (3)a multi-leaf collimator or, in some cases, a patient collimator makes the lateral shape of the radiation field identical to the cross sectional shape of the target volume. The second process works for superposing Bragg peaks with changing of the ranges. Consequently it forms a spread Bragg peak, so that the highest LET region covers the target volume. A range compensators modulates the deepest ranges, furthermore, to finely shape the distal part of the radiation field to meet the distal surface of the target volume. The range compensators and the patient collimator are manufactured for each patient. These procedures result in forming biologically homogeneous dose distribution all over the target volume.

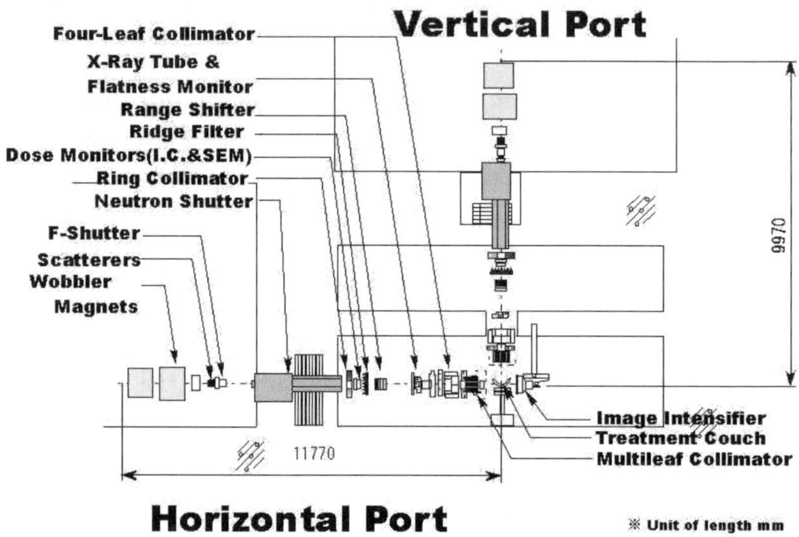

Figure 3. Beam delivery system of treatment room B.

4. Clinical Trials

Since the clinical trials of the carbon-ion radiotherapy began in 1994, a total of 1,954 patients were treated until August 2004. The details of the number of the patients are shown in Figure 4.

The clinical trials were initiated by conducting Phase I/II study to certify non-toxicity of the radiotherapy and to optimize dose for various diseases (dose escalation trials). The trials were started with 18 fractions per six weeks as the standard. In the study, the fraction numbers has been successfully decreased with increasing dose per fraction. For example, the first stage of non-small cell lung cancer (NSCLC) is one of cancers those are not suited to operation. For the NSCLC, only one fraction has been achieved via the trials with four fractions given in a week. Bone and soft tissue tumor is also a typical case that is highly resistant to photon radio-

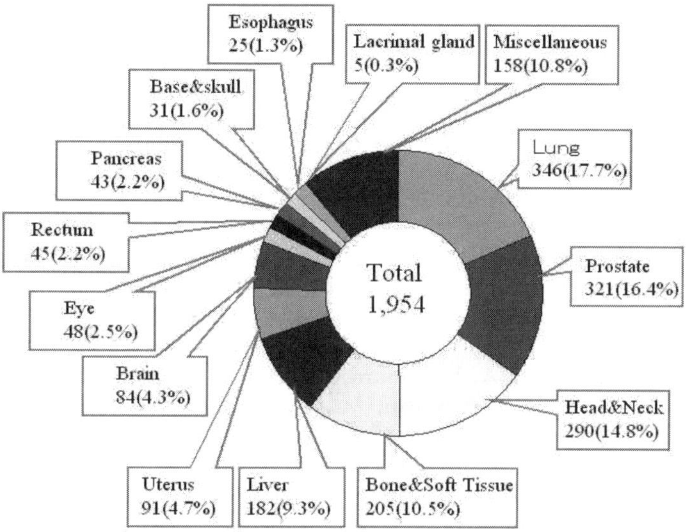

Figure 4. Patient distribution registered in carbon-ion radiotherapy at NIRS (June 1994–August 2004).

therapy. It has been proven, however, that it is well suited to carbon-ion radiotherapy. This fact shows that the high-LET radiations also have great effectiveness for tumors that are not ever suited to photon radiotherapy. The clinical results show that eligible tumors would be those that show biological resistance to the low-LET radiations.

5. Summary

HIMAC was completed in October 1993, and the clinical trials of the carbon-ion radiotherapy were started in June 1994 after the beam commissioning and some biological experiments lasting for about six months. The advantages of carbon-ion that is one of high-LET radiations are summarized as follows: the well localized dose distribution, high RBE and low OER. These advantages have appeared in the clinical trials such as: comparatively small number of fractions in treatment and high effectiveness for radioresistant tumors such as bone and soft tissue tumor.

HIMAC also provides beams for basic experiments of physics, biology, chemistry, material science and so forth. In these years, about 150 research proposals have been approved. HIMAC has been working with excellent stability and reliability with alomost no trouble since beam commissioning.

References

1. G. W. Barendsen, *Curr. Top. Radiat. Res.* **Q4**, 293 (1968).
2. Y. Furusawa, K. Fukutsu, M. Aoki, H. Itsukaichi, K. Eguchi-Kasai, H. Ohara, F. Yatagai, T. Kanai, K. Ando, *Rad. Res.* **154**, 485 (2000).
3. ICRU Report 30, (1979).
4. K. Noda, M. Kanazawa, A. Itano, E. Takada, M. Torikoshi, N. Araki, J. Yoshizawa, K. Sato, S. Yamada, H. Ogawa, H. Itoh, A. Noda, M. Tomizawa and M. Yoshizawa, *Nocl. Instrum. Meth.* **A374**, 269 (1995).
5. T. Kanai, M. Endo, S. Minohara, N. Miyahara, H. Koyama-Ito, H. Tomura, N. Matsufuji, Y. Futami, A. Fukumura, T. Hiraoka, Y. Furusawa, K. Ando, M. Suzuki, F. Soga and K. Kawachi, *Int. J. Radiation Oncology Biol. Phys.* **44**, 201 (1999).

THIN UC2 TARGET FOR SPES PROJECT

A. ANDRIGHETTO[1], S. CEVOLANI[2], C. PETROVICH[2]

1) INFN Laboratori Nazionali di Legnaro, Viale dell'Università 2, 35020 Legnaro (Pd), Italy

2) ENEA, Via M.M.Sole 4, 40129 Bologna, Italy

The study of a possible solution for a target system with a high intensity (up to 10 mA) proton primary beam, aimed at the production of exotic nuclei as a result of high energy fissions in ^{238}U, is reported. The work is inserted in the framework of the SPES project (Study for the Production of Exotic Species) of the Legnaro INFN Laboratories (Italy). An ideal configuration with the low energy primary beam (E<50 MeV) directly impinging on a UC_2 target device has been here considered. The isotopes fragments yields and the power depositions have been studied by means of the Monte Carlo code MCNPX as a function of the primary beam energy. The proposed configuration has been analyzed from a thermal point of view. A parameterization of the disk radius has also been performed.

1. Introduction

Large part of the experiments using exotic beams, performed at the existing first-generation facilities, are at present limited by the poor beam intensity (of the order of μA). Nevertheless, the development of high intensity (up to tens of mA) primary beam accelerators (*e.g.* TRASCO Project [1]) opens a new field in RIB technologies (Radioactive Ion Beams). The present study is inserted in the framework of the R&D of the SPES project (Study for the Production of Exotic Species) [2], which consists of an accelerator facility providing intense neutron-rich radioactive ion beams of highest quality, in the range of masses between 80 and 160. This is achieved by means of high energy fissions in ^{238}U compounds. This facility is planned to be constructed in Italy, at the Legnaro INFN Laboratory (the italian institute for nuclear physics). The first design project [3] was based on a high intensity 100 MeV (100 kW) proton LINAC driver.

Here the optimization of the nuclei production is investigated: a configuration with the low energy proton primary beam (E<50 MeV) directly on a Uranium Carbide target device (UC_2 in the present work) has been considered. In target isotopes fragments yields and power depositions are studied by means

of the Monte Carlo radiation transport code MCNPX as a function of the primary beam energy.
The thermal analysis of the proposed system was performed in ideal conditions, *i.e.* by considering the disk irradiating to a low temperature environment. The target high temperature (around 2000 °C) is an essential condition for the extraction of the fission products. This is the reason for the choice of uranium carbide instead of natural uranium as target material.
Only the on-target isotopes production and the thermal status were analyzed here, without considering other technological problems (*e.g.* diffusions, effusions, ionisations, etc).

2. Target Configuration

Three different parameters can be chosen as the most important for setting the target configuration: the number of fission yields, the fission fragment distribution and the power released in the materials (window and targets). Even if the 2-step configuration (proton beam on a production target used to produce fast neutrons for fissioning the uranium target) is the most exploited solution in the RIB projects (*e.g.* RIA, EURISOL, SPIRAL 2), the 1-step target configuration is chosen here (proton beam directly on the uranium target, as shown in Fig.1). This represents a better solution for a low energy proton beam driver (less than 50 MeV) as far as the on target fission rates are concerned [4].

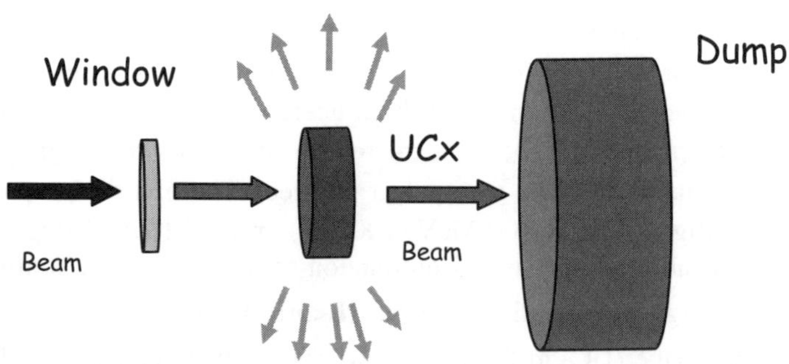

Fig.1. Scheme of the 1-step configuration: the primary beam impinges directly on the production target

The main problem in the 1-step configuration is the high energy loss of the incident beam through the medium (window and target), mainly due to the electromagnetic interactions. A direct high intensity beam fully impinging directly on the production target, although previously assessed as a possible solution, was rejected because beam powers of the order of 10 kW give rise to problems in the design of the target cooling system. Such problems are concerned with the necessity to dissipate the heat by means of thermal radiation; and with the low thermal conductivity of the target materials.
However, a solution to this problem might be to drive only the protons with higher fission cross sections in a thin production target and then to drive the outgoing beam inside a passive dump. In this way the number of fission residuals is maintained high and the power deposited in the target is lowered considerably.
As far as the choice of the energy of the primary beam is concerned, it is important to point out that even if high energy beams involve less dissipated energy per length, they increase sensitively the cost of the apparatus. Moreover, the ^{238}U fission cross sections [5], as reported in Fig.2, does not increase considerably for energies higher than 35 MeV: between 35 and 50 MeV particle energy, there is almost the same values (about 1.6 barns) [6], [7]. The availability of a high intensity proton LINAC for the isotope production will be sufficient at about 35 MeV, since higher energies do not seem to bring a clear advantage in the number of the fragment yields.

3. The MCNPX Code

Simulations trials reported in this work have been performed by means of the Monte Carlo radiation transport code MCNPX [8] (version 2.5.e). The code allows a detailed 3D definition of the system to be analyzed and a full transport calculation, starting from the proton particles. The interaction physics in MCNPX is determined in two ways: through table-based cross-section data and through on-line calculation by means of physics models. For neutrons below 20 MeV, the nuclear reactions are accounted for by means of cross-section evaluations (such as ENDF/B-VI). Whenever evaluated cross-section libraries are missing, MCNPX offers different physics models describing the nuclear interactions by the transition of different stages: in the first stage the incident particle interacts with the individual nucleons of the nucleus via particle-particle cross-sections emitting high-energy particles and light ions. This phase is called Intra-Nuclear Cascade (INC) and is followed by a pre-equilibrium stage. In the second stage, the residual nucleus either undergoes evaporation, releasing neutrons and light ions, or fissions. In the final stage the excited nucleus decays by gamma emission.

Among these physic models the Bertini-Dresner Model [9] and the CEM2k (Cascade-Exciton-Model) [10] have been chosen for our final calculations. The fission model used to describe the fragmentation distribution is the RAL (Rutherford Appleton Laboratory) fission model [11]. The fission yields are calculated by means of the HTAPE3X code [8].

Since there are no evaluated cross section data for protons interacting with ^{238}U (the most important interaction for this target system), a validation of the fission process described by the above mentioned models has been performed. The proton fission cross-section, obtained from the experimental data [5]-[6] and from the MCNPX calculations using the Bertini model and the CEM2k model, is shown Fig.2. CEM2k is in good agreement (discrepancies below 15%) with the experimental data of [5] in the whole energy range. It has some discrepancies (up to 35%) with the data of [6] in the range 35-50 MeV. The Bertini model has discrepancies with the experimental data and with CEM2k in the range 10%-35%. These comparisons are considered here to be good enough for an analysis of the target system by means of MCNPX.

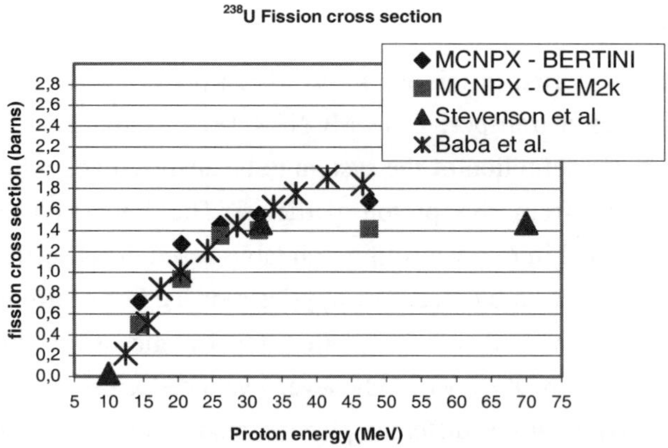

Fig.2. ^{238}U experimental and calculated fission cross section for protons.

4. Target optimization

The simulation trials were performed in a 1-step configuration varying the proton beam energy from 20 to 50 MeV. The beam shape is assumed to be gaussian with $\sigma = 2.5$ cm. A thin (400 μm) pyrolytic graphite window foil has been introduced, just before the UC_2 target, in order to separate the beam line with the target void regions. The target itself is constituted by a disk of 3 cm radius and 2 mm thickness. The thickness is a compromise between the need of low energy deposition (thinner target are preferred) and mechanical resistance of the material (thicker target are preferred). The target material is UC_2 with a density of 2.5 g/cm^3.

From a thermal point of view, the target is assumed to be cooled by thermal radiation to an environment at about room temperature.

The total fission fragment yields, the ^{132}Sn isotope yield, and the power dissipated in the medium have been calculated for each beam energy and reported in Table 1. ^{132}Sn is taken here as reference because, being a double-magic nucleus and thus interesting for nuclear physics, is one of the radioactive nuclei to be produced.

As expected, the energy dissipated in the medium by the charged beam (mainly due to electromagnetic interactions) is high for low energies and it decreases with higher beam energies: the decrease in the production target is about 3 MeV going from 20 to 35 MeV, while it is only 1 MeV going from 35 to 50 MeV.

Table 1. Fission yields and energy losses for different incident proton energies.

Incident proton energy (MeV)	Fission residuals (atoms per incident proton) BERTINI	Fission residuals (atoms per incident proton) CEM2k	^{132}Sn Yield (atoms per incident proton) BERTINI	Energy loss in C window (MeV)	Energy loss in UCx target (MeV)	Average outgoing proton Energy (MeV)
20	1.6E-03	1.1E-03	$6\cdot 10^{-6}$	2.2	7.2	10.4
25	2.9E-03	2.1E-03	$8\cdot 10^{-6}$	1.8	5.8	17.4
30	3.4E-03	3.1E-03	$1.\cdot 10^{-5}$	1.6	4.9	23.5
35	3.6E-03	3.2E-03	$1\cdot 10^{-5}$	1.4	4.3	29.2
50	3.9E-03	3.3E-03	$1.3\cdot 10^{-5}$	1.1	3.4	45.2

The fission fragment mass distribution for the different proton bombarding energies is shown in Fig.3. There also fissions induced by secondary neutrons, but they are negligible (below 1%) if compared to those induced by protons. As expected, the fission residual production does not increase sensitively if the beam proton energy is raised above 30 MeV. Moreover, it appears that at 35 MeV incident energy the symmetric valley of the mass distribution has been already filled and the only difference with the 50 MeV incident energy is that in this case the distribution is slightly wider.

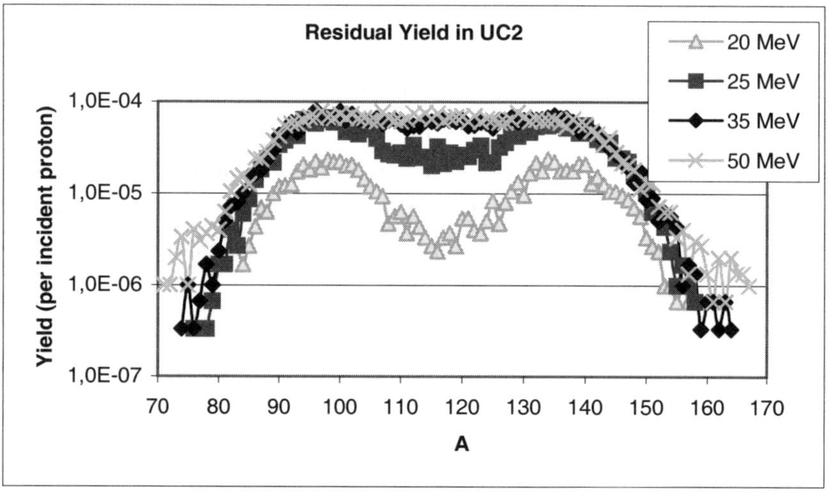

Fig.3. Fission fragment mass distribution for different incident proton energies.

It might be concluded that since a 50 MeV proton beam involves a higher cost bringing only low gain in the isotope production and in the dissipation of energy in the medium, it appears that a good compromise is to employ a proton beam with energy of about 35 MeV.

As far as the choice of the beam current is concerned, Table 2 shows for a 35 MeV primary proton beam the same quantities reported in Table 1, varying the current intensities from 0.1 to 2 mA. When a 1 mA of beam current is used, the total fission residuals in the UC_2 target is about $2.2 \cdot 10^{13}$ atoms per second with this target system configuration. A large amount of the power impinges into the dump (29 kW), 1.4 kW hits the window, while the remaining thermal power (4.4 kW) is dissipated in the production target. This corresponds to a power density averaged in the whole target volume of approximately 800 W/cm^3. As it will be discussed in the next section, preliminary calculations show that these values do not represent a serious problem for the target, because the mean temperature remains below the UC_2 melting point if an appropriate target disk radius is used.

Table 2. Fission residuals and power losses with E_p=35 MeV varying the beam current.

Incident proton current (mA)	Fission residuals (s^{-1})	^{132}Sn yield (s^{-1})	Power loss in the C window (kW)	Power loss in the UC_2 target (kW)	Power loss in the beam dump (kW)
2	$4.4 \cdot 10^{13}$	$1 \cdot 10^{11}$	2.8	8.7	58.4
1	$2.2 \cdot 10^{13}$	$6 \cdot 10^{10}$	1.4	4.4	29.2
0.5	$1.1 \cdot 10^{13}$	$3 \cdot 10^{10}$	0.7	2.2	14.6
0.1	$2.2 \cdot 10^{12}$	$6 \cdot 10^{9}$	0.1	0.4	2.9

Thus a good solution might be based on a 35 MeV 1mA (35 kW) proton beam, which provides enough fission rates together with the possibility to dissipate the power in the target and in the window. In this way, a proton beam power of 35 kW produces about $2 \cdot 10^{13}$ s^{-1} nuclei, the same order of magnitude as that obtained with a deuteron beam power of 200 kW (5 mA, 40 MeV) in the framework of the SPIRAL 2 proposal.

5. Target Thermal Analysis

In this section the thermal behaviour of the target system is analyzed, when the average total deposited power in the target is 4.4 kW (obtained from a 35 MeV, 1 mA proton beam).

The disk constituting the target is heated by the beam and, in steady state, the same amount of power has to be transmitted to the environment by thermal radiation. Purpose of the thermal analysis is to determine the temperature field within the disk; by means of the obtained results, some indications on the target design are obtained.

The heat deposition in the disk is assumed here to be uniform along the disk thickness (the beam direction); in radial direction, the power profile can be assumed to follow the beam shape and thus to be Gaussian; in the circumferential direction it can be also considered to be uniform.

For the thermal analysis, in order to face the variable radial power profile, it is necessary to subdivide the disk in radial meshes of annular shape, each one conducting with the adjacent ones and radiating toward the environment. No mesh subdivision is necessary in the circumferential direction, where no heat deposition gradient is supposed to exist. In the thickness direction some heat conduction takes obviously place but, due to the uniformity of the power profile, a mesh subdivision is not considered necessary: the temperature drop across the disk thickness can be computed *a posteriori* basing on the analytical solution of the Fourier law in a parallel slab.

Basing on such assumptions, a small one-dimensional computer code has been developed. The core of the solving system is constituted by the application of the Fourier law to the disk, performed by means of the *Control Volume Method* [12] [13]; the thermal radiation to the environment will be introduced as a boundary condition at the mesh surface. Due to the fact that the radiation is proportional to the fourth power of the temperature, the system of linear equations deriving from the Fourier law becomes non-linear. As a consequence, the system has to be solved be means of an iterative procedure, basing on the classical linearization of the radiation term. In this way, the solution is constituted by the radial temperature profile at the disk surface. As mentioned above, the temperature profile at the centre of the disk (in the thickness direction) can be determined by superposing to the obtained result the analytical solution of the Fourier law in a parallel slab.

The described model has been applied to the target proposed in the previous sections: a disk radius of 30 mm (2 mm thick) with a beam Gaussian with σ = 25 mm. The total power in the target is 4.4 kW and we consider a environment temperature of about 50 °C.

The disk material physical properties were assumed as follows: a thermal conductivity of 1 W m^{-1} °C^{-1} [14], a thermal emissivity of 0.6 [14] and a UCx melting point with the value between 2315 °C [15] and 2527 °C [16].
It has to be pointed out that both emissivity and conductivity are referred to UC_2 in porous form. The main obtained results are shown in the Fig.4 in form of temperature profiles in radial direction at the disk surface and at the disk centre. It appears clearly that for great part of the disk the temperature is above the melting point, even by trusting the most favourable of the collected data for the melting point.

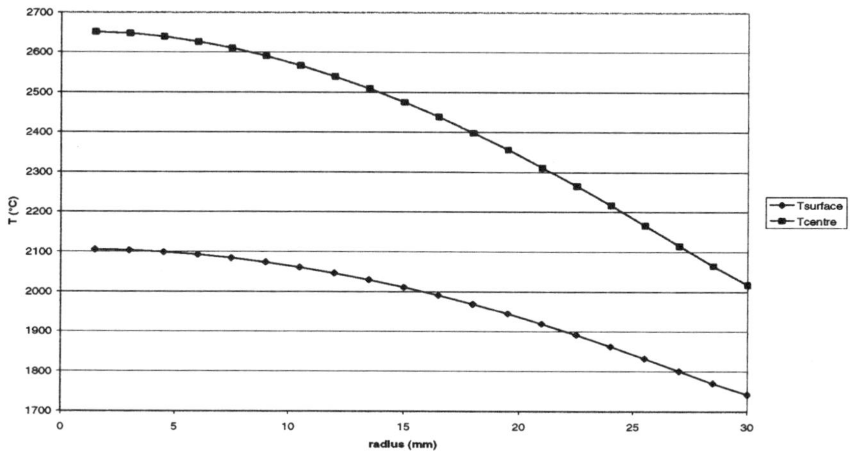

Fig.4. Temperature profiles along the disk (3 cm of radius) at the surface and at the centre.

In order to reduce the temperature level, there exists several possibilities. In this preliminary phase, only the increase of the disk radius will be taken into account. For this purpose, parametric calculations were performed, by obtaining the results shown in Table 3.

Table 3. Temperature at the centre and at the border of the disk for different disk radii.

Radius [cm]	Disk surface		Disk centre	
	Tmax [°C]	Tmin [°C]	Tmax [°C]	Tmin [°C]
3	2105	1742	2650	2017
4	1911	1353	2298	1468
5	1813	1047	2136	1095
6	1767	790	2063	809

It appears that by adopting a suitable disk radius, even with a small increase with respect to the up to now proposed value, the system becomes feasible; once more, the disk radius shows to be a good tool in order to drive the temperature level in the disk. Figure 6 represents the thermal behaviour of a target with a radius of 5 cm. This target has, in all the considered points, a temperature well below the UC_2 melting point (around 2400 °C).

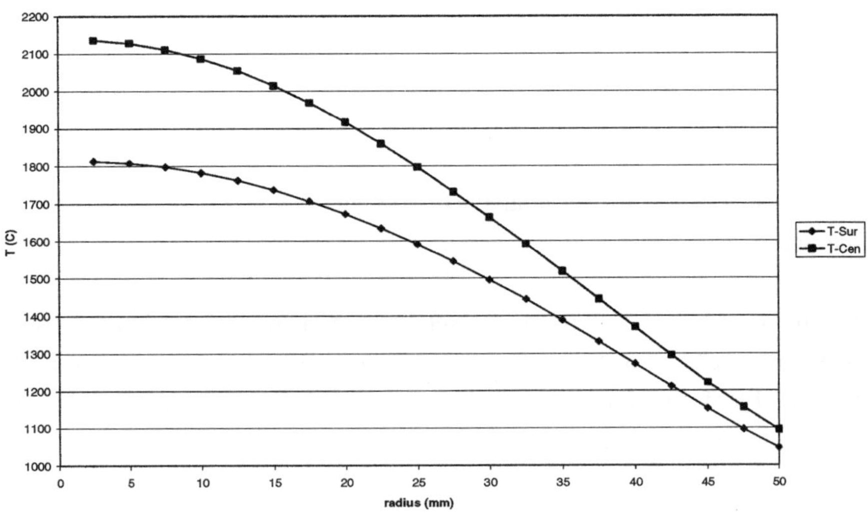

Fig.5. Temperature profiles along the disk (5 cm of radius) at the surface and at the centre.

More difficult will be the control of the thermal gradients: probably they have to be reduced for sake of the fission products migration. In the radial direction, the temperature gradient can be lowered by flattening the power deposition profile either by increasing the Gaussian width or by allowing a periodical movement of the beam. In the thickness direction, the only possibility seems to be the reduction of the thickness itself, for instance by subdividing the target in several disks disposed in serial. A further problem could arise from the necessity to hold the enclosure where the target is placed (the heat well) at a temperature higher than the room one. Also in this case it could be necessary to use several disks disposed in serial.

6. Conclusions

A possible solution for producing exotic nuclei is a configuration with a proton beam with energy of 35 MeV and a current of 1 mA impinging directly on a UC_2 target. In this way about $2 \cdot 10^{13}$ s^{-1} fission residuals are formed as a result of high energy fissions in ^{238}U. The fission fragment distribution has been calculated by means of the MCNPX code, here validated as far as the fission cross section is concerned. Preliminary numerical calculations suggest that the power dissipated by the beam inside the production target (< 5 kW), might not represent a critical engineering issue. In fact, using a UC_2 target disk with thickness 2 mm and radius above 4 cm, the thermal radiation is enough (in ideal conditions) to keep the temperature of the target below the melting point. Further and detailed R&D study on the thermo-mechanical behaviour of the Uranium Carbide target is in any case required.

References

[1] Status of the High Current Proton Accelerator for the Trasco Program – INFN/TC 00/23.

[2] SPES Technical Design for an Advanced Exotic Ion Beam Facility – LNL-INFN (REP)181/02.

[3] A. Andrighetto, J. Li, C. Petrovich, Q. You, Nucl Instr. Meth. **B204** (2003) 205.

[4] A. Andrighetto, Proceeding of *Exon 2004 Internatioal Conference (2004).*

[5] P. C. Stevenson et al , Phys. Rev. **111**, 3 (1958) 886.

[6] S. Baba et al, Nucl Phys **A175** (1971) 177.

[7] B. L. Tracy et al, Phys Rev **C5** (1972) 22.

[8] L. S. Waters, Editor, "*MCNPX™ User's Manual*", Version 2.4.0, LA-CP-02-408, September 2002 (http://mcnpx.lanl.gov/).

[9] H. W. Bertini, *Low-Energy Intranuclear Cascade Calculation*. Phys. Rev. **131** (4) 1801 (1963).

[10] Stepan G. Mashnik and Arnold J. Sierk, *Recent Developments of the Cascade-Exciton Model of Nuclear Reactions*, International Conference on Nuclear Data for Science and Technology, October 7-12, 2001, Tsukuba, Japan. Published in "Journal of Nuclear Science and Technology", Supplement 2, pp. 720-725 (2002).

[11] F. Atchison, *Spallation and Fission in Heavy Metal Nuclei under Medium Energy Proton Bombardment,* in *Targets for Neutron Beam Spallation Sources,* Jul-Conf-34, Kernforschungsanlage Julich Gmmh (January 1980).

[12] S. V. Patankar, *Numerical Heat Transfer and Fluid Flow* McGraw-Hill (1980).

[13] S. Cevolani, Dissertation, Universität Karlsruhe. KfK 3148, EUR 7051d (1981).

[14] J.A. Nolen, M. Petra, J. Green: R&D Related to a Future Rare Isotope Accelerator Facility.

[15] R M.M. El-Wakil *Nuclear Power Engineering* McGraw-Hill (1962).

[16] S. McLain, J.H. Martens, Editors *Reactor Handbook* Interscience Publishers (1964).

APPLIED NUCLEAR PHYSICS FOR CULTURAL HERITAGE AND ENVIRONMENTAL STUDIES: THE NEW LABEC LABORATORY IN FLORENCE

FRANCO LUCARELLI

Department of Physics of the University and Sezione INFN of Florence

A well established activity of the group in the field of applied nuclear physics was present in Florence already since the Eighties. This activity was carried out at an accelerator laboratory based on a single-ended 3 MV Van de Graaff. In this paper first I will explain the main features of Ion Beam Analysis (IBA) techniques and I will describe the experimental set-up at our laboratory; then I will give some examples of our work to show which kind of information may be obtained both in cultural heritage and environment field; finally, I will present the new INFN LABEC Laboratory in Florence.

1. IBA techniques

The Ion Beam Analysis (IBA) techniques are a powerful tool to investigate in a fully non-invasive way the composition of a material. To this purpose, the object to be analysed is used as a target for a beam of accelerated particles. The interactions of the beam particles with the atoms (or the nuclei) of the target material induce from the latter the emission of secondary radiation (X-rays, gamma rays, particles), having an energy characteristic of the emitting atom or nucleus. Suitable nuclear detectors are used to collect and discriminate in energy the emitted radiation and make it therefore possible – in a single measurement - to detect and quantify the presence of the different elements in the analysed material. Particle Induced X-ray Emission (**PIXE**) exploits the X-rays emitted by the atoms. Particle Induced Gamma-ray Emission (**PIGE**) exploits the gamma-rays emitted by the target nuclei. Rutherford Backscattering Spectrometry (**RBS**) exploits the beam particles, elastically scattered from the target nuclei, whose energy is a function of the mass of the target nuclei and of the depth. Among them PIXE is by far the most commonly used. In general,

IBA analysis are: multi-elemental, quantitative, fast (a single measurement takes few minutes at most), sensitive down to trace level, information on the stratigraphy can be obtained, by using RBS and differential PIXE, i.e. PIXE measurements performed at different beam energies (see below), non-destructive: due to the high cross section values (for PIXE in particular), extremely low beam currents can be used, without any risk of damage for the sample. Furthermore, external analysis can be performed by extracting the beam into atmosphere. In this way, the object to be analysed can be kept in its "natural" air environment, thus, avoiding the need of picking up of samples and keeping to negligible values the beam-induced heating; in summary, no damage at all is procured to the work. Fig. 1 shows a picture of our setup with a manuscript or atmospheric aerosol samples under analysis. Specific supports have been designed for handling the different kind of samples, which are sometimes large and heavy (manuscripts, paintings etc.); this stage is equipped with a system of remotely controlled micrometric movements. The beam comes out from the vacuum line into air through a thin Upilex foil. The sample is kept at a distance of a few mm from the exit window. For the X-ray detection 2 Si (Li) detectors are used to optimize the sensitivity over a large range of energies. A He flow in front of the detectors and the sample avoids the absorption of lower energy X–rays (we are able to detect from sodium (included) upwards in the periodic table). A high-purity Germanium detector is used for PIGE measurements. A laser-based alignment system allows us an accurate aiming at the detail to be analyzed. A TV-camera system is used both in the phase of positioning, and to remotely monitor the "target" during beam bombardment. For thin targets a Faraday cup behind the sample is used to measure the beam current; for thick insulated samples a rotating chopper vane (a Ni layer evaporated on a graphite substrate) is placed between the exit window and the target, to monitor the beam current intensity by measuring the Ni X-rays yield.

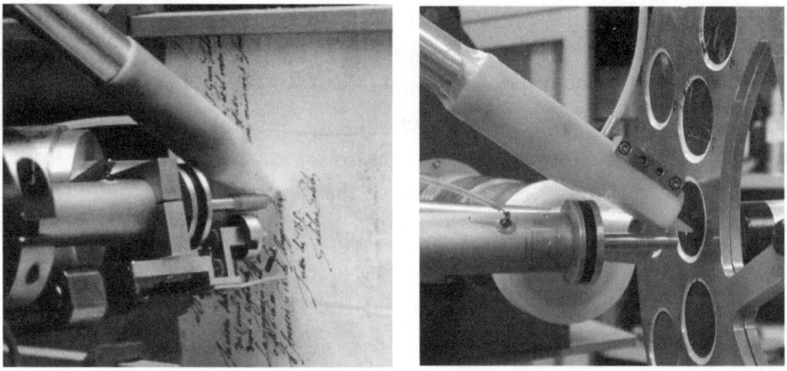

Fig.1 : PIXE-PIGE experimental set-up for the measurement of art objects (left) or atmospheric aerosol samples (right).

2. Cultural heritage applications

Which are the possible outcomes from doing IBA analysis on objects of relevance for the cultural heritage? Plenty of precious historical information on the cultural development at different times is obtained by the knowledge of the materials used for manufacturing art objects; it can provide an insight into the technical development in ancient times; it can lead to the discovery of the sources of supply of raw materials, thus contributing to the solution of historical problems, e.g. about the existence of certain trade channels at the time of manufacture of the ancient object under analysis; it can constitute a way of indirect dating, through compositional analogies with dated materials; under lucky circumstances, it can even lend strong support to the attribution to a specific artist, thus providing a criterion for authentication (or refutation); it is sometimes a useful tool in the diagnosis of deterioration processes of materials; it may be an essential preliminary step to guide decisions about *compatible* and *reversible* restoration procedures.

IBA can certainly play an important role, especially since it is often fully non-destructive. Of course, for these investigations to be really productive an essential prerequisite is an intense collaborative work of scientists with archaeologists, art historians, museum curators, librarians, etc. In the following, I will describe with some details examples taken from the experience of our group in Florence.

2.1. *Ceramics*

By PIXE measurements, both major and minor elements concentrations in ceramics, porcelains and glasses can be determined with reasonable accuracy. Our laboratory has been involved in an international collaboration (including the Louvre museum in Paris, the Bargello museum and the Opificio delle Pietre Dure in Florence, and the Physics Departments and INFN groups in Genoa and Catania) aimed at analyzing the glazed terra-cotta sculptures produced during the Florentine Renaissance, and in particular the works of the family Della Robbia. This kind of measurements were mainly intended to solve problems of authenticity - actually the enormous success of these terra-cottas stimulated the production of many contemporary artists like Rustici, Sansovino and the family Buglioni - but also to get important information for restoration. [1].

2.2. *Iron gall inks*

Iron gall inks of ancient manuscripts are characterized not only by iron itself but also by other metals, like zinc, copper and lead: their relative abundances represent a sort of fingerprints of a particular typology of ink. PIXE is an ideal technique to study these materials since it can provide all the required information without risk of damage to the paper. The elemental analysis of inks can give us also the possibility for an indirect dating of manuscripts. In the framework of “Galileo project” we have analyzed lots of notes and demonstrations written by the scientist on folios without a date: science historians are very interested to determine the various steps in the evolution of Galileo’s thought, but, sometimes, establishing the chronology of the various folios can be very difficult only from a textual analysis. Inks in many of these folios have been analyzed and their composition compared with the ones of dated documents (like letters) [2].

2.3. *Pigments and paintings*

Most of the pigments are characterized by inorganic elements and can be thus studied by PIXE. For example, pigments used in miniatures in medieval manuscripts have been analyzed. More than one hundred manuscripts or lose pages, belonging to the Italian environment and dating from the XI to the XV century, have been analyzed with an external PIXE set-up. The aim was to build a satisfactory database and hence establish trends in the technologies for illuminating, infer the possible sources of supply for the raw materials, confirm or deny attributions to an artist [3]. Perhaps the most striking of the general results of our research, is the widespread use of ultramarine blue, which we detected in the decorations of most of the examined manuscripts from the XII and XIII century, even in those cases where the decoration was very simple.

So far, worldwide relatively little PIXE activity has concerned the study of the materials of traditional paint on wood or canvas. For these objects a limitation for PIXE is the presence of the protective varnish, an organic layer that actually drastically absorbs soft X-rays emitted from the paint layers below, so that elements with the lowest Z detectable by PIXE (Na, Mg, Al...) become practically invisible. However, information about light elements can be determined by means of PIGE, because γ-rays have a much higher energy than X-rays [4]. Moreover, the contributions of the different layers and thus the discrimination of the depth sequence of elements in this kind of samples can be obtained using the so-called differential PIXE (PIXE measurements using incident ion beams of decreasing energy). A first application of both PIGE and differential PIXE to the study of a painting is represented by the analysis of the Madonna dei Fusi by Leonardo da Vinci (this analysis was the pilot study for the international collaboration of the “Universal Leonardo Project”, aimed at discovering the “secrets” of the painting and drawing techniques of the artist by non-destructive analysis) [4, 5]. The analysis allowed us to characterize the pigments and the substrate composition. In particular, PIXE analysis of the blue areas revealed the presence of a lot of restorations (mainly in the mantle of the Virgin), which were identified by the presence of zinc white (ZnO) and cobalt blue ($CoO{\cdot}Al_2O_3$), pigments used only since the beginning of XIX century. The use of lapis-lazuli for the original areas (mountains for example) was pointed out by PIGE measurements, detecting sodium 441 keV γ-rays from the (p,p'γ) reaction.

3. Environmental applications

IBA techniques and mainly PIXE, are a powerful tool to investigate environmental problems: they are indeed particularly suitable for the study of atmospheric aerosol, which nowadays is the pollutant of major concern in many Italian urban areas. Aerosol sampling campaigns produce huge amounts of few μg particulate samples: fast, quantitative, high-sensitivity and multi-elemental analytical methods are thus required. Non-destructiveness is also an important issue to extend the range of detectable elements by complementary techniques applied to the same sample. Scanning possibility may also be very useful, allowing time trend reconstruction by the analysis of time-sequence aerosol deposits collected by continuous samplers. All these requirements put severe difficulties in the path of standard chemical methods and open up opportunities for IBA techniques.

All the elements with $Z > 11$ are simultaneously detectable in few minutes of beam time (one measurement may last 5-10 min, so the samples collected on a daily base for one year may be analyzed in only 3 days of analysis!), including several important tracers of peculiar aerosol sources and potentially harmful elements, with minimum detection limits ranging between below 1 and 10 ng/m^3. Before laboratory analysis, atmospheric particulate has to be collected, by means of different kind of samplers, on filter membranes or impaction plates. Collected aerosol mass can be obtained gravimetrically.

LABEC is one of the few laboratories in the world employing forward and backward Particle Elastic Scattering Analysis (PESA) for the detection of light elements, which are the main aerosol constituents [6]. The application of both PIXE and PESA allows a complete mass reconstruction of aerosol samples.

The identification of the sources is obviously one of the most important issues of particulate matter studying: it is necessary for any pollution abatement strategy. Since aerosol particles retain elemental composition characteristic of their sources, even at long distances, the simultaneous detection of groups of elements may be a signature of pollution sources. Absolute Principal Component Analysis (APCA) groups the detected elements into "*factors*" according to the similarity of the time behavior of their concentrations, thus succeeding in pointing out the sources of pollution and in determining the contribution of each of the identified sources to the total aerosol mass. In fig 2 are reported the results of a study on the elemental composition of the aerosol in Florence [7]. The average contribution of traffic over a period of one year (its contribution is higher during winter) is dominant not only in a heavy traffic area, but also inside an urban park, while in the residential suburban area the main source is oil combustion. The situation is completely different in an industrial district not far from Florence. Here two different kinds of industries (the one producing ceramics and the one producing artistic glass) are the main aerosol sources. Looking at the PM_{10} (particles with diameter below 10 μm) source apportionment day by day, we can look what happen in the days in which PM_{10} levels are higher than the European limit of 50 μg/m^3. In the above mentioned industrial area, in many of these days the "soil" source, which is natural dust, gave the main contribution. Why dust is so important? Again PIXE is useful to give an answer. We observe an increase in PM_{10} concentration with a parallel increase the concentrations of all soil related elements like Si but no increase of all the anthropogenic elements and at the same time there is also change in the ratio between the soil elements, with ratio typical of Saharan dust; so these "soil" episodes are due to Saharan dust transport. These results are confirmed by air masses backward trajectories.

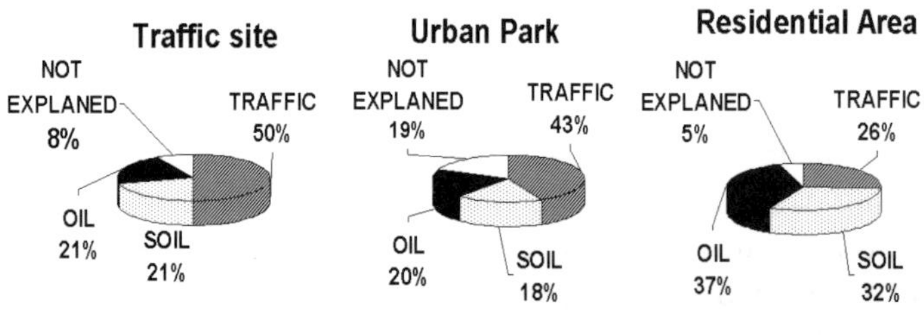

Fig. 2 : Aerosol source apportionments in the three sampling sites inside Florence urban area.

4. The new LABEC laboratory

LABEC (LAboratorio BEni Culturali) is a laboratory in the Sezione di Firenze of INFN, established with the main purpose of performing applications of nuclear techniques in the field of problems related to the Cultural Heritage. However, applications of nuclear techniques also to other fields are currently being performed: studies on environmental problems (mainly, but not only, air quality monitoring), applications to geological studies such as chronological reconstruction of past volcanic eruptions or geochemical modeling of magma evolution, applications to material science in general, and others.

The main equipment of LABEC is a Tandem accelerator, 3 MV terminal voltage, which has been constructed by High Voltage Engineering Europe. It is conceived, from its initial design, for both Ion Beam Analysis and Accelerator Mass Spectrometry operations. Although IBA and AMS share the same track inside the machine, all the essential settings needed to the beam transport can be easily retrieved for both operating modes at any time. A fast switching between IBA and AMS operations is thus made possible, since the system is equipped with two fully independent injecting lines - each provided with its own source(s), focusing elements, steerers, slits and vacuum segmentations. On the high energy side a switching magnet is present to serve up to 9 independent IBA lines, while the AMS line is provided with the facilities to perform measurements on isotopes of carbon, beryllium, aluminum and iodine.

Three independent beam lines are already active (as to January 2005): external beam line dedicated to Archaeometry, external beam line dedicated to atmospheric aerosol measurements, external micro-beam line (~ 10 μm FWHM) which give the possibility to perform beam scanning on the analyzed sample in order to reconstruct maps of elemental distribution over the scanned area through PIXE, PIGE and RBS. In particular, μPIXE maps can provide elemental distributions of even trace elements in the sample. Three more beam lines will be completed within the next months (fast pulsed beam line, a beam line dedicated to measurements of particle scattering cross sections, a beam line with a multi-purpose chamber for in vacuum IBA).

Acknowledgments

I am deeply indebted to all the members of the LABEC laboratory and mainly to Prof. P.A. Mandò, S. Nava, N. Grassi, M. Chiari, M.E. Fedi, L. Giuntini, M. Massi, F. Taccetti, L. Carraresi who have been involved in some of the projects described in the paper.

References

1. A. Zucchiatti, A. Bouquillon, G. Lanterna, F. Lucarelli, P. A. Mandò, P. Prati, J. Salomon, M.G. Vaccari, *Nucl.Instr.&Meth.**B*** **189** (2004), 358.
2. F.Lucarelli, P.A. Mandò, *Nuclear Physics News*, **Vol.6, N.2** (1996), 24.
3. M. Bernasconi, R. Cambria, L. Dal Poz, P. Del Carmine, M. Grange, F. Lucarelli, J.D.MacArthur and P.A. Mandò, *Ancient and Medieval Books Materials and Techniques, Studi e Testi* **357-358**, Biblioteca Apostolica Vaticana (1993), 57.

4. P.A. Mandò, M.E. Fedi, N. Grassi and A. Migliori, accepted for publication on *Nucl.Instr.&Meth.B*.
5. N. Grassi, A. Migliori, P. A. Mandò and H. Calvo del Castillo, *Nucl.Instr.&Meth.* ***B*** **219-220** (2004), 48.
6. M. Chiari, P.Del Carmine, F. Lucarelli, G. Marcazzan, S. Nava, L. Paperetti, P. Prati, G. Valli, R. Vecchi and A. Zucchiatti, *Nucl.Instr.&Meth.* **B219-220** (2004), 166.
7. F. Lucarelli , P.A. Mandò, S. Nava, P. Prati and A. Zucchiatti, *Journal of the Air & Waste Management Association*, **54** (2004), 1372.

HYDROGEN ANALYSIS FOR GEOSCIENCE BY USING NUCLEAR MICROBEAM

K. KOMATSUBARA, K. SASA, S. ISHII, Y. YAMATO, K. MIYAKAWA AND K. SATOU

*Applied Accelerator Division,
Research Facility Center for Science and Technology,
University of Tsukuba,
Tsukuba, Ibaraki 305-8577, Japan
E-mail: komatsubara@tac.tsukuba.ac.jp*

M. KUROSAWA

*Institute of Geoscience, University of Tsukuba,
Tsukuba, Ibaraki 305-8577, Japan*

In order to analyze hydrogen content in magmatic samples nuclear microbeam irradiation system with elastic recoil coincidence spectrometry (ERCS) has been developed. This new method of analysis is only sensitive for hydrogen. The analysis is non-destructive and quick in which depth profile of hydrogen concentration can be deduced from measured energy spectrum.

1. Introduction

Nuclear microbeam technique can provide utility to wide range of studies in analytical research, cell irradiation surgery, material science, and semiconductor devices. In many facilities of particle accelerator science, research activities with microbeam irradiation are increasing to be important subjects[1,2]. In our lab at University of Tsukuba, Research Facility Center for Science and Technology, Applied Accelerator Division, microbeam facility has been constructed and being operated for hydrogen analysis to determine water content in geological samples. In this paper, our recent technical developments on the irradiation and detection system for the hydrogen analysis will be presented.

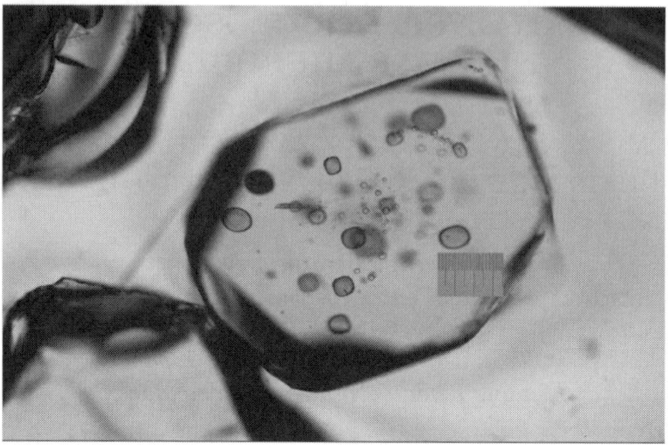

Figure 1. Melt inclusions trapped in a quartz crystal.

2. Geological aspect

The aim of the nuclear microbeam analysis is determination of hydrogen content in melt inclusions which are tiny objects trapped in quartz or ore. Figure 1 shows microscopic view of a slice of a quartz crystal that includes many small sub-structures called as melt inclusions which are slightly dark color. A scale in this figure indicates the size of the inclusion that is about 180 μm. The quartz crystal is about 3 mm in diameter. The sample was found from deposited layer at Enda near Zao volcano and the crystal is considered as a part of evacuated magma from Zao volcano.

The quartz is a crystal of SiO_2 contained no hydrogen and the melt inclusions are regarded as glass of rhyolite which can solve certain amount of water content under a high pressure. If a sample has mass more than a gram, it can be analyzed by gravimetric method or Karl Fisher volumetric method to evaluate water content. However, the nuclear microprobe can analyze the tiny objects one by one without any destruction.

Hydrogen analysis of the melt inclusions is quite important to understand magmatic dynamics and evolution, since the inclusions are regarded as specimens of the pre-eruptive magma[3,4]. Figure 2 shows a schematic view of growth of magma and transition of water. It shows movement of ocean plate which is colliding to Japan island and shift down under the land. This movement of the plate can be assumed to introduce migration of seawater to mantle. Under a condition of high pressure, the mantle can solve a certain amount of water and change its phase to be melt, i.e. incident of magma. Because of its low density, the magma can drift up. As the

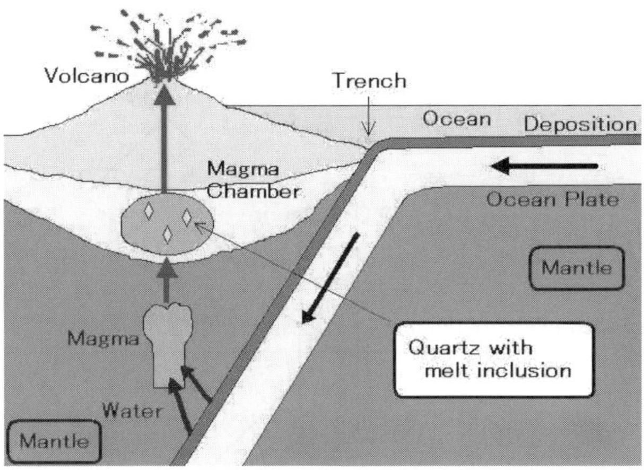

Figure 2. Schematic view of a volcano and magma. Melt inclusions are trapped in quartz crystal in the magma chamber.

final event of the magmatic evolution, volcanic eruption occurs in which the steam pressure in magma chamber acts as the motive power of the vigorous dynamics.

Before the eruption of volcano, the quartz crystals are growing up in the magma chamber and trap tiny parts of melting magma. They are similar to bubbles in ice as shown in Fig. 1. In this figure, inclusions appear as clear droplets of glass with negative crystal which has the same axis as the matrix. In some cases, melts are evolved and shown with sub-structure or micro-crystals when the cooling proceeds slowly.

3. Proton-proton elastic scattering coincidence spectrometry

In order to analyze hydrogen in side of the geological samples, a detection system with proton-proton elastic recoil coincidence spectrometry (ERCS) has been developed. This system has been combined with nuclear microbeam facility focusing tiny object in the crystal[5,6,7]. The proton-proton ERCS is powerful analytical method for hydrogen since the coincidence method is only sensitive to proton under the kinematic condition of scattering and recoil particles.

The ERCS is not only derivative analytical method from elastic recoil detection analysis (ERD) and Rutherford back scattering (RBS). Since Cohen[8] developed this ERCS, there are many efforts have been paid to

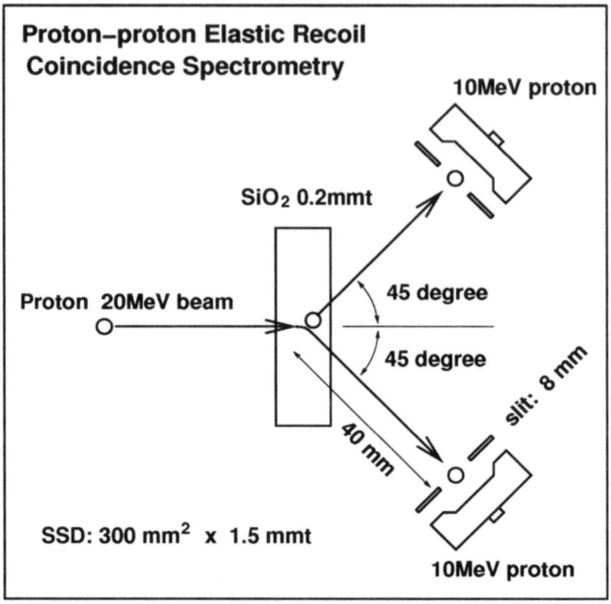

Figure 3. Schematic view of detectors for proton-proton ERCS. Scattered protons of incident beam and recoil from target are detected with coincidence technique.

improve and extend this spectrometry[9,10,11,12]. Recently, extensive works by using multi-detector array have been reported[13].

Fig. 3 shows schematic view of our detection system in which two Si-detectors are located at $\pm 45°$ with respect to the beam axis. Since the kinematic condition, scattered two protons always possess invariant separation angle of $90°$.

When a mineral sample is irradiated by high energy proton beam, one can imagine quite complex spectrum of scattered particles from elastic and inelastic collisions at silicon, oxygen, aluminum, carbon and so on. Fig. 4 shows the single and two-fold coincidence spectra measured by the same detector. Scattering yield from proton is less than one tenth of the total yield, however, the coincidence condition can only select scattering between two protons. It should be noted that the chance coincidence is one of the most disturbing background which can be reduced by better time resolution.

Scattering yield of the coincidence measurement is consistent to the previous report of singles measurement[14]. Since nucleus-nucleus collision is dominated, the cross section is about three times of Rutherford cross section.

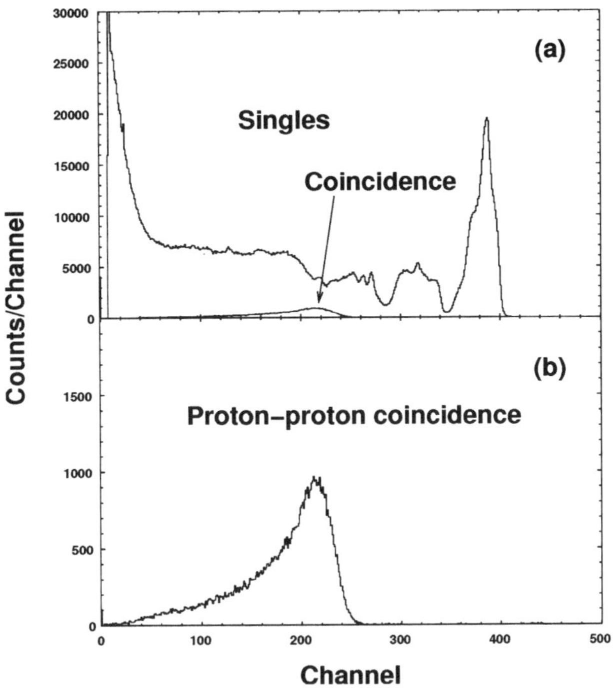

Figure 4. (a) A singles and a two-folded spectra of scattered protons. (b) The same spectrum of the coincidence with expanded scale.

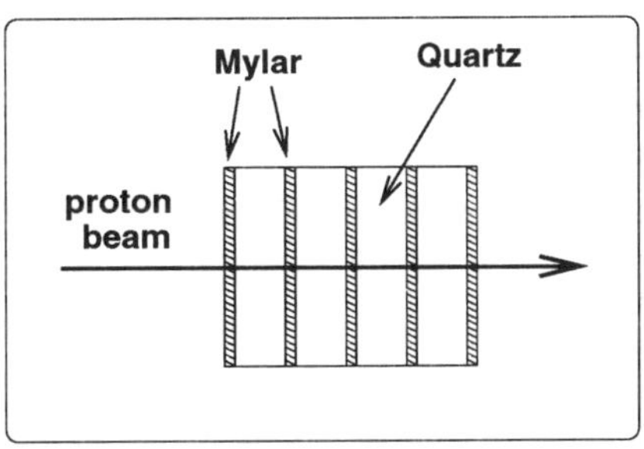

Figure 5. A stacked target consists of five mylar films and four quartz glass plates.

One of the most distinguished advantages of the ERCS is depth information of the scattering position. Difference of stopping powers between incident beam and scattered proton causes position dependence of sum energy. In Fig. 5, a stack of organic sheets made of mylar and quartz glass plates is shown as a target of irradiation. The mylar films are thin and

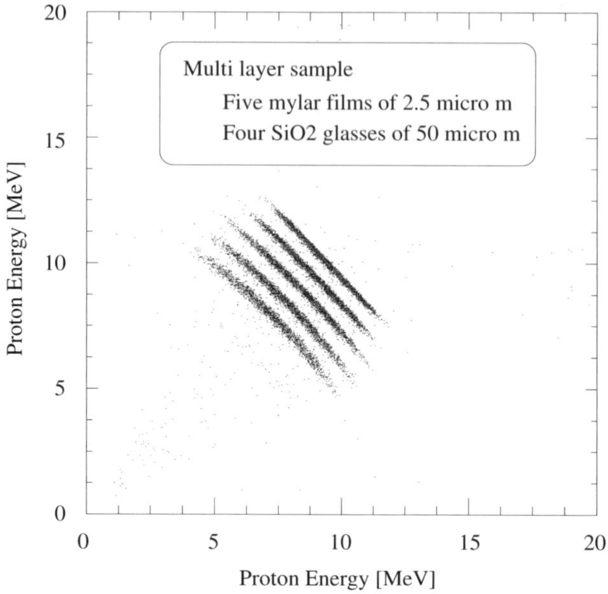

Figure 6. 2-dimensional spectrum for the stacked sample shown in Fig. 5

possess certain amount of hydrogen, however, the plates of quartz glass contain no hydrogen. Measured spectrum is shown in Fig. 6 where x and y axes show proton energies measured by the two detectors located at the scattering angles of $\pm 45°$, respectively. The clearly appeared five lines are corresponding to five mylar films.

For the determination of the hydrogen content, we have developed empirical method to reproduce hydrogen depth profile from the measured sum energy spectrum on the basis of stopping process which reduces proton energies through rather thick target material. In our analysis, the energy scale in the sum energy spectrum can be converted to the depth of the scattering position in a unit of mg/cm^2. Furthermore, the second chance collisions occur in the passage from scattering position to detectors. The unfavorable another scattering reduces the yield by a factor of $e^{-\mu x}$ which was initially pronounced by Cohen[8].

4. Result

For the analysis of the melt inclusions in quartz crystal, the quartz is grounded to become a thin plate. Both surfaces should be well polished

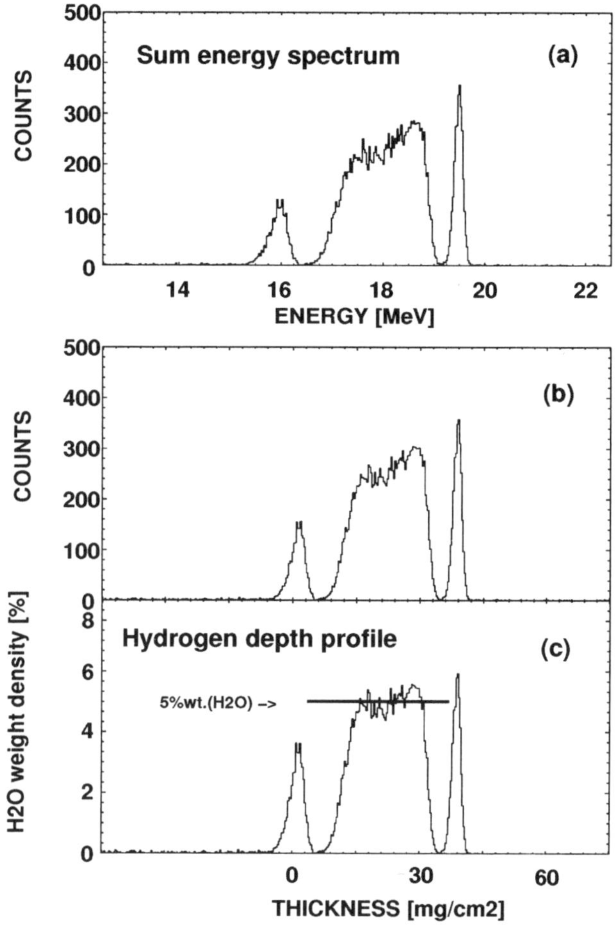

Figure 7. (a) Measured sum energy spectrum for a quartz crystal with a a melt inclusion. Two peaks correspond mylar films attached to both sides of the quartz crystal. (b) Reduced spectrum where abscissa indicates depth position in a unit of mg/cm^2. (c) Depth profile of hydrogen content which is modified by a factor of $e^{-\mu x}$. Horizontal line indicate 5 wt.% (H_2O).

like as mirror and thickness of the plate is less than 0.2 mm. In order to sustain the tiny plate at the irradiation position, two mylar films were used to hold it. The measured spectrum of sum energy is shown in Fig. 7(a). Two sharp peaks corresponds to the hydrogen in the thin mylar films. The yields of the peaks are quite available for calibration as standard and the estimation of the second chance collision rate.

When we consider stopping process depending on the proton energy, abscissa of the energy spectrum can be converted to become depth position in a unit of mg/cm^2 where a small square term of the relation can recover linearity in the weight-surface-density scale, as shown in Fig. 7(b). Next, exponential decrease by the second chance collision is compensated by a multiplier factor of $e^{\mu x}$. In the depth profile in Fig. 7(c), height of the spectrum can indicate water content as a weight ratio. The horizontal line indecates 5 w.t. % (H_2O) which is a measure of water content in the melt inclusion.

5. Summary

Hydrogen analysis with ERCS is successfully progressing for geological samples of volcanic melt inclusions and combined with microbeam irradiation system. Range of the hydrogen analysis is quite wide from several ppm to a few dozen percent. Practical methods have been developed to convert sum energy spectrum to become hydrogen depth profile. Hydrogen density has been determined to be $\sim$ 5 w.t.%(H_2O) for melt inclusions in quartz crystal taken from Zao region.

Acknowledgements

The authors thank Emeritus Professor K. Furuno for his fruitful discussions. This work is supported in part by the Grant-in-Aid for Scientific Research of Japan Ministry of Education, Culture, Sports, Science and Technology, under grant no. 16540436.

References

1. H. Watanabe and Y. Sudo, *Nucl. Inst. and Meth.* **B210**, 1 (2003).
2. G. Dollinger, G. Datzmann, A. Hauptner, H.J. Körner, P. Reichart and B. Volckaerts, *Nucl. Inst. and Meth.* **B210**, 6 (2003).
3. S.H. Sie, G.F. Suter, A. Chekhmir and T.H. Green, *Nucl. Inst. and Meth.* **B104**, 261 (1995).
4. M. Kurosawa, S. Sueno, K. Shima, H. Oshima, S. Ishii, H. Kamiya, S. Kimoto, H. Ohyi and K. Hayashi, *Nucl. Inst. and Meth.* **B142**, 599 (1998).
5. K. Sasa, H. Oshima, Y. Yamato, T. Komatsubara, T. Katabuchi, K. Shima, K. Furuno, M. Kurosawa and N. Yanagisawa, *Nucl. Inst. and Meth.* **B190**, 287 (2002).
6. K. Furuno, T. Komatsubara, K. Sasa, H. Oshima, Y. Yamato, S. Ishii, H. Kimura and M. Kurosawa, *Nucl. Inst. and Meth.* **B210**, 459 (2003).

7. K. Furuno, M. Kurosawa, T. Komatsubara, K. Sasa, Y. Yamato, S. Ishii and H. Ohshima, *Japanese Magazine of Mineralogical and Petrological Sciences* **33**, 51 (2004).
8. B.L. Cohen, C.L. Fink and J.H. Degnan, *J. Appl. Phys.* **43**, 19 (1972).
9. P. Paduschek, P. Eichinger, *Appl. Phys. Lett.* **36**, 62 (1980).
10. H.C. Hofsäss, N.R. Parikh, M.L. Swanson, W.K. Chu, *Nucl. Inst. and Meth.* **B45**, 151 (1990).
11. D. Dujmić, M. Jakšić, N. Soić, T. Tadić, I. Bogdanović, *Nucl. Instr. and Meth.* **B111**, 126 (1996).
12. K.A. Sjöland, P. Kristainsson, M. Elfman, K.G. Malmqvist, J. Pallon, R.J. Utui, C. Yang, *Nucl. Instr. and Meth.* **B124**, 639 (1997).
13. P. Reichart, G. Dollinger, A. Bergmaier, G. Datzmann, A. Hauptner and H.J. Körner, *Nucl. Inst. and Meth.* **B197**, 134 (2002).
14. J.W. Burkig, J.R. Richardson and G.E. Schrank, *Phys. Rev.* **113**, 290 (1959).

SECTION VII

INNOVATIVE INSTRUMENTATION: DETECTORS AND ELECTRONICS

NEW LINES OF RESEARCH WITH THE MAGNEX LARGE-ACCEPTANCE SPECTROMETER

A.CUNSOLO[1,2], F.CAPPUZZELLO[1], A.FOTI[2,3], A.LAZZARO[1,2], S.E.A.ORRIGO[1,2], J.S.WINFIELD[1], C.NOCIFORO[1], A. KHOUAJA[1], H. LENSKE[4]

1. INFN - Laboratori Nazionali del Sud, Catania, Italy
2. Dipartimento di Fisica e Astronomia, Università di Catania, Catania, Italy
3. INFN - Sezione di Catania, Catania, Italy
4. Institut für Theoretische Physik, Universität Giessen, Giessen, Germany

A new design of magnetic spectrometer, MAGNEX, is under construction for the INFN-LNS, Catania. The unique features of MAGNEX are its solid angle of acceptance (51 msr), momentum acceptance (±10%), overall momentum resolution of 1/2000 and mass resolution of 1/200, together with a focal plane detector having a low detection threshold (0.5 MeV/A). The spectrometer is based on a 55° bend angle dipole magnet with mean radius of 1.6 m. It is designed for a maximum rigidity of 1.8 Tm. Despite the large acceptance, a good momentum resolution is achieved by a combination of careful ion-optical design and software ray-reconstruction. The latter depends on three things: the availability of detailed field maps, the precise measurement of position and angle by the detection system, and the solution to high order of the equation of motion based on, in our case, the program COSY INFINITY. The MAGNEX spectrometer, connected with the broad choice of both stable and radioactive beams at the LNS, will provide new opportunities for, e.g., spectroscopy of weakly-bound nuclei by direct reactions, reaction mechanisms with large isospin and nuclear astrophysics.

1. Introduction

The concept and layout of the MAGNEX spectrometer has been described in refs. [1-3]. In brief, it is a large-acceptance device (50 msr) based on a vertically-focussing quadrupole and 55° bend-angle dipole. The angles and profiles of the dipole entrance and exit pole faces are used to correct partly the aberrations in the ion-optics. Further corrections are performed in software by a ray-reconstruction technique, resulting in an expected average momentum resolution of about 1/2000. The ray-reconstruction is based on a solution to high orders of the equation of motion using, in our case, the program COSY INFINITY of Berz [4]. Detailed measured field maps in five vertical planes are used. A position-sensitive timing detector (PSD) between the target and quadrupole gives both the angle of the scattered particles and a start signal for the time-of-flight. The focal plane detector (FPD) measures positions and angles and provides particle identification [5].

The spectrometer will be commissioned at the beginning of 2005. In this paper we focus on some of the experimental opportunities opened by the advent of MAGNEX at the LNS, including its use with radioactive ion beams (RIB) from the ISOL-type EXCYT facility [6]. In particular a deeper analysis of the MAGNEX capabilities is shown in reference to a specific RIB experiment simulated with our Cosymag routine libraries [7].

2. Direct reactions with Tandem stable beams

Recent studies have shown how the (^{7}Li,^{7}Be) reaction at Tandem energies is a powerful tool to explore the excited states of light neutron rich nuclei [8]. In ref. [9] an energy resolution of 50 keV has been obtained in the excitation energy spectrum of ^{11}Be by detecting the ^{7}Be ejectiles with the Split-Pole magnetic spectrometer at IPN-Orsay. The observation of narrow resonances embedded in the continuum (BSEC), well beyond the ^{11}Be neutron emission threshold, has raised the question whether this phenomenon is connected to the exotic properties of the ^{11}Be structure. It has stimulated similar studies for different ions such as ^{12}B, ^{14}B, ^{15}C and ^{7}He. A clear indication of the BSEC has been obtained [10,11] for the ^{15}C nucleus, which presents many similarities with ^{11}Be. It is worthwhile to note that both ^{11}Be and ^{15}C have a structure of three neutrons coupled to an integer number of α particles. Such results have determined the development of sophisticated microscopic theories based on QRPA. The microscopic model of Dynamical Core Polarisation (DCP) of refs. [12,13], which accounts for the correlation of core phonons with single particle excitations of the external neutron, predicts the existence of narrow resonances at low excitation energy for light neutron rich odd nuclei. The calculated response functions have shown how these narrow resonances cannot be explained by two quasi-particle (2QP) excitations and need a broader phase space including at least 4QP configurations. This is due to the presence of a weakly bound, and thus easily polarizable core, e.g., ^{10}Be for the ^{11}Be nucleus and ^{14}C for the ^{15}C, that effectively exchange energy with the unpaired neutron. Consequently the energy produced in a collision, proceeding through a direct mechanism, can be directly transferred to the core, the valence neutron being weakly influenced. In the extreme case of BSEC the valence nucleon remains bound to the core even when the energy transferred to the whole nucleus by the direct process would be enough to extract it.

A systematic study of the Nα + 3n nuclei up to the Iron-Nickel region via the (^{7}Li,^{7}Be) reaction at Tandem energies would allow us to follow the evolution of this phenomena as a function of the mass, charge and binding energy. This would greatly help to understand the microscopic origin of the BSEC and to clarify whether this phenomenon can be described within the general framework of mean field theory or many-body correlations are unavoidable. This study has been hindered up to now by the reduction of the cross sections for heavier systems at the low incident energy necessary for achieving the high resolving power in the energy spectra. The use of the MAGNEX spectrometer with its large solid angle (more than 25 times the Split-Pole) will open new opportunities in this field. The energy resolution achievable will be of the order of 50 keV, thus allowing a clean separation of most of the excited states.

3. Direct reactions with EXCYT RIB's

The (^{7}Li,^{7}Be) studies described above may be extended to the use of RIB's from the EXCYT facility. Inverse kinematic reactions and the large acceptance of MAGNEX would be used to overcome the low secondary beam intensity. One of the first beams from EXCYT will be ^{8}Li [14]. We can use this beam to study the neutron-rich nucleus ^{8}He. The proposed experiment is ^{7}Li(^{8}Li,^{7}Be)^{8}He, i.e., the ^{8}Li beam bombards a ^{7}Li target and ^{7}Be reaction products are detected in the focal plane of the spectrometer. Besides the possibility of exciting bound states in the continuum, in a similar manner as for ^{11}Be, the low-lying level structure of ^{8}He could be more firmly established. In the most recent evaluation of mass 8 nuclei, three excited states in ^{8}He (S_{2n} = 2.14 MeV) are listed at ~ 3.1, 4.36 and 7.16 MeV [15] with evidence for another at 6.03 MeV [16]. In fact, it is not clear whether the 2^+ first-excited state at ~ 3.1 MeV is a single state or two states: some groups have reported a level near 2.7 MeV [17,18], others give the excitation energy as about 3.6 MeV [19-21] (in addition, Belozerov et al. [17] report a level at 1.3 MeV). With a high resolution spectrum from a generally unselective reaction as (^{7}Li,^{7}Be), one might expect either to excite both levels (if there are two) or determine a more precise energy of a single level.

For the above ^{8}He experiment a complete simulation accounting for the ray reconstruction technique [7] is presented in the following. In the simulation a beam of 57 MeV incident energy and of 3×10^5 pps intensity is assumed, according to the preliminary predictions for the EXCYT facility [14]. From the systematics for light nuclei of the (^{7}Li,^{7}Be) reaction at 57 MeV, the cross section for the Gamow-Teller transition from the ^{8}Li$_{gs}$(2^+) ground to the ^{8}He(2^+) excited state is estimated to be about 100 ÷ 200 μb/sr at forward angles. The weak intensity of the beam and the low value of the cross section put severe constraints on the solid angle and target thickness, in order to perform the experiment in a reasonable time. For this purpose the large solid angle (50 msr) of MAGNEX, connected to the precise reconstruction of the scattering angle and the effective compensation of the kinematic effect, are key elements for the feasibility of this experiment. The kinematic broadening of the peaks in the energy spectra due to the large scattering angle interval (from 1° to 15°) is more than 2 MeV when the spectrometer aperture is fully open. Under these conditions the counting rate per each micron of target thickness is about 0.8 counts / 1 μm × hour, while the effect of target on the energy resolution is about 28 keV / 1 μm. To limit this contribution to the energy resolution to within 200 keV a counting rate of about 6 counts / hour is achievable. This leads to the necessity to fix the spectrometer in the same conditions for at least 100 hours to get enough counts in the GT transition peak. In Fig. 1 the initial conditions of the simulations are shown for a sample of 20000 particles distributed over 12 simulated levels, which corresponds to about 300 hours of measurement. The ground and the known excited states at 2.7, 3.6, 6.03, 7.16 MeV are visible. A broad resonance known at 4.5 MeV is also included. In the simulation the 2.7 and 3.6 MeV states are both considered to exist, and the experimental goal is to resolve them. For each ^{8}He excitation two lines are present in the spectra, arising from either the population of the ground or the 0.429 MeV bound first excited state of the ^{7}Be ejectiles. The strong kinematic effect is evident in the plot, appearing as a noticeable curvature of the kinetic energy lines as a function of the scattering angle. In the right panel of Fig. 1 the projected kinetic energy spectrum is shown, emphasizing the need to measure the scattering angle with good precision. It is important to bear in mind that any possible angular segmentation of the data is hindered by the very low counting rate of this experiment.

The distribution of particles along the focal plane after tracking through the spectrometer is shown in Figure 2. In the simulations all the active and dead layers are included realistically. To reduce the kinematic effect, the detector has been shifted to the predicted location of the focal plane (as allowed by MAGNEX), and the quadrupolar and sextupolar surface correction coils are used. The scatter plot and the one dimensional spectrum give an idea of the difficulties with the large acceptance condition if trajectory reconstruction is not employed, even with an optically-refined spectrometer such as MAGNEX [3]. The position resolution is obviously not enough to distinguish the peaks at 2.7 and 3.6 MeV.

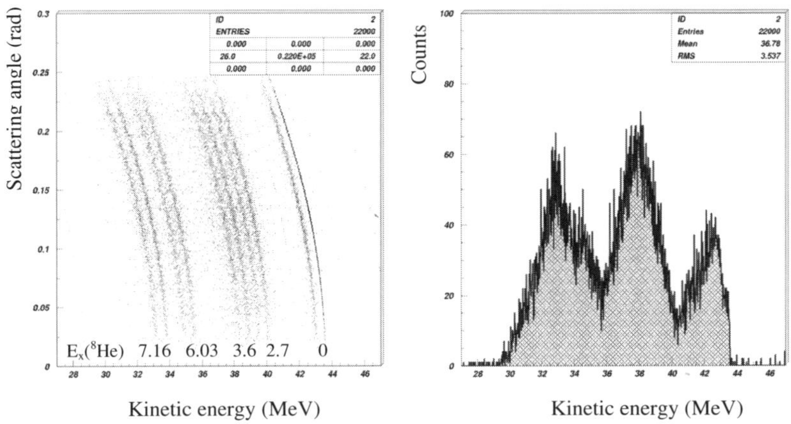

Figure 1. Left panel; initial conditions (after the target) for the simulation of the ^{7}Li(^{8}Li,^{7}Be)^{8}He reaction at 57 MeV. Right panel; initial energy spectrum.

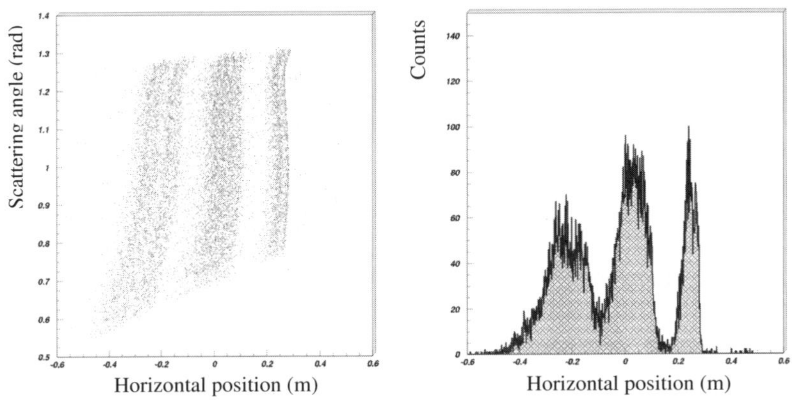

Figure 2. Left panel; scatter plot at the focal plane for the ^{7}Li(^{8}Li,^{7}Be)^{8}He reaction at 57 MeV. Right panel; focal plane position spectrum.

In Figure 3 the result of the application of the ray reconstruction method is shown. The order of reconstruction has been set to the 11th. The reconstructed kinematic scatter plot clearly indicates the power of this technique in compensating both the kinematic effect and the effects of residual aberrations that were observed in Figure 2. The excitation energy spectrum shows the clear separation of the peaks in the region of interest around 3 MeV. The broadening of the peaks is almost entirely due to the effect of target thickness which is unavoidable and not dependant on the instrument itself.

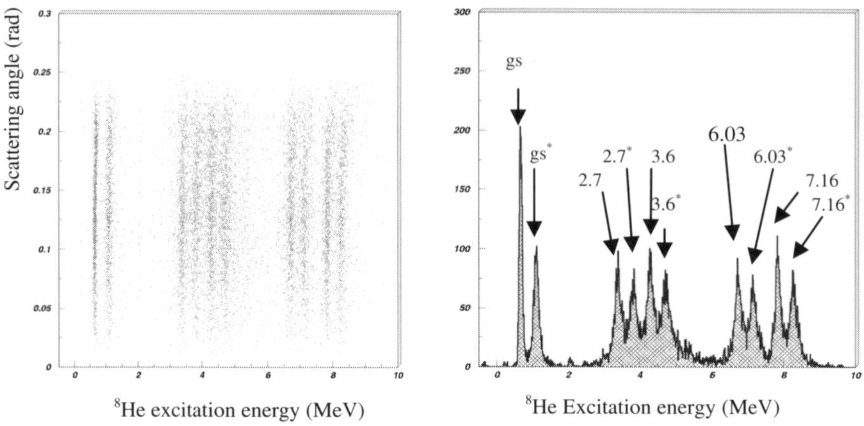

Figure 3. Reconstructed scatter plot (left) and energy spectrum (right) at the target for the ^{7}Li(^{8}Li,^{7}Be)^{8}He reaction at 57 MeV. The 11th order algorithm of Cosymag has been used.

References

1. A. Cunsolo et al., Proc. Workshop Giornata EXCYT, Catania (1996) pp. 143-161, Proc. Workshop II Giornata EXCYT, Catania (1997) pp. 71-80.
2. A. Cunsolo et al., Proc. 9th Intl. Conf. on Nucl. Reaction Mechanisms, Varenna, 2000, p. 661.
3. A.Cunsolo et al., Nucl. Instr. and Meth. A 481 (2002) 48; 484 (2002) 56.
4. M. Berz, Nucl. Instr. and Meth. A 298 (1990) 473.
5. A.Cunsolo et al., Nucl. Instr. and Meth. A 495 (2002) 216.
6. G.Ciavola et al., Nucl. Phys. A616 (1997) 69c.
7. A. Lazzaro, PhD thesis, University of Catania, (2003), A. Lazzaro et al., Proc. 7th Int. Computational Accelerator Physics Conf., East Lansing, Michigan, Oct. 2002 (M. Berz, ed., IOP Publishing, in press).
8. F. Cappuzzello et al., Proc. Int. Conf. on Nuclear Physics at Border Lines, Lipari, May 2001 (G. Fazzio, G. Giardina, F. Hanappe, G. Immè, N. Rowley, eds., World Scientific, Singapore, 2002) p. 64.
9. F. Cappuzzello et al., Phys. Lett. B 516 (2001) 21.
10. C. Nociforo, PhD thesis, Univ. of Catania (2002) and Proc. XXXVII Zakopane School of Physics, 2002, Acta Physica Polonica B 34 (2003) 2387.
11. S. Orrigo et al., these proceedings, and F. Cappuzzello et al., EuroPhys.Lett. 65 (2004) 766.
12. G. Baur, H. Lenske, Nucl. Phys. A 282 (1977) 201.
13. H. Fuchs et al., Nucl. Phys. A 343 (1980) 133.
14. G. Cuttone, Workshop on forthcoming facilities at LNS, Catania, 2003.

15. J.H. Kelley et al., Energy levels of light nuclei A = 8, preliminary version #1, TUNL Nuclear Data Evaluation Group, February 2002.
16. H.G. Bohlen et al., Prog. Part. Nucl. Phys. 42 (1999) 17.
17. A.L. Belozerov et al., Izv. Akad. Nauk. SSSR Ser. Fiz. 52 (1988) 100.
18. K. Markenroth et al., Nucl. Phys. A 679 (2001) 462.
19. A.A. Korsheninnikov et al., Nucl. Phys. A 588 (1995) 123c.
20. W. von Oertzen et al., Nucl. Phys. A 588 (1995) 129c.
21. Y. Iwaka et al., Phys. Rev. C 62 (2000) 064311.

DEVELOPMENT OF POSITION SENSITIVE Ge DETECTOR

T. FUKUCHI, S. SHIMOURA, E. IDEGUCHI, H. BABA,
M. TAMAKI AND M. NIIKURA

Center for Nuclear Study, Graduate School of Science
University of Tokyo
2-1 Hirosawa, Wako, Saitama 351-0198, Japan
E-mail: fukuchi@cns.s.u-tokyo.ac.jp

S. OTA

Department of Physics, Kyoto University
Sakyou-ku, Kyoto 606-8502, Japan

M. KUROKAWA

RIKEN,
2-1 Hirosawa, Wako, Saitama 351-0198, Japan

We have been developing three-dimensional position sensitive Germanium semiconductor detector using an Artificial Neural Network(ANN). For training training of the ANN, pulse shape sampling was performed using flash-ADCs. In order to reduce the time for sampling, several method were proposed and tested.

1. Introduction

We have been developing the Gamma-Ray detector Array with Position and Energy sensitivity (GRAPE) which is mainly used for detecting γ-ray from fast moving nuclei. The GRAPE consists of 18 high-purity Germanium (Ge) detectors. All detectors have two planar-type Ge crystals (60 mm in diameter and 20 mm thickness). The electrode of Ge crystal is segmented in 3 $\times$ 3. These Ge detectors were manufactured by the company Eurisys Measures. Figure 1 shows schematic view of the Ge detector using in the GRAPE. The interacting positions of γ-ray in perpendicular with electrodes are extracted by the analogue modules [1]. This one-dimensional position sensitivity enable to correct the Doppler shifted γ-ray energy accurately. For more precise correction and so-called γ-ray tracking, we start to develop the three-dimensional position sensitive Ge detector. The three-dimensional extraction of interaction positions are based on pulse shape digitization of not only true signal from detector but also the transient signal. We are presently developing Artificial Neural Network (ANN) algorithm to deduce the γ-ray interaction positions. We are also developing the pulse shape sampling method for training of the ANN.

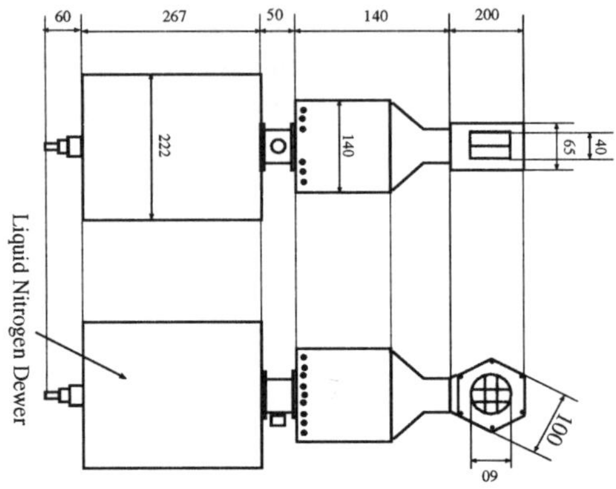

Figure 1. Schematic view of the Ge detector using in the GRAPE.

2. Pulse shape sampling

In order to train the ANN, the pulse shape sample in each detector position is need for a supervisor. One method of three-dimensional pulse shape sampling is measure a detector position using colimator and another detector with slit. In this method, γ-ray which is injected in remitted direction and measuring position is restricted by correction of Computon event scattered angle is fixed by slit and another detector. However scanning whole region of detector needs long time. To shorten the sampling time, we perform the sampling by combination of γ-ray injection from two perpendicular direction. Practically, (1)Gamma-ray is injected into the detector in parallel with electrodes. (2)Gamma-ray is injected in perpendicular with electrodes. (3)The sampling data of a point is made from gating by the data (1) on the data(2). In this measurement, an automatic measured system was used. The total and 9 segment signals were taken. The signals from the crystal are amplified by the charge-sensitive pre-amplifiers. After pre-amplifiers, operational amplifiers were used for matching with dynamic range of flash-ADCs. The rising pulse shape were acquired in the time range of 1 μs. The

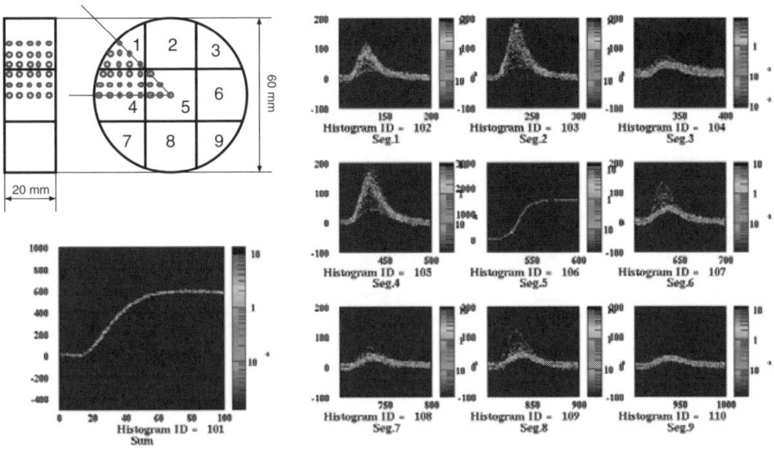

Figure 2. The sampling data positions (red circles) and the example of the pulse shape at center position in 10 mm depth.

data set of 59 different positions and produce 145 sampling data positions by combination of two direction injection. Figure 2 shows sampling data positions (red circles) and example of the pulse shape data at the center position in 10 mm depth.

3. Algorithm and result

We had attempted to apply an ANN algorithm to determine the γ-ray interaction positions. The ANN algorithm we tried is based on the Kohonen's method [2]. In this algorithm, first, a distance between input and output d_j is calculated by a following equation,

$$d_j = \sum (x_i(t) - w_{ij}(t))^2 \tag{1}$$

where i and j are the data number of the flash-ADCs and detector position respectively. Parameter t is event number. Parameter x_i and w_{ij} are input value and pulse shape corresponding to the each detector position respectively. Second, minimum search of d^j is executed and interaction position is deduced. For the last parameter w_{ij} is modified according to the deduced position and the sampling position by a following equation,

$$w_{ij}(t+1) = w_{ij} + g(t)(x_i(t) - w_{ij}(t)), \tag{2}$$

where g is gain parameter. In this modification, the parameters of the circumference are also modified. Figure 3 shows example of the ANN results. For the input, sampling data was arranged in random order of each

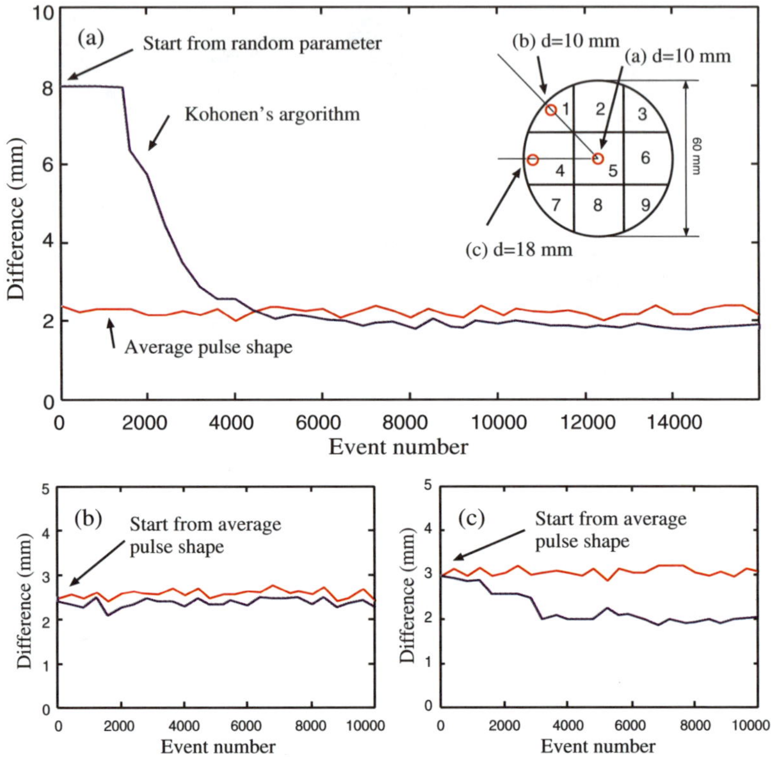

Figure 3. The example of the ANN results. The ANN training is started form random parameter and average pulse shape parameter for (a) and (b)(c) respectively. Tested positions are also shown by red circles.

detector positions. The result of center position is shown in fig.3(a). The vertical axis is difference between interaction position which is derived form the ANN and sampling position. The horizontal axis is the event number. The red and blue line indicate result using the ANN and result using an average pulse shape respectively. The average pulse shape is constant parameter. The ANN result in 3(a) is started from random parameter. The ANN results of fig. 3(b) and (c) are started from average pulse shape. Starting from random parameter and average parameter case need about 8000 events an 3000 events need to converge respectively. The ANN makes good result in some positions.

4. Future plan

In future developments, we plan sampling method using the Compton scattering for more effective one. In Compton scattering, the energy of the scattering angle θ is related to the energy of the incident photon E_0 and the energy of the scattered photon E_1 by the formula,

$$E_0 = E_1 + E_2, \quad cos\theta = 1 + m_e c^2 (\frac{1}{E_0} - \frac{1}{E_2}). \tag{3}$$

In this method, the event scattered in one crystal and deposited full energy in another crystal are stored. This sampling is performed by minimum search of distance d_j in eq.(1) through the all possible combination of interaction positions based on scattering angle. We plan two step sampling, (1)About 5 mm interval using collimated γ-ray on a test-bench. (2)The sampling of 1-2 mm resolution by the Compton scattering (Starting from the parameter made by the sampling(1)). The goal of development is perfection of the all sampling just putting a standard γ-source on the target position.

Acknowledgments

This work was supported by Center for Nuclear Study (CNS), Graduate School of Science, University of Tokyo.

References

1. M. Kurokawa et al., IEEE **50**, 1309 (2003).
2. T. Kohonen, Self Organization and Associative Memory, third edition, Spinger-Verlag 1990.

STATUS OF THE AGATA PROJECT

DINO BAZZACCO
INFN Padova, Via Marzolo 8, I-35131, Padova, Italy

New accelerator facilities for radioactive-ion beams and high-intensity stable beams will start operation in a few years. They will provide interesting opportunities for exploring unknown territories of the nuclear landscape but the harsh experimental conditions require the construction of a new generation of detector arrays for gamma-ray spectroscopy built fully from germanium detectors and based on the emerging technique of gamma-ray tracking. The "Advanced GAmma Tracking Array" (AGATA), proposed in Europe, will be built out of 180 highly segmented high purity germanium crystals operated in position sensitive mode by means of digital data techniques and pulse shape analysis of the segment signals. AGATA will be capable of measuring gamma radiation in a large energy range (from ~10 keV to ~ 10 MeV), with the largest possible photopeak efficiency (25 % at $M_\gamma = 30$) and with a good spectral response. In particular, its very good Doppler correction and background rejection capabilities will allow to perform "standard" γ-ray spectroscopy experiments using fragmentation beams with sources moving at velocities up to $\beta \sim 0.5$. The talk reviews the status of development of the γ-ray tracking technique and the present design of AGATA.

1. Introduction

In the last decade, EUROBALL and GAMMASPHERE have been the highest efficiency 4π detector arrays available for in beam γ-ray spectroscopy. These arrays are built out of ~100 Compton-suppressed high-purity germanium (HPGe) spectrometers arranged in a spherical configuration around the reaction point and provide, for 1 MeV γ-rays, a total peak efficiency ε_{ph} of about 10 % and a P/T-ratio of about 60 %. These figures combine to a good selectivity for identification of weak reaction branches and, in fact, the two detectors have played a major role in experimental nuclear structure studies, in particular in the study of high spin phenomena.

The γ-ray spectroscopy community has now the opportunity to extend the field of nuclear structure studies exploiting the radioactive ion beams provided by the next generation of accelerator facilities of ISOL and in-flight fragmentation type. It is well realized that, due to low beam intensities, large Doppler broadenings and high backgrounds, the experimental conditions at such facilities will be very challenging. To cope with them, both the total peak efficiency and the selectivity of our arrays must be improved and it is a matter of fact that this cannot be obtained by simply increasing the number of detectors. The reason is that, to obtain a good P/T ratio the germanium crystals must be surrounded by BGO suppression shields, which, however, take a consistent fraction of the solid angle and limit the total peak efficiency of the so-called 4π arrays to no more than 10 % (at 1 MeV). The solution of the problem will come from recent advances in crystal segmentation technology and digital signal processing, which make it possible to operate large-volume germanium detectors in a position sensitive mode. This means that besides the total energy released in the Ge crystal we will also know the energy and the position of the individual interaction points within the detector. With a position resolution of a few mm it will be possible reconstruct the interaction history of the absorbed gamma radiation, realizing the so-called gamma ray tracking concept [1]. According to realistic Monte Carlo simulations, an array consisting of a relatively small number (100-200) of such detectors would have an efficiency of up to 50 % and a P/T ratio of ~70 %. It should also be remarked that an important aspect of γ-ray tracking is the possibility to determine with very good precision (~ 1°) the emission direction of the reconstructed transitions, allowing for an almost perfect correction of Doppler broadening effects for gammas emitted by nuclei recoiling at velocities as high as $\beta = 0.5$.

The unique capability of the γ-ray tracking arrays will allow the study rare reaction channels produced in experiments with weak beam intensities (e.g. radioactive beams), addressing many questions that are still open in nuclear structure, astrophysics and fundamental interactions.

Finally, it is clear that, besides the perspective use for fundamental research, the principle of γ-ray tracking is also extremely interesting for its possible application in the field of γ-ray imaging.

2. Gamma-ray Tracking Methods

The feasibility of γ-ray tracking has been the subject of extensive R&D performed in the last seven years by the TMR network "Gamma Ray Tracking Detectors" [2] in Europe and the GRETA group in the USA [3]. The main lines of development are outlined in the following.

Highly-segmented HPGe detectors. The development of highly segmented germanium detectors, both in cylindrical and planar configuration, is pursued in several laboratories. The following brief summary is limited to closed-end coaxial detectors, which are believed to be more suited for the construction of 4π tracking arrays. The 36-fold segmented "hexa-conical" detector for GRETA at Berkeley, was the first prototype to be studied in details [4]. In Italy, we are using a 25 fold segmented detector (called MARS) that has 6 angular sectors, 4 transversal slices and a small segment on the front face [5]. More recently the36-fold segmented prototypes detectors of AGATA have also been delivered.

The goal of these developments is to produce the building blocks of highly-packed 4π detector systems. The issues of reliability and maintainability of the crystals is solved by exploiting the encapsulation technology developed for the cluster detectors of EUROBALL and MINIBALL [6] where (Fig. 1) the germanium crystal is kept in permanent vacuum by a thin aluminum can.

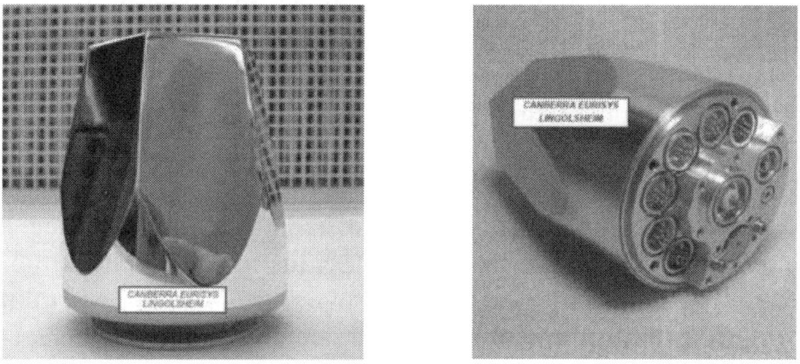

Figure 1. The prototype (symmetric) HPGe crystal of AGATA before and after encapsulation. The 36-fold segmentation of the outer electrode is not shown.

Digital electronics and DAQ. To preserve the full signal-shape information that is needed to extract energy, timing and position of the interaction points, the electronics has to sample the preamplifier signals with fast ADCs (e.g. 100 Ms/s, 14 bit) located as close as possible to the detectors. This gives 200 MB/s/channel and the impressive value of 7.4 GB/s for a 36-fold segmented detector. Several successive steps of real time data processing have to be employed to reduce this figure to a more reasonable value. The first reduction comes from the fact that energy and timing of the signals can be determined in real time using efficient digital signal processing techniques [7]. After this, only a small number (50÷60) of samples relative to the leading edge of each pulse has to be transferred to the data processing step where the position of the interactions is determined and a further rate reduction is achieved. Considering a singles counting rate of 10 kHz, the final data rate per detector is down to ~1 MB/s. The total data flow into the DAQ of a 4π γ-ray tracking array is of the order 100 Mbytes/s. Event building of such data is achieved by using a common "time-stamped" clock for all the ADCs.

Pulse shape analysis (PSA). The input for the tracking algorithms consists of energy and three-dimensional position and of each point of interaction. This information is obtained from an analysis of amplitude and shape of the signals induced at the detector electrodes by the charge collection process. The transient signals seen in the segments neighbors to those collecting the net charge released by the interaction are an important feature of segmented detectors. These signals have no charge at the end of the collection process and their amplitude and shape depend on the distance of the interaction point to the border of the segment. Signal shapes and amplitudes can be calculated rather accurately for almost arbitrary crystal geometries using finite element analysis and the method of weighting fields [8].

For the development and test of the algorithms for pulse shape analysis, simplified problems with just a few interactions within the segments are usually considered. The resulting composite signals from the segments are analyzed (decomposed) to extract the number, position, and energy of the original point(s) in various ways. Programs using e.g. genetic algorithms (GA) can determine the number of interactions correctly for ~ 90 % of the events with variances between reconstructed and true positions of typically ~ 2 mm. The performance of the PSA algorithms has been tested extensively with data from the GRETA and MARS prototypes using tightly collimated γ-sources. More recently, some test-beam experiments have been performed to see how they behave in more realistic situations. In this case, the idea is to check the achieved position resolution by looking at the improvement of energy resolution of a severely Doppler broadened peak when the detected energy is corrected using the position of the first interaction point as determined by the PSA procedure.

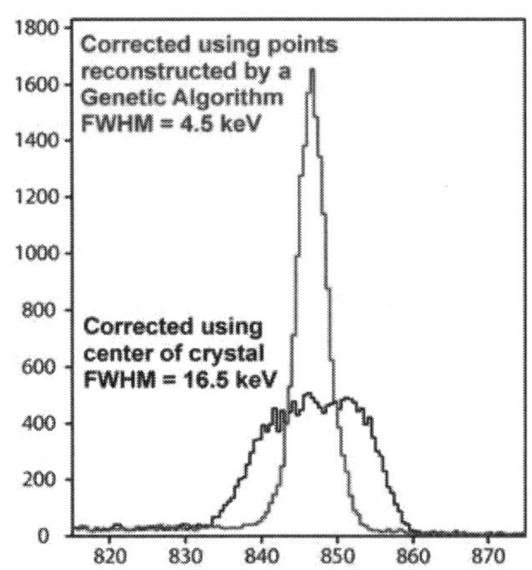

Figure 2. Energy resolution obtained in the in-beam experiment. The Doppler shift correction of the sharp peak is done event-by-event using the position of the first interaction point given by the PSA.

The result reported here was obtained with the MARS detector in a Coulomb excitation experiment using a 240 MeV ^{56}Fe beam on a thin ^{208}Pb target. At this bombarding energy, the spectrum contains, practically, only the ^{56}Fe $2^+ \rightarrow 0^+$ transition at 846.7 keV. The direction of the scattered ^{56}Fe ions, which is needed to perform the Doppler correction, was detected with a precision of ~ 2° by a set of 14 well collimated silicon detectors. The ~ 10^5 collected events have been decomposed into interaction points using a Genetic Algorithm [9], and the final result of this analysis is shown in Fig. 2. The improvement in energy resolution is reproduced by a detailed simulation of the experiment assuming a position resolution of 5 mm FWHM. In calculating the performance of a realistic γ-tracking array, it seems therefore safe to assume that pulse shape decomposition can be done with at least a 5 mm position resolution. Further R&D will most likely improve this value so that the figures given in the following should be taken as safe values.

Tracking algorithms. The main task of γ-ray tracking is to identify the individual transitions of an event and to reconstruct their scattering sequence inside the detector using the energy and the spatial coordinates of the interaction sites. The characteristic features of γ-ray tracking can be best illustrated considering single gamma events: **(i)** Low energy γ-rays end-up into isolated interaction points that are recognized as photoelectric absorption events upon checking the compatibility between γ-ray energy and interaction depth in the detector. **(ii)** Gamma rays with energy from a few hundred keV to some MeV are absorbed in the detector with a sequence of (a few) Compton scattering interactions and a final photoelectric effect (see e.g. Fig. 3).

Such events are reconstructed using a figure of merit that quantifies how well the scattering angles determined form the position of the involved interaction points agree with the values obtained inserting the pertinent energies into the Compton scattering formula. The figure of merit is calculated for all permutations of the interaction points and the event is accepted if the merit of the best permutation is compatible with an empirically defined limit. **(iii)** Pair production events become important above a few MeV. A strong signature of this interaction mechanism is given by the fact that, at least for not too high γ-ray energies, the first point of interaction collects the total energy of the γ-ray minus the 1022 keV needed to create the e^+e^- pair, while the two 511 keV annihilation photons generate their own clusters of interaction points in the vicinity of this site.

The γ-ray tracking algorithms have been developed and tested using data from Monte Carlo simulations where the finite resolution of actual detectors is taken into account by means of position and energy smearing procedures applied to the calculated interaction points. It is important to remark that, due to experimental uncertainties but also because of fundamental limits, every conceivable algorithm will always accept some background events (corresponding e.g. to partial energy release in the detector) and reject good ones. The amount of accepted background can normally be reduced at the cost of an increased rejection of good events: i.e. better P/T implies lower efficiency. Real events involve in general several coincident γ-ray and are reconstructed exploiting the fact [10] that the interaction points of transitions emitted into sufficiently separated directions tend to "cluster" into spatially isolated groups. These clusters of interactions, identified using different algorithms, are validated as individual transitions with the methods explained above.

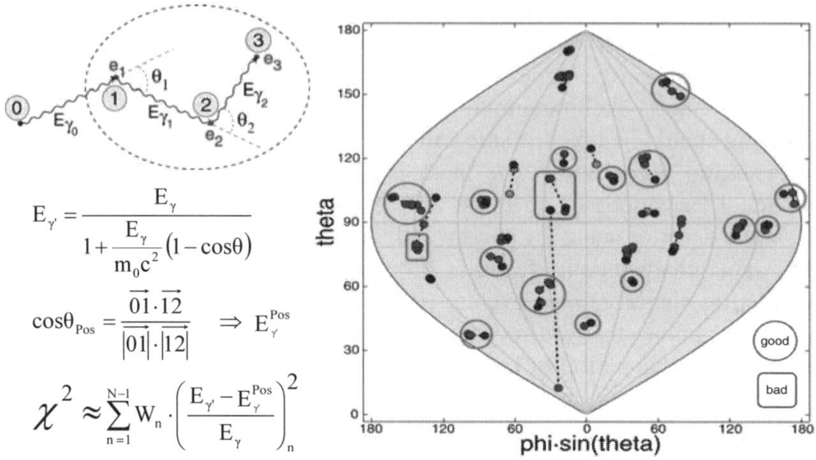

$$E_{\gamma'} = \frac{E_\gamma}{1+\frac{E_\gamma}{m_0c^2}(1-\cos\theta)}$$

$$\cos\theta_{Pos} = \frac{\overrightarrow{01}\cdot\overrightarrow{12}}{|\overrightarrow{01}|\cdot|\overrightarrow{12}|} \Rightarrow E_{\gamma'}^{Pos}$$

$$\chi^2 \approx \sum_{n=1}^{N-1} W_n \cdot \left(\frac{E_{\gamma'}-E_{\gamma'}^{Pos}}{E_\gamma}\right)_n^2$$

Figure 3. Tacking of a gamma ray absorbed in the detector material (indicated by dashed contour) by means of a sequence of 2 Compton scattering events followed by a photoelectric interaction. The right panel shows the reconstruction of a simulated high multiplicity event using the cluster-tracking method.

The reconstruction performance of our algorithms is often benchmarked using a reference detector consisting of a shell of germanium with an inner radius of 15 cm and a thickness of 9 cm. Assuming a position resolution of 5 mm, the cluster-tracking yields $\varepsilon_{ph} = 36$ % and P/T = 60 %, for a gamma ray multiplicity $M_\gamma = 30$ (Fig. 3). The 4π Tracking Array

The "Advanced Gamma Tracking Array" AGATA [11] in Europe and the "Gamma Ray Energy Tracking Array" GRETA [3] in the USA will be the first full implementations of the γ-ray tracking concept.

The specific events triggering these projects are the planned radioactive beams facilities RIA in the US and GSI-future in Europe. However, the detectors are being designed for a more general range of applications, e.g. also for experiments with high intensity stable beams. As already mentioned, peculiar features of radioactive beams are: limited intensities (particularly for the most exotic nuclei); wide range of recoil velocities (from stopped to $\beta = 50$ %); high gamma and particle backgrounds; large γ-ray multiplicities (up to $M_\gamma = 30$). To cope with these conditions, a 4π γ-array with the highest efficiency, selectivity, energy resolution and capability to handle high counting rates is required. These features can only be achieved with a close packed arrangement of germanium detectors, i.e. a 4π shell built from large, highly segmented large volume HPGe crystals.

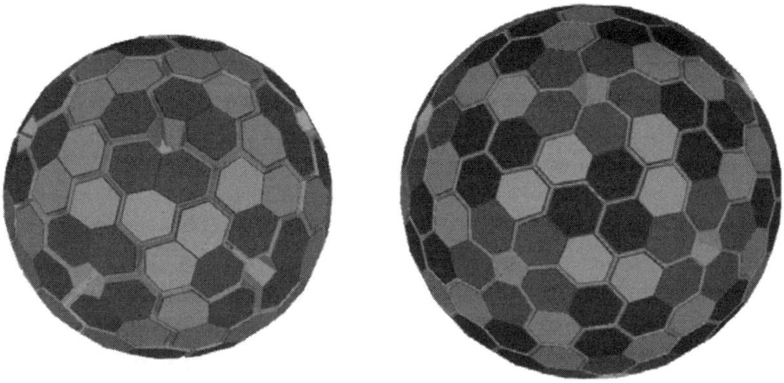

Figure 4. Schematic view of the two 4π configurations considered as optimal for the new γ-ray tracking arrays. The smaller configuration has 120 hexagonal crystals of 2 different shapes (colour coded) that can be packed into 40 triple clusters of two different types as in the figure or into 30 all-equal quadruple clusters. In the larger configuration 180 crystals of three different shapes are packed into 60 all-equal clusters. AGATA has recently chosen the bigger configuration while GRETA has decided for the smaller configuration with quadruple clusters.

The geometric structure of the arrays is based on a geodesic tiling of the spherical surface with hexagons and 12 pentagons. Two possible configurations have been identified (see Fig. 4): a smaller one with 120 hexagons of two different shapes and a bigger one with 180 hexagons of three different shapes. The pentagons will, most likely, not be used and, to minimize inter-detector space losses, the hexagons will be packed into clusters of a few crystals. The 120-configuration will therefore have 40 triple-clusters (of two different types) or 30 all-equal quadruple clusters while the 180-configuration will have 60 (all-equal) clusters. Using the largest available HPGe crystals (80 mm diameter, 90 mm length), the inner radius of the two arrays turns out to be ~18 and ~23 cm (large enough for ancillary detectors), while the total solid angle covered by germanium can be as high as 80% in both cases.

The performance of the candidate arrays has been simulated within the GEANT4 [12] framework considering the actual hexa-conical shape of the crystals and the dead materials from the encapsulation and the canning into clusters [13]. With the conservative assumption of a position resolution of 5 mm at 100 keV, the realistic detectors achieve an efficiency of about 40 % and 25% at γ-ray multiplicity of 1 and 30 respectively. Although this is only ~ 60% of the performance of the ideal shell, it means an increase in selectivity of several orders of magnitude, in particular for high γ-multiplicity experiments. The germanium crystals will be 36-fold segmented and the resulting 4440 or 6660 total segments will provide unprecedented position sensitivity. A key feature of the arrays is the capability to determine the emission direction of the detected γ-ray with a precision of ~ 1°. This corresponds to an effective solid angle granularity of more than 5 10^4 and ensures an energy resolution better than 0.5 % for transitions emitted by nuclei recoiling at velocities as high as 50 % of the speed of light. This value is only a factor of two worse than the intrinsic resolution of HPGe detectors and corresponds to what the current arrays provide for ~ 10 times smaller recoil velocities.

3. The AGATA Demonstrator

The first step of the construction of AGATA will be the so-called Demonstrator, a small array of 5 triple-clusters complete of digital electronics, DAQ and full on line data processing. As the name says, the target of this project is to verify with a real, although reduced in size, system whether the so-far only calculated performance of the γ-ray tracking principle can be achieved in practice.

Figure 5. Schematic view of the AGATA Demonstrator with 5 triple-clusters

The Demonstrator is built by a collaboration of 12 European countries (Bulgaria, Finland, France, Germany, Hungary, Italy, Poland, Romania, Sweden, Turkey, and UK). To speed-up the development of the detectors, it was decided that the first triple-cluster is built out of symmetric (i.e. simpler) crystals while the other 4 are of the exact asymmetric shape that is needed by the 180-crystal ball. As of this Symposium, two out of the 3 ordered symmetric crystals have been delivered by the manufacturer Camberra-Eurisys. Their energy resolution is well within specifications and one of them is being scanned at Liverpool to check quantitatively the agreement between real and calculated pulse shapes. Most of the orders for three asymmetric clusters have already been placed and the first deliveries are expected by the end of 2005. According to the present plans, the symmetric triple cluster should be tested in beam by mid of 2005, the development of the electronics should be completed in 2006 and the Demonstrator should be ready for commissioning around the middle of 2007.

The AGATA community is now planning how to proceed after the Demonstrator. It is reasonable that the full AGATA will be built in successive phases, the first one (15 triple-clusters covering a solid angle of 1π) likely to be ready for experimentation with radioactive beams in 2010.

Acknowledgments

This presentation was done on behalf of the European collaboration AGATA. Some of the reported results come from the R&D of GRETA and the TMR Research Network "Gamma Ray Tracking Detectors", which was supported by the EC under contract ERBFMRX-CT97-0123.

References

1. 1.I-Y. Lee, *Nucl. Instr. Meth.* **A422** (1999) 195.
2. R.M. Lieder et al., *Nucl. Phys.* **A682** (2001) 279c.
3. M.A. Deleplanque et al., *Nucl. Instr. Meth.* **A430** (1999) 292.
4. K. Vetter et al., *Nucl. Instr. Meth.* **A452** (2000) 105 and 223.
5. D. Bazzacco et al., *LNL Ann.Rep 2000* (LNL-INFN-Rep-178/01) 166.
6. J. Eberth et al., *Prog. Part. Nucl. Phys.* 38(1997) 29.
7. W. Gast et al., *IEEE Trans. Nucl. Sci.* **48** (2001) 2380.
8. Th. Kröll, D. Bazzacco *Nucl. Instr. Meth.* **A463** (2001) 227.
9. Th. Kröll, to be published.
10. G.J. Schmidt et al., *Nucl. Instr. Meth.* **A430** (1999) 69.
11. AGATA Technical proposal, J.Gerl and W.Korten eds. (2001)
12. S. Agostinelli et al., *Nucl. Instr. Meth.* **A506** (2003) 250.
13. E. Farnea, to be published.

DEVELOPMENT OF CHARGED PARTICLE DETECTORS AND ANCILLARY FOR GAMMA ARRAYS

G. PRETE, R. PONCHIA, G. BASSATO
INFN Laboratori Nazionali di Legnaro Italy

D. MENGONI, G. AMBROSI, C. PETRACHE
INFN and University Perugia, Italy

G. CARUGNO G. GALEAZZI, C. BRAGGIO, E. FELTRIN
INFN and University Padova, Italy

G. BRESSI
INFN Pavia, Italy

The use of exotic beams of low intensity asks for high efficiency experimental set-ups. Light charged particles are important tools in studying nuclear reactions with exotic beams, mainly to study direct reactions in inverse kinematics and to develop powerful ancillary detection systems to be coupled to gamma arrays. Large solid angle coverage, precise measurement of energy, timing and scattering angle are the main features of these new generation detectors. The number of electronic channels associated to these high segmentation detectors is quite large: in the order of 1000 or more. To reduce the amount of electronic components, ASIC technology is required and in the next future an effort will be devoted to design new high-density electronics with adequate resolution and dynamic range. Several actions to coordinate the efforts at European level are on the way. The aim is to develop new instrumentation and related electronics suitable to be used in the above experiments. Italian Nuclear Physics community actively participate to design new instrumentation and to perform the preliminary tests.

1. Introduction

Nuclear reactions using secondary radioactive beams opened a new possibility for studying the structure of exotic nuclei far from the valley of stability. These studies have been at the forefront of nuclear physics research during the last years and lead to the discovery of Neutron Halos in several neutron rich nuclei such as ^{11}Be and ^{11}Li through measurements of the interaction cross sections [1] and the momentum distributions [2] of the projectile fragments of these nuclei. Density distributions were determined from the energy and the target mass dependence of the interaction cross-sections [3].

A relevant aspect related to changes in size and diffusivity encountered in neutron rich nuclei is the modification of the average field experienced by a single nucleon. This is a basic ingredient in the many-body theories used to describe nuclear properties. Therefore it is very important to study experimentally how the single-particle levels shift or re-order with neutron excess, inducing changes in the standard magic numbers and, possibly, even the breakdown of shell gaps and magicity. A new generation of radioactive beam facilities is under design in Europe, USA and Japan. In Italy two ISOL facilities are under construction: Excite at LNS with a light ion beam of 100 AMeV and 1 pμA, is at debugging level and SPES at LNL was financed for the first stage of a high intensity proton and deuteron beam of 100 MeV and 5 mA. In Europe GSI is starting with an in-flight fragmentation facility and Spiral2, in France, is proposed for an ISOL facility based on a deuteron beam of 40 MeV and 0.5 mA. The long range project in Europe is the design of EURISOL, an highly performing ISOL facility based on an high energy and high intensity proton driver (1GeV, 5 mA). These new facilities are very expensive (in the order of 100 Meuro for Spiral2 and Spes) but nevertheless the beam quality is expected worse then normal stable beams, both on the point of view of the intensity and of the beam purity. To overcome these problems and to make the best use of the radioactive beam, coordinated efforts among the European nuclear physics community are on the way both to develop more intense unstable beams and to study new detection techniques, detectors and related electronics.

To take full advantage of the radioactive beams, high efficiency detectors are needed, able to cover, with a high granularity, a large part of the solid angle and possibly 4π.

2. Direct reactions

To sketch the main requirements of an up-to-date detection system one can analyze the needs for the study of direct reactions, a basic tool for nuclear structure studies based on the interaction of nuclei with light particles like protons and deuterons. Direct reactions, involving the transfer of one or more nucleons between beam and target nuclei, have been used in the past as a source of nuclear structure information. The availability of radioactive nuclear beams at energies up to few tens of MeV per nucleon and with sufficient intensity, will now make possible to extend such technique to nuclei far from stability addressing therefore basic problems of the nuclear structure of loosely bound systems. In such energy range the cross section for single nucleon exchange is found to be of the order of 1-10 mb/sr at small scattering angles. Similar cross sections are observed for alpha particle transfer, while ten times lower cross sections are observed for multi-nucleon transfer reactions. Elastic scattering and inelastic nucleon scattering are the main probes to study the interaction potentials and the density distribution of halo nuclei such as ^{11}Be, ^{11}Li discovered in the last decade [4, 5]. Due to the presence of the neutron halo, new low-energy resonances named "soft modes" are predicted [6] which could be studied by inelastic nucleon scattering. Transfer and break-up reactions give access to the nuclear structure and could be used to study the shell modifications: with the increasing number of neutrons, nuclei are expected to modify the shell structure [4] and the disappearance of magic numbers for nuclei approaching the neutron drip line is predicted.

While studying stable nuclei, direct reactions were performed using as target the nuclei of interest and a beam of protons or deuterons. In the case of unstable nuclei we have to produce a beam of the nuclei of interest (exotic beam) and an inverse kinematic reaction should be performed on a proton or deuteron target; a thin foil of CH2 or CD2 in the case of proton or deuteron scattering, or a cryogenic target of hydrogen or helium. Considering a two-body reaction, the forward going heavy projectile yields the full characteristics of the reaction, excitation energy and scattering angle. For a proper detection a magnetic spectrometer is necessary, nevertheless for heavy projectiles with A = 20-30 amu the outgoing nuclei are very forward focused, and the angular resolution obtainable is not sufficient to measure meaningful angular distributions. In addition if the heavy ion is excited above the particle emission threshold, in-flight decay destroys the two-body kinematics and complete kinematics coincidence experiments between the ejectile and the decay particles are necessary.

The alternative method is to measure the energy and the angle of the recoiling light particle. This technique has already been performed in several experiments [7,8,9]. Analyzing the kinematics of such reactions we observe that the heavy ions are focused in a small cone in the forward direction and the light particles are scattered at large angles up to the backward direction. In order to measure down to very forward center of mass angles a low energy threshold of approximately 500 KeV protons is necessary. The excitation energy resolution is strongly affected by the angular and energy resolutions, in order to have 1 MeV resolution, an angular resolution in the laboratory frame of about 0.5° and an energy resolution not worse than about 100 KeV are crucial. Particle identification is also important in order to suppress background arising from break-up or fusion reactions, as well as from reactions on Carbon nuclei contained in the target.

To summarize, a successful implementation of the detection of light recoils requires a system able to identify light charged particles over a large energy range (500 KeV-70 MeV for protons) with energy resolution of 100 KeV and, if positioned at 15 cm from the target, with 1-2 mm spatial resolution. Moreover, a large solid angle must be covered to overcome the low intensities of exotic beams. As the angle measurement is a crucial parameter, to perform these experiments the beam direction should be defined carefully, by beam tracking if necessary.

Another very useful application of high granularity light particle detectors is their coupling to gamma arrays. The use of ancillary detectors in gamma spectroscopy is crucial to improve the reaction channel selection. To fulfill this requirement the particle detector should have a large efficiency to detect the particles emitted in the nuclear reaction and a small impact on the characteristics of the gamma array itself, mainly on the gamma-ray efficiency.

In the last years the interest in using ancillary detectors with available gamma arrays is strongly increased and several set-ups were developed not only to detect light charged particles but also neutrons, fission fragments, recoils. The use of ancillary detectors increases the sensitivity of the gamma array, allowing the cleaning of the gamma spectra according to the reaction channel selection that they can perform.

3. High granularity detectors

To study direct reactions with radioactive beams several new detection systems have been developed. According to the parameters described in the previous section, the main characteristics of these devices are the high granularity and the capability to detect light charged particles. In the following two examples of high granularity detectors for direct reaction studies with radioactive beams and an ancillary detector for gamma arrays are presented with the aim to underline the state of the art and to point to future developments.

A new and innovative array, MUST [10], based on silicon strip technology was developed by a French collaboration. The detector consists of 8 silicon strip – Si(Li) telescopes used to identify recoiling light charged particles through time of flight and ΔE-E measurements and to determine precisely their scattering angle through *X*, Y position measurements. Each 60x60mm^2 double sided silicon strip detector with 60 vertical and 60 horizontal strips yields an *X*–Y position resolution of 1 mm, an energy resolution of 50 keV, a time resolution of around 1 ns and a 500 keV energy threshold for protons. The backing Si(Li) detectors stop protons up to 25MeV and can be followed by CsI crystals with photo-diode read out to stop protons up to 70MeV.

VXI electronics was developed to instrument and read-out almost 1000 channels. Isotope identification of light charged particles over the full energy range has been achieved and the capability of the system to measure angular distributions of states populated in inverse kinematics reactions has been demonstrated. A further development for a larger solid angle detector, MUST 2, is under progress. It is based on the same detection technique but with an area 3 times larger and a total number of parameters larger than 4000. In this case the system will be equipped with ASIC electronics to overcome the complexity of manage this large number of electronic channels.

Another high granularity detector, named EXODET, to study direct reactions in inverse kinematics, was developed at Naples. It is also based on silicon micro-strip technology and it is presented with more details elsewhere in this workshop [11]. It consists of 8 modules of ΔE-E telescopes packed in a very compact quasi-cylindrical geometry with the target in the center. 60 + 600 μm micro-strip silicon detectors make the telescope. Each detector has 100 strips 500 μm wide on one side and a continuum electrode on the other, the telescope is mounted with the strips orthogonal on the two layers obtaining a position resolution of 500 μm in X and Y for charged particles which go through the ΔE detector. This set-up has a total amount of more than 1600 channels that are managed using the AToM chip, an ASIC circuit developed for the BaBar Silicon Vertex Tracker [12] and adapted to read the charged particle telescopes. This set-up has been successfully used to study several reactions. The main characteristic is the compactness and the use of high integration electronics.

The last example we should mention is the coupling of high granularity charged particle detectors to gamma arrays. Two ΔE-E silicon balls: ISIS and EUCLIDES, developed by Padua-LNL [13], have been extensively used during the past years in conjunction with the GASP and EUROBALL gamma arrays respectively. The main structure is a 7 cm radius ball of 40 ΔE-E silicon telescopes of 140 μm - 1000 μm. A careful design to optimize the coupling to EUROBALL was performed using the GEANT code. The forward detectors were segmented to improve the reaction rate capability and to increase the angular resolution, a useful parameter for the Doppler correction of the gamma spectra. The total number of channels is in the order of 200 and a CAMAC electronics was developed for the read-out to allow the best low threshold particle identification by pulse shape analysis [14].

The use of this ancillary detector with the GASP array highly improves the selection power as shown in ref. 13.

4. Future developments

According to the needs for a charged particle detector of high granularity, several tests and design studies are under development at LNL. The main use is in studying direct reactions with radioactive beams and as an ancillary for gamma arrays. The ΔE-E identification method was chosen as the geometry can be very compact and the identification threshold can be reduced applying the pulse shape analysis at the first element if a silicon detector mounted in flipped mode is used [15]. A possible configuration is a barrel of about ~10 cm target-to-detector distance with end-caps, covering as much as possible the solid angle around the target. Modules of simple rectangular shapes with pixel detectors of 4x4 or 2x2 mm^2 can be used to obtain angular resolution in the order of 2° or 1°.

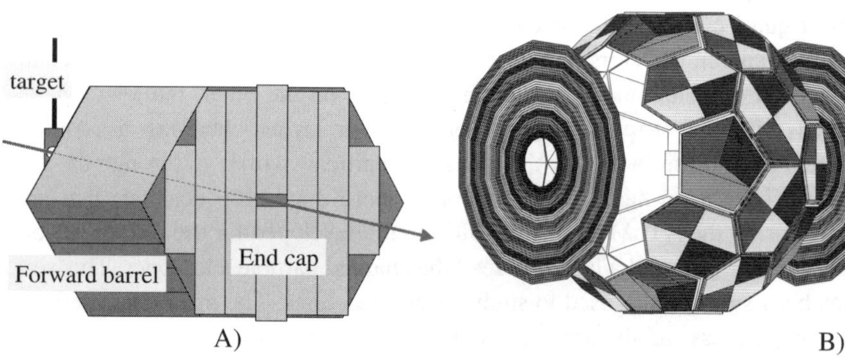

Figure 1. A) Half barrel and forward end-cap.
B) Ball-like geometry with strip silicon detectors end-caps

A second candidate is a ball-like geometry with a radius of 10-15 cm which follows the gamma array shape to minimize the impact on the gamma detection, similar to ISIS-EUCLIDES already mentioned, with two end-caps of 1 mm strip detectors to ensure the good angular resolution of 1°. A sketch of the two designs is shown in fig. 1.

The number of electronic channels evaluated for both geometries are 8500 and 1000-2000 respectively for 4x4 mm^2 pixeled barrel or ball solution. (If pixel of 2x2 mm^2 are used the number of channels exceed 34000). For the next future the development of an adequate high-density electronics is mandatory to properly instrument the new detectors. ASICS have been extensively developed for High Energy Physics but they seldom cover our needs of dynamic range, energy and timing resolution.

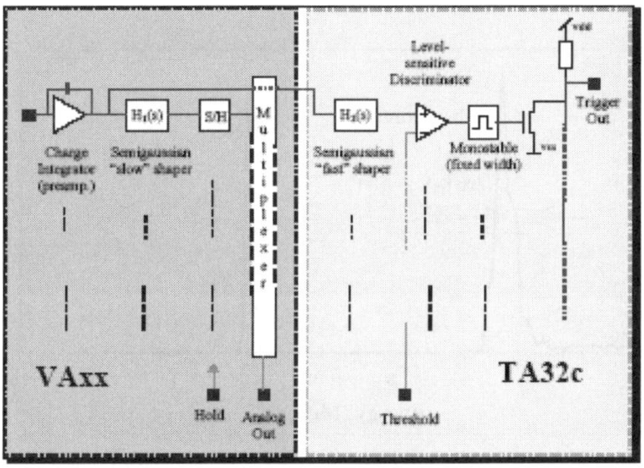

Figure 2. Operating principle of VA and TA modules.

With the aim to study the possible application of available ASIC's, preliminary tests were performed on the VA32-HDR11 produced by IDEAs [16]. It is a CMOS chip integrating 32 channels of low power (2.3 mV/channel) charge sensitive preamplifiers followed by CR-RC shapers and sample-and-hold circuitry. The outputs of the sample-and-holds are multiplexed on one analog output buffer in which the amplitude of each channel is serialized for subsequent digital conversion. An input multiplexer allows a test-pulse to be injected in each channel for calibration and diagnostics. To allow the connection with trigger electronics each preamplifier output is also available on external pins. Two chips are connected together for a total of 64 input channels and one analog serial output. For trigger purpose the two VA32_HDR11 are connected to two TA32 which generate the trigger for the sample-and-hold circuitry and the clock for the output multiplexer. The physical size of each ASIC is ~ 4x4 mm^2. The principle of operation of the VA and TA chips is shown in fig.2. The dynamic range for the VA32-HDR11 is 25pC corresponding to an energy loss of 400MeV in a silicon detector. Due to the high dynamic range this device is suitable for the detection of heavy ions too. Preliminary tests [17] were performed to check the integral linearity and the noise response as a function of the input capacitance. Good linearity was measured up to an energy of 350 MeV. Preliminary results to alpha detection, using a 1 cm^2 300 μm thick silicon, is shown in fig. 3, a resolution of 500KeV FWHM was measured with ^{241}Am source in air.

The next step in this development will be the coupling of a module of design A (10x2 cm2 with 120 pixels of 4x4 mm2) to the VA32-HDR11 for an in-beam test.

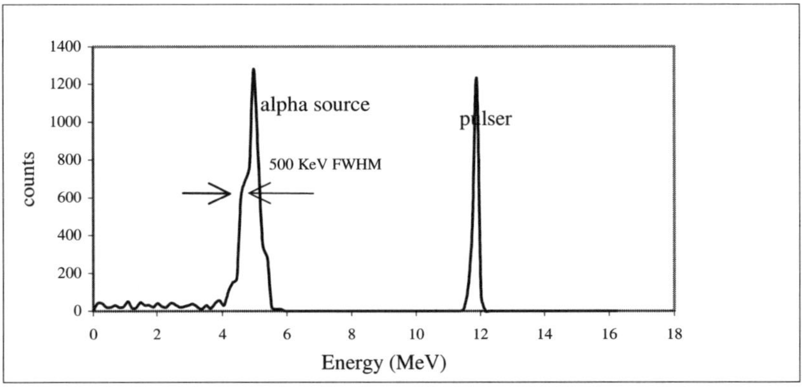

Figure 3. Alpha spectrum from a ^{241}Am source in air at 1 cm from the silicon detector

In the framework of the EURISOL working groups the nuclear structure and reaction communities have proposed new concepts for charged particle arrays. The GRAPA concept [18] was proposed with the aim to fulfill the best compromise between charged particle and gamma detection. It is based on a ball configuration with multi-layer detectors able to detect charged particles of low and high energy as well as X-rays, soft γ-rays and electrons. This configuration optimizes the detection of charged particles and has a minimum effect on the gamma efficiency when used as ancillary detector inside a gamma-array.

The first requirement is the detection of both charged particles and electromagnetic radiations in the same detector. To study the possibility of the application of this concept at affordable costs some work was done to investigate the performances of detectors based on Si or Ge crystals operated at cryogenic temperatures as low as 10K. At this temperature a freeze-out effect reduce drastically the number of carriers and a low voltage can deplete thick detectors of modest purity [19]. Several tests were performed with Schottky barrier detectors made both of Si and Ge. Preliminary results show the possibility to operate these detectors inside a cryostat with a complete charge collection and good energy resolutions.

An Al-Si-Al Schottky barrier detector of diameter 10 cm and thickness 5 mm was operated at 10°K with a Dark Current ~ 10pA up to 700V. The measured minimum voltage for the total charge collection was 500 V. A test was performed for a high purity Germanium diode also. The detector had a diameter of 3 cm and was 1 cm thick, it was operated at 4°K. In this case with a bias of 30 V the full charge collection was reached for the ionization produced by a ^{137}Cs gamma source.

Linearity and energy resolution of the Schottky Germanium diode were measured with calibration gamma sources and the results are shown in fig. 4 and 5 demonstrating the good capability of the detector.

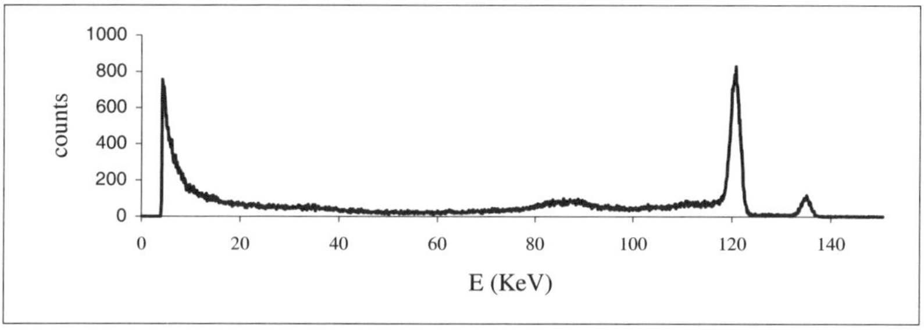

Figure 4. Gamma spectra for ^{57}Co source obtained with a Schottky Germanium diode of diameter 3 cm and thickness 1 cm operated at 4°K.

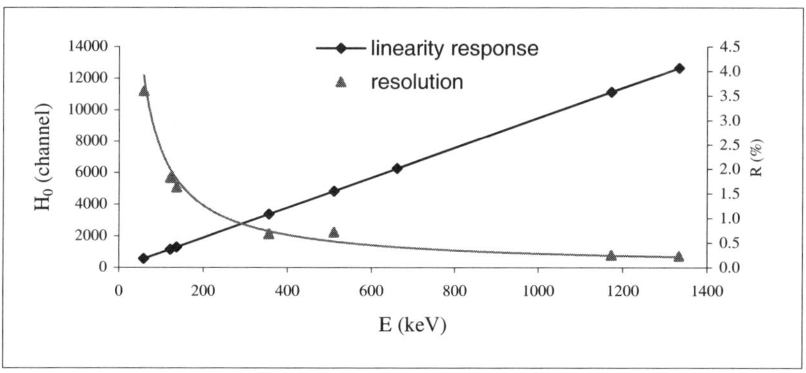

Figure 5. Resolution and linearity obtained with a Schottky Germanium diode of diameter 3 cm and thickness 1 cm operated at 4°K. Standard nuclear electreonics was used.

This device can detect gamma as well as charged particles if the cryostat is properly designed. The next step in the experimentation will be to modify the cryostat allowing the input of alpha particles from a source.

5. Conclusions

The use of charged particle detectors is of grate interest in the study of Nuclear Physics. New frontier for the development of high granularity telescopes to be used with radioactive beams ask for the study of complex systems with thousands of electronic channels; high density electronics based on ASIC components is mandatory to instrument the next generation set-ups. Several developments on new detectors and associated electronics are on the way in Italy to fulfill the requirements of the next generation experiments.

References

[1] I. Tanihata et al., Phys. Rev. Lett. 55 (1985) 2676
[2] T. Kobayashi et al., Phys. Rev. Lett. 60 (1988) 2599.
[3] I. Tanihata et al., Phys. Lett. B 287 (1992) 307.
[4] I. Tanihata, J. Phys. G22 (1996) 157
[5] M.D. Cortina-Gil et al., Phys. Lett. B 401 (1997) 9
[6] P.G. Hansen, B. Jonson, Europhys. Lett., 4 (1987) 409
[7] J.H. Kelley et al., Phys. Rev. C 56 (1997) R1206.
[8] G. Kraus et al., Phys. Rev. Lett. 73 (1994) 1773.
[9] A.A. Korsheninnikov et al., Phys. Rev. C 53 (1996) R537.
[10] Y. Blumenfeld at al., Nucl. Instr. and Meth. A 421 (1999) 471
[11] M. Romoli at al., this conference
[12] http://costard.lbl.gov/LBNL_BaBar_Web
[13]A.Gadea et al., Nucl. Instr. and Meth. A 400 (1997) 87
[14] A.Gadea at al., LNL-INFN Rep 160-00 (2000) 151
[15] E.Fioretto at al., IEEE Trans. Nucl. Sci., vol.NS-44, N.3 (1997) 1017
[16] IDEAs : http://www.ideas.no
[17] LNL-INFN (Rep) 202 (2004)
[18] http://www.ganil.fr/eurisol/Final_Report.html
[19] M.J. Penn. At al., Nucl. Instr. and Meth. A 364 (1995) 118.

EXODET: A NEW DETECTOR ARRAY FOR CHARGED PARTICLES WITH INTEGRATED ELECTRONICS FOR THE POSITION READ-OUT

M. ROMOLI*, M. DI PIETRO, E. VARDACI, A. DE FRANCESCO, A. DE ROSA, G. INGLIMA, M. LA COMMARA, B. MARTIN, V. MASONE, P. PARASCANDOLO, D. PIERROUTSAKOU, M. SANDOLI

INFN, Sezione di Napoli and Dipartimento di Scienze Fisiche, Università di Napoli "Federico II", Compl. Univ. MSA, Via Cintia, I-80126, Napoli, Italy

T. GLODARIU†, M. MAZZOCCO, C. SIGNORINI

Dipartimento di Fisica, Università di Padova and INFN, Sezione di Padova, Padova, Italy

R. BONETTI, A. GUGLIELMETTI

Dipartimento di Fisica, Università di Milano and INFN, Sezione di Milano, Milano, Italy

F. SORAMEL

Dipartimento di Fisica, Università di Udine and INFN, Sezione di Udine, Udine, Italy

The low intensity (about 10^5-10^6 pps) of the presently available RNBs (Radioactive Nuclear Beams) requires detection apparatuses with large solid angle coverage and high granularity. Moreover, the utilization of highly segmented detectors needs unconventional read-out systems appositely designed and the development of electronic boards making use of highly miniaturized and integrated circuitry (ASIC chip). The EXODET (EXOtic DETector) apparatus has been designed and constructed following these guide lines. It consists of large-area silicon detectors segmented in 100 strips on a single side. The particle energy loss information is obtained from the unsegmented side using standard electronic chains, while for the read-out of the incident position information we found suitable a chip already developed for high-energy particle physics experiments. EXODET has been already successfully used in exotic nuclei experiments devoted to the study of the scattering of ^{17}F on a ^{208}Pb target and of ^{11}Be on a ^{209}Bi target at energies around the Coulomb barrier, performed at the Argonne National Laboratory (USA) and at the RIKEN Laboratory (Japan), respectively. The development of these modern and compact detection systems opens new and interesting perspectives for experimentation with RNBs.

* Electronic address: mauro.romoli@na.infn.it

† On leave from: NIPNE-HH, Romania

1. Introduction

The requirements for the project of a new detector system designed to be used with RNB (Radioactive Nuclear Beam) are solid angle coverage as large as possible, high energy and position resolution and low cost. The EXOTIC Collaboration devoted a considerable effort in the last years to develop a new detector apparatus satisfying the previous requirements. The EXODET (EXOtic DETector) system consists of solid state detectors arranged to form two-layer telescopes and placed around the target in the forward and backward hemispheres to subtend a solid angle of about 70% of 4π sr. The position information is obtained from the front side of the detectors that is segmented in strips, while the energy information comes out from the rear side that is unsegmented. The obtained energy resolution is 1% to 6% depending on the detector thickness while the position indetermination for the particles passing through the first layer is 0.5x0.5 mm^2. The granularity of the system (80'000 position pixels) and the detector segmentation (1'600 position channels) have required the use of a highly integrated and miniaturized electronics (ASIC chip) for the read-out. The EXODET apparatus is now fully operational and first experiments have been performed at ANL, Argonne, USA and at RIKEN, Japan. Details of the EXODET features, of the read-out electronics and of the data acquisition system will be given in the following sections. A discussion about new developments and perspectives for this kind of detection systems will conclude the report.

2. Detector features and geometry

The detectors used for the EXODET apparatus are silicon single-sided strip detectors having a large active area (50x50 mm^2) and the front side segmented in 100 strips with pitch size of 0.5 mm and inter-strip separation of 50 μm. The rear side of the detectors is unsegmented and used to bias the detector and to get the information concerning the energy released by the incident particles inside it.

EXODET consists of 16 such detectors, delivered by MICRON Semiconductors Ltd., arranged to form 8 two-layer telescopes placed around the target in two boxes, as shown in fig. 1. The strips of the first layer are perpendicular to the beam direction and those of the second layer are parallel to the beam direction and orthogonal to the strips of the first layer. Thus, for particles passing through the first layer, a position pixel is defined of 0.5x0.5 mm^2. For these particles a Z-identification is also possible using the usual ΔE-E technique.

The two boxes are placed as close as possible to the target and the whole system covers a solid angle of about 70% of 4π sr.

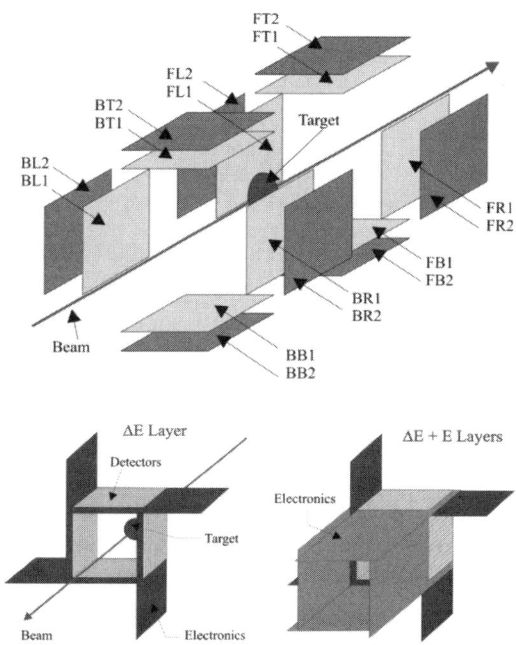

Figure 1. EXODET detectors placed around the target and assembling of the two telescope layers.

The θ_{lab} angular ranges covered by the EXODET apparatus are [26°, 82°] and [98°, 154°]. The distance between the ΔE layer and the beam axis is about 26 mm and the closest approaching distance between one of the two boxes (active area of its detectors) and the target plane is 2.5 mm (4.0 mm). The E layer is placed about 5 mm far from the ΔE layer, then the total solid angle is reduced by about 10% if the ΔE-E coincidence condition is imposed.

The thickness of the ΔE layer (40 μm at backward angles and 60 μm at forward angles with respect to the beam direction) ensures a low energy threshold for particles detectable with position high resolution, but also for particles stopped inside the ΔE layer it is possible to reconstruct the angular distribution, although with lower resolution. EXODET

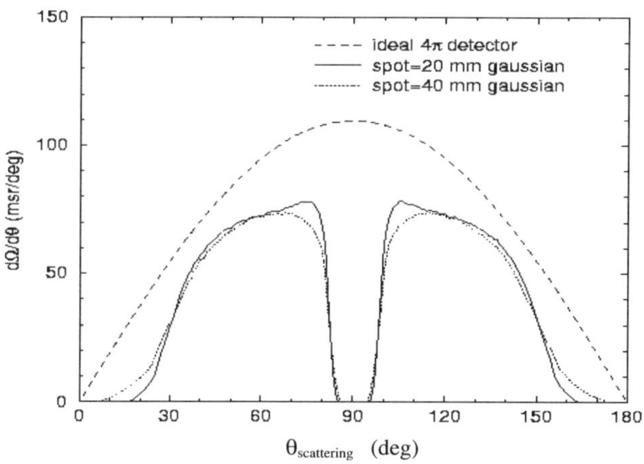

Figure 2. The solid angle covered by EXODET, considering beam spots on target with diameter of 20 and 40 mm. For comparison, the solid angle for an ideal 4π sr detector is also shown.

geometrical features have been evaluated by using Monte Carlo calculations and tested in laboratory with α-sources.

Fig. 2 reports the solid angle covered by the EXODET apparatus when a beam with finite spot diameter of 20 mm and 40 mm is impinging on the target, simulating real and common experimental situations encountered treating with RNBs. In the same figure, the solid angle covered by an ideal 4π sr detector is also shown.

The total number of pixels of the EXODET apparatus is 80'000 (obtained crossing the ΔE strips with the E ones) and the channels to be analyzed, in order to get the incident position information, are 1'600, so the utilization of standard read-out chains is prohibitive. Thus, we designed a read-out system using highly integrated and miniaturized electronics that means an ASIC chip.

3. ASIC chip and read-out electronics

The ASIC chip we used for the EXODET position read-out is a chip originally developed for high-energy particle physics experiments (BABAR) [1, 2]. It has 5.7x8.3 mm^2 size and can analyze in parallel up to 128 input channels (only the first 100 are connected to the detector strips). In fig. 3, the scheme of the operations performed for each single channel is reported. The input signals are pre-amplified and shaped, and then a continuous comparison between the input voltage level and an externally settable threshold is performed at a clock frequency of 15 MHz.

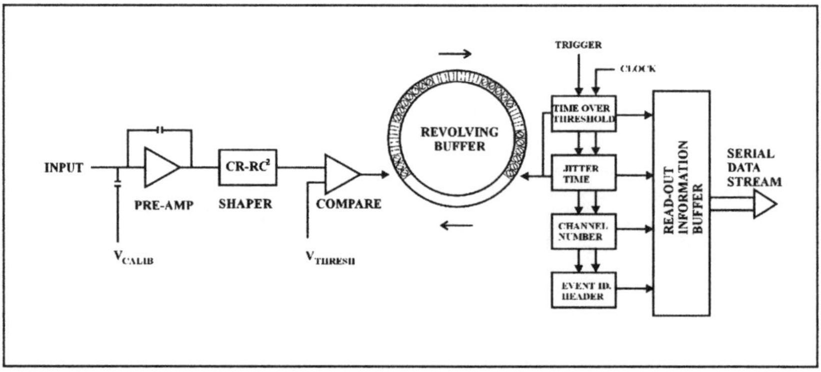

Figure 3. The scheme of the ASIC chip used for the position information read-out.

When there is an input signal higher than the threshold, a bit of the shift register, that is the input memory buffer, is set to 1. When a valid trigger assertion command arrives to the chip, the analogue treatment of the input signals is stopped and the analysis of the input memory buffer starts. If a signal is present in the buffer, a digitized stream of information is given as output. It contains a header with the chip identification number and the trigger time stamp followed, for each strip hit, by the strip identification number, the Jitter Time (JT) and the Time over Threshold (ToT).

The JT is the time distance between the signal threshold crossing and the trigger command arrival, measured in clock cycles, and it is a sort of correlation time useful to distinguish events correlated to the trigger from spurious non-correlated ones.

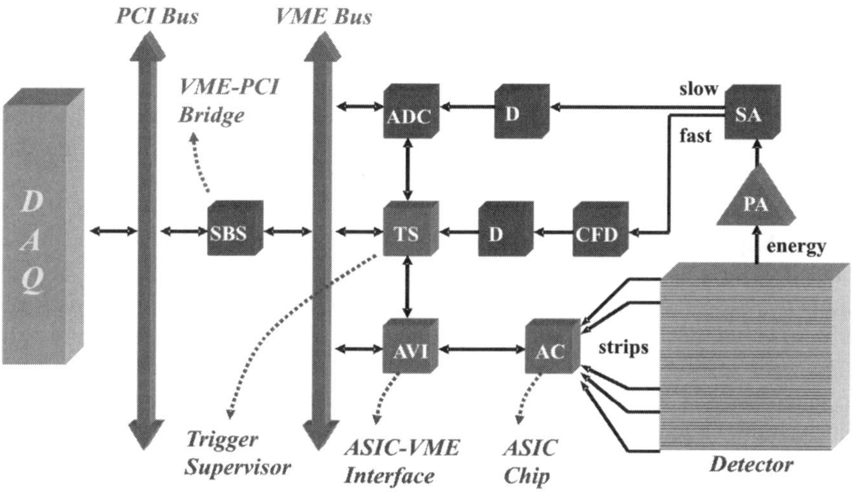

Figure 4. The EXODET read-out system.

The ToT is the time spent by the input signal over the threshold, measured in clock cycles, and it is roughly proportional to the energy released by the incident particle into the detector strip. This information can be used to disentangle the position of particles with sufficiently different energy ranges, when hitting different strips of the same detector.

Other important features of the chip are the conversion gain of 150 mV/fC, the time resolution of 67 ns (corresponding to one clock cycle) and the time length of the input memory buffer (12 μs) and of the input signal analysis window (2 μs).

In order to use the above described chips in our experimental set-up, it has been necessary to design an appropriate detector-chip interface board. Such board is directly bonded to the detector strips and contains the ASIC chip, a resistive attenuator in order to reduce the amplitude of the input signals (with an attenuation factor of 37 dB) for matching the chip input dynamic range, a pitch size adapter necessary to reduce the pitch size from the 0.5 mm of the strips to the 80 μm of the chip input pads, the connectors for the chip I/O, for the detector and electronic components bias and for the energy signal output and finally an hybrid circuit devoted to the pre-amplification of the energy signals.

The read-out system is described in fig. 4. We decided to develop the new front-end boards using the VME standard because of its widespread use in nuclear physics laboratory and for the large variety of modules commercially available (ADC, TDC, QDC, scalers,...). The energy information read-out chain is standard, except for the homemade hybrid preamplifier placed on the detector-chip interface board (28 mV/MeV sensitivity and 300 e rms noise), while for the read-out of the position information coming from the chips we developed a new ASIC-VME Interface (AVI). The AVI board, one card for each chip, allows in diagnostic mode the set-up of the chip parameters and the chip calibration. In runtime mode, it sends the trigger command to the chip and makes the data stream received from the chip available to the VME bus.

We have also developed another highly versatile VME-based board, named Trigger Supervisor (TS). The TS board checks the status of all the chips and of their AVI boards, reports on occurrences of errors or FIFO half-full signals, filters the trigger and delivers it to the whole front-end. It is possible to mask and to sample one or more of the input channels and to force the assertion of a valid trigger for testing purpose. Moreover there are counters for the proposed and accepted triggers in order to evaluate the dead time and dividers by 10/100/1000 for the input channels.

4. Data acquisition system

A commercial VME-PCI bridge is used to connect the VME crate to a PC with Linux as operating system. This PC runs all the tasks necessary to the data acquisition (DAQ) and is the interface of the entire system to the external world. The DAQ is organized as a client/server architecture in order to be accessible to external users, and all the main tasks are distributed as POSIX threads spawned by a single and supervising process, the System Manager process. The advantages of the multi-threads architecture lie in several points. The threads much less demanding on the operating system resources than processes and provide a natural ground to take advantage of a multiprocessor environment (potential parallelism). Moreover, threads share various resources (i.e., memory, file descriptors,...) and this eliminates the need to introduce methods to make communications between the processes. All of these features allow building a very compact and portable system which can take naturally advantage of the growing processing power available commercially. The DAQ developed is a general purpose data acquisition system and is not connected to a specific technology. For instance, it can be transferred over a CPU based VME system only by changing the modality of accessing the VME address space.

The setup and control of the DAQ is achieved through a set of client graphical applications developed using the Motif GUI. The user can prepare the setup for the specific application in a friendly environment by using graphical objects and can define the set of rules for the readout. The setup of each front-end module can also be achieved graphically.

Finally, one of the tasks is to perform on-line data analysis and histogramming. The System Manager provides a default data histogramming for the data coming from each of the front-end modules. The user has also the possibility to define a custom set of 1D and 2D histograms. Any client with the proper set of client applications can connect to the server via Ethernet and control the DAQ and monitor the data.

5. First measurements

The first experiment successfully performed using a part of the EXODET apparatus was the measurement of the scattering of a ^{17}F beam, produced at the ANL, Argonne (USA), by a ^{208}Pb target at 90.4 MeV of incident energy. The very encouraging results have already been published elsewhere [3] and showed the capability of the detection system for scattering measurements with RNBs.

The EXODET performances have been confirmed in the second experiment performed at the RIKEN Laboratory (Japan), where the whole apparatus was used to study the scattering of ^{11}Be by a ^{209}Bi target at incident energies in the 40-50 MeV range. The beam energy and spatial spread was taken into account implementing the EXODET read-out system with a commercial TDC devoted to the collection of the data coming from the beam tagging system of the RIKEN Laboratory, consisting of a couple of position-sensitive PPAC (Parallel Plate Avalanche Counter) and a plastic scintillator.

As an example of the data given in output by the EXODET read-out system, fig. 5 shows, in the upper panel, the JT spectra of the scattered ^{11}Be for the detectors of the first layer (ΔE) and for the second layer ones (ER) without the coincidence condition with the corresponding ΔE one (middle panel) and with it (lower panel). It is possible to see how the background of uncorrelated events, present in the middle panel spectrum and probably due to the delayed decay of the ^{11}Be ions implanted into the E detectors, disappears when the coincidence condition is imposed. In fig. 6a) and b) the ToT spectra of the ΔE and of the E detectors are reported, respectively; in fig. 6c) and d) ^{11}Be energy spectra are shown, obtained gating on different values of the ToT of the E detectors and the relationship between ToT and energy can be clearly evinced.

Details of the analysis concerning the ^{11}Be scattering data experiment have been reported elsewhere [4].

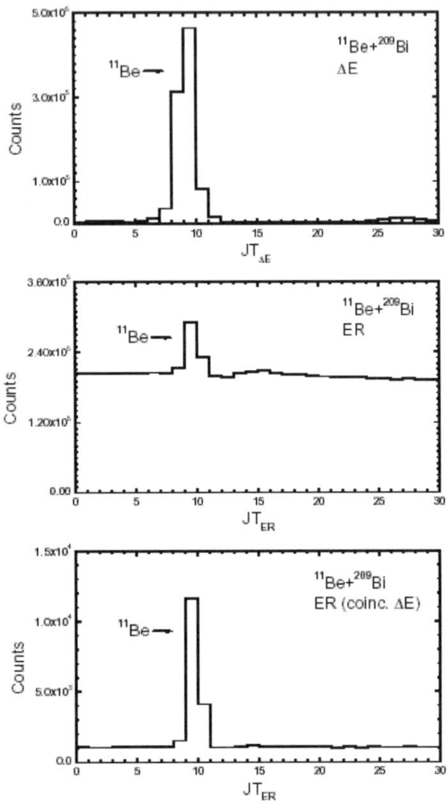

Figure 5. JT spectra of scattered ^{11}Be ions for ΔE detectors (upper) and for E ones without the coincidence condition with the corresponding ΔE (middle) and with it (lower).

6. New developments and future perspectives

Detection systems as EXODET, taking advantage of the low intensities of the RNBs available at the present and in the next future, can be used extensively in nuclear physics experiments. EXODET is at the moment fully operational, it is now mounted at the Laboratori Nazionali di Legnaro (LNL) and it will be mainly used, in the next years, to study the scattering of the RNBs produced by the EXOTIC beam line, at energies around the barrier. But EXODET represents only the first step in the development of such kind of detector arrays.

Important improvements can be obtained, for example, using double-sided strip detectors instead of telescopes of single-sided strip ones. Thus, only one layer of detectors can be used to get the incident particle position with good angular resolution, decreasing the economical effort and increasing the total solid

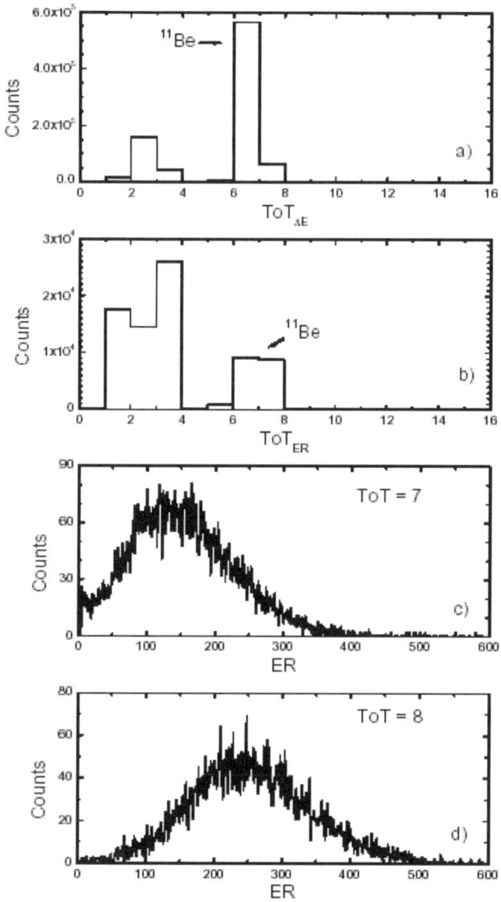

Figure 6. a) ToT spectrum of the ^{11}Be for the ΔE detectors. b) ToT spectrum for the E detectors and energy spectra gated with ToT=7 and ToT=8 respectively c) and d).

angle covered by the system with respect to that covered by EXODET when the ΔE-E coincidence is required. The particles to be detected often do not have enough energy to pass through the first layer of a telescope (about 40 μm thick), therefore it is very important to reduce the energy threshold to get high-resolution position information that means to use the pixel structure of the apparatus. The particle Z-identification could be eventually obtained, if necessary, using less expensive unsegmented detectors as second layer or improving the performances of the read-out chips, that is the better choice.

In fact, another field of research and development concerns the study and design of new ASIC chips appositely developed for experimentation with RNBs. The main goals of the project of this new electronics are a proper analogue treatment of the expected input signals (i.e. appropriate dynamic range, gain and shaping), an accurate sampling of input signals, in order to find the peak, and an analogue or digital multiplexing of the energy information coming from all the channels, with zero suppression. Moreover, the chip should allow the generation of a fast trigger to drive the acquisition logic and possibly a better timing information regarding the rise time of the signals, which could be used to disentangle different particle species with the same energy lost. Finally, the energy information coming from all the detector elements should have high resolution (we estimate sufficient an overall energy resolution of 150 keV FWHM for elastic scattering and break-up cross-section measurements with RNBs at the barrier) and that means a very low intrinsic noise of the chip and of the related electronics.

The geometry of the new apparatuses, probably similar to the EXODET one, should allow the better arrangement of the read-out electronic boards, in order to increase the total solid angle covered by the active area of the system, to reduce the occupancy of the apparatus and to make easy the eventual replacement of the electronic cards. A suggestion could be the utilization of flexible flat cables in Kapton to connect the detectors to the read-out electronic boards, ensuring the possibility of various configurations for the electronics set-up. The detector support frames should be moreover chosen taking into account the possibility of an easy reconfiguration also of the detectors, independently from the electronics, to use the same apparatus elements in very different experimental situations.

In conclusion, we can summarize saying that EXODET has been a successful achievement of our collaboration work, well performing and satisfying the requirements of its original design. We have used and will use it in the next future to perform studies of RNBs reaction mechanisms at the barrier, however, the increasing interest in this research field stimulates also the study of new and better performing apparatuses and gives propulsion to the project of new custom highly integrated and miniaturized ASIC chips more appropriate for the read-out of the typical physical quantities involved in the RNBs physics.

References

1. A. Perrazzo and N. Roe, *Babar note # 501* (1999).
2. J. Beringer et al., *Babar note # 518* (2000).
3. M. Romoli at al., *Phys. Rev.* **C 69**, 064614 (2004).
4. M. Mazzocco et al., *"Scattering process for the system ^{11}Be+^{209}Bi at Coulomb barrier energies with the EXODET apparatus"*, this symposium.

HARDWARE AND SOFTWARE DEVELOPMENT FOR FAST DIGITAL PROCESSING OF SIGNALS FROM NUCLEAR PHYSICS DETECTORS

L.BARDELLI, G.POGGI, M.BINI, R.CIARANFI, G.PASQUALI, N.TACCETTI

I.N.F.N. and Department of Physics, University of Florence
Via G.Sansone 1, Sesto Fiorentino, 50019 Italy

An application of digital sampling techniques is presented which can greatly simplify experiments involving sub–nanosecond time-mark determinations and energy measurements with nuclear detectors, used for Pulse Shape Analysis and Time of Flight measurements in Heavy-Ion experiments. In this work a 100 MSample/s, 12 bit analog to digital converter has been used. A modular system well suited for applications to large detector arrays is also described. Examples of application of these devices to various detector types (i.e. silicon, germanium, CsI(Tl), Gas, ...) in experiments involving particle identification via Pulse Shape analysis and Time of Flight measurements are presented. The achievable digital timing resolution is much smaller than the converter sampling period, for example a resolution of 100 ps FWHM has been obtained with a 12 bit, 100 MSample/s converter. The system is suited for applications to large detector arrays and to different kinds of detectors.

1. Introduction

The use of fast digitizing systems for identification of charged particles using various kind of detectors is discussed. Modern electronic sampling techniques (for example pipelining) have made it possible to design commercial high resolution ($\gtrsim$10 bit) fast sampling analog to digital converters (ADC) which permit to retain the high precision of the standard analog methods (for instance for the energy measurement), while the detailed information achievable with signal sampling can be used in newly designed pulse shape discrimination applications: an example of these applications has been reported in refs.[1,2,3,4]. While the feasibility of high-quality energy measurements, given the high resolution of the converter, represents a somewhat obvious application of the digitizer, the time-mark determination capability of these devices deserves a more detailed illustration because

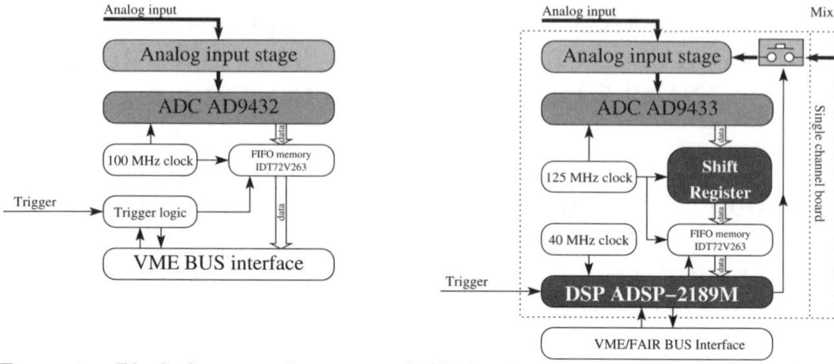

Figure 1. Block diagram of a custom DSP hardware targeted to nuclear physics applications. On the left: block diagram of a first prototype without signal processing capabilities. On the right: block diagram of an improved version with on-board DSP and signal mixing capabilities. See text for details.

this novel application may constitute a breakthrough for experimental setups involving a large number of detectors, possibly of different types: in ref.[2] a detailed study of the timing resolution achievable with these digital systems has been presented, and a 100 ps FWHM timing resolution with a 100 MSample/s, 12 bit digitizer is reported and discussed.

This presentation is organized as follows: in the next section a brief description of the sampling system used in our experimental tests is given. In section 3 the algorithms used for high resolution energy measurements are presented, and experimental results obtained with standard Si-CsI ΔE–E telescopes and CsI scintillators are discussed. In section 4 some experimental results regarding high resolution timing measurements with digital sampling systems will be shown, and results of digital PSA in a reverse mount silicon detector will be presented, whereas in Sec. 5 the time-coincidence test using a large-volume germanium detector is presented. Section 6 will present the results obtained using a "Single Chip Telescope" detector coupled to such a sampling system.

2. The digitizer

The digitizer already described in ref.[1] has been used: the system is provided with 4 independent input channels, each sampled by a fast Analog to Digital Converter (ADC, AD9432 from Analog Devices) with a resolution of 12 bit and 100 MSample/s. Specific measurements have determined the effective number of bits as ENOB = 10.8. The analog input stage of each channel is a 3-poles active Bessel filter, that operates as antialiasing

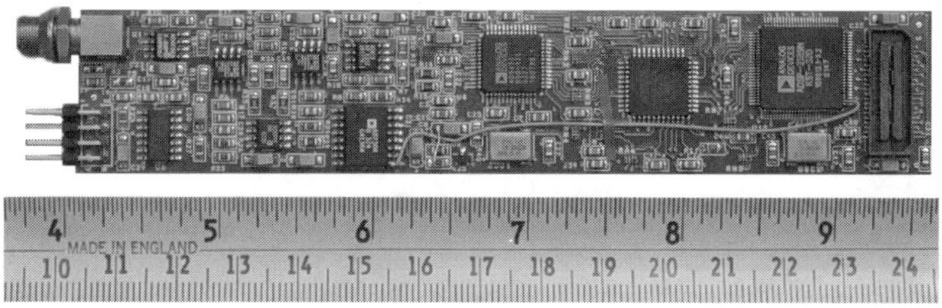

Figure 2. Photo of a single digital acquisition channel of the custom modular system described in the text. The board size is about $14{\times}2.5$ cm^2. A on-board DSP is included.

filter while minimizing signal distortion. The sampled data is stored on a temporary memory (FIFO) and then transferred to the acquisition system through the VME bus for off-line analysis. As discussed in ref.[1] and in the following, all the proposed algorithms are simple and fast enough to be used for on-line analysis with Digital Signal Processors (DSPs).

The configuration used in the prototype board (fully discussed in ref.[1]) is shown in Fig. 1 (left). The prototype does not provide any on-line signal processing capability, because after each data acquisition it transfers the complete acquired waveform to the acquisition system. This solution is clearly unfeasible for a large scale application of the device (the requirements on data transmission and storage would be unacceptable even for a relatively small experiment), but it has allowed for detailed off-line studies of the characteristics of the acquired signal waveforms. Various algorithms have been developed and optimized analyzing data obtained from different detectors and then it is now possible to implement them into an on-line signal processing system.

A new modular system has thus been developed [5] including a on-board DSP for on-line signal processing. The system is meant to be used in the GARFIELD[6] experiment for the NUCL-EX collaboration. Its block diagram is shown in Fig. 1 (right).

The main differences with respect to the original prototype are the use of a faster AD converter (AD9433, 125 MS/s 12 bit) and the presence of a DSP processor (fixed point ADSP2189 from Analog Devices). Data coming from the ADC is stored in a temporary memory (the shift register is needed for acquiring some samples before the actual start of the signal) that is subsequently read by the DSP. At boot time a suitable DSP software is loaded into the processor (through the acquisition interface) depending on the information that needs to be extracted from the collected waveforms. The few extracted values are then transferred to the general acquisition system of the experiment. Each channel of the modular system is also

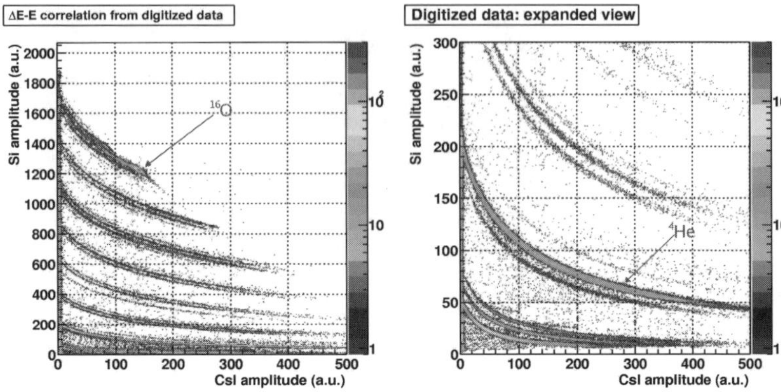

Figure 3. ΔE–E correlations obtained with a standard Si-CsI telescope and digital sampling techniques (see text). Both energy ranges have been obtained using only one converter for the silicon detector.

provided with a common mixing analog input in order to apply the Time of Flight or coincidence method discussed in ref.[2]. Both the prototype and the modular system use a trigger based acquisition logic. Whereas the AD converter is continuously running[a], the signal waveform is stored (and possibly processed) only when a general trigger signal is received by the board. More details about the used trigger logic are presented in ref.[1,7].

From the electronic/mechanical point of view, the new system employs a highly modular design: each acquisition channel is implemented on a small sized board that houses the whole electronics needed for a complete signal acquisition and on-line processing (a photo of a single channel PCB is presented in Fig. 2). Each of these single channel boards is connected to a "motherboard" card able to house up to 8 channels. The motherboard acts as a complex interface between the DSPs and the acquisition system, and makes it possible to integrate the card with standard VME [8] or FAIR [9] acquisition systems. The FAIR acquisition interface will be used inside the GARFIELD[6] apparatus, that planned an upgrade based on these sampling systems.

3. Energy measurements

The use of a high resolution (12 bit) fast sampling analog to digital converter makes it possible to perform high resolution energy measurements with an

[a]This is the recommended solution in order to maintain a good stability of the AD conversion against temperature and counting rate.

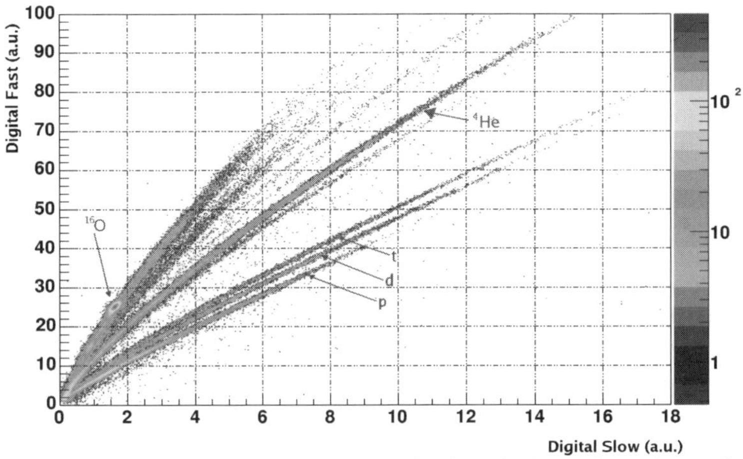

Figure 4. Pulse shape in a CsI scintillator with photodiode readout obtained using a single sampling system. Data have been processed using two different digital shaping time constants.

intrinsic resolution that well compares with standard analog equipment: for example it is possible to apply to the collected samples a Digital Filter (see for example [10,11]) in order to extract the relevant amplitude information and to achieve an optimal signal to noise ratio.

In order to check these assumptions we have performed an experimental test at the Laboratori Nazionali di Legnaro of INFN (LNL, Padova, Italy) using the $^{16}O+^{116}Sn$ reaction at a beam energy of 250 MeV. A first test has been performed on a usual detector configuration: a standard Si-CsI ΔE–E telescope of the Garfield apparatus[6]. The silicon detector was 300 μm thick with a $\sim$ 2 cm^2 area, and the CsI scintillator fluorescence was read using a light guide and a photodiode. The outputs of the two charge preamplifiers connected to the detectors were directly fed into two channels of the digitizer system. On both signals, a digital semigaussian-like shaping has been performed, and the results reported in Fig. 3. The system achieves full isotopic resolution for all the detected particles in the whole dynamic range.

Signals coming from the CsI scintillator ($\sim$30 cm^3) have been also processed with a different algorithm in order to study the pulse shape capabilities of the system using the CsI scintillator only: each signal has been digitally shaped with a semigaussian-like filter ($\tau_{fast} \simeq 700$ ns, $\tau_{slow} \simeq 2$ μs). The results are reported in Fig. 4: the isotopic resolution achieved for $Z \lesssim 3$ is apparent.

These tests are a clear indication of how a high resolution sampling system can be effectively used as a replacement of the standard analog techniques for energy measurements and PSA: the digital resolution and identification performances well compare with the ones of analog systems, with the advantage of significant reduction of costs and system complexity.

4. Digital timing measurements

A very common PSA technique for particle identification is the study of signal risetime (or similar quantities) as a function of particle energy and type: an investigation of the timing properties of sampling systems is thus required to understand whether these methods can be used for this kind of applications.

In ref.[2] a full discussion of this topic is reported as well as a characterization of the properties of digital sampling systems that allows to perform high resolution timing measurements (for example 100 ps FWHM timing resolution with a 100 MSample/s, 12 bit digitizer). Sampling system are usually controlled by a sampling clock (for example a quartz oscillator) that is characterized by a very good stability over time and temperature: in our case errors are in the 1-10 ps range. Thus, as long as the signal risetime is not too fast with respect to sampling period (see Fig.4 of ref.[2]), the timing resolution achievable is determined by the Effective Number of Bits (ENOB) of the converter and the overall electronic noise. It is possible to interpolate between the acquired samples (cubic interpolation is far superior than linear one, see ref.[2]), and then reach a digital timing resolution much smaller than the sampling period of the system.

As an example of this application, we have performed a Digital Pulse Shape Analysis on a reverse mount silicon detector (for details see ref.[2]): the results are reported in Fig. 5. The amplitude versus digital zero-crossing time correlation allows the identification of particles stopped in silicon with full charge resolution (partially for mass): the sub-nanosecond resolution attainable with this digital sampling system is evident. From the data reported in Fig. 5 a digital timing resolution of 100 ps FWHM can be extracted.

5. Digital timing using germanium detectors

The electric field inside a large-volume coaxial germanium detector is not constant and thus many signal shapes correspond to a fixed γ energy, depending on the interaction point. In a coincidence measurement between

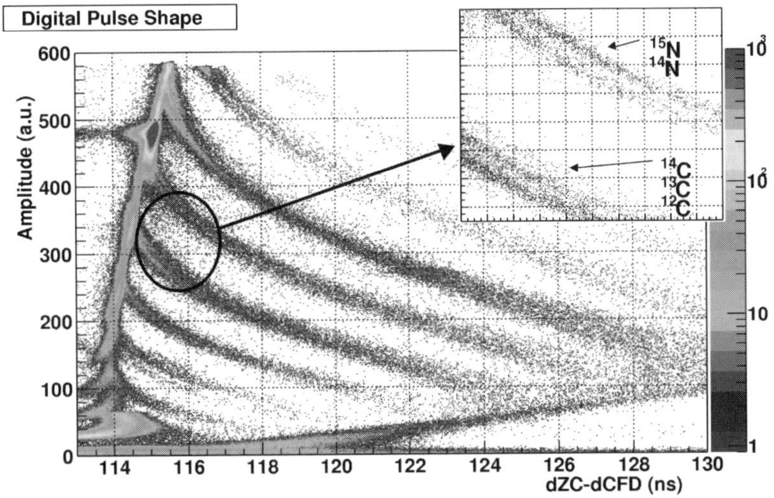

Figure 5. Identification plot for a reverse mount silicon detector obtained using digital sampling techniques: charge identification is apparent. Data have been obtained using a 12 bit 100 MSamples/s converter. In the inset: detail of the isotope separation of Carbon and Nitrogen.

two or more germanium detectors a timing algorithm insensitive to the different shapes is needed in order to achieve a good timing quality.

A test has been performed with a large-volume (30% efficiency) coaxial germanium detector. The achieved timing resolution with such a detector using standard state-of-the-art analog electronics is about 3.2 ns FWHM @ 1.332 MeV, in agreement with the manufacturer specifications[12]. The two cascaded γ rays coming from the decay of a ^{60}Co source are measured in coincidence using the specified germanium detector and a Fast Plastic Scintillator (that is used as reference time). Two sampling AD converters are used, controlled by the same clock and the same trigger: with this configuration a time synchronization between the two channels is trivial, because there is a sample-by-sample time correspondence between the two acquired waveforms. Note that this electronic setup is very effective and easily implemented only when few digital channels are involved (two in this test), but it would be unfeasible for applications with many acquisition channels (an original solution to this problem is presented in ref.[2]). The standard analog electronic chain for extracting a timing information from the germanium output (characterized by a non-fixed shape) is the use of a Amplitude and Risetime Compensated CFD (ArcCFD, [13,14]). It is possible to employ the very same operations performed by a analog ArcCFD module

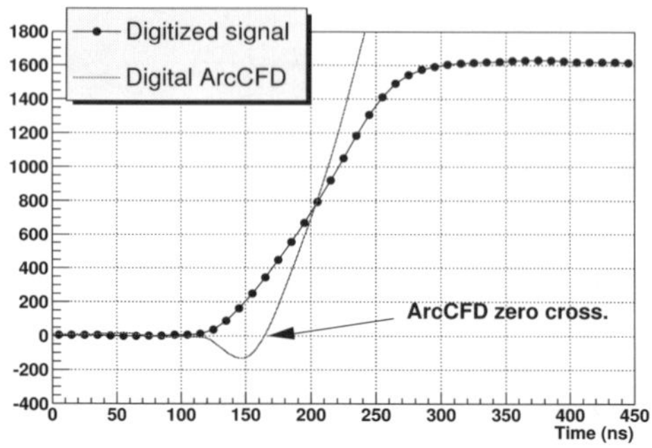

Figure 6. Example of digital ArcCFD timing. Starting from the digitized samples the digital equivalent of a ArcCFD signal is used for timing determination with a cubic interpolation of samples.

on the digital samples, thus building a digital ArcCFD, or "dArcCFD": an experimental example is shown in Fig. 6.

The first step of the computation is a time shift δ of the original sequence $S[k]$. For the case of fractional delay (i.e. when δ is not a multiple of τ_{clk}) an interpolation of the signal can be used to reconstruct the signal $S_\delta[k] = S(k\tau_{\text{clk}} - \delta)$. As discussed in ref.[2] the choice of an interpolation algorithm is critical in determining the achievable timing performances. In this analysis a cubic interpolation of samples has been used. The obtained sequence $S_\delta[k]$ is then multiplied for the desired fraction f and subtracted from the original signal. The result of this operation is presented in Fig. 6 (the final signal is shown as continuous for better presentation, although it is not the case). Using a proper interpolation the zero-crossing of the final signal can be extracted and used as time reference. This algorithm (as its analog counterpart) is characterized by two parameters, namely the delay δ and the fraction f. A search for the optimal parameter choice has been performed in the $\delta - f$ plane: the best timing resolution is achieved using δ=30 ns and f=0.3.

The obtained digital resolution for the 1.332 MeV line is 1.9 ns FWHM, to be compared with its analog counterpart of 3.2 ns FWHM.

The superior performances of the digital sampling system can be traced back to the zero-crossing algorithm used: the cubic interpolation method employed uses four samples for the determination of the zero-crossing, so that the whole information available ~20 ns before and after the zero-crossing is used. On the contrary, the wide-band analog system uses only the information near to the zero-crossing, and thus it is more sensitive to high frequency noise, finally worsening the achievable resolution.

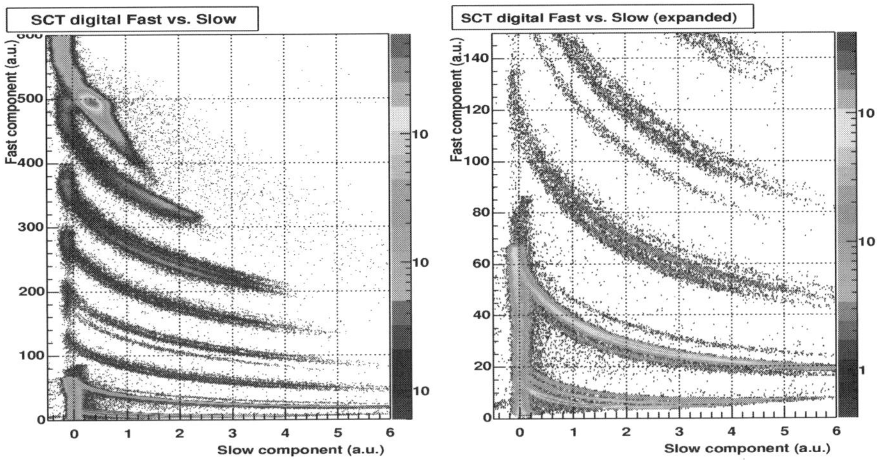

Figure 7. ΔE–E-like correlation obtained using a Single Chip Telescope coupled to a single digital sampling channel: both energy ranges have been obtained using only one converter.

The presented result has been restricted to a very small dynamic range (few keV near the 1.332 MeV line). More tests are planned in order to check the performance of the system (both hardware and software) over a wider dynamic range.

6. The single chip telescope

In the previous sections the main characteristics of high resolution energy and timing measurements using digital sampling techniques have been discussed and their performances experimentally tested using common detector configurations. It has to be noted that the flexibility of digital sampling techniques allows also a simple and compact implementation of new detectors types: for example we have tested[3] the performances of a Single Chip Telescope (SCT[15]) coupled to such a digital sampling system. With respect to the standard analog implementation described in ref.[15], the digital one provides an important reduction of the required electronics, making the SCT an attractive device for high granularity experiments.

For particles stopped in the silicon detector the same analysis described in the previous section can be applied: identification of particle can be achieved using Pulse Shape Analysis of silicon signals. Particles punching through the silicon detector can be identified using a standard Fast vs Slow identification method (the same method used in [15] with analog electronics): the results obtained using digital shaping are presented in Fig. 7. The achieved identification results of this very first digitized SCT prototype are indeed encouraging: further work is in progress to improve the isotopic resolution of the system.

7. Conclusions

An extensive evaluation of the properties of digital sampling systems has been performed, focusing on high-resolution energy and high-resolution timing measurements. A custom digital sampling system has been developed, well suited to application to large experiments, due to its modular design and low cost. Various experimental energy resolution tests have been performed: examples regarding ΔE–E particle identification as well as Pulse Shape analysis are reported. It is important to note that an adequate number of effective bits of the converter (i.e. low quantization noise) allows the extraction of a time mark information with a resolution much smaller than the converter sampling period (for example a resolution of 100 ps FWHM has been obtained with a 12 bit, 100 MSample/s converter). As discussed in ref.[2] this result can be reached if a proper interpolation procedure is used. Examples regarding Pulse Shape in reverse mount silicon detectors and timing with large-volume germanium detector have been presented. In the latter case a 1.9 ns FWHM experimental resolution has been achieved.

References

1. L.Bardelli *et al.*, Nucl. Instr. and Meth. **A491** (2002) 244.
2. L.Bardelli *et al.*, Nucl. Instr. and Meth. **A521** (2004) 480.
3. L.Bardelli *et al.*, Proc. of RNB6 Conf., 2003, Nucl.Phys. **A** (2004) 746.
4. L.Bardelli *et al.*, *Application of a fast digital sampling system to $\Delta E/E$ identification and subnanosecond timing*, LNL Annual Report 2002.
5. M.Bini, R.Ciaranfi, G.Pasquali, *to be published*, 2004.
6. F. Gramegna *et al.* Nucl. Instr. and Meth. **A389** (1997) 474.
7. M. Bini, *et al.*, Nucl. Instr. and Meth. **A515** (2003) 497.
8. Motorola, *The VMEbus Specification Manual*, revision C.1, October 1985.
9. A.Ordine *et al.*, IEEE Trans. Nucl. Science, vol.45, no.3, June 1998, 873.
10. A.V.Oppenheim R.W.Shafer, *Digital Signal Processing*, Prentice Hall, 1975.
11. Steven W. Smith, *The Scientist and Engineer's Guide to Digital Signal Processing*, available on `www.dspguide.com`.
12. Timing performances of GMX30p series, ORTEC, WEB page `http://www.ortec-online.com/`.
13. R.L.Chase, Rev.Sci.Intrum. **38** (1968) 1318.
14. J.J.Kozyczkowski and J.Bialkowsky, Nucl. Instrum. Meth., **137** (1976) 75.
15. G.Pasquali *et al.*, Nucl. Instr. and Meth. **A301** (1991) 101.

DAQ SYSTEM FOR SMALL-MIDDLE SIZE NUCLEAR PHYSICS EXPERIMENTS PRESENT AND FUTURE

Y. WATANABE

RIKEN (The Institute of Physical and Chemical research),
2-1 Hirosawa, Wako, Saitama 351-0198, JAPAN
E-mail: watanaby@riken.jp

We are starting to develop a next generation DAQ system which consists of autonomous data taking modules distributed over DAQ network. The each module is tightly connected to its partner detector and has memorizing capability of the detector's characteristic. The framework of present DAQ systems for nuclear physics experiments which stands on modular structure is not changed for 30 years, because it is universal and flexible. Advanced research works require more precise measurement and more detector channels, which caused difficulty of complicated cabling, increasing noise etc. The biggest difficulty is time consumed calibration work which is mandatory on each experiment and makes the analysis period so long. On the other hand, IT (Information Technology) field is growing so quickly and we can take fast CPU, big memory, network technology etc. with so small expense. It is no wonder to imagine autonomous data taking modules distributed over DAQ network as more universal and flexible DAQ system.

1. Introduction
- Pros and cons of the present DAQ systems -

"The data acquisition (DAQ) has a lot of difficult problems to tackle", which was claimed for more than ten years in the experimental nuclear physics field. Figure 1 shows the essence of present DAQ systems for small-middle size nuclear physics experiments. It was developed about 1970's then keep the framework without big change until now, as:

- Processing analog signal before digitization.
- Connecting (with coaxial or twist pair cables) modules, each of which has single function.
- Applying fast trigger signal to start digitization.
- Only a few bus lines are shared with many digitizing modules for digital data output.
- A few intelligent modules (PC or mini computer...) control all

electronics.

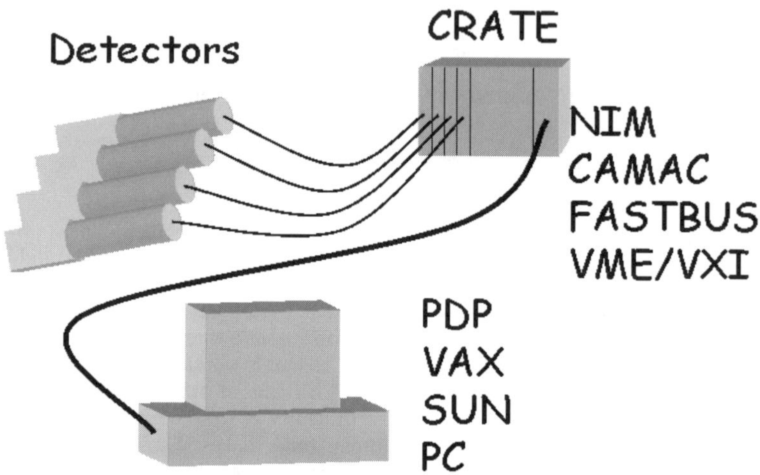

Figure 1. The framework of prsent DAQ systems for small-middle size nuclear physics experiments.

The most important feature is flexibility which stands on modular structure of the framework. Even complicated function can be realized by connecting several simple modules designed with standardizations of signal (NIM, TTL), connector (BNC, LEMO), bus (CAMAC, VME) etc. Moreover thanks to the modular structure, each module (include detector, amplifier and digitizer etc.) can be improved or developed independently to satisfy requirements of the experiment. It is obvious that these flexibilities cause longevity of the framework.

On the other hand, nuclear physics experiments were being bigger and more complicated. Figure 2 shows how complicated the electric wiring of recent experiment is, which was done recently at the RIKEN RIPS beam line. Each module has only single function. It causes the complications because connecting many modules is needed to realize required functions. For example, a linear signal from a detector is divided to three destinations as a high gain ADC, a low gain ADC, and a timing module. Not only visible complications are there. We nuclear physics experimenters often disassemble these modules and rearrange again for next experiments. Every time of rearrangement, the ratio between input signal and output data must be

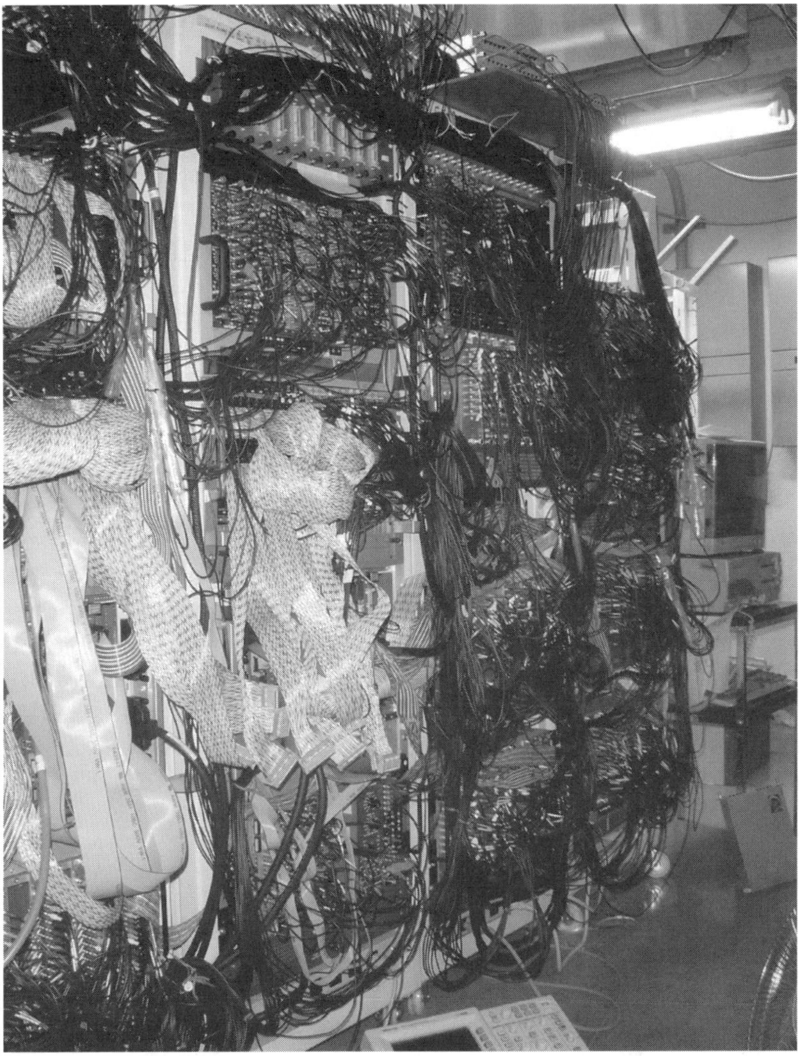

Figure 2. An example of complicated wiring for recent nuclear physics experiment.

calibrated, because the connection of modules include detectors would be changed always. There is another weak point of the framework for small experiments. It is always needed to prepare many things (e.g. detector, HV, NIM module/crate, CAMAC(VME) module/crate, computer, interface, software... as shown in Figure 1) even taking only one channel data.

The number of digitizers hidden in Figure 2 is about 700 (i.e. 700 channels), and it will reach several thousands at next generation experiments. We should figure out a new framework of DAQ which satisfy;

- Flexibility - Easily can be rearranged for next experiment.
- Scalability - Not only "possible" but also "practicable/feasible".
  - Very easy for small system: Plug & Play.
  - Not so complicated for larger system (~10 k channels).
- Availability.
  - Robustness.
  - Normalcy verification of each detector/module.

2. An idea sketch of future DAQ system - Autonomic detector -

In order to satisfy the requirements as discussed in the previous section, we think of "autonomic detector", which puts out digital data instead of analog signal. Figure 3 shows an example of "autonomic detector" which consists of a NaI scintillator, a photo multiplier, and an "autonomic box"; and its simplest usage with Plug & Play. There is an important point that the overall calibration parameters between physical inputs and digital outputs can be managed by each of "autonomic detector" because these components are tightly connected and hard to be disassembled.

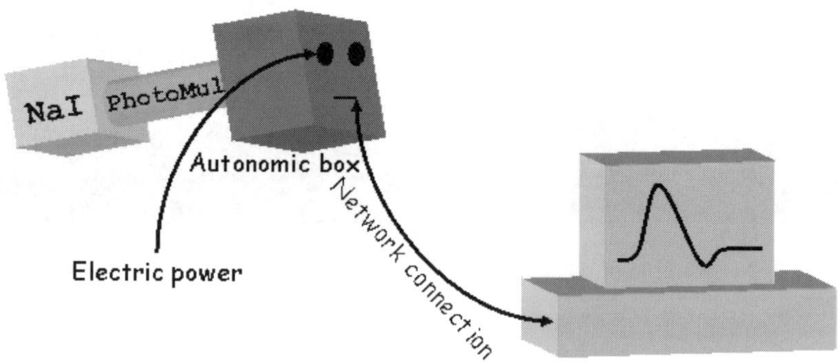

Figure 3. An idea sketch example of "autonomic detector" and it's simplest usage.

The detector should work as self triggered when putting electric power then a PC connected network can easily takes data with using *http* protocol for example. It looks like a multi channel analyzer which takes spectra with just connecting a signal cable instead of the network cable.

2.1. *The autonomic box*

Obviously the "autonomic box" is a key component of "autonomic detectors". Figure 4 shows a basic block diagram example of the box. It is almost same diagram of the heart of digital oscilloscope, and also takes digitized waveform data and extracts intrinsic information with pulse shape analysis using digital signal processing technique by FPGA or DSP or CPU. Once a digitized waveform data is taken with enough precision, the extracted information such as pulse height or timing with digital technique is equivalent to ones with analog technique, for example using shaping amplifiers or constant fraction discriminators.

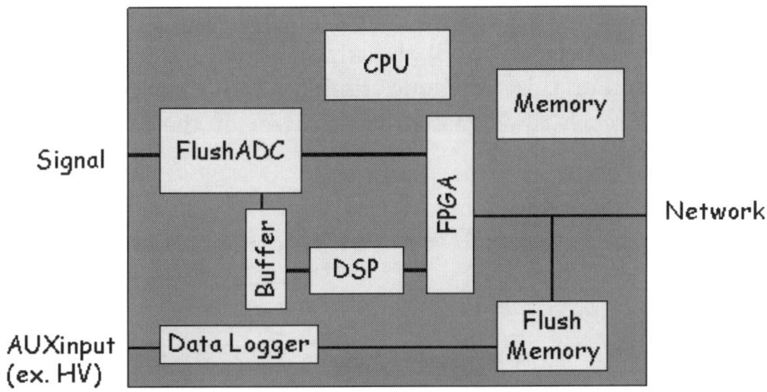

Figure 4. A block diagram of "autonomic box", which has basically same architecture of digital oscilloscope. The data logger function is needed to collect parameters of measurement condition and memorize with other calibration parameters.

It is another important point that the box has non-volatile memory device which enable to keep the history of past calibration parameters. Of course the last calibration parameter set is used for current measurement as initial values. Moreover the history of calibration parameters represents a characteristic of the "autonomic detector" itself, it also can be used to estimate damage and life time of the detector. It is an essential condition for autonomy to keep the past memory and making good use of it at present and future.

The following list shows the minimum requirements of "autonomic box" and those should be pre-installed on the box as Plug & Play functions:

- Waveform recording and outputting.
- Simple pulse shape analysis of waveform data.
- Some parameters to memorize includes calibration history.
- Network protocol for control and data transfer.

2.2. *Multi channel usage of "autonomic detectors"*

Since each "autonomic detector" works independently, simplest multi channel measurement is so easy as just putting many detectors on a network. The PC can take data from each detector labeled with unique number (e.g. MAC address of ETHERNET).

When you take the coincidence measurement, each detector must work simultaneously. In the present coincidence measurement, the trigger technique is the usual way to assure the simultaneous work of detectors: Each detector waits measurement until the trigger signal issued which identify the target event. For this technique, timing adjustment of each detector is mandatory to reach the signal to its digitizer at the same time of the trigger issued.

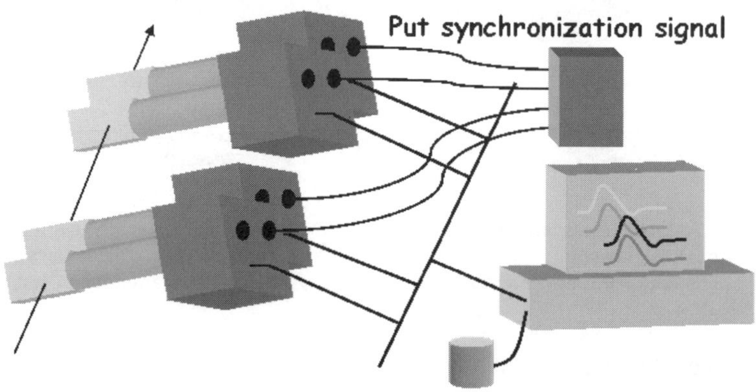

Figure 5. An idea sketch of multi channel system using "autonomic detectors". Another signal for synchronization work of "autonomic detectors" might be needed.

Meanwhile thanks to the digital memory, the "autonomic detector" is able to memorize any data includes waveform, and extract them after the target event is identified. However it is impossible to point out the memorized data which corresponded the target event, unless the timing difference among detectors are well determined. Figure 5 shows a solution that distributing an accurate timing signal over detectors to work synchronously. Once all detectors work synchronously, the timing difference are easily determined. The distribution mechanism of accurate timing signal is one of the technical challenge to develop the "autonomic detector".

There is also rather trivial but practical merit for checking or diagnosing every detector remotely because the waveform data of each "autonomic detector" can be collected easily through network. You don't need to bring an oscilloscope to the area at the detector sitting nor set signal cables to every detector each for diagnosing.

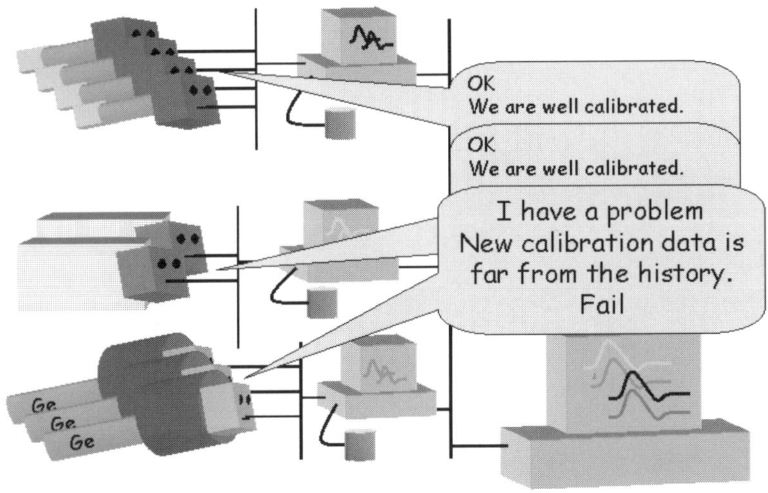

Figure 6. An idea sketch of larger experiment using "autonomic detectors". Each "autonomic detector" can do self-watching by comparing the run-time calibration data with its history.

3. Summary

The "autonomic detector"; this idea was gotten from live cell of organism. Each cell develops independently and it also cooperates with other cells as an individual of organism. As the cell in organism, each "autonomic detector" is designed to work independently and also cooperate with other detectors. It should be much easier and robust that a large experiment setup consists of “autonomic detectors’ as shown in Figure 6, because any trouble of a detector is detected by its autonomic function and influence of the trouble is isolated in the detector .

The “autonomic detector” which distributed over network should solve difficulties of recent and future nuclear physics experiments.

3D-SCANNING OF nTD-SILICON DETECTORS: PRELIMINARY RESULTS

A.Boiano[1], G.Giordano[1], D.Grassi[2], A.Ordine[1], E.Rosato[1,2], G.Spadaccini[1,2], M.Vigilante[1,2]
and the AZ4π and CHIMERAPS collaborations

[1] *INFN, Sezione di Napoli*
[2] *Dipartimento di Scienze Fisiche*
Università di Napoli "Federico II" , I80126 Napoli (Italy)

A dedicated beam line has been installed at the Naples tandem accelerator aiming to verify the homogeneity of nTD silicon detectors. The experimental technique relies on pinhole proton beams, x- and y-translations by remotely controlled stepping motors, tuning the bombarding energy to explore different (z) depths. Comparing the pulses delivered by a special preamplifier, which allows studying current signals induced by ionizing particles, allows checking the detector homogeneity point by point.

1. Introduction

Nucleus-nucleus collisions around the Fermi energy produce nuclear matter in extreme conditions of excitation energy, spin, pressure, temperature and isotopic composition, far away from equilibrium and very similar to the early universe [1,2]. The presently available exotic beams and the future facilities planned to accelerate nuclei farther and farther from stability should allow to investigate the N/Z degree of freedom over a wide range [3]. In particular, experimental measurements should constrain the nuclear equation of state and the asymmetry term, which depends on the proton and neutron densities [4].

As a consequence there is an increased need of multidetector array able to measure and identify all the products of nuclear reactions, to lower detection thresholds, to cover the overall solid angle, to have high granularity and angular resolution. The detection at same time of neutrons, gamma's and their energy has a crucial relevance; however the reduction in geometrical efficiency and the absorption and (re)scattering by charged product detectors introduce severe difficulties [5]. The simultaneous identification of both atomic and mass numbers of reaction products is a serious challenge. The mass measurement is the most ambitious goal; in fact, the usual methods suffer several limitations especially for slow or very energetic particles.

The pulse shape discrimination (PSD) applied to silicon and scintillator detectors is a very promising technique. In fact, the signal profile depends on the density and the spatial distribution of charge carriers produced by the ionizing particles.

Due to the different drift velocities of electrons and holes, the charge-collection time and the rise time vary with the charge and the mass of the detected ion [5-9]. Further, by using both on- and off-line numerical and computational algorithms, the complicated circuitry and the associated hardware techniques are reduced and simplified. Nowadays PSD is becoming a basic tool for signals of charged particles in silicon detectors and both charged and neutral products in light scintillators [10-12].

The sensitivity to charge and mass of the ionizing particle is higher for reverse mount, i.e. when particles are injected into the rear side of totally depleted silicon detectors. The entrance face is at a higher voltage and the electric field strength increases with the penetration depth. If two isoenergetic particles hit the crystal, the heavier one will generate a shorter track and a higher ionization density. As a consequence, the Bragg peak and the whole ionization track are located within the region of weaker electric field: the path and the drift time of the holes towards the cathode increase while the electron behavior is exactly the opposite with respect to the lighter particle. As a final result, for a given energy, the total charge-collection time regularly increases with the charge and the mass of the detected particle. Anyway several factors, all of which can be attributed to the lack of doping homogeneity [13-16], limit the resolution and large fluctuations have been found for detectors cut from the same silicon ingot or wafer and with the same nominal resistivity, thickness and surface. This has been ascribed to the heterogeneous concentration of impurities added for doping the bulk silicon, which is inherent in the growth of the high resistivity float-zone silicon [17,18]. The homogeneity of the doping concentration seems to be the key parameter, which controls the quality and the performance of the particle identification by PSD.

Neutron transmutation doped (nTD) silicon detectors, obtained by exposing crystals to the neutron flux from a nuclear reactor are a very appealing alternative. The thermal neutrons are captured by ^{30}Si and the following β-decay induces nuclear transmutation of ^{31}Si into ^{31}P: the net result is a n-type semiconductor of lower resistivity with little variance from bulk material.

A few years ago data on isotopic identification have been published obtained by PSD applied to nTD silicon detectors. The claimed performances are excellent both for isotopic identification and charge-collection time and hence for the accuracy of energy measurement [19]. These results have been obtained for too small detectors, 250 mμ thick and 1 cm^2 surface, which prevent them from being used in multidetector arrays of large geometrical coverage.

In the literature [20] results have also been reported concerning both planar and nTD silicon detectors with larger surface values, S ~ 10 cm^2, by adopting a different PSD strategy. In fact, Pausch and coworkers [20] measure the rise time of the shaping amplifier signals; instead Mutterer et al. [19] extract the PSD information already at the preamplifier level. Furthermore it is not clear if the increased capacity, due to the larger area, reduces the advantages of the homogeneity. Good

results have been reported on isotopic identification for ion-implanted strip detectors by measuring a combination of time of flight and pulse shape effects [21]. The CHIMERAPS collaboration has got very promising charge identification by PSD for particle stopped in large area Chimera silicon detectors [11].
This contribution will report on the experimental apparatus installed at the Laboratorio dell'Acceleratore (LdA) of the Università di Napoli and on preliminary results concerning nTD silicon detectors by means of charge- and current-sensitive preamplifiers, specially designed for improving PSD technique and studying current signal waveforms [10].

2. Experimental

A dedicated beam line has been installed at the 3 MV tandem accelerator hosted by LdA (fig. 1). Its main purpose is to produce a parallel beam of both reduced cross section and low intensity for bombarding well-identified small portions of the detectors and avoiding radiation damages at the same time. The system consists in two ∅=0.5 mm collimators, 320 cm apart from each other. The first collimator is a pierced Faraday cup mounted at the exit of the switching magnet, the second one is placed in the scattering chamber just in front of the detector position at a 2.5 cm distance from it.

Fig. 1- The experimental apparatus. In the background the beam pipe, the switching magnet and the insertion device of the first collimator are visible on the left. The –45° beam line, the scattering chamber and the pumping system are in the foreground.

By aligning the two collimators a parallel beam is produced except for lateral divergence due to the scattering from the borders of last diaphragm. Tests have been performed by means of a 25 mm^2 silicon detector aligned with the two collimators. An almost constant counting rate has been observed when displacing the detector ± 2.5 mm along both the x- and y-axis perpendicular to the incident beam; the rate considerably drops when the detector is further moved up to ±3 mm and no counts are detected outside the [-3,3] mm range from the centre. These results lead to conclude that the transmitted beam is parallel, its dimensions are comparable to the last collimator diameter and remain unchanged when travelling from diaphragm to the detector. Furthermore a counting rate of 100-150 particles per second (p/s) has been measured, that has to be compared to an incident current of 10-20 nA on the first collimator corresponding to 10^{11} particles per second, in quite good agreement with the geometry. In fig. 2 a typical spectrum is shown for E_p=3.5 MeV protons. The lower energy tail amounts to ≈ 15% of the peak and is due to scattering by the borders of the 2 mm thick Al collimator in front of the detector. To reduce asymmetry the central part of the diaphragm has been replaced with a 100 μm thick Tantalum foil so that the thinner thickness and the higher atomic number reduce the scattering yield and the energy spread of detected protons.

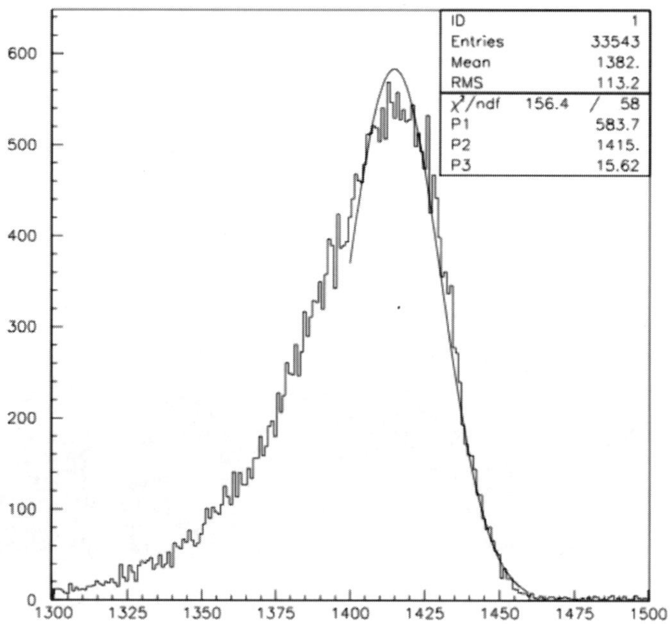

Fig. 2 – A typical spectrum of E_p = 3.5 MeV protons.

The detector is located immediately behind the collimator. It can translate along both x- and y-axis by means of 3 nested frames (fig. 3) and 2 stepping motors, which are remotely controlled and have a 5 μm sensitivity. By displacing the detector in front of the collimator well-identified regions of its rear face can be accurately selected and exposed to the incident beam. Changing the proton energy allows exploring different depths along the z- axis, from 16 μm for E_p=1.0 MeV protons up to 290 μm for E_p=6.0 MeV, which is the highest available energy [22]. This way a point-by-point scanning of the entire detector volume V(x,y,z) is made possible.

Fig. 3 – System of x-y translation. The outermost frame is fixed to the scattering chamber while the inner ones can freely translate with respect to each other and to the first one. The 1700 mm² nTD silicon detector and the PACI preamplifier are also visible.

The scheme of electronics is shown in fig. 4. The core of the detection system is the PACI preamplifier [10]. It has especially been conceived for applying PSD to the current pulses produced by detected particles and provides two independent outputs. The first waveform, *charge output*, is the voltage signal proportional to the collected charge; it is split and fed into a classical spectroscopic chain to measure the energy released into the detector and into a second chain for rise time measurement. This part is managed by the FAIR acquisition system [23].
The second signal, *current output*, is the time derivative of the first one and it is an accurate image of the hole- and electron-current. A digital oscilloscope separately analyzes this current output and records the corresponding waveforms, averaged over 512 pulses.

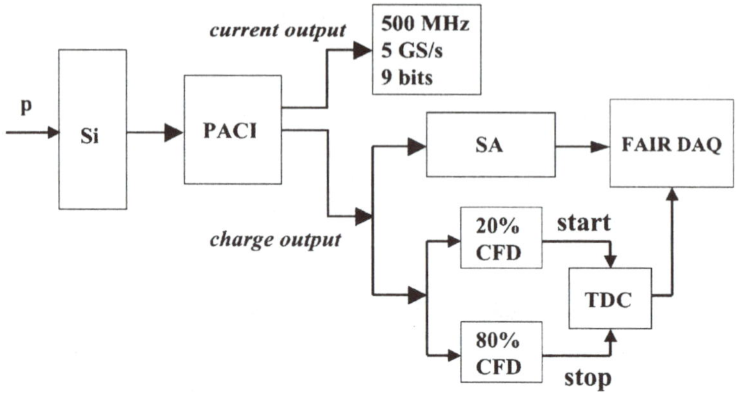

Fig. 4 – Block diagram of the electronics. The characteristics of the digital oscilloscope and the CFD fractions are also shown.

3. Results

A series of tests have been carried out aiming at setting up the apparatus, ascertaining the feasibility of the pinhole beam and verifying the capabilities of the method. A nTD silicon detector (S=1700 mm^2 active area, z=300 μm thickness, ρ=2.5 kΩcm resistivity, R=5.2 Ω resistance, C=630 pF capacitance, L=62 nH inductance, V_d= 140 V and 170 V full depletion and operating bias voltage, respectively) has directly been exposed to proton beams in the energy range E_p= 1.0-4.5 MeV and measurements have been performed all over the rear face in 5 mm steps.

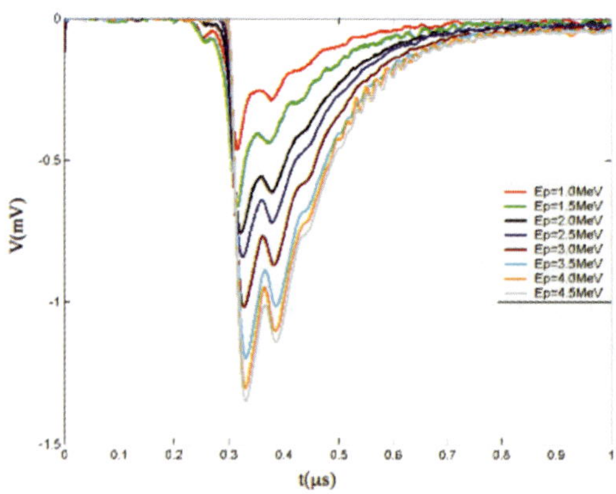

Fig. 5 – Typical current waveforms averaged over 512 pulses.

The intensity of beam current on the first collimator has been kept at 10-20 nA and a counting rate as low as 100-150 particles/second has been measured at the detector level. The stability of the beam alignment has been monitored by inspecting the spectrum of the spectroscopy amplifier and the signal fall time. The PACI current signals, averaged over 512 pulses, have separately been recorded and analyzed. Typical waveforms are shown in fig. 5. The signals have been superimposed for making easier the comparison, but the abscissa has not to be considered an absolute time. Amplitudes as low as 450 μV have been measured and it has been necessary to reduce the bandwidth of the oscilloscope and to select the 1 MΩ coupling for allowing the current pulses to emerge from the noise. As a consequence the coupling impedance mismatch has sparked off signal reflections as it can be seen from fig. 5.

Moreover the detector capacitance and the preamplifier gain (C=630 pF and 13 mV/MeV, respectively) make the system unstable and suddenly oscillations develop, as it is clearly visible in the signal tail of fig. 5. Present results confirm and explain previous measurements performed by the AZ4π collaboration on an identical detector both at GANIL and Orsay tandem. A modification of the preamplifier has been envisaged and is presently under test [24].

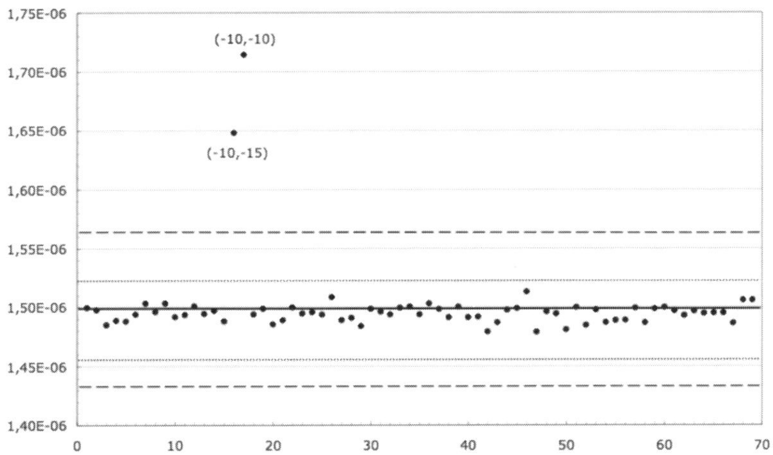

Fig. 6 – Mean values of the current pulses for E_p=4.5 MeV.

Therefore, in order to check the capabilities of the method, the time moments of the current waveforms have been evaluated up to the 4th order for each incident energy aiming at verifying the doping homogeneity point by point. Since the recorded current pulses are *ab initio* average values, the experimental samples must obey the normal distribution unless systematic deviations or other spurious effects are present. Comparing the distribution of the calculated moments with expectation values allow to conclude about systematic effects. In figs. 6-9 the

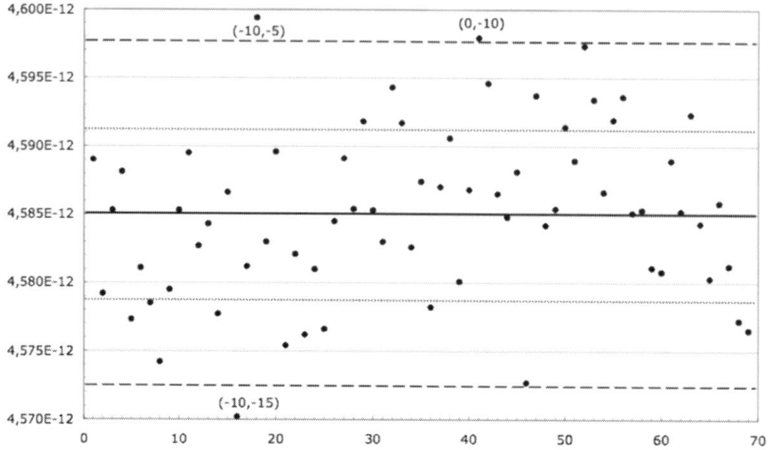

Fig. 7- Same as fig. 6 but for 2nd order moments.

control charts of 1st to 4th moments of current waveforms are displayed for E_p=4.5 MeV. On the ordinate axis time is shown in units of second, while the abscissa is the sequence number. Full line represents the global average μ of the mean values; the dotted lines are drawn at $\mu\pm\sigma$ and the chain lines indicate the range at the 95% confidence level. In parentheses the coordinates are shown of points outside the $\mu \pm 1.96\,\sigma$ interval.

The graphs corresponding to the other bombarding energies are completely similar to those shown in the figures and, due to space shortage, they are not displayed. By figure inspection it is evident that the data are in any case compatible with the hypothesis of a random sample. In fact the number of points falling outside the 95% confidence level is equal to 3 at most. Moreover it has to be considered that scattering from the borders of the collimator could modify the current signal profile and increase the deviation from the mean of single average values, so the previous figure of 3 has to be assumed as an upper limit.

One could worry about the statistical significance and the intrinsic meaning of moment distributions bearing in mind the above mentioned problems connected with noise, instabilities and coupling troubles in measuring current waveforms. In fact, at a first glance it could seem that oscillations, reflections and consequent lengthening of the signal would mask all fluctuations including any deviation due to possible lack of homogeneity. That is not the case as confirmed by results shown in fig. 10, which displays the average values of the fall time. In fact, it has to be considered that reflections simply add a delayed and attenuated image of the signal to itself, whose attenuation and delay depend on the impedances at both ends of the cable and its length.

From the figure the same conclusions as above can be drawn concerning randomness of the experimental results.

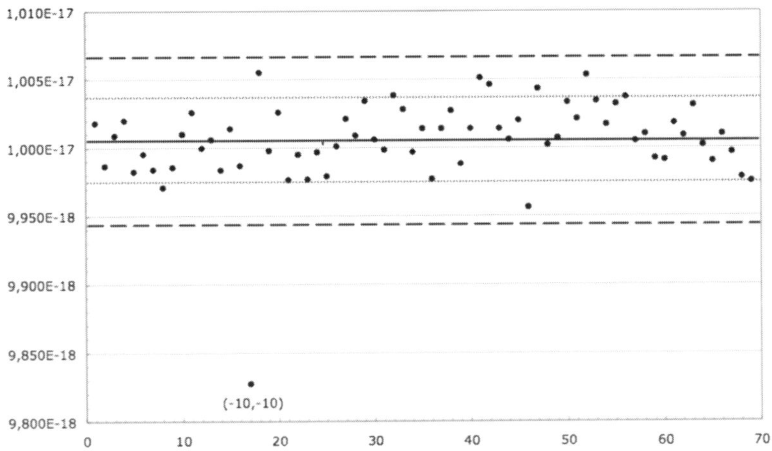

Fig. 8 - Same as fig. 6 but for 3rd order moments.

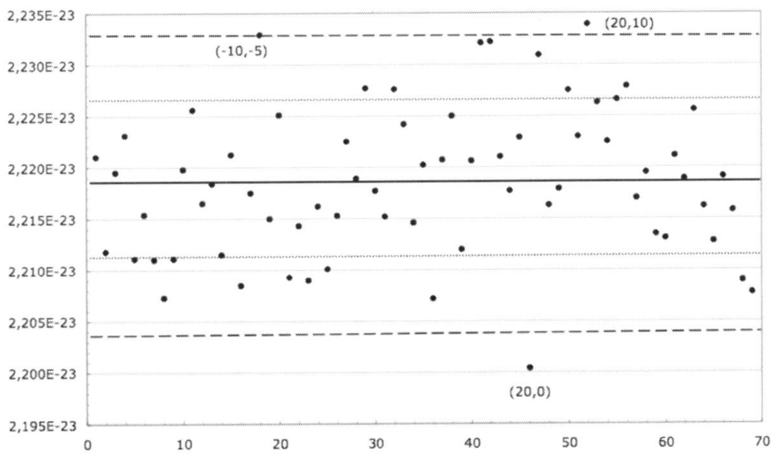

Fig. 9 - Same as fig. 6 but for 4th order moments.

The measured values, centered on 18 ns, are entirely compatible with typical values of analogous detectors. Furthermore, as it is argued from fig. 5, the front of the current waveforms exhibits an almost regular shape and it is not affected by the above-cited problems. Therefore the extracted results have to be considered accurate and complete confidence can be given to their statistical significance. The same results with exactly identical accuracy and to the same confidence level are obtained concerning the fall time measurement from the charge signals of the PACI preamplifier (see fig. 4). Furthermore it has to be pointed out that the charge output suffers only to a least extent from instabilities and coupling mismatch.

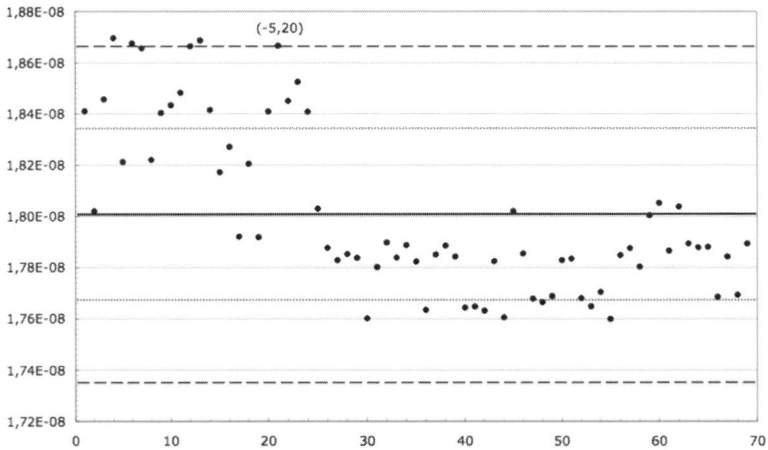

Fig. 10 – Fall time in units of second. The meaning of the symbols is the same as in fig. 6.

4. Conclusions

A dedicated beam line, installed at the tandem accelerator of Università di Napoli, has been presented. It allows producing a well-collimated "pinhole" beam with a $\approx 0.2\ mm^2$ spot size. As a consequence the beam intensity on the target is reduced by a factor of 10^9 in accordance with geometry: for protons 100-150 particles/second have been counted compared to a 10-20 nA incident current. nTD silicon detectors are directly exposed to the pinhole beam with no risk of radiation damage. A remotely controlled x-y translation system and the energy of the bombarding particles together with such a beam allow selecting well-defined volumes V(x,y,z) of the detector. A specially designed charge- and current-sensitive preamplifier has been used. A series of preliminary measurements, aimed at setting-up the experimental technique, have bee performed. Protons in the E_p=1.0-4.5 MeV range have been injected through all over the rear face of the nTD detector with a 5 mm scanning step. From the analysis of the current waveforms produced by bombarding protons there is no evidence of any heterogeneity in doping concentration. This result is confirmed by fall time of both charge and current signals. In summary at the "Laboratorio dell'Acceleratore" of the University of Naples an effective system has been built, installed and tested able to perform a 3-D scanning of nTD silicon detectors for checking their doping homogeneity.

Acknowledgements

Dr. L.Campajola and Mr. M.R.Borriello are deeply acknowledged for their support during the installation of the experimental apparatus.
This work has been supported by the Istituto Nazionale di Fisica Nucleare and by the Ministero dell'Istruzione, dell'Università e della Ricerca under Grant PRIN 2002 No. 2002028413_002.

References

1) Proc. II Int. Workshop on Multifragmentation and Related Topics, GANIL (France) November 5 - 7, 2003;
2) Proc. 18th Int. Nucl. Phys. EPS Conf. "Phase Transition in Strongly Interacting Matter", Prague (Czech Republic), August 23-29, 2004, Nucl. Phys. A749(2005)
3) W.Henning, Nucl. Phys. A746(2004)3
4) M.Di Toro, talk to this workshop and references therein;
5) G.Prete, talk to this workshop and references therein;
6) C.A.J.Ammerlaan et al., Nucl. Instr. and Meth. 22(1963)189
7) W.D.Emmerich et al., Nucl. Instr. and Meth. 83(1970)131
8) T.Kitahara et al., Nucl. Instr. and Meth. 178(1980)201
9) J.B.A.England et al., Nucl. Instr. and Meth. A280(1989)291
10) H.Hamrita et a., Nucl. Instr. and Meth. A531(2004)607
11) A.Pagano, talk to this workshop and references therein;
12) L.Bardelli, talk to this workshop and references therein;
13) G.Pausch et al., Nucl. Instr. and Meth. A365(1995)176
14) F.Z.Henari et al., Nucl. Instr. and Meth. A288(1990)439
15) G.Pausch et al., IEEE Trans. Nucl. Sc. 43(1996)1097
16) G.Pausch et al., IEEE Trans. Nucl. Sc. 44(1997)1040
17) W.Keller, A.Muhlbauer, Float-zone Silicon (Dekker, New York, 1981)
18) S.M.Sze, Physics of Semiconductor Devices (Wiley, New York, 1985)
19) M.Mutterer et al., IEEE Trans. Nucl. Sc. 47(2000)756
20) G.Pausch et al., Nucl. Instr. and Meth. A443(2000)304
21) J.Lu et al., Nucl. Instr. Meth. A471(2001)374
22) J.F. Ziegler, et al., The Stopping and Range of Ions in Solids, (Pergamon Press, New York, 1985)
23) A.Ordine et al., IEEE Trans. Nucl. Sc. 45(1998)873
24) E.Rauly, private communication

The 3 MV tandem accelerator of the "Laboratorio dell'Acceleratore" in Naples

L.Campajola[1,2], L.D'Ambrosio[1], M.Fortunato[1], P.Pedicini[2] and E.Rosato[1,2]

[1] *Dipartimento di Scienze Fisiche*

[2] *INFN, Sezione di Napoli*

Università di Napoli "Federico II", I80126 Napoli (Italy)

An overview of the "Laboratorio dell'Acceleratore" (accelerator laboratory), which hosts a 3 MV tandem accelerator, is presented.

1. Introduction

The Naples 3 MV tandem accelerator is in operation since 1977. At the beginning it was equipped with a RF source, a 90° analysing magnet and a general-purpose scattering chamber, mounted at the end of the unique beam line.

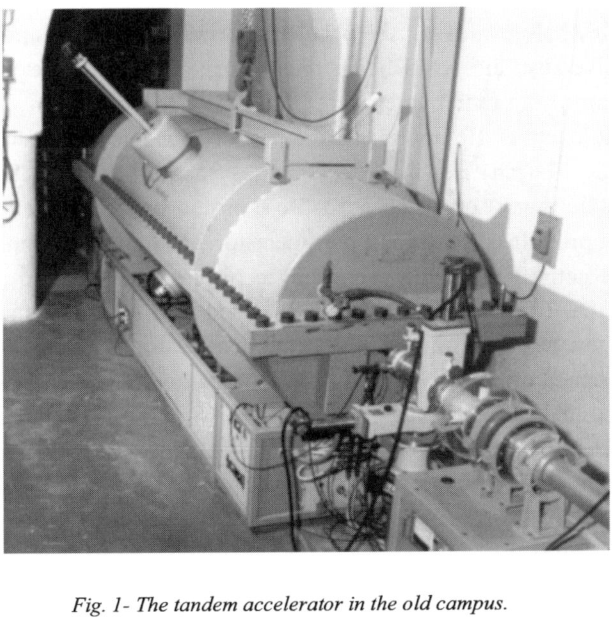

Fig. 1- The tandem accelerator in the old campus.

Since the very beginning a very important mission of the laboratory has been training of students both in making simple experiments for laboratory courses and in preparing their degree. This educational activity continues still today. Over the years the number of beam lines has increased and several improvements have been made aiming to get better performances, which will be briefly described in the following.

The laboratory, officially named “Laboratorio dell’Acceleratore” (LdA), is a division of the Dipartimento di Scienze Fisiche. It is jointly funded by INFN and the University of Naples.

2. The AMS system

During the 1983-1985 years the LdA layout was deeply redesigned. An important improvement was achieved by replacing the injection magnet, by adding a new sputtering source and a 7-port switching magnet.

A Wien filter and a telescope detector for heavy ion identification were installed on a dedicated beam line aiming to perform radiocarbon dating. It became fully operational in 1987 and was the first AMS (Accelerator Mass Spectrometry) facility in Italy [5]. Several hundreds of ^{14}C dating were produced both within the scope of research programs, mainly in archaeology and geology, and on behalf of a third party [6, 7].

2.1 Radioactive ion beam

In 1996, before moving the accelerator into the new campus buildings of Naples University, the AMS beam line was modified in order to measure the cross section of the radiative capture $p(^7Be, ^8B)\gamma$ reaction in inverse kinematics [8]. The radioactive ^7Be nuclides were produced inside a pill-shaped Li_2O matrix, via the $^7Li(p,n)^7Be$ charge-exchange reaction at the KIZ-cyclotron of Karlsruhe (Germany) by using a 13 MeV proton beam with a 10 mA current. Typically a pill, after proton bombardment for 230 hours, accumulated 6.0 GBq of total ^7Be activity. The pills were sent to Naples and were introduced into the sputter source. The ^7Be nuclides were extracted from the Li_2O matrix as a $^7BeO^-$ molecular ion beam and injected into the accelerator [9]. The radioactive ^7Be were accelerated up to E = 8.0 MeV and bombarded a windowless H_2 gas target [9,10]. The incident beam had a typical intensity of 18 pA in the target region.

The ^7Be beam has been the first and unique accelerated radioactive ion beam in Italy.

3. Present layout

In 1997, when the measurements of the ^7Be+p reaction were completed, the accelerator was dismantled and definitively moved to its own hall in the new campus. Advantage was taken of this opportunity to introduce major changes to the layout of the laboratory:

- a new injection system coupled to a pre-accelerator was designed and installed;

- improvement of the pumping system;
- fast sequential injection system;
- computer control of vacuum;
- automatic conditioning of the accelerator;
- new beam lines.

Fig. 2 – The accelerator in its definitive hall.

3.1 The injection and the low-energy side

The transmission efficiency, defined as the percentage of negative ions measured at the low energy Faraday-cup with respect to the positive ones (with a definite charge state) detected at the image point after the analysing magnet depends on terminal voltage, stripper choice and the nature of the negative ions. Good transmission depends on the beam optics and on the vacuum conditions. The beam optic in the low-energy side of the accelerator has been improved with the new injector especially designed for ions as heavy as ^{14}C. The sputtering source used in AMS measurements has been equipped with a home made pre-acceleration stage able to deliver an injection energy up to 90 keV. For a ^{12}C beam the transmission efficiency increases by a factor of ~ 3, being higher than 0,5 to be compared to the previously measured values without pre-acceleration, which, in any case, were less than 0.2.

3.2 The pumping system

The original pumping system was based on manually operated diffusion pumps. They have been completely replaced with turbo molecular pumps. In addition the accelerator has been equipped with two ion pumps at both ends, the low and the high-energy side. The operating vacuum is typically 8 x 10^{-8} and 6 x 10^{-7} mbar at the end of the accelerator and on the beam lines, respectively.

Fig. 3 - The sputtering sources: without (left) and with pre-acceleration (right).

3.3 The sequential injection of AMS

The fast sequential injection of ^{13}C and ^{14}C isotopes is achieved by applying different voltages leading to the same magnetic rigidity for the different isotopes without changing the injection energy. The voltage switching takes ~ 0.1 ms, but stable beam currents require about 100 ms.

The AMS measurements are supervised by a system control based on a PC. The system sets and measures the fields applied to the injection magnet and velocity filter (fig. 4). It also applies the pulsed voltage to cycle between the carbon isotopes and activates the proper beam stops at the due time. Finally the control system enables the acquisition of ^{13}C counts in the Faraday-cup and ^{14}C in the particle detector.

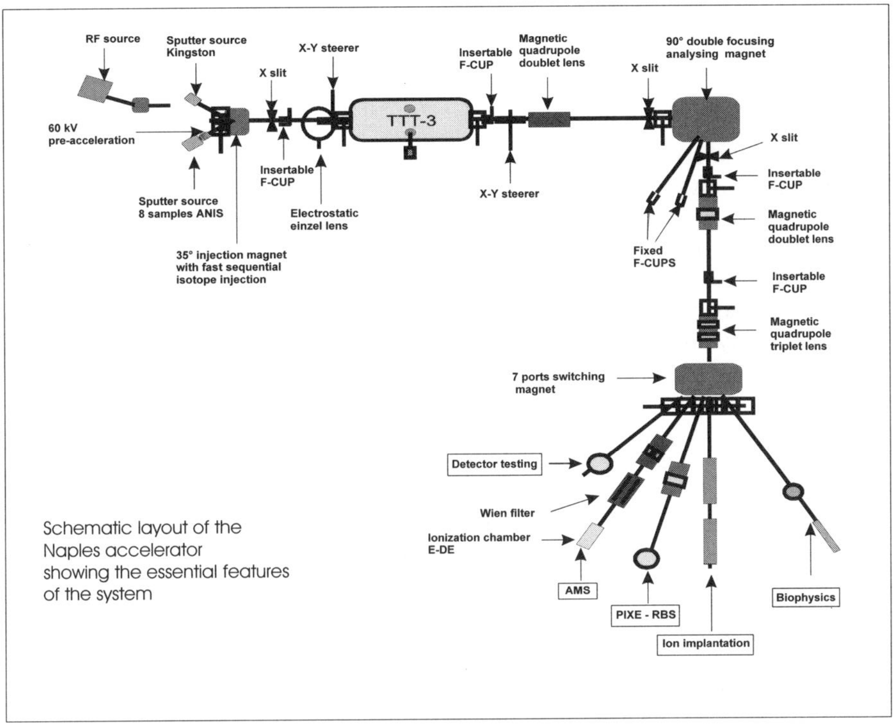

Fig. 4 - Schematic layout of the LdA with beam lines

3.4 The control system

The status of the vacuum conditions from the sources up to the switching magnet is continuously monitored by another control system based on a dedicated PC. It acquires and controls the pressure and the status of turbo molecular pumps, gate valves and ion pumps.

A third PC controls the Van de Graaff generator, acquires all the parameters of the electrostatic generator and performs the automatic conditioning.

3.5 The beam lines

A general layout of the LdA is shown in fig. 4. In the same figure are also sketched the five beam lines, which are in operation at the moment. They are listed from left to right

- Detector testing
- AMS

- Ion Beam Analysis
- Ion implantation
- Radiobiology

Besides the research activity the LdA plays a very important role from an educational point of view. In fact, each year many students, under the guidance of their teachers or experienced tutors, perform simple experiments of nuclear physics, PIXE, RBA, radiobiology, etc. as a part of their laboratory courses. Furthermore several students have done the degree and/or the PhD thesis at LdA. As a confirmation of this fact, the entire computer control system has been almost done by undergraduate students.

References

1. P.Cuzzocrea et al., Lett. N. Cimento **32**(1981)33
   E.Rosato, *Phys. Rev.* **A28**(1983)2759
   P.Cuzzocrea et al., J.Phys. **B18**(1985)711
2. A.M.Mancini et al., *Thin Sol. Films* **89**(1982)407
   A.M.Mancini et al., *J. Appl. Phys.* **53**(1982)5785
3. G.Gialanella et al., *Radiat. Res.* **96**(1983)462
4. P.Cuzzocrea et al., Lett. N. Cimento **28**(1980)515
   P.Cuzzocrea et al., Report INFN/BE-80/5
5. L.Campajola, et al., Nucl. Instr. and Meth. in Phys. Res. **B29**(1987)129
6. F.Terrasi et al., L'Italia Scacchistica, Numero speciale, Maggio 1994
7. C.Albore-Livadie et al., Hommes et Volcans dans la Prehistoire - Quaternaire (1997)
8. L.Gialanella et al., Eur. Phys. J. **A7**(2000)303-305
9. L.Campajola, et al., Z. Phys. A-Hadron Nucl. **356**(1996) 107
10. L.Gialanella et al., Nucl. Instr. and Meth. in Phys. Res. **A376**(1996)174

SECTION VIII

RI BEAMS

EXPERIMENTAL NUCLEAR PHYSICS IN ITALY

E. CHIAVASSA

Dipartimento di Fisica Sperimentale, Università di Torino
and
I.N.F.N. - Sezione di Torino
Via P. Giuria 1, I-10125 Torino, Italy
E-mail: Emilio.Chiavassa@to.infn.it

A. FELICIELLO

I.N.F.N. - Sezione di Torino
Via P. Giuria 1, I-10125 Torino, Italy
E-mail: Alessandro.Feliciello@to.infn.it

The experimental activity in the field of the Nuclear Physics is briefly outlined: about 470 Italian researchers are involved in 32 experiments, supported by INFN and carried out both at national and international facilities.

1. Introduction

The experimental Nuclear Physics program of the INFN is devoted to the investigation of the structure and of the dynamics of many-body systems. It spans over a wide spectrum of topics, ranging from the structure of the nucleon, described in terms of quarks and gluons, to state and phase transitions of the nuclear matter; moreover it includes the study of the properties of nuclei under extreme conditions. About 470 Italian Nuclear Physicists are working on these subjects in several national and international laboratories, by using a large variety of probes, different in energy and nature.

Electromagnetic probes of both high and intermediate energy allow to investigate the nucleon spin structure,to assess the properties of baryonic resonances and to look for a possible multi-quark state.

Charged K mesons produced at LNF have been intensively exploited to produce and to study single Λ-hypernuclei and kaonic atoms.

Ion beams of both low and intermediate energy represent the key tool for studying nuclear structure, reaction mechanisms and phase transitions of nuclear matter.

Nuclear structure under extreme conditions of isospin and momenta has been extensively explored using multi-array γ-detectors.

It is worth noting that many of these activities have been carried out at the INFN National Laboratories LNF, LNL and LNS.

Recently the possibility of producing and using radioactive ion beams has been attentively pursued: they represent the main interest of this branch of the Nuclear Physics in the next future.

A very successful experimental program has been performed in the few keV energy domain with the aim of measuring the cross section for a set of nuclear reactions of astrophysical interest: data have been collected in several laboratories and in particular at the INFN LNGS, profiting of the very low background environment provided by this unique underground site.

Few years ago there was a strong participation in the ultra-relativistic heavy ion program at the CERN SPS; today the Italian Groups are deeply involved in the construction of the ALICE detector for LHC.

In all the experiments in which the Italian researchers are present, they give important contributions both by developing advanced detectors with the associated electronics and by actively participating to the analysis and to the interpretation of the experimental data. The best proof of this statement is represented by the large number of papers published on referred international journals and of talks given at several international conferences.

Finally it is important to underline that INFN National Laboratories have successfully delivered beams to the scheduled experiments: among other positive, obvious consequences, this fact attracted many foreign researchers. Moreover the high-quality beams allowed the realization of several medical and cultural applications.

The complete list of both concluded and running experiments, their main achievements, their publication list and other information can be found at the INFN web site[a]. In Appendix A table 1 gives the list of the recently approved experiments cited in this paper.

2. Quark-hadron physics

The HERMES Collaboration at DESY is studying the nucleon spin structure through the deep inelastic scattering of 27.5 GeV, longitudinally polarized electrons on an internal target (unpolarized, longitudinally or transversely polarized). Among its results it is worth quoting the advanced

[a] go to http://www.infn.it, then follow the links “Activities” and “Nuclear Physics”

knowledge of the structure functions, the first direct measurement of pPDF for single flavors, the first estimation of the gluon contribution to the nucleon spin and the signals for transversity and for GPD's. The measurements will be continued with the aim of addressing exclusive processes, having added a recoil detector to the spectrometer.

At JLab the AIACE Group is performing experiments with the CLAS spectrometer: the goal is to obtain precise information on the nucleon structure and on the nature of the strong interaction. The main measurements under Italian responsibility refer to: production of baryon resonances in meson electro-production channels; hard production processes and quark gluon structure; higher twist and quark structure in the semi-inclusive hard meson production processes. A search for exotic baryons (pentaquark) is currently carried out. Among other topics, the large AIACE scientific program includes the studies of the GDH sum rule on proton and of the polarized function at very low momentum transfer.

ELETTRO is the second Italian Group working at JLab with a high resolution spectrometer in the Hall A. Also in this case the experiment is producing a lot of good results: among others it is worth reminding the decreasing ratio of the electric and magnetic proton form factors and the change in the GDH integral behavior at low Q^2. The installation of septum magnets and of a proximity RICH allowed to begin the studies on the electro-production of hypernuclei that will be completed in the next years. In addition measurements on GDH integral and on parity violation are foreseen.

The CTT Group, working at MAMI and at ELSA, confirmed the validity of the GDH rule on the proton in the energy range explored. This result is an experimental proof of the validity of the theoretical model for the photon-nucleon interaction. The Group will continue its activity at the MAMI-C accelerator: it will focus its interest on the study of the properties of baryonic resonances.

The review of the activities based on the use of electromagnetic probes is completed by the GRAAL experiment: it is producing interesting results on baryonic resonances, obtained using Compton backscattered polarized photons.

Measurements with hadronic projectiles are mainly performed at LNF, where two experiments use charged K mesons following ϕ decays at DAΦNE.

FINUDA collected about 30 millions of events of single Λ-hypernuclei formation on several targets: ^{6}Li, ^{7}Li, ^{12}C, ^{27}Al and ^{51}V. The Collaboration will continue the measurements to reach a total integrated luminosity of the order of 1 fb^{-1}: the chosen targets will range from light to heavy nuclei and the experiment will push the production of hypernuclei and the study of their decay modes at unprecedented level.

The second experiment completed at LNF is DEAR: it measured the shift and width of transition on kaonic nitrogen and hydrogen; data will allow to obtain new precise information on the kaon-nucleon system. The upgraded, triggerable apparatus SIDDARTHA will allow more precise measurements of kaonic hydrogen and the first measurements on kaonic deuterium and helium. A determination of the K mass at a precision level of 10 keV will be possible as well.

The review of the hadron physic experiments is completed by the PAINUC experiment, running at Dubna and investigating the interaction of low energy pions on nuclei.

3. Relativistic and ultrarelativistic heavy ion physics

In this sector the activities of the Italian researchers are mainly concentrated at CERN where measurements with the SPS heavy ion beams have been successfully performed and the heavy ion apparatus for LHC (ALICE) is now under construction.

The NA57 experiment found an enhancement of the strange hadron production in Pb-Pb interaction (with respect to p-p) at ion energies of 40 and 158 GeV per nucleon.

An anomalous suppression of the J/Ψ production has been observed in central Pb-Pb collisions by the dimuon spectrometer of the NA50 experiment: this results could hardly be interpreted by means of non-QGP models.

NA60, the upgraded version of NA50, is now studying the D charmed meson production together with the low mass spectra. Moreover first data on In-In collisions have been collected.

Today the main engagement of the INFN physicists in this field is represented by the ALICE experiment: more than 120 researchers are involved and the Italian Groups play a key role in the management of the whole project and in the construction of the apparatus. As it is well known the ALICE experiment consists of a set of sub-detectors, immersed in a magnetic field, plus a forward dimuon spectrometer and some additional forward detectors. In the following the contribution of the Italian Groups to this experiment is briefly summarized.

- Inner Tracking System (ITS)
  6 Si layers form the ITS; the main tasks that it must accomplish are:

  - determination of the primary and secondary vertices for the charm and hyperon decay reconstruction
  - identification and tracking of the low momentum particles
  - improvement of the TPC momentum measurement

  The first two ITS layers consist of pixel detectors; they cover an area of 0.27 m^2 for a total of 15.7×10^6 channels. The realization of the detectors is shared between CERN and INFN.
  The 3rd and 4th layers are made of Si drift detectors; the corresponding area is of 1.3 m^2, for a total of 13.3×10^3 channels. The project and the realization are mainly due to the Italian Groups.
  Finally the 5th and 6th layers are instrumented with double sided microstrips; the sensitive area is of 5 m^2 with 1.7×10^6 channels. In respect to this device the Italian Groups are engaged in the construction of a sub-set of the detectors and in the module assembly. The skilfulness and the deep involvement of the Italian physicists in this ALICE sub-detector is acknowledged by the fact that they have the leadership of the whole ITS project.

- Time Of Flight system (TOF)
  The TOF aims to identify particles in all the ALICE acceptance. It covers an area of 140 m^2 with 160×10^3 channels and is made of an array of Multigap Resistive Plate Chambers; each element of the detector is a long strip of 7×120 cm^2 with 96 readout pads of 3.5×2.5 cm^2. The detector has proven to be very performing with excellent efficiency and with a time resolution better than 50 ps. The apparatus has been completely designed and assembled in Italy.

- High Momentum Particle IDentification (HMPID)
  A 1.5×1.5 m^2 proximity focusing Čerenkov ring detector covers part of the barrel, with the goal of recognizing particles with momentum greater than 2.5 GeV/c; a prototype of the detector has been successfully operated in the STAR experiment at RICH.

- 2 muons spectrometer
  The Italian Groups collaborate to the construction of the tracking chambers and of the RPC needed for the trigger of this spectrometer that aims to measure the forward production of heavy "quarkonia"

- Zero Degree Calorimeters (ZDC)
  A set of four calorimeters made of quartz fibers is totally build in Italy. Its purpose is to give an evaluation of the centrality of the interaction by means of the measurements of the non interacting nucleons. Moreover the calorimeters will be a powerful device to estimate the luminosity of the collider.

- Computing
  Italian physicists are deeply involved in computing; a pilot Tier2 center has been installed and a test on data production has been carried out by simulating events using the ALIROOT code developed by the ALICE software Group.

A new, interesting initiative is represented by the FRIBS experiment at LNS: its main idea is to produce radioactive ion beams at intermediate energy, by exploiting the in-flight method.

The activities in this field are completed by the participation to the HADES2 experiment at GSI with the design and the construction of the TOF detector electronics and with an active presence in the data taking and analysis.

4. Nuclei under extreme conditions

The classical Nuclear Physics focuses its interest on the studies of nuclei produced under extreme conditions of isospin, angular momentum and temperature. In this context it is possible to create new shapes and pairing of nuclei thus allowing to look at the shell structure, at the isospin effects and at the nuclear matter transition from liquid to gas. Italian physicists work on these subjects mainly at the two dedicated INFN national laboratories (LNL and LNS); however important contributions come from measurements performed in foreign laboratories like GSI, GANIL and TAMU. Important information on the nuclear structure have been obtained by the GASP Collaboration at LNL and in the framework of the EUROBALL program. Measurements with the GASP apparatus and with the PRISMA2 spectrometer, operated in coincidence with the CLARA γ-detector array, are presently carried out at LNL, where the PIAVE-ALPI complex will produce new stable ion beams. Use of radioactive ion beams is pursued at GSI in the framework of the RISING project, aimed to the study of nuclei of astrophysical interest. To complete the review of the activities in the field of spectroscopy, it is worth reminding the measurements of TRARE on homologous nuclei produced in transfer reactions and the strong participation to the development of the new segmented HPGe detector (AGATA project).

The understanding of the dynamics and of the thermodynamics of the nuclear reactions represented the goals of several experiments (NUCL-EX, ISOSPIN, N2P, FIESTA, TRASMARAD) carried out mainly at LNL and LNS by using sophisticated apparatuses like GARFIELD, 8πLP, MACISTE and CHIMERA that has been successfully completed last year. Among the large number of topics addressed, it is worth quoting the evolution of the fission mechanism, the isospin dependence of the dissociation of nuclei far from the stability, the phase transition of warm nuclei and the reaction mechanisms in peripheral collisions where a large and fast emission of complex fragments is observed.

The EXOTIC Collaboration put in operation a ^{17}F beam at LNL to study the elastic diffusion cross section, the break-up and the stripping on light and heavy nuclei around the Columbian barrier.

The N2P experiment will study nuclear reactions induced by neutrons at LNL and TAMU, where the emission of delayed neutrons in the decays of exotic nuclei produced in the U-U interactions will be measured.

It is important to remark that at LNS the CATANA facility is now operative and that several patients, affected by ocular melanoma, have already been irradiated. It must be also underlined that at LNS new exotic beams will be soon produced: this will be done in the framework of the EXCYT project that will be fully exploited by means of the new MAGN-EXP spectrometer.

Finally at LNL the first phase of the SPES project, aiming to produce a high intensity p beam as an injector for the European RIB generator EURISOL, has started. The project includes as well a part devoted to the application of the BNCT therapy.

5. Nuclear astrophysics

The interest for measurements of nuclear reactions at energies around the Gamov peak is greatly grown up in the last years; a well coordinate program has started in Italy with measurements at LNS and LNGS INFN National Laboratories and at Bochum. The LUNA2 facility at LNGS represents the first worldwide underground experiment to measure low energy cross sections profiting of the very low cosmic ray background. Cross section measurements of the ^{3}He(^{4}He,γ)^{7}Be and d(p,γ)^{3}He have been done down to the Sun Gamov peak; from the measurement of the ^{14}N(p,γ)^{15}O reaction a new estimation of the Globular Cluster age has been inferred.

The ASFIN2 Collaboration, working mainly at LNS, developed and proved the efficacy of the indirect method (Trojan Horse) in which the cross section of the $A+x \rightarrow C+c$ reaction is deduced from the measurement of the $A+a \rightarrow C+c+b$ (where $a = x+b$). This method allows the determination of the cross sections of the investigated reactions at energies considerably lower than that of the projectile, provided that the Fermi momentum of the nucleon x in the nucleus a is known. By using deuterium beams the ^{7}Li(d,α^{4}He)n and ^{6}Li(d,α^{3}He)n cross sections have been measured: the results gave important hints on the electron screening potentials.

The Italian researchers belonging to the ERNA Collaboration are involved at Bochum in the measurement of the $^{12}\mathrm{C}(\alpha,\gamma)^{16}\mathrm{O}$ reaction cross section by using a highly selective spectrometer to reject C ions.

Contributions to nuclear astrophysics came also from the n_TOF facility operating at the CERN PS: the primary goal of the apparatus is the neutron cross sections measurements for ADS realization, but it is now obtaining values on capture reactions of astrophysical interest like, for example, the ^{151}Sm(n,γ)^{152}Sm.

6. Conclusions

The researches endowed by INFN cover all the fields of the modern Nuclear Physics and are completely inserted in the context of the European Nuclear Physics as it is illustrated in the NuPECC long range plane 2004. Following a consolidated tradition, all the experimental activities are developed in close cooperation with the Italian Universities and all the experiments are carried out in the framework of international Collaborations. The studies on nuclear structure and reaction mechanism are mainly carried out at the two Nuclear Physics Laboratories LNL and LNS, where also medical applications have successfully (and will be) developed. Moreover environment and cultural applications are intensely pursued and a new dedicated laboratory was recently settled out. The intensive use of INFN National Laboratories is also proved by the hypernuclei and kaonic atoms experiments running at LNF and by the low energy measurements performed at LNGS. Nucleon structure and nucleon resonances are studied in foreign laboratories by using electromagnetic probes. New experiments (PANDA and PAX) are now being proposed at the new GSI FAIR facility. In respect of the very high energy sector Italian Groups work at CERN: in particular they are deeply involved in the construction of the ALICE apparatus to study QGP with the LHC collider. The Italian contribution in this experiment is very remarkable: in this moment it represents the main engagement of the community, especially for what concerns the resources.

Appendix A.

Table 1.: List of the Nuclear Physics experiments supported by INFN.

experiment	laboratory	accelerator	probe
Quark-hadron physics			
HERMES	DESY (Germany)	HERA	$e^{\pm}$
AIACE	JLab (USA)	CEBAF	e^{-}, γ
ELETTRO	JLab (USA)	CEBAF	e^{-}
CTT	Mainz (Germany)	MAMI-C	γ
GRAAL	Grenoble (France)	ESRF	γ
	BNL (USA)	NSLS	γ
FINUDA	LNF (Italy)	DAΦNE	$e^{\pm}$, $(K^{\pm})$
SIDDHARTA	LNF (Italy)	DAΦNE	$e^{\pm}$, $(K^{\pm})$
PAINUC	JINR (Russia)	phazotron	$\pi^{\pm}$
PANDA	GSI (Germany)	HESR	$\bar{p}$
PAX	GSI (Germany)	HESR	$\bar{p}$

Table 1.: List of the Nuclear Physics experiments supported by INFN (cont'd).

experiment	laboratory	accelerator	probe
Relativistic and ultrarelativistic heavy ion physics			
ALICE	CERN (Switzerland)	LHC	*u.r. ions*
NA57	CERN (Switzerland)	SPS	*u.r. ions*
IPER (NA60)	CERN (Switzerland)	SPS	*u.r. ions*
FRIBS	LNS (Italy)	CS	*ions*
HADES2	GSI (Germany)	SIS	*r. ions*, p, $\pi^{\pm}$
Nuclei under extreme conditions			
GAMMA	LNL (Italy)	ALPI/PIAVE	*ions*
	GSI (Germany)	SIS	*ions*
PRISMA2	LNL (Italy)	ALPI/PIAVE	*ions*
TRARE	Garching (Germany)	tandem	p
NUCL-EX	LNL (Italy)	ALPI	*ions*
	LNS (Italy)	CS	*ions*
	GANIL (France)	Spiral	*ions*
ISOSPIN	LNS (Italy)	CS	*ions*
N2P	LNL (Italy)	ALPI	*ions*
	TAMU (USA)	SC	*ions*
FIESTA	LNL (Italy)	ALPI	*ions*
	JYFL (Finland)	SC	*ions*
TRASMARAD	LNS (Italy)	CS	*ions*
	UCL (Belgium)	cyclotron	*ions*
EXOTIC	LNL (Italy)	ALPI	*ions*
	LNS (Italy)	CS	*ions*
	RIKEN (Japan)	ring cyclotron	*ions*
	ANL (USA)	ATLAS	*ions*
	GANIL (France)	Spiral	*ions*
MAGN-EXP	LNS (Italy)	EXCYTE, CS	*ions*
	CERN (Switzerland)	REX-ISOLDE	*ions*
	IFUSD (Brazil)	PELLETRO	*ions*

Table 1.: List of the Nuclear Physics experiments supported by INFN (cont'd).

experiment	laboratory	accelerator	probe
Nuclear Astrophysics			
LUNA2	LNGS (Italy)		p, α
	Bochum (Germany)		p, α
ASFIN2	LNS (Italy)	tandem	*ions*
	Bochum (Germany)	tandem	*ions*
	TAMU (USA)	cyclotron	*ions*
ERNA	Bochum (Germany)	tandem	*ions*
n_TOF	CERN (Switzerland)	PS	n

THE EXCYT PROJECT

G.CUTTONE, M.MENNA, M.RE, L.CALABRETTA, L.CELONA,
L.COSENTINO, P.FINOCCHIARO D.GARUFI, G.RAIA, D.RIFUGGIATO

Laboratori Nazionali del Sud,
Via S. Sofia, 62
95123 Catania
E-mail: cuttone@lns.infn.it

The EXCYT facility (EXotics with CYclotron and Tandem) at the INFN-LNS is based on a K-800 Superconducting Cyclotron injecting stable heavy-ion beams (up to 80 MeV/amu, 1μA) into a Target-Ion Source assembly (TIS) to produce the required nuclear species, and on a 15 MV Tandem for post-accelerating the radioactive beams. Since the previous TIS had been radically modified in its shape, dimensions and geometry, it became necessary to verify off and on-line the TIS behaviour with respect to the mechanical and thermal stresses at 2300 K as well as to obtain information about the production and the release processes. In order to save time, this was successfully achieved in May 2003 at the SIRa test-bench of SPIRAL, GANIL, by shooting a ^{13}C primary beam (60 MeV/amu) on a ^{12}C target under the same operational conditions that will be initially used at EXCYT. The measured yields and production efficiencies for ^{8,9}Li were compatible with the ones obtained at SIRa. We decided to deliver ^{8}Li as the first EXCYT radioactive beam taking into account the requests and the first results obtained by the Big Bang collaboration as well as the availability of the large magnetic spectrometer MAGNEX in late 2004. The commissioning of the EXCYT facility is foreseen by the beginning of 2005 together with the start of nuclear experiments programme.

1. Superconducting Cyclotron

Following the upgrading process of the Superconducting Cyclotron (CS) accomplished in the last few years [1], in 2003 an accelerator stop was scheduled to test the cooled deflector by shooting high-intensity beams on it. The tests were also aimed to the extraction of a light ion beam sufficiently intense to be used as a primary beam for EXCYT. When shooting a ^{20}Ne beam (45 MeV/amu), we found a power limitation at 40 W: the beam caused mechanical damages in the Ta septum of the electrostatic deflector. To solve this problem we designed and mounted:

(1) A new septum modified to a V-notch for an improved power dissipation;
(2) A small chopper installed in the injection line allowing to decrease the beam current in the longitudinal way;
(3) Two internal current probes provided with a cooling circuit, to find the best conditions for minimum beam dissipation in the septum;
(4) Phase slits to increase the extraction efficiency up to 60%.

New tests with ^{13}C, the prospective primary beam for the production of ^{8}Li, resulted in an extracted beam power of about 100 W corresponding to 680 enA (fig. 1). This figure is estimated to be sufficient for an adequate production of ^{8}Li (10^4-10^6 pps on target) and at the moment is the maximum achievable for radioprotection issues.

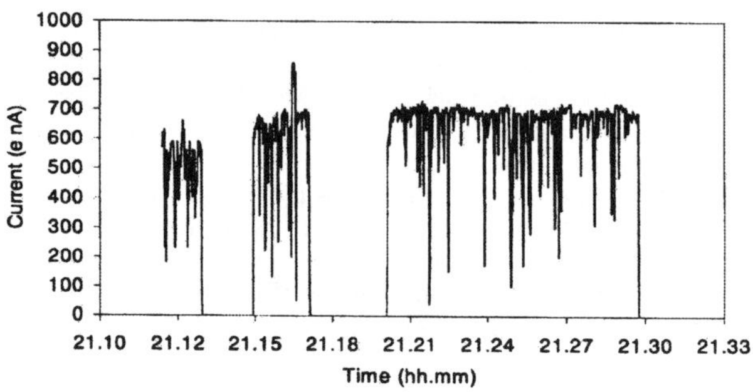

Figure 1. ^{13}C (60 MeV/amu) intensity versus time. The extracted power was 100 W.

2. Primary Beam Line and HV Platform

The vertical segment of the primary beam line has been redesigned and additional diagnostic elements have been introduced through the whole line. Since the TIS is at 50 kV with respect to the high-activity platform and its operating temperature is around 2000 °C, to check the correct beam positioning a remote-controlled linear actuator with one alumina disc and three collimators (diameter 6, 9, 12 mm respectively) had to be placed on the top of the platform. The disc can withstand 10 W maximum, thus it will be used only during the beam preparation and replaced with one of the collimators during target irradiation. The outer halo of the primary beam will constantly collide with the collimator and an increase or decrease of the relevant measured current outside the fixed thresholds automatically triggers the safety system to stop the beam. During target irradiation, beam shape and positioning will be provided by a special diagnostic element placed at a few centimetres above the TIS: it consists of a 10μm scintillating layer deposited on a 10-μm window, and of a second 6-μm window to shield the first one from the TIS thermal irradiation. Studies and tests on prospective windows indicated that -as for thermal dissipation, irradiation heating and radioactivation- Al has better properties with respect to graphite, Ti, Cu, Ag, Au. The scintillator must have good radiation hardness, therefore at LNS we deposited inorganic materials on the Al window: CsI(Tl) by thermal evaporation and Al_2O_3(Cr) by electron-gun bombardment. Alternative techniques like Ion Beam Assisted Deposition e Plasma Spray are currently under study. The thermal dissipation and the energy decrease on the windows are about 0.6% of the primary beam power while the minimum detectable beam power is around 1W. Computer simulations show that the temperature increase is negligible under those conditions (fig. 2).

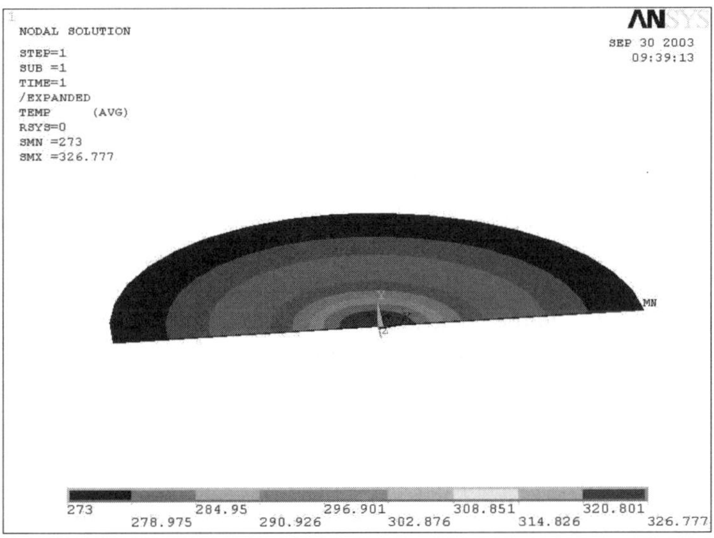

Figure 2. Temperature distribution (in Kelvin) for a window made of Al_2O_3(Cr)/Al (10+10 μm) with diameter = 25 mm.

A video camera with zoom will be placed on the TIS vertical axis to look at both diagnostic elements through a dedicated view port on the 90 ^{o} bending magnet. Finally, two linear actuators have been placed soon after the 18 ^{o} magnetic dipole in the secondary beam line: one at the 0 o exit, the other at the 18 ^{o} exit. Both actuators can host alumina discs for beam monitoring by means of video cameras connected to the video multiplexer on the low-activity platform.

3. Target Material and Geometry

The main characteristics of an ideal target are:

(1) High production rates;
(2) Large diffusion coefficient of the products;
(3) Low product hold-up time on surfaces;
(4) Large values of dissipated power;
(5) High melting point;
(6) Low vapour pressure to preserve ion-source efficiency ($< 10^{-8}$ mol/s);
(7) Low Z to reduce activation;
(8) No sintering;
(9) No reaction with the container;
(10) No damage due to the radiations.

By taking into account some of these characteristics (points 4-10), graphite has been chosen as a suitable material. To comply also with favourable production and release properties (points 1-3), it has been necessary to select among the many kinds of commercially available graphite the ones with higher open porosity, lower closed porosity, smaller grain size, higher thermal conductivity, higher density. After an extensive market research, we chose the graphite type UTR146 manufactured by XYCARB with a few parts per million of impurities. Experiments with radiotracers in collaboration with CERN-ISOLDE clearly showed a better release from this kind of graphite compared to other materials [2]. As for the production rates, we were less restricted on the choice since at EXCYT the radioactive nuclides are obtained mainly by projectile fragmentation. Nevertheless, we obtained good yields of ^{18}F, ^{7}Be and ^{22,24}Na by shooting ^{19}F (48 MeV/amu) on a XYCARB UTR146 graphite target [3]. To get an efficient heat dissipation for simple heat radiation loss, the target is tilted by a certain angle with respect to the incoming beam while keeping constant its volume, thus reducing the effective target thickness and increasing the area of two of the heat-dissipating surfaces [4]. The previous TIS assembly [4] proved to be unreliable and was therefore radically modified in its shape, dimensions and geometry as shown in (fig. 3).

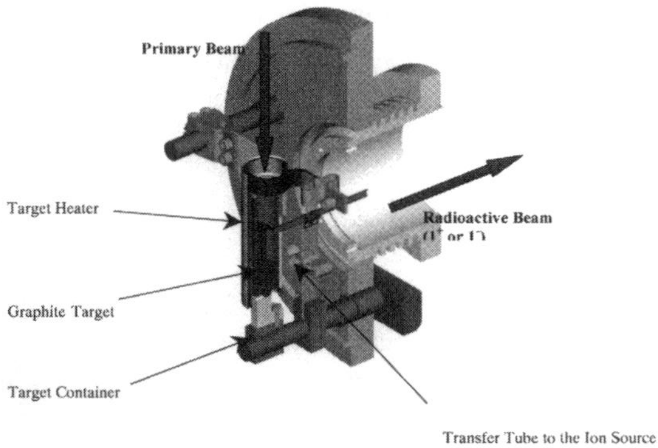

Figure 3. The new TIS. The real target area is mainly constituted by its upper part and is tilted to enhance both power dissipation and desorption

In the new TIS the graphite target is standing in a tantalum container, which in turn is inside a tantalum heater. In this configuration, the heater does not touch the container and the primary beam impinge on the target from the top. Heating is achieved by means of direct current while several tantalum layers on the outside of the heater shield the aluminium external container (not shown) from high temperature. The target container is heated by irradiation; the target is heated by: irradiation, contact at its bottom, primary beam power. The effective target volume is mainly its upper part, the rest constituting a simple mechanical support. In fact, the thickness of the upper part is chosen according to the range of the primary beam and of the produced radioactive nuclides. By a judicious choice, in principle it is possible to obtain that most of the products are located into the upper part of the target close to the transfer tube.

4. Positive Surface Ion Source (PIS)

Given its physical and chemical properties, the ionisation efficiency of lithium can be conveniently studied by implanting stable ^{6}Li into the graphite substrate. The substrate is then positioned nearby the ion source and heated to allow for the escape of the implanted atoms to be ionised. Then the heating temperature is varied in order to keep the extracted beam in the range 1-10 nA (operational values for radioactive beams), the total extracted charge for the selected mass is recorded and converted into the number of extracted ions. The ratio of the latter to the implanted atoms gives the lower limit for the ionisation efficiency. By using an ISOLDE-type Positive Ion Source with a hollow W tube, we obtained an efficiency of 75% for ^{6}Li$^+$, very close both to its theoretical value and to on-line extrapolations [5].

5. Isobaric Mass Separator and Beam Diagnostic

The isobaric mass separator has been mounted and aligned with the required precision. Tests to measure its mass resolving power are in progree. During 2003 we installed and aligned 4 LEBI (Low Energy Beam Imager/Identifier) devices on the low-intensity platform. In order to measure and monitor also the radioactive beams, the devices were modified by introducing a beta-detector with 50% efficiency. The detector consists of a 6x6x5cm^3 plastic scintillator coupled to a photomultiplier, both controlled via a pneumatic actuator. By using stable and collimated radioactive sources, the measured current sensibilitythreshold was 10^6pps for light stable beams up to 300keV and 10^4pps for beta particles with average energy of 1MeV [6,?]. A graphic pattern superimposing to the beam image has been assigned to each LEBI: it contains all the information about the alignment and the relevant spatial calibration. The LEBI devices will be completely operated by means of two personal computers placed on the low-intensity platform and connected to Ethernet. In addition, their video signals are input into a multiplexer and from there directly to the console through glass-fibre cabling. Identification of radionuclides will be performed with a counter board (National Instruments 6601) and a multichannel analyser (Ortec MAESTRO) installed into a personal computer in the low-activity platform. As for the post-accelerated beams, currently we are purchasing ten Glass-Fibre Based Beam Sensors (GFIBBS); in addition telescopes and microchannel plates for identifying the radionuclides by time-of-flight techniques will be mounted within 2004.

6. The Experiment at Ganil

Since the TIS had been radically modified, it became necessary to verify off-line its behaviour with respect to the mechanical and thermal stresses at 2300 K as well as to obtain information about the on-line production and the release processes. In order to save time, it was decided to run an experiment at the SIRa test-bench of SPIRAL, GANIL, by shooting a ^{13}C primary beam (60 MeV/amu) on a ^{12}C target under the same operational conditions that will be initially used at EXCYT.

6.1. *Interface and Off-line Test*

Since the primary beam at SIRa is horizontal, we needed to:

(1) Rotate by 90 ^{o} the target, its container and its heater;
(2) Design and test an interface between EXCYT assembly and SIRa.

This implies that in the GANIL configuration the heater does touch the container, unless one puts something in between. In order to make the minimum electrical contact we put a thin tantalum zigzag structure between the target heater and the target container, thus keeping their temperatures almost independent. The same configuration problem, to a minor extent, exists between target and container: at high temperatures the graphite target reacts with the tantalum container and the transfer tube can be obstructed by misplacement of the target. Therefore spacing between the target and its container was obtained by inserting a few tiny porous graphite lumps. Off-line tests at LNS showed that the assembly was mechanically and thermally stable up to 2300 K.

6.2. *ANSYS Simulation*

The second task consisted in simulating by means of the ANSYS code the target behaviour in terms of mechanical and thermal stresses under the simultaneous influence of the heating power and of a 400 W primary beam. For radioprotection purposes this was done long before conducting the real experiment. The results allowed to determine the target and the heater temperatures as functions of input heater current and primary beam power. In addition, we could extrapolate the best operational parameters and ascertain that under those conditions the target was not going to break. Details of the simulations can be found in [8]

6.3. *The Production of $^{8,9}Li$*

After all the preliminary work reported above, in May 2003 the TIS was mounted and outgassed at SIRa. During the off-line outgassing the TIS withstood three abrupt thermal cycles from about 1700 K to room temperature due to failures of the power supplies. We started the on-line experiment and on May 14th at 2:20 am we detected the first radioactive beams of ^{8,9}Li. The run lasted three days and the system proved to be robust and reliable (fig. 4): because of radioprotection constraints the maximum primary beam power was limited to 370 W but there was no sign of leaks or breakdown, despite five additional on-line failures of the primary beam (i.e. unwanted thermal cycles).

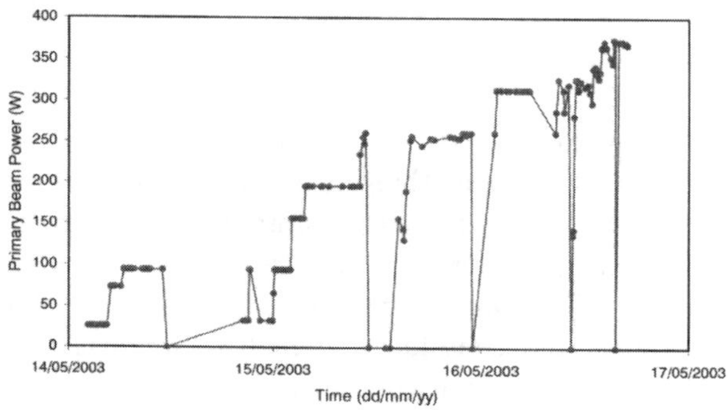

Figure 4. Primary beam power versus time. The target withstood five on-line plus three off-line abrupt thermal cycles due respectively to power and primary beam failures

The graphs in (fig. 5) show the production efficiencies for ^{8,9}Li: these are defined as the ratios of the extracted radioactive ions to the atoms produced by nuclear reactions in the target core, as estimated via the EPAX code. It is clear from the plots that the efficiencies increase by passing time. This is mainly due to two factors: increase of the target temperature by the primary beam power and increase of the heater temperature by the Joule effect. Currently we are estimating the influence of the first factor on the second one as well as trying to unfold the three stages of the production process: diffusion, effusion and ionisation, thus getting the efficiency for each stage.

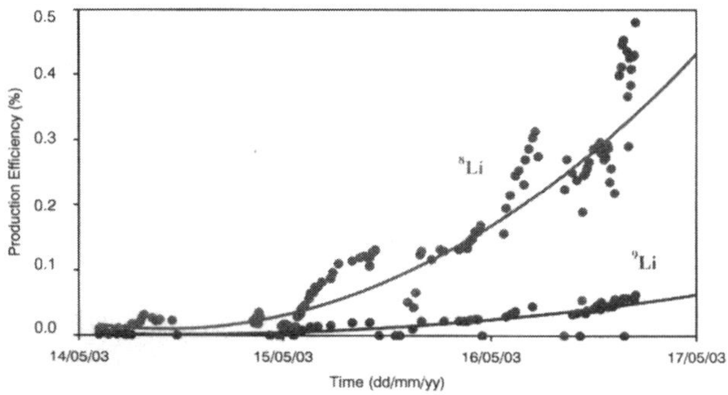

Figure 5. Production efficiencies for ^{8,9}Li versus time. The production efficiency is the ratio of the extracted radioactive ions to ones theoretically produced into the target (estimation by the EPAX code)

The yields for ^{8,9}Li were compatible with the estimation made at the beginning of the project (table 1).

Table 1. Beam intensities for ^{8,9}Li pre and post-accelerated by the Tandem. Values in the first row were estimated at the beginning of the project while in the second and third row are reported the expected on-line rates deduced from the experiment at GANIL.

Radioactive beam	*Primary beam*	*Energy MeV/amu*	*Target*	*Intensity pre-accelerated (pps/pμA)*	*Intensity post-accelerated (pps/pμA)*	*Intensity post-accelerated 500W primarybeam (pps)*
^8Li	^{15}N	50	C	6.7*10^7	2.4*10^6	1.6*10^6
^8Li	^{13}C	60	C	1.4*10^7	5.0*10^5	3.3*10^5
^9Li	^{13}C	60	C	3.2*10^5	1.2*10^4	7.4*10^3

However, they can be improved by adding a Re liner inside the ioniser and by increasing the target and ion source operational temperatures. A comparison with SPIRAL graphite target in table 2 shows that its efficiency[9] is about twice compared to EXCYT.

Table 2. Maximum production efficiencies achievedrespectively with EXCYT and GANIL graphite targets.

EXCYT efficiency (%) ^8Li	*EXCYT efficiency (%)* ^9Li	*GANIL efficiency (%)* ^8Li	*GANIL efficiency (%)* ^9Li
0.48	0.063	1.20	0.14

Some tentative explanations for this difference are:

(1) A better performance of GANIL assembly with respect to EXCYTs one;
(2) We could not push the TIS to its real limits in terms of duration (schedule), target temperature (power supplies limitations) and primary beam intensity (radioprotection constraints);
(3) The figures for the production efficiencies are affected by a large error (100%) coming from the uncertainties in the EPAX code.

We remind that the figures for SPIRAL are based on another nuclear reaction, namely 95 MeV/amu ^{36}Ar on graphite. By taking into account the third point, the values are consistent and their overlap seems to indicate an order of magnitude for the efficiency of lithium produced from graphite matrices: around 1% and 0.1% respectively for ^{8}Li and ^{9}Li.

6.4. *The TIS Post-mortem at Ganil*

The post-mortem inspection showed that the target unit was intact (fig. 6) and reusable after a suitable radioactive cooling period. The TIS was partly dismounted inside a glove box and put in a special stainless steel container designed and built at LNS. Low-activity items were put in the bottom of the container, adequately shielded by layers of 1 cm lead.

High-activity items (e.g. target) were put in a small stainless steel cylinder; the cylinder was sealed by screwing its cap on a metallic gasket, put on a dedicated shelf inside the container and additionally shielded by lead bricks. The container was sealed by screwing its cap on a plastic gasket and put into a wooden box for radioactive transportation back to LNS. The last two years signed two milestones for the EXCYT activity: the radioactive beams ^{8,9}Li were successfully obtained at GANIL with the EXCYT TIS and a 100 W ^{12}C primary beam was extracted at LNS [10]. In 2005, EXCYT will be completed and authorised to work with a primary beam power of 500 W. The commissioning of the facility with stable beams is planned for September 2004, while the start of the nuclear experiments with ^{8}Li is foreseen by the end of 2004. The experimental programme takes into account the availability of the MAGNEX detector [11,12], the requests and the first results obtained by the Big Bang collaboration [13] and the RSM experiment [14].

Figure 6. The target still intact after irradiation

References

1. D. Rifuggiato et al., *Nukleonika* **Vol. 48 Supplement 2**, pp. S131-S134 (2003).
2. M. Menna et al., Release of implanted fluorine from C and SiC matrices, (in preparation).
3. M.Menna, *NIM B*, **184**, 446 (2001).
4. G. Ciavola et al., LNS Report 1996-99, 225.
5. Ravn et al., *NIM B*, **88**, 441 (1994).
6. L.Cosentino and P.Finocchiaro, Ion beam imaging at very low energy and intensity, *NIM B*, **211**, 443 (2003)
7. L.Cosentino and P.Finocchiaro, Recent developments of the EXCYT radioactive beam diagnostics, Proceedings of the Dipac 2003, Mainz, Germany, May 5-7, 2003.
8. M. Re et al., Thermal simulations for the EXCYT target assembly, LNS Report 2003, 170.
9. S. Gibouin, Ph.D. Thesis N^o T 03 02, 2003, University of Caen, France.
10. G. Cuttone et al., The EXCYT facility towards its commissioning, LNS Report 2003, 138.
11. A.Cunsolo et al., *NIM B*, **481**, 48 (2002).
12. A.Cunsolo et al., *NIM B*, **484**, 56 (2002).

13. S. Cherubini et. al., *Eur. Phys. Journ.* **A20**, 355 (2004).
14. A. Di Pietro et al., Using the resonance scattering method to study Li-He cluster states in boron exotic isotopes, LNS Report 2003, 30.

PRESENT STATUS OF THE KEK-JAERI JOINT RNB PROJECT

H. MIYATAKE, S. ARAI, Y. ARAKAKI, Y. FUCHI, Y. HIRAYAMA, N. IMAI, H, ISHIYAMA, S. C. JEONG, I. KATAYAMA, H. KAWAKAMI, K. NIKI, T. NOMURA, M. OKADA, M. OYAIZU, M. H. TANAKA, E. TOJYO, M. TOMIZAWA, Y. X. WATANABE, AND N. YOSHIKAWA

High Energy Accelerator Research Organization (KEK),
Oho 1-1, Tsukuba, Ibaraki, 305-0801 JAPAN
E-mail: hiroari.miyatake@kek.jp

S. ABE, M. ASAI, S. HANASHIMA, K. HORIE, S. ICHIKAWA, H. IIMURA, H. IKEZOE, T. ISHII, N. ISHIZAKI, H. KABUMOTO, S. KANDA, T. KANEKO, M. KOIZUMI, M. MATSUDA, S. MITSUOKA, Y. NAGAME, T. NAKANOYA, K. NISHIO, I. OHUCHI, A. OSA, M. OSHIMA, M. SATAKA, S. TAKEUCHI, H. TAYAMA, K. TSUKADA, Y. TSUKIHASHI, AND T. YOSHIDA

Japan Atomic Energy Research Institute (JAERI),
Shirakata-Shirane 2-4, Tokai, Ibaraki, 319-1195 JAPAN

High Energy Accelerator Research Organization (KEK) and Japan Atomic Energy Research Institute (JAERI) are jointly constructing the Tokai Radioactive Ion Accelerator Complex (TRIAC) facility at Tokai site of JAERI . This facility in a final goal consists of an Isotope Separator On-Line (ISOL), a charge-breeding 18 GHz ECR (CB-ECR), and a linac complex such as a split-coaxial RFQ (SCRFQ-) linac, an interdigital-H type (IH-) linac, three rebunchers, and a superconducting (SC-) linac. Radioactive nuclei are produced with using primary beams from the 20 MV Tandem accelerator. The output RNB energy is variable between 0.1 to 8 MeV/u. At the end of 2004, the TRIAC facility will open up the RNB science with 1.1 MeV/u beams supplied from IH-linac. The higher energetic (5 - 8 MeV/u) RNBs will be available in the near future.

1. Introduction

Research and development on an isotope separator on-line (ISOL)-based radioactive nuclear beam (RNB) facility in Japan had mainly been carried out at the Institute for Nuclear Study (INS), University of Tokyo[1]. The first RNB acceleration was realized with the split-coaxial RFQ (SCRFQ) and interdigital-H (IH) type linacs in March 1997. A joint RNB facility, TRIAC (Tokai Radioactive Ion Accelerator Complex) facility, is partly based on the re-installation of the above INS-facility in Tokai site of the Japan Atomic Energy Research Institute (JAERI)[2]. It will open up the RNB science with 1.1 MeV/u RNBs from FY2005.

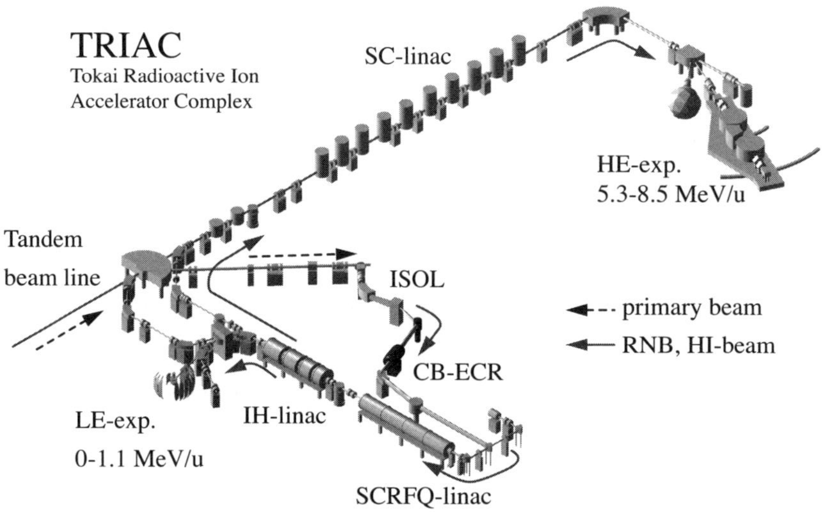

Figure 1. Layout of the TRIAC.

The facility consists of the ISOL, a newly designed charge-breeding electron cyclotron resonance ion-source (CB-ECR), and a linac complex such as the SCRFQ-linac, the IH-linac. Connecting beam line from IH-linac to a superconducting (SC-) linac through newly designed rebunchers (see Fig. 1), the output energy will be variable from 0.1 to 8 MeV/u. The primary protons as well as heavy-ions are supplied from the 20 MV Tandem accelerator to produce radioactive nuclei via the nuclear reactions, namely, nuclear fusions, transfers and fission processes. These radioactive nuclear atoms are ionized and are mass-separated to form low-energy RNBs (2 keV/u). The CB-ECR, which is an ionic charge state converter of the mass-separated RNB from 1^+ to q^+, makes it possible to accelerate heavy radioactive nuclei by the linac complex.

The output energy of IH-linac can be varied from 0.14 to 1.09 MeV/u. The 1.1 MeV/u RNBs from the IH-linac will be further accelerated to 8.5 MeV/u by the SC-linac as mentioned in the following section. There are two experimental halls. One is for the experiments with low-energy RNBs up to 1.1 MeV/u, and the other is for the use of high-energy RNBs up to 8 MeV/u. It is noted that an another 10 GHz ECR for the production of the stable nuclear beam is also available not only for a tuning of the linac complex, but also for some research subjects with the intense heavy-ion beams independent from Tandem beams. In this paper, we will report the present status of the research and developments concerning to the ion-source and the CB-ECR together with the plan of the energy up-grade.

2. RNBs development

Various kinds of neutron rich radioactive nuclei produced via the proton induced fission or the heavy-ion transfer or the fusion reaction have been ionized so far in R&D works at Tokai site[3]. The RNB intensities of fission fragments at the exit of the ISOL have been measured by utilizing the surface ionization type ion-source and FEBIAD type ion-source. In this measurement, parameters of the proton beam energy, intensity, and the thickness of UC-target were 20 MeV, 100 nA, and 0.33 g/cm^2, respectively.

As can be seen in Fig. 2, the extraction efficiency decreases from several 10 th % to the order of 1 % due to the short lifetime of each extracted fragment. The UC-target in the graphite fiber (11 μm^{ϕ}) indicates higher efficiency than one in the graphite block (porosity 50 %). So far, 98 isotopes in 16 elements are ionized and mass-separated. In the actual production condition, 30 MeV protons with 3 μA will yield 1.5×10^{11} fissions / sec for the 2.6 g / cm^2 UC-target. Based on these measured data, the intensity of ^{143}Cs-RNB is, for example, expected to be 5×10^5 pps.

RNBs of ^{8}Li, ^{18}F, ^{20}F, and ^{111}In have been also produced via the heavy-ion transfers and fusion reactions and were mass-separated with those intensities of 10^7, 10^6, 3×10^5, and 3×10^5 pps, respectively.

The ion-source and the production target are contained in one module, which can be seen at the center of the photograph in Fig. 2. The tandem beam comes through the beam line of the back wall in this picture. Ac-

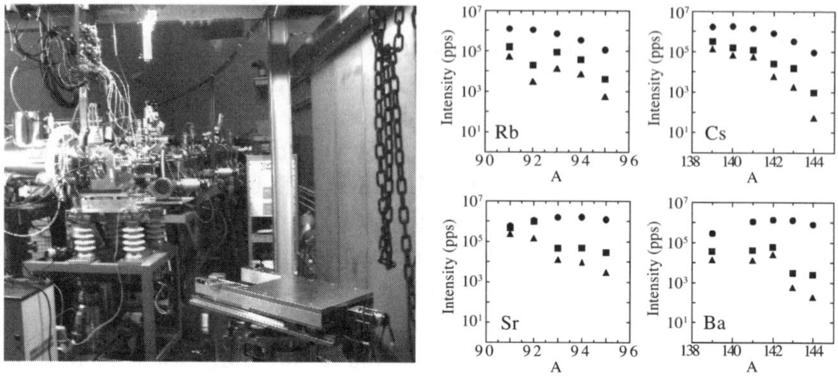

Figure 2. Photograph of the ion-source of the ISOL and measured intensities of mass separated Rb, Cs, Sr, and Ba isotopes. The solid circles indicate the expected production rates, and squares and triangles, showing measured intensities, correspond to the different graphite materials, fiber (11 μm^{ϕ}) and block (porosity 50%), respectively, in which UC_2 is deposited.

cording to an operation plan of the TRIAC, after one weak machine time, the used module will be stored during 3 weeks in order to reduce its high radioactivity of 40 mSv / h at the 1 m distance. It will be removed remotely from High-voltage platform of the ISOL and transferred to an Pb-shielded box by an compact robot system. The shielded box is shown at the right side of the picture together with the robot system.

3. Performance of the CB-ECR

The 18 GHz CB-ECR converts singly charged radioactive ions, axially injected from the ISOL, to higher charged ones in its ECR plasma in-flight. It aims at realizing the small mass to charge ratio (A / q) required from the acceleration by the linac complex. The maximum acceptable A/q-values for linacs are 29 for SCRFQ-linac, 10 for IH-linac, and 7 for SC-linac, respectively. So far, the charge breeding efficiencies for the externally injected 1^+-ions were measured at the KEK-test bench before the installation[4]. The highest efficiencies reached to be 13.5% for Ar^{9+}, 10.4% for Kr^{12+}, and 6.8% for Xe^{20+}, respectively. It is noted that these high charge states fulfill the overall acceleration condition ($A / q \leq 7$) for the linac complex.

A charge-breeding time was also measured with Xe-ions. This quantity is characterized as a delay time of the extracted q^+-ion beam from the beginning of the injection of the 1^+-ion beam and a time constant of a growth curve of the extracted beam. The singly charged Xe-ions were

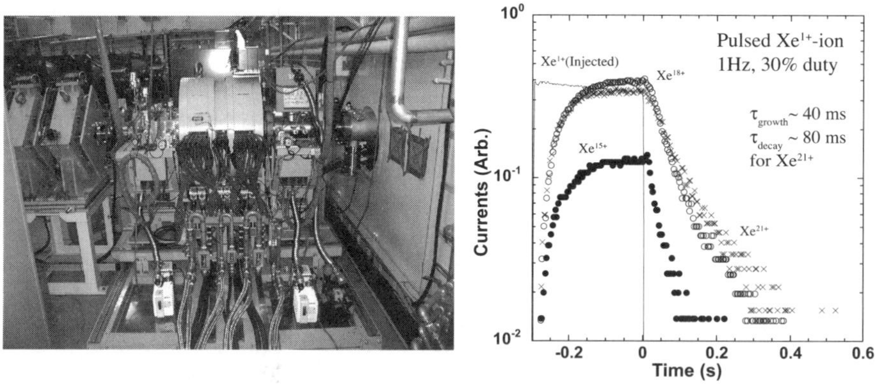

Figure 3. Photograph of the CB-ECR and the time spectrum of extracted Xe-ions from the CB-ECR. Solid circles, open circles, and crosses indicate extracted ions with $q = 15^+$, 18^+, and 21^+, respectively, while the solid line shows the 1^+- injection beam.

injected to the CB-ECR for 300 ms with repetition frequency of 1 Hz. Fig. 3 shows the time structures of the extracted charge-bred ions with $q = 15^+$, 18^+, and 21^+. From this measurement, it is found that the charge-breeding time is as short as 60 ms including the delay time of 20 ms for the conversion from $q = 1^+$ to 21^+. This value is enough short with respect to the half-lives of almost all the heavy neutron-rich fission fragments available in TRIAC.

4. Linac complex

The linac complex in a final goal consists of the SCRFQ-linac, IH-linac, and SC-linac. In order to accelerate the output beam of the SCRFQ/IH-linacs with the SC-linac, which has a resonant frequency of 129.8 MHz, the frequency of SCRFQ must be changed from 25.5 to 25.96 MHz and that of IH from 51 to 51.92 MHz. It was successfully done before the installation in the following way. In the case of the SCRFQ linac, the resonant frequency and inter-vane voltage along the beam axis were well tuned by changing locally inter-vane capacitance and inductance of stems used for supporting vane electrodes. Values of the capacitance and inductance were estimated by using an equivalent circuit analysis. As for the IH cavities, the resonant frequencies were tuned by increasing the gap lengths between drift-tubes, that is, by decreasing the capacitance. All drift-tubes were thus replaced with the modified ones. The gap lengths were determined by model tests after MAFIA calculations. The measured resonant frequency of each modified cavity agreed with a goal value within a variable range of the frequency due to the tuners.

The left side figure of Fig. 4 shows the SCRFQ/IH-linacs installed in the LE-experimental hall. The charge-bred radioactive ions come through a dipole magnet as a charge-state separator at the left side in this figure, and they are transported in the low-energy beam line to inject the SCRFQ-linac placing the upper side of the figure. Accelerated RNBs are delivered to the experimental hall from the IH-linac placing the center of this figure.

Figure 4. Photographs of the linac complex. Left: The mass-separated radioactive ions with appropriate charge state after CB-ECR are transported through the low-energy beam line and are accelerated by SCRFQ/IH-linacs. Right: The RNBs from IH-linac will be further accelerated by this SC-linac from the back to forward direction in near future.

RNBs from IH-linac will be more accelerated by the SC-linac (see the right side figure of Fig. 4), which was constructed in 1993 as a booster of the Tandem accelerator[5]. This linac comprises 10 cryostats and inter-cryostat quadrupole doublets. Each cryostat has 4 cavities. This linac has a large energy acceptance, since there are only 2 accelerating gaps in a cavity and the rf-phase of each cavity can be independently adjusted. The optimum incident velocity is designed to be $\beta = 0.1$, which is, however, still high for the output energy of IH-linac (1.1 MeV/u). Therefore, we will modify the upstream 8 cavities to accept the low-β beams by replacing new ones, each of which will be optimized for the beam velocity of $\beta = 0.06$. Moreover, one cryostat including the original 4 cavities will be added to the last cryostat to achieve the higher output energy.

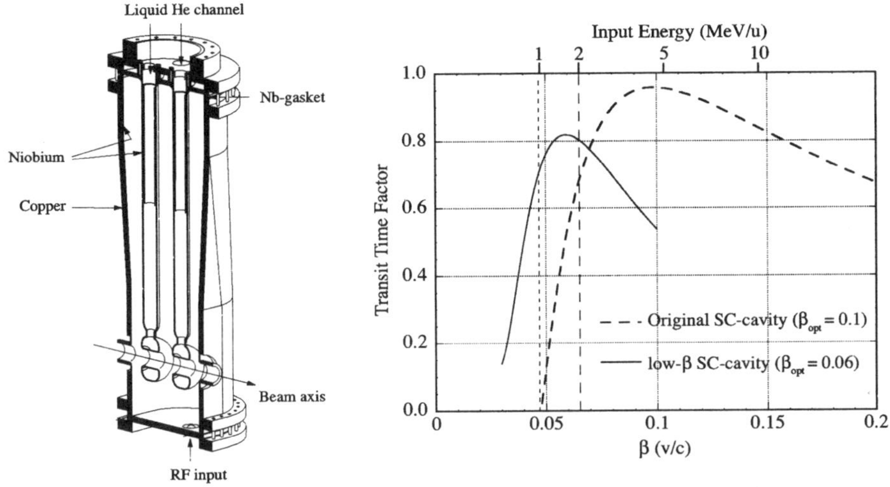

Figure 5. A cut-view of a low-β SC-cavity as a superconducting twin-quarter wave resonators and a transit time factor of the cavity.

Figure 5 shows a newly designed low-β cavity having 3-gaps. Based on the estimated acceleration efficiency (time transit factor) as shown in this figure, the output energy of the modified SC-linac, consisting of 8 low-β cavities and 36 original cavities, is expected to be[6] 5.25 MeV/u for A / q = 7 and 8.52 MeV for A / q = 4. The test of the modified cryostat with low-β cavities has been performed since the year of 2004.

Between the modified SC-linac and the IH-linac, 2 rebunchers with 26 MHz will be installed to match a longitudinal emittance. Each structure of them is double coaxial like an already existing rebuncher (26 MHz) between the SCRFQ-linac and IH-linac[7]. The performance test with using a half-scale model as shown in the Fig. 6 has been finished. All the above devices together with the beam-line components will be installed within 2 to 3 years.

5. Summary

KEK-IPNS (Institute of Particle and Nuclear Studies) and JAERI-Tokai have been collaborating to construct the radioactive-nuclear-beam facility based on the ISOL and post-acceleration scheme. The facility is called as TRIAC (Tokai Radioactive Ion Accelerator Complex). From FY2005, the low-energy RNB having its energy up to 1.1 MeV/u will be available at the LE-experimental hall for various scientific subjects of nuclear astrophysics,

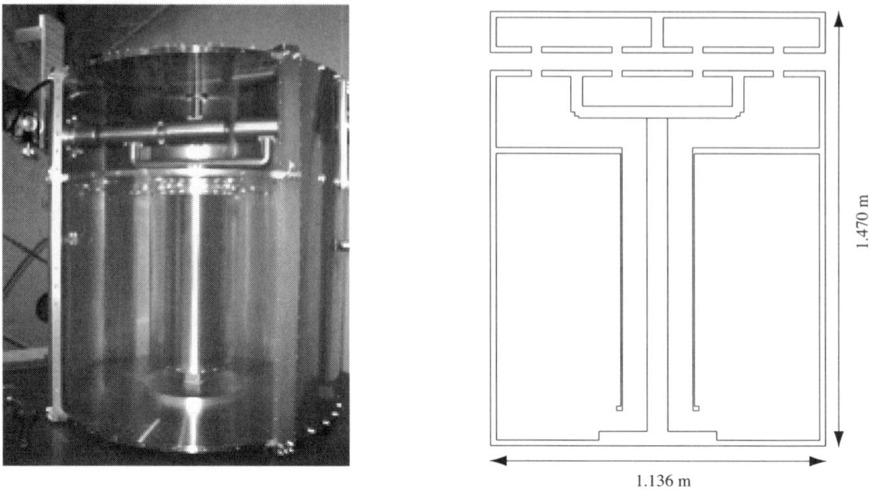

Figure 6. A photograph of the half-scale model for the rebuncher and its cross section.

nuclear physics, material science, and related research fields.

Along with the installation of the components of TRIAC, following developments have been performed. (1) Development of UC-target for producing RNBs by proton-induced fission. (2) Measurement of the charge-breeding efficiency and its breeding time in the ECR plasma. (3) Modification of the resonance frequencies of SCRFQ/IH-linacs and the design work of the low-β superconducting cavity together with the model test of the rebuncher.

References

1. T. Nomura, *Proc. 1st Int. Conf. Radioactive Nuclear Beams*, California, USA, 16-18 Oct., 1989, World Scientific (1990).
2. H. Miyatake et al., *Nucl. Instrum. Meth.* **B204**, 746 (2003) and references therein.
3. S. Ichikawa et al., *Nucl. Instrum. Meth.* **B204**, 372 (2003).
4. S. C. Jeong et al., *Rev. Sci. Instrum.* **75**, 1631 (2004).
5. T. Ishii et al., *Nucl. Instrum. Meth.* **A328**, 231 (1993).
6. S. Takeuchi, private communication.
7. K. Yoshida et al., *Nucl. Instrum. Meth.* **A430**, 189 (1999).

THE RADIOACTIVE ION BEAM PROJECT SPES AT LNL

A. PISENT
On behalf of SPES study group

INFN - Laboratori Nazionali di Legnaro, I-35020 Legnaro (PD), Italy

At LNL it has been proposed an Advanced Exotic Ion Beam facility, that will allow a frontier program in RIBs Physics. In particular beams of neutron rich nuclei in the medium heavy mass region will be produced. The structure properties of these nuclei are interesting also for the understanding of the astrophysical problem of the stellar nucleosyntesis. We illustrate here the concept of the SPES facility, and how this proposal is interconnected with the longer term European project EURISOL. The first part of the SPES facility, SPES-1, is presently under construction.

1. Introduction

In the last years the availability of new intense radioactive ion beams has been recognized as a fundamental tool for future research in Nuclear Physics. For the future development GSI has presented, with FAIR project, the proposal of a major facility based on fragmentation method, while for the ISOL (Ion Separation On Line) method, after a preliminary feasibility study, a European design study, named EURISOL-DS [1], has been funded by EU as a four year research program. Therefore the size of EURISOL is comparable to the American project RIA, but it takes advantage of the complementarily with the GSI present and future capabilities.

The time scale for EURISOL is likely to be behind fifteen years from now. In the meantime an important role for RIB's physics will be played by European national laboratories, with facilities like SPES (Study and Production of Exotic nuclear Species) at LNL (Laboratori Nazionali di Legnaro). This facility, with an intermediate size between existing first generation facilities and EURISOL, will allow, together with an extensive physics program, to boost the technological development (accelerator, production targets, detectors etc) in view of the European project. Indeed at LNL the design work for EURISOL is perfectly integrated in the design and R&D work for SPES, and with the accelerator development for future high intensity linacs (TRASCO project). LNL has an important participation to EURISOL-DS scientific and management board. For what concerns SPES a first project report was published in '99 [2] while a detailed Design Report has been published in 2002 [3]. In October 2003 the

INFN has decided the construction of a first phase of SPES (SPES-1) corresponding to the first part of the driver linac, equipped with a neutron source for interdisciplinary applications.

2. SPES and EURISOL

SPES is a new generation ISOL facility proposed at LNL, able to represent a competitive intermediate step between the existing facilities, and the longer-range high performance facility EURISOL. Between EURISOL and SPES there will be one or two orders of magnitude in beam power on target, in the production of specific isotopes and likely in financial investment. The nominal fission rate is 10^{15} s^{-1} for EURISOL and 10^{13} s^{-1} for SPES, while the beam power on target is 5 MW and few 100 kW respectively.

In both facilities the driver accelerator will be a high power superconducting linac, and the post accelerator will be a heavy ion superconducting linac. In particular for the post-accelerator the choice of a superconducting linac allows "tandem quality" beams with tunable energy, CW operation determining maximum transmission of thermally released ions and flexible stripping schemes with multi charge state transmission. For EURISOL it has been proposed a high energy experimental area (100 MeV/u), a medium energy area (about 20 MeV/u), a few MeV/u area for experiments of astrophysical interest and a low energy area for experiments "à la ISOLDE". The LNL facility SPES will be able to cover (with lower intensity) the last three energy ranges using the superconducting linac (ALPI), able to accelerate exotic ions above 15 MeV/u.

An efficient way to produce a large variety of neutron rich isotopes is the use of the nuclear fission induced by high-energy neutrons. The fission target consists of ^{238}U, under uranium carbide form. The neutrons are produced by a primary proton beam impinging on a thick beryllium or carbon target. The main advantage of the two targets method is that the beam power is dissipated in the first target (converter), while the second target (production target) only withstands the fission power (few hundred W for 10^{13} f/s).

The SPES superconducting driver in an open machine, able, with the implementation of upgraded injector, to accelerate deuterons up to 100 MeV for an increased production of neutrons so that the rate of 10^{14} f/s can be exceeded.

As an alternative to the two target method the use of a fission target directly invested by a proton beam is under study with very promising results[4]. The process is in this case more efficient, and 10^{13} f/s can be reached with a 40 MeV proton beam current lower than 1 mA, and few tenths of grams of uranium target. The main problem is to withstand the high stopping power density in the target, formed by few thin uranium carbide discs. An intense experimental program on this concept has started within SPES project.

The driver linac design based on independently phased superconducting cavities (ISCL) and with a capability of 5 mA, is the same as for the corresponding part of EURISOL project. The beam current choice is convenient for the linac, since a power level of the order of 10 kW per amplifier can be achieved with solid state technology.

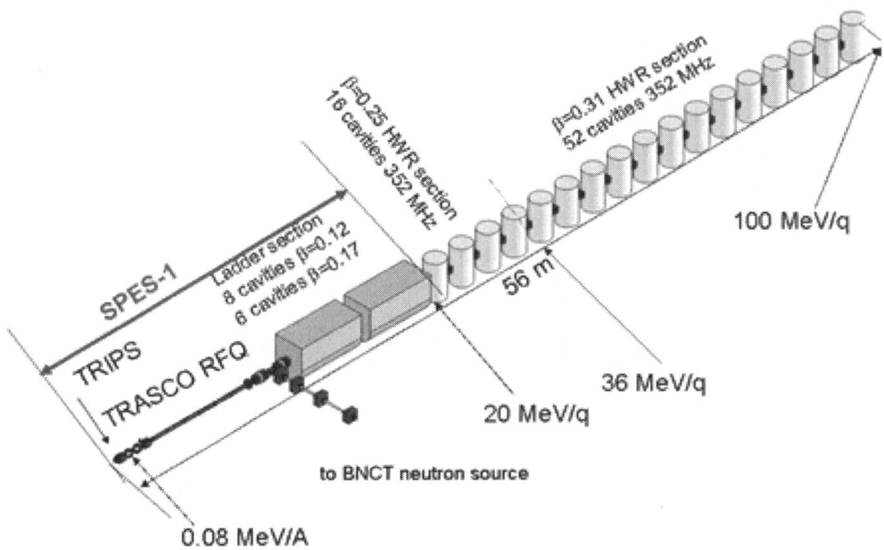

Figure 1. Possible lay-out of SPES driver linac (100 MeV protons).

The main linac components (Fig. 1) are the off resonance rf source TRIPS [5], the TRASCO RFQ [6] and the Independently phased Superconducting Cavity Linac (ISCL). The source and the RFQ have been developed within the TRASCO research program, aimed to the development of a high intensity linac for nuclear waste transmutation. The source TRIPS is being commissioned at LNS, while the RFQ is under construction at LNL (Fig. 2). The source and the RFQ, installed at LNL, will represent a unique facility, able to deliver 30 mA 5 MeV beam. The RFQ is a 7.13 m long structure composed by 6 brazed modules, fed by one klystron of LEP kind.

The following ISCL is the natural extension of the superconducting heavy ion linac technology, where LNL has a recognized role in Europe. Respect to the accelerator ALPI now operating at LNL [7], there will be a similar scheme, with many independent resonators working in continuous wave mode (CW), but at higher frequency and much higher beam current. As a consequence new RF amplifiers (keeping the reliability of solid state RF sources) and new cavities (with large beam hole and good field quality) are being developed. In fig.3 half wave [8] resonator and ladder cavity [9] are shown. The cavities are built using Nb sheets electron beam welded. The half wave resonator prototype has already been built and tested up to its design field.

Finally is should be noted that a high current proton beam requires a very good control of beam losses so to avoid an unacceptable activation of the accelerator components. In particular one has to control the mechanisms able to push few particles at a large distance from beam core (the so called beam halo).

The distribution of fission products in A achieved in this scenario is more homogeneous than for a reactor and simulations show for example 2 x 10^{11} ^{132}Sn per second produced in target. From the UCx target heated to more than 2000 °C the fragments diffuse into the source, where they are ionized by an electron or laser beam.

A first bending magnet selects the ions with a resolution of about 1/300 in q/A and cuts the largest part of unwanted radioactive species. The charge state is therefore boosted in an ion source, of ECR or EBIS type, working in breeder mode. A high-resolution spectrometer, able to resolve isobars in specific cases, can be located at this point. A line of bypass should give the possibility to the user of choosing between higher resolution and higher transmission.

The ions are then bunched and accelerated in an RFQ section, in ALPI and finally impinge into the experimental target. Again for ^{132}Sn typically 10^8 ions/s (0.02 pnA) can be expected. A final energy of about 15 MeV/u can be achieved in ALPI, with the completion of both the low β and the high βsections. A bunching RFQ, plus a couple of superconducting RFQs, identical to PIAVE, have to be installed in the same ALPI vault. The three experimental halls presently used at LNL will all be reached by accelerated RIBs.

3. The funded facility SPES-1

Taking into account the available resources, and taking advantage from the European framework for the development of RIBs, it was decided to launch SPES-1 that will be a first important step in the direction of SPES and

EURISOL, a very good test for the high intensity community (ADS), and will be able to serve a community of interdisciplinary physics and medical users.

The main interdisciplinary user of the facility is the Boron Neutron Capture Therapy (BNCT). The 5 MeV, 30 mA proton beam produced by the first acceleration step, the RFQ, through (p,n) reaction on ^{9}Be and a heavy water moderator, generates the neutron flux of thermal neutrons (about 2.5 x 10^{9}/(s cm^{2})) necessary for patient treatment. A concentration of B in correspondence of neoplastic tissues is achieved pharmacologically, so that the α particles, emitted in correspondence of n capture, determine a very localized damage of these tissues. In particular, at LNL the BNCT facility is foreseen to explore the cure of extended skin melanoma with this method [10]. The main ingredients of the research program carried out by an interdisciplinary research group, formed by medical doctors, biologists, physicists and nuclear engineers, are the development of the neutron source based on an accelerator, the molecule (boron carrier) and the microdosimetry.

SPES-1 will include the completion and installation of the 5 MeV 30 mA proton injector, the construction of the thermal neutron facility for BNCT, with a 150 kW beryllium target, the realization of the superconducting p linac up to 20 MeV and the continuation of the R&D program on RIB production targets.

In fig. 1 a schematic lay-out of SPES-1 is shown, with the source TRIPS, the TRASCO RFQ, and the transport line that allows to deliver the beam either to the superconducting linac or to the BNCT complex. SPES-1 is integrated in infrastructures that will allow the implementation of the entire SPES project.

Figure 2. TRASCO-SPES RFQ: view of the first module before and after brazing.

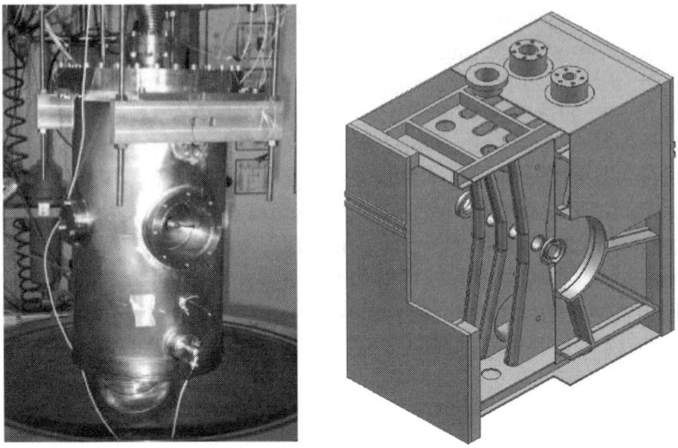

Figure 3. Superconducting cavities operating at 352 MHz under development at LNL for the driver linac: half wave and ladder prototypes.

4. Conclusions

The long range ISOL facility will necessarily be on European scale (EURISOL), based on a high intensity linac as driver and a superconducting linac for the post-acceleration. The technology of the driver (and of the converter) is common to other applications like material science, nuclear waste transmutation and High Energy Physics.

It exists an intermediate phase with an essential role for National Laboratories, like SPES project at LNL, described in this paper, or SPIRAL II at GANIL, that is also based on a superconducting driver. There is therefore an integrated plan for the development of ISOL facilities in Europe, and good perspectives for the implementation of the relative Physics programs. The first step in Italy will be the construction of the first phase of SPES in the next five years.

References

1. The EURISOL Report (Co-ordinated by Prof. J. Vervier, edited by J. Cornell) (2003) and http://www.ganil.fr/eurisol/index.html
2. SPES Study Group "Project Study of an advanced facility for Exotic Beams at LNL" LNL-INFN (REP) 145/99 report

3. SPES Technical design report (A. Bracco and A. Pisent editors) LNL-INFN(REP) 181/2002 and http://www.lnl.infn.it/~spes/
4. A. Andrighetto, S. Cevolani, C. Petrovich, Eur. Phys. J. A (2005) DOI 10.1140/epja/i2005-10064-8
5. G. Ciavola, L. Celona, S. Gammino: "First Beam from the TRASCO Intense Proton Source (TRIPS) at INFN-LNS" Proceedings of the 2001 Particle Accelerator Conference, Chicago, June 18 - 22, 2001
6. A. Pisent, et al. "TRASCO RFQ", Proceedings of the XX International Linac Conference, Monterey, California, August 21 - 25,p. 902 (2000).
7. A. Porcellato et al., "Operation experience with ALPI resonators",SRF2001 proceedings
8. A. Facco et al. "Contruction and testing of the b=0.31, 352 MHz Superconducting Half Wave Resonator for the SPES project" Proceedings of EPAC04, 2004.
9. V. Andreev et al. "Study of a Novel Superconducting Structure for the very low beta part of high current linacs" Phys. Rev. ST Accel. Beams 6, 040101 (2003)
10. S.Agosteo et al. "Advances in the INFN-Legnaro BNCT Project for Skin Melanoma". Proceedings of the international physical and clinical workshop on BNCT. Held in Candiolo (Torino) on the 17th of February 2001.

RESEARCH PROGRAM IN RIKEN RIBF PROJECT

T. MOTOBAYASHI

RIKEN, 2-1 Hirosawa, Wako, Saitama 351-0198, Japan
E-mail: motobaya@riken.jp

Since 1990 the RIKEN Accelerator Research Facility has provided variety of fast radio-isotope (RI) beams based on the projectile-fragmentation scheme. Various studies have been made and are in progress using these exotic beams, the intensities of which are highest in the world for many light unstable nuclei. The RIKEN RI Beam Factory, which is now being built, will greatly extend the region of study to more exotic and heavier nuclei. The first beam is expected in the end of the year 2006. Installations of experimental equipments are being considered to fully exploit the new opportunity of nuclear research.

1. Current Research with RI Beams at RIKEN

1.1. *RIKEN Accelerator Research Facility*

The current RIKEN Accelerator Research Facility (RARF) has an accelerator complex that provides intermediate-energy light and heavy ions. It consists of a main ring cyclotron (RRC) with K=540 MeV with two injector machines, an AVF (azimuthally varying field) cyclotron for ions up to Kr and a linear accelerator (RILAC) for heavier ions. The schematic view of the accelerators is shown in Fig. 1 including the part of the Radioactive Isotope Beam Factory (RIBF) discussed later. The RARF started its routine operation in 1987. It provides light heavy-ions with energies up to 135 MeV/nucleon, including 270 MeV polarized deuterons. Due to various improvements so far, such as developments of an 18-GHz ECR ion source (ECRIS-18)[1], a frequency variable RFQ (radio frequency quadrupole) linac[2], and a booster linac system called CSM (Charge State Multiplier) installed in collaboration with CNS (Center for Nuclear Study, the University of Tokyo), performance of the facility has been much improved, and various experiments requiring high beam intensities become possible.

Intense lower-energy ions at several MeV/nucleon energies are also available in the RILAC experimental hall. They are used for super-heavy element search by heavy-ion fusion reactions[3]. The most recently, one event has been observed for the production of the isotope $^{278}113$ by the $^{209}Bi(^{70}Zn,n)$ using a intense beam of 5 MeV/nucleon ^{70}Zn [4].

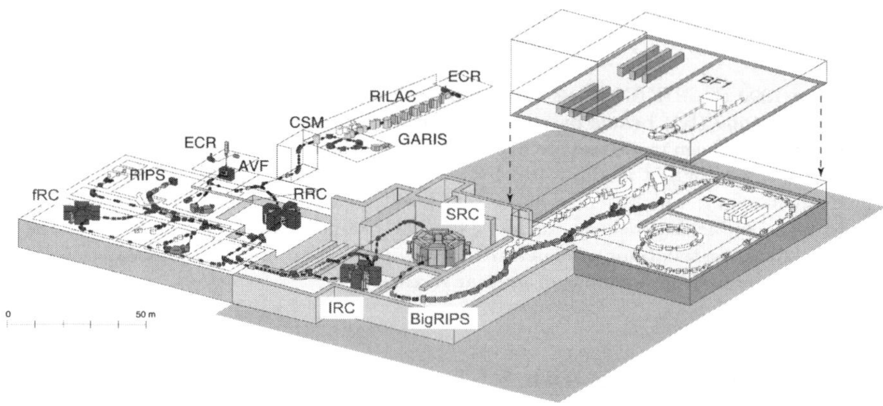

Figure 1. Schematic view of the RIKEN accelerator complex. The part with gray background is the RIBF facility being built.

1.2. *Studies with Intermediate-Energy RI Beams*

RI Beams of light unstable nuclei have been produced by the projectile fragmentation scheme since 1990. The primary beams are mainly light heavy ions with energies at around 100 MeV/nucleon accelerated by the RRC with the AVF injector. Due to the large angular and momentum acceptance and high bending power of the fragment separator RIPS[5], RI beams with reasonable intensities for many nuclei far from the stability line are available. These intensities are highest in the world for many light unstable nuclei. Due to the upgrade of the RILAC injector system mentioned before, available RI beams have been extended to the $A \approx 70$ region, which could not be realized by the AVF-RRC acceleration scheme, where the intensity are too low for efficient production of RI beams.

Besides the intermediate-energy RI beams, beams of light unstable nuclei at lower energies, typically 5 MeV/nucleon, are also available in the CRIB facility[6] constructed by CNS. Several experiments of astrophysical reactions and spectroscopy have been performed[7]. Low energy RI beams are also available with the RIPS using an energy-degrader. Recently an experiment for the ^{11}Be+^{209}Bi elastic scattering at Coulomb barrier energies has been performed under the RIKEN-INFN collaboration [8].

Direct reactions are useful for nuclear spectroscopy. However, new methods were necessary for studies with the fast RI beams because of difficulties due to the reversed kinematics, poor beam-energy resolution, and low beam intensity. We have developed the following two methods of nuclear spectroscopy to overcome the difficulties. The invariant mass is measured if the excited state of interest is particle unbound. For bound states, deexcitation γ rays are measured to identify the excited levels.

In the invariant mass method, particles decaying in flight is measured. The excitation energy of their parent nucleus is obtained from the invariant mass measured by their momentum vectors. The energy resolution is almost free from the energy spread of the incident secondary beams. We have been studying, by this method, astrophysical (p,γ) processes with the Coulomb dissociation. Breakup processes, such as ^{8}B$\rightarrow^{7}$Be+p, ^{9}C$\rightarrow^{8}$B+p, and ^{14}O$\rightarrow^{13}$N+p have been measured with Pb targets for studying astrophysical hydrogen burning and solar neutrino production processes[9]. Recently we extended the Coulomb dissociation study to a few rp-process nuclei as ^{22}Mg$\rightarrow^{23}$Al+p and ^{26}Si$\rightarrow^{27}$P+p. Another application of the Coulomb dissociation is to measure the E1 strength function of very neutron-rich nuclei. Coulomb dissociation experiments with halo nuclei as ^{11}Li[10] and ^{11}Be[11] beams are such examples.

Coulomb excitation to particle-bound states has also been studied for various unstable nuclei. Comparing the yield of deexcitation γ rays and theoretical prediction of the Coulomb excitation cross section, the transition probability of the relevant state can be extracted. We built arrays of NaI(Tl) scintillators called DALI and DALI-2, and measured the probabilities of the 2^+-0^+ transitions in ^{32}Mg, ^{34}Mg, ^{56}Ni, etc, and the E1 transitions in ^{11}Be, ^{12}Be and ^{15}O. The same method was extended recently to (p,p') and (d,d') reactions. Efficient measurements were realized by using a liquid hydrogen target. The location of the first 2^+ state in ^{30}Ne[12], the most proton deficient N=20 even-even nucleus, has been determined for the first time by the (p,p') experiment. Recently, we have performed another (p,p') experiment, and found two bound excited states in ^{27}F[13], which are not expected in any of recent theoretical calculations.

Studies of nucleon transfer reactions, (α,t) and (α,^{3}He) [14], have been started with a liquid helium target coupled with the segmented Ge array GRAPE [15]. One of the recent highlights is the 2^+-0^+ transition in ^{16}C. A lifetime measurement and a ^{16}C+Pb inelastic scattering experiment point to a surprising nature of the transition: very much hindered E2 and enhanced neutron excitations[16]. This indicates almost complete decoupling of proton- and neutron-motions in this nucleus.

2. RI Beam Factory Project

2.1. *Overview*

These successfull achievements in RARF together with those in other RI beam facilities with the fragmentation scheme triggered a new project. RIKEN plans to extend the research with RI beams by building a new experimental facility called "RI Beam Factory (RIBF)". The goal of the project is to provide a wide range of experimental opportunities by increasing the variety of RI with higher intensities.

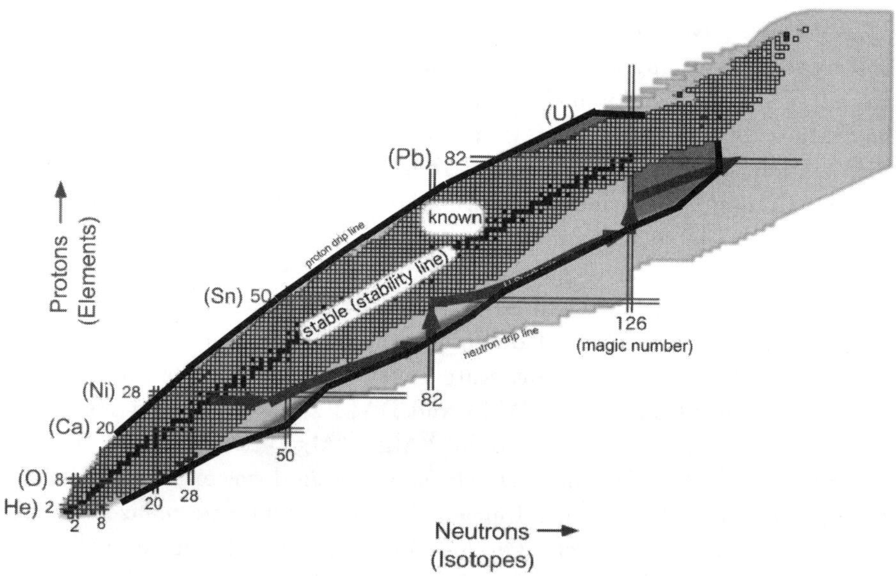

Figure 2. Nuclear chart covered by the RIBF project. The thick solid curves indicate the limit of RI productions of 1 particle per day.

The RIBF facility is illustrated in Fig. 1 together with the present one. Three ring cyclotrons in cascade accelerate heavy ions up to uranium to 350 MeV/nucleon with high intensities up to 1 pμA. RI beams will be produced via the projectile fragmentation or in-flight fission of uranium ions. The limit of production in the rate of one particle per day is indicated by the shick solid curves in Fig. 2. As shown in the figure, most of the expected path of the r-process nculeosynthesis is covered. In view of its outstanding performance on the variety and intensity of RI beams, the RIBF is one of the next-generation facilities competitive to the ones planned in the United State (RIA) and Europe (GSI FAIR), and is their first realization in the in-flight fragmentation scheme. In order to fully exploit this scientific opportunity, we set three major categories of research that should be pursued in the RIBF: i) establishment of a comprehensive picture of atomic nucleus by studying new dynamics in asymmetric nuclei, ii) understanding of element genesis by studying properties and reactions of unstable nuclei, and iii) new applications in multi-disciplinary fields such as medicine, environmental researches, and material science, hoping that some of them will create new industrial domains. We plant to build the facility of its phase I by the end of 2006.

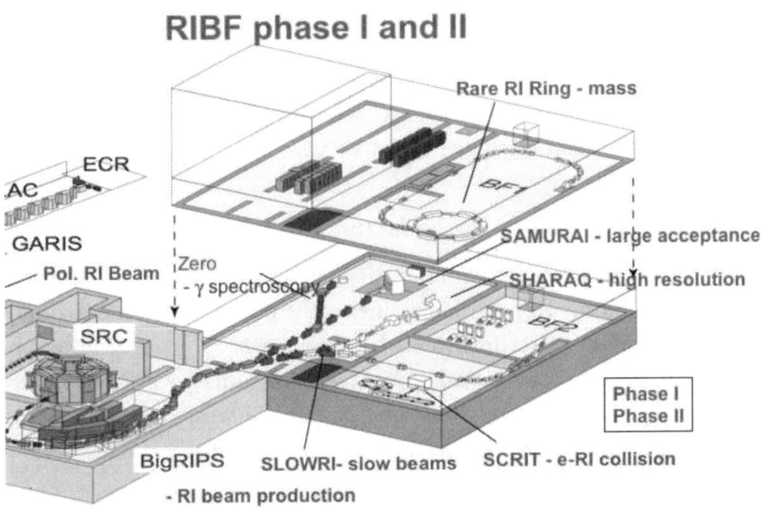

Figure 3. Experimental equipments in the RIBF phase I and II.

2.2. *Phase I*

The enlarged view for the experimental equipments of the RIBF is shown in Fig. 3. In the phase I, a magnetic analyzer called Zero-degree spectrometer and a few beam lines will be constructed together with the two-stage fragment separator BigRIPS[17]. First experiments, which will start in the year 2007 after the RIBF comes into operation, should be in this condition.

Experimental possibilities without the Zero-degree spectrometer in the first few years are: interaction cross-section measurement, β-decay study in the vicinity of the "r-process path", β-γ and/or isomer spectroscopy, proton elastic scattering in inverse kinematics, spectroscopy with degraded RI beams, studies with stopped RI's, unbound state spectroscopy by correlation measurements, and spectroscopy of pionic atoms.

With the Zero-degree spectrometer, search for new isotopes using the long TOF line realized by coupling the Zero-degree spectrometer and BigRIPS, in-beam γ spectroscopy with direct reactions (inelastic scattering including Coulomb excitation, charge exchange and fragmentation reactions), Missing mass spectroscopy, proton elastic scattering / giant resonance studies, and so on will be possible.

2.3. *Phase II*

The phase II of RIBF is in the planning state at present. It contains constructions of the following devices. The budget request will be made in 2005, and the construction will be in five years.

2.3.1. *A large-acceptance superconducting spectrometer (SAMURAI)*

Its large angular and momentum acceptance allows for multi-particle correlation measurements. Experiments currently planned are electromagnetic dissociation for nuclear structure and nuclear astrophysics studies, proton elastic / inelastic scattering and knockout reactions to study density distribution and single-particle orbit of nucleus, polarized-deuteron induced reactions to explore 2-3N forces and short-range correlation, and multi-particle measurement for the EOS study.

2.3.2. *Gas catcher and rf ion guide system (SLOWRI)*

As illustrated in Fig. 4, RI beams from the BigRIPS are stopped in a gas volume and efficiently extracted by the rf ion guide scheme to conduct various experimentssuch as mass measurements with the multi reflection time-fo-flight (MR-TOF) spectrometer and determination of charge radii of unstable nuclei by the collinear laser spectroscopy.

It is also expected as a possible ion-source of the SCRIT system.

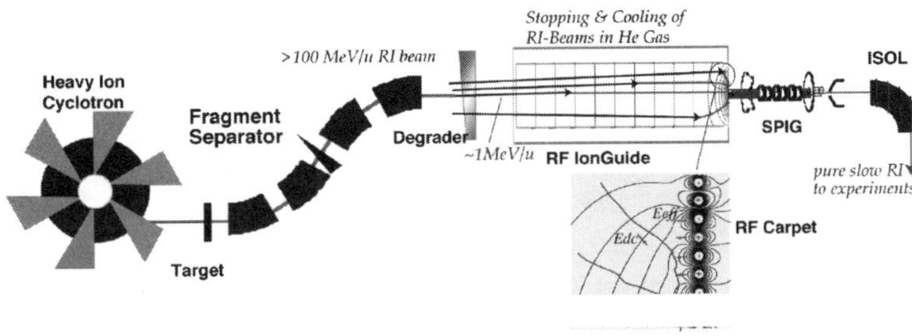

Figure 4. Scheme of the SLOWRI.

2.3.3. *Polarized RI beams*

Polarized RI beam with high-intensity and low to medium-energy will be developed as a new probe for material science as well as for nuclear structure study. They are generated with the existing RIPS separator which is newly connected to a return beam line delivering heavy-ion beams from the IRC.

2.3.4. *High resolution RI beam spectrometer (SHARAQ)*

A high-resolution spectrometer coupled with a specially designed beam line with dispersion matching is proposed. Exploiting the advantage of the high momentum resolution, experiments with RI beams as reaction probes rather than the object are planned such as study on double Gamow-Teller states and production of neutron nuggets by RI-induced double charge-exchange reactions. Choice of spin- and isospin-transfers and Q-value of the reaction can control the kinematical conditions which cannot be realized by a beam of a stable nucleus.

2.3.5. *Ring for rare RI ions*

Conceptual design of an isochronous ring with individual injection of rarely produced RI ions is proposed (Fig. 5). It is mostly dedicated for new precision mass measurement scheme for short-lived nuclei very far from the valley of stability. Time-of-flight measurements with a long flight path with

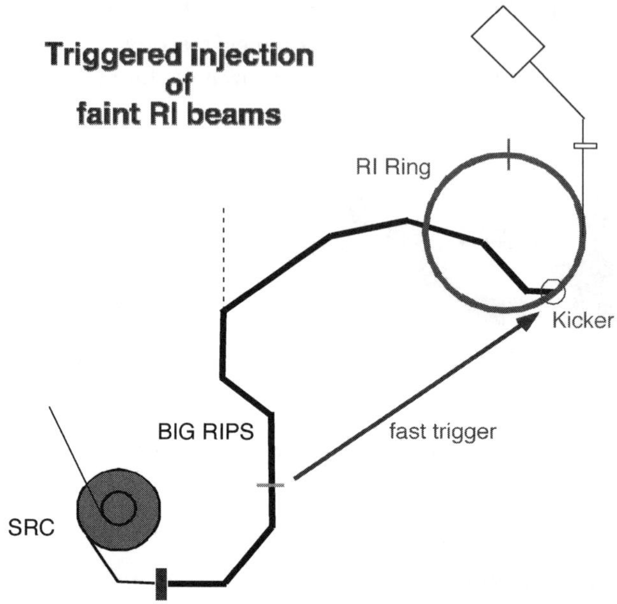

Figure 5. Scheme of the rare RI ring.

an accurate isochronous field enable one to determine the ion mass with a great precision. Almost 100% injection efficiency is realized with cyclotron-based (rare) RI beams with the individual injection method. The ring is equipped with a kicker for ion extraction as well as the one for injection for the mass measurement.

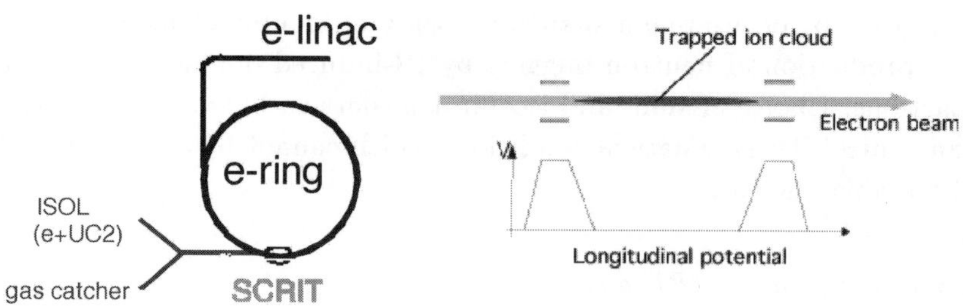

Figure 6. Scheme of elestron scattering experiments using the Self Confining RI Target (SCRIT).

2.3.6. *Self confining RI target (SCRIT)*

Conceptual design of new electron-scattering experiment scheme for RI ions are proposed (Fig. 6). RI ions produced by an ISOL apparatus are delivered to the area of strong electron beams running. These ions are trapped transversely with the help of the attractive force caused by the electrons. With longitudinal confinement by appropriate potentials, the RI ions form a target in the electron beam, and electron-RI scattering measurements can be performed.

3. Summary

The RIKEN RI beam factory (RIBF) will provide various possibilities for experimental studies using radioactive beams. Construction of the three ring cyclotrons (fRC, IRC and SRC) will be completed in 2006. Then the first experiments will start with 200-300 A MeV beams of various unstable nuclei from 2006. To extend further the research with RI beams, the "phase II" of RIBF is being considered. International collaboration in various aspects is highly welcome to exploit the research opportunity provided by the RIBF.

References

1. T. Nakagawa *et al., Jpn. J. Appl. Phys.* **33**, 378 (1996).
2. O. Kamigaito *et al., Jpn. J. Appl. Phys.* **33**, L537 (1996).
3. K. Morita, this symposium.
4. K. Morita *et al., J. Phys. Soc. Jpn.* **73**, 2593 (2004).
5. T. Kubo *et al., Nucl. Instr. Meth.* **B70** 309 (1992).
6. Y. Yanagisawa *et al., Nucl. Instr. Meth.* **A539** 74 (2005).
7. S. Kubono, this symposium.
8. M. Mazzocco, this symposium.
9. T. Motobayashi, *Nucl. Phys.* **A693** 258 (2001), and references therein.
10. S. Shimoura *et al., Phys. Lett.* **B348** 29 (1995).
11. N. Fukuda *et al., Phys. Rev.* **C70** 054606 (2004).
12. Y. Yanagisawa *et al., Phys. Lett.* **B566** 84 (2003).
13. Z. Elekes *et al., Phys. Lett.* **B599** 17 (2004).
14. S. Shimoura, this symposium.
15. M. Kurokawa *et al., IEEE Trans. Nucl. Sci.* **50** 1309 (2003); T. Fukuchi, this symposium.
16. N. Imai *et al., Phys. Rev. Lett.* **92** 062501 (2004); Z. Elekes *et al., Phys. Lett.* **B586** 34 (2004);
17. T. Kubo, this symposium.

LIST OF PARTICIPANTS

Akiyama Takahiro	Department of Physics, Saitama University, Sakura-ku, Saitama 338-8570 Japan Akiyama@ne.phy.saitama-u.ac.jp
Andrighetto Alberto	Laboratori Nazionali di Legnaro Via Romea 4, 35020 Legnaro (Padova), Italy alberto.andrighetto@lnl.infn.it
Asakawa Masayuki	Department of Physics, Osaka University Toyonaka, 560-0043 Japan yuki@phys.sci.osaka-u.ac.jp
Bardelli Luigi	Dipartimento di Fisica, Università di Firenze Via G. Sansone 1, 50019 Sesto Fiorentino (Firenze), Italy bardelli@fi.infn.it
Bazzacco Dino	Istituto Nazionale di Fisica Nucleare, Sezione di Padova Via Marzolo 8, 35131 Padova, Italy dino.bazzacco@pd.infn.it
Benzoni Giovanna	Dipartimento di Fisica, Università di Milano Via Celoria 16, 20133 Milano, Italy benzoni@mi.infn.it
Bizzeti Pier Giorgio	Dipartimento di Fisica, Università di Firenze Via G. Sansone 1, 50019 Sesto Fiorentino (Firenze), Italy bizzeti@fi.infn.it
Bonaccorso Angela	Istituto Nazionale di Fisica Nucleare, Sezione di Pisa Largo B. Pontecorvo 3, 56127 Pisa, Italy bonac@df.unipi.it
Bracco Angela	Dipartimento di Fisica, Università di Milano Via Celoria 16, 20133 Milano, Italy angela.bracco@mi.infn.it
Bressani Tullio	Dipartimento di Fisica Sperimentale, Università di Torino Via Pietro Giuria 1, 10125 Torino, Italy tullio.bressani@to.infn.it

Brondi Augusto	Dipartimento di Scienze Fisiche, Università di Napoli "Federico II" Via Cinthia, 80126 Napoli, Italy augusto.brondi@na.infn.it
BrunoMauro	Dipartimento di Fisica, Università di Bologna Via Irnerio 46, 40126 Bologna, Italy bruno@bo.infn.it
Capuzzello Francesco	Istituto Nazionale di Fisica Nucleare, Laboratori Nazionali del Sud Via Santa Sofia 62, 95123 Catania, Italy capuzzello@lns.infn.it
Cherubini Silvio	Istituto Nazionale di Fisica Nucleare, Laboratori Nazionali del Sud Via Santa Sofia 62, 95123 Catania, Italy cherubini@lns.infn.it
Chiavassa Emilio	Dipartimento di Fisica Sperimentale, Università di Torino Via Pietro Giuria 1, 10125 Torino, Italy chiavassa@to.infn.it
Covello Aldo	Dipartimento di Scienze Fisiche, Università di Napoli "Federico II" Via Cinthia, 80126 Napoli, Italy aldo.covello@na.infn.it
Cunsolo Angelo	Dipartimento di Fisica ed Astronomia, Università di Catania Via Santa Sofia 64, 95123 Catania, Italy angelo.cunsolo@ct.infn.it
De Angelis Giacomo	Istituto Nazionale di Fisica Nucleare, Laboratori Nazionali di Legnaro Via Romea 4, 35020 Legnaro (Padova), Italy deangelis@lnl.infn.it
Di Toro Massimo	Dipartimento di Fisica ed Astronomia, Università di Catania Via Santa Sofia 64, 95123 Catania, Italy massimo.ditoro@ct.infn.it

D'Onofrio Antonio	Dipartimento di Scienze Ambientali, II Università di Napoli Via Vivaldi 43, 81100 Caserta, Italy donofrio@na.infn.it
Feliciello Alessandro	Istituto Nazionale di Fisica Nucleare, Sezione di Torino Via Pietro Giuria 1, 10125 Torino, Italy feliciello@to.infn.it, Italy
Fioretto Enrico	Istituto Nazionale di Fisica Nucleare, Laboratori Nazionali di Legnaro, Via Romea 4, 35020 Legnaro (Padova), fioretto@lnl.infn.it
Foti Antonino	Dipartimento di Fisica ed Astronomia, Università di Catania Via Santa Sofia 64, 95123 Catania, Italy antonino.foti@ct.infn.it
Fukuchi Tomonori	Center for Nuclear Study, University of Tokyo 2-1 Hirosawa, Wako, Saitama 351-0198, Japan fukuchi@cns.s.u-tokyo.ac.jp
Gadea Andres	Istituto Nazionale di Fisica Nucleare, Laboratori Nazionali di Legnaro, Via Romea 4, 35020 Legnaro (Padova), Italy andreas.gadea@lnl.infn.it
Gargano Angela	Istituto Nazionale di Fisica Nucleare, Sezione di Napoli Via Cinthia , 80126 Napoli, Italy angela.gargano@na.infn.it
Garibaldi Franco	Istituto Nazionale di Fisica Nucleare, Sezione Sanità Viale Regina Elena 299, 00161Roma, Italy franco.garibaldi@iss.infn.it
Gelli Nicla	Istituto Nazionale di Fisica Nucleare, Sezione di Firenze Via G. Sansone 1, 50019 Sesto Fiorentino (Firenze), Italy gelli@fi.infn.it, Italy
Gialanella Lucio	Istituto Nazionale di Fisica Nucleare, Sezione di Napoli Via Cinthia , 80126 Napoli, Italy lucio.gialanella@na.infn.it

Grieco Giovanni — CAEN Electronics for Nuclear Physics
Via Vetraia 11, 55049 Viareggio, Italy
g.grieco@caen.it

Gulino Marisa — Istituto Nazionale di Fisica Nucleare, Laboratori Nazionali del Sud
Via Santa Sofia 62, 95123 Catania, Italy
mgulino@dmfci.unict.it

Guaraldo Carlo — Istituto Nazionale di Fisica Nucleare, Laboratori Nazionali Frascati
Via E. Fermi 40, 00044 Frascati (Roma), Italy
guaraldo@lnf.infn.it

Hamagaki Hideki — Center for Nuclear Study, University of Tokyo
7-3-1 Hongo, Bunkyo-ku,Tokyo 113-0033, Japan
hamagaki@cns.s.u-tokyo.ac.jp

Ikezoe Hiroshi — Japan Atomic Energy Research Institute
Tokai, Ibaraki 319-1195, Japan
ikezoe@popsvr.tokai.jaeri.go.jp

Imbriani Gianluca — Dipartimento di Scienze Fisiche, Università di Napoli "Federico II"
Via Cinthia, 80126 Napoli, Italy
imbriani@na.infn.it

Ishiyama Hironobu — Institute of Particle and Nuclear Studies, High Energy Accelerator Research Organization (KEK),
Tsukuba, Ibaraki, 305-0801 Japan
hironobu.ishiyama@kek.jp

Ishikawa Tomoko — Institute of Particle and Nuclear Studies, High Energy Accelerator Research Organization (KEK),
Tsukuba, Ibaraki, 305-0801, Japan
tomoko@post.kek.jp

Itaco Nunzio — Dipartimento di Scienze Fisiche, Università di Napoli "Federico II"
Via Cinthia, 80126 Napoli, Italy
itaco@na.infn.it

Iwasaki Masahiko — DRI, RIKEN- The Institute of Physical and Chemical Research
Wako-shi, Saitama 351-0198, Japan
masa@riken.jp

Kimura Sachie	Istituto Nazionale di Fisica Nucleare, Laboratori Nazionali del Sud Via Santa Sofia 62, 95123 Catania, Italy kimura@lns.infn.it
KuboToshiyuki	RIKEN- The Institute of Physical and Chemical Research 2-1 Hirosawa, Wako, Saitama 351-0198, Japan kubo@rarfaxp.riken.jp
Kubono Shigeru	Center for Nuclear Study, University of Tokyo Wako Branch at RIKEN, 2-1 Hirosawa, Wako, Saitama, 351-0198, Japan kubono@cns.s.u-tokyo.ac.jp
La Commara Marco	Dipartimento di Scienze Fisiche,Università di Napoli "Federico II" Via Cinthia , 80126 Napoli, Italy marco.lacommara@na.infn.it
Lanzalone Gaetano	Istituto Nazionale di Fisica Nucleare, Laboratori Nazionali del Sud Via Santa Sofia 62, 95123 Catania, Italy lanzalone@lns.infn.it
La Rana Giovanni	Dipartimento di Scienze Fisiche, Università di Napoli "Federico II" Via Cinthia, 80126 Napoli, Italy giovanni.larana@na.infn.it
Lucarelli Franco	Dipartimento di Fisica, Università di Firenze Via G. Sansone 1, 50019 Sesto Fiorentino (Firenze), Italy lucarelli@fi.infn.it
Lunardi Santo	Dipartimento di Fisica, Università di Padova Via Marzolo 8, 35131 Padova, Italy lunardi@pd.infn.it
Martin Brunella	Dipartimento di Scienze Fisiche,Università di Napoli "Federico II" Via Cinthia , 80126 Napoli, Italy martin@na.infn.it
Marzari Chiesa Alberta	Dipartimento di Fisica Sperimentale, Università di Torino Via Pietro Giuria 1, 10125 Torino, Italy marzari@to.infn.it

Mazzocco Marco	Dipartimento di Fisica, Università di Padova Via Marzolo 8, 35131 Padova, Italy mazzocco@pd.infn.it
Menna Mariano	Istituto Nazionale di Fisica Nucleare, Laboratori Nazionali del Sud Via Santa Sofia 62, 95123 Catania, Italy menna@lns.infn.it
Miyatake Hiroari	High Energy Accelerator Research Organization (KEK), Oho 1-1, Tsukuba, Ibaraki, 305-0801 Japan hiroari.miyatake@kek.jp
Morita Kosuke	RIKEN- The Institute of Physical and Chemical Research Wako, Saitama 351-0198, Japan morita@rarfaxp.riken.jp
Moro Renata	Dipartimento di Scienze Fisiche, Università di Napoli "Federico II" Via Cinthia, 80126 Napoli, Italy renata.moro@na.infn.it
Motobayashi Tohru	RIKEN- The Institute of Physical and Chemical Research 2-1 Hirosawa, Wako, Saitama 351-0198, Japan motobaya@riken.jp
Musumarra Agatino	Istituto Nazionale di Fisica Nucleare, Laboratori Nazionali del Sud Via Santa Sofia 62, 95123 Catania, Italy musumarra@lns.infn.it
Nagame Yuichiro	Japan Atomic Energy Research Institute Tokai, Ibaraki 319-1195, Japan nagame@popsvr.tokai.jaeri.go.jp
Nappi Eugenio	Istituto Nazionale di Fisica Nucleare, Sezione di Bari Via G. Amendola 173, 70125 Bari, Italy eugenio.nappi@ba.infn.it
Nociforo Chiara	Istituto Nazionale di Fisica Nucleare, Laboratori Nazionali del Sud Via Santa Sofia 62, 95123 Catania, Italy nociforo@lns.infn.it

Nomura Toru	Institute of Particle and Nuclear Studies, High Energy Accelerator Research Organization (KEK), Tsukuba, Ibaraki, 305-0801 Japan tnomura@post.kek.jp
Pagano Angelo	Dipartimento di Fisica ed Astronomia, Università di Catania Via Santa Sofia 64, 95123 Catania, Italy angelo.pagano@ct.infn.it
Petrascu Catalina	Istituto Nazionale di Fisica Nucleare, Laboratori Nazionali Frascati Via E. Fermi 40, 00044 Frascati (Roma), Italy petrascu@lnf.infn.it
Pierroutsakou Dimitra	Istituto Nazionale di Fisica Nucleare, Sezione di Napoli Via Cinthia, 80126 Napoli, Italy pierroutsakou@na.infn.it
Pisent Andrea	Istituto Nazionale di Fisica Nucleare, Laboratori Nazionali di Legnaro Via Romea 4, 35020 Legnaro (Padova), Italy pisent@lnl.infn.it
Poggi Giacomo	Dipartimento di Fisica, Università di Firenze Via G. Sansone 1, 50019 Sesto Fiorentino (Firenze), Italy poggi@fi.infn.it
Porto Francesco	Dipartimento di Fisica ed Astronomia, Università di Catania Via Santa Sofia 64, 95123 Catania, Italy francesco.porto@ct.infn.it
Prati Paolo	Dipartimento di Fisica, Università di Genova Via Dodecaneso 33, 16146 Genova, Italy prati@ge.infn.it
Prete Gianfranco	Istituto Nazionale di Fisica Nucleare, Laboratori Nazionali di Legnaro Via Romea 4, 35020 Legnaro (Padova), Italy prete@lnl.infn.it
Ricco Giovanni	Dipartimento di Fisica, Università di Genova Via Dodecaneso 33, 16146 Genova, Italy giovanni.ricco@ge.infn.it

Romano Mario	Dipartimento di Scienze Fisiche, Università di Napoli "Federico II" Via Cinthia, 80126 Napoli, Italy mario.romano@na.infn.it
Romano Stefano	Dipartimento di Fisica ed Astronomia, Università di Catania Via Santa Sofia 64, 95123 Catania, Italy romano@ct.infn.it
Romoli Mauro	Istituto Nazionale di Fisica Nucleare, Sezione di Napoli, Via Cinthia, 80126 Napoli, Italy mauro.romoli@na.infn.it
Rosato Elio	Dipartimento di Scienze Fisiche, Università di Napoli "Federico II" Via Cinthia, 80126 Napoli, Italy elio.rosato@na.infn.it
Sakai Hideyuki	Department of Physics and Center for Nuclear Study, University of Tokyo Hongo 7-3-1, Bunkyo, Tokyo 113-0033, Japan sakai@phys.s.u-tokyo.ac.jp
Sandoli Mario	Dipartimento di Scienze Fisiche, Università di Napoli "Federico II" Via Cinthia, 80126 Napoli, Italy mario.sandoli@na.infn.it
Santonocito Domenico	Istituto Nazionale di Fisica Nucleare, Laboratori Nazionali del Sud Via Santa Sofia 62, 95123 Catania, Italy santonocito@lns.infn.it
Satou Kenichiro	Applied Accelerator Division, Research Facility Center for Science and Technology, University of Tsukuba, Tsukuba, Ibaraki 305-8577, Japan ksatou@tac.tsukuba.ac.jp
Scomparin Enrico	Istituto Nazionale di Fisica Nucleare, Sezione di Torino Via P. Giuria 1, 10125 Torino, Italy scomparin@to.infn.it

Shimoura Susumu	Center for Nuclear Study, University of Tokyo Wako branch at RIKEN, 2-1 Hirosawa, Saitama 351-0198, Japan shimoura@cns.s.u-tokyo.ac.jp
Signorini Cosimo	Dipartimento di Fisica, Università di Padova Via Marzolo 8, 35131 Padova, Italy signorini@pd.infn.it
Sona Anna Maria	Dipartimento di Fisica, Università di Firenze Via G. Sansone 1, 50019 Sesto Fiorentino (Firenze), Italy annamaria.sona@fi.infn.it
Spitaleri Claudio	Dipartimento di Fisica ed Astronomia, Università di Catania Via Santa Sofia 64, 95123 Catania, Italy spitaleri@ct.infn.it
Sulignano Barbara	G.S.I. D-64291 Darmstadt, Planckstr. 1, Germany b.sulignano@gsi.de
Tamura Hirokazu	Department.of Physics, Tohoku University, Aramaki, Aoba-ku, Sendai 980-8578, Japan tamura@lambda.phys.tohoku.ac.jp
Terrasi Filippo	Dipartimento di Scienze Ambientali, II Università di Napoli Via Vivaldi 43, 81100 Caserta, Italy filippo.terrasi@unina2.it
Torikoshi Masami	National Institute of Radiological Sciences, 4-9-1 Anagawa, Inage-ku, Chiba-shi 263-8555, Japan torikosi@nirs.go.jp
Trotta Monica	Istituto Nazionale di Fisica Nucleare, Sezione di Napoli Via Cinthia , 80126 Napoli, Italy monica.trotta@na.infn.it, Italy
Ur Calin Alexandru	Dipartimento di Fisica, Università di Padova Via Marzolo 8, 35131 Padova, Italy ur@pd.infn.it
Utsuno Yutaka	Japan Atomic Energy Research Institute Tokai, Ibaraki 319-1195, Japan

utsuno@nsbox.tokai.jaeri.go.jp

Vardaci Emanuele — Dipartimento di Scienze Fisiche, Università di Napoli "Federico II"
Via Cinthia, 80126 Napoli, Italy
emanuele.vardaci@na.infn.it

Vitturi Andrea — Dipartimento di Fisica, Università di Padova
Via Marzolo 8, 35131 Padova, Italy
andrea.vitturi@pd.infn.it

Watanabe Yasushi — RIKEN- The Institute of Physical and Chemical Research
2-1 Hirosawa, Wako, Saitama 351-0198, Japan
watanaby@riken.jp

AUTHOR INDEX